Springer INdAM Series

Volume 64

This series will publish textbooks, multi-authors books, thesis and monographs in English language resulting from workshops, conferences, courses, schools, seminars, doctoral thesis, and research activities carried out at INDAM - Istituto Nazionale di Alta Matematica, http://www.altamatematica.it/en. The books in the series will discuss recent results and analyze new trends in mathematics and its applications.

THE SERIES IS INDEXED IN SCOPUS

Biagio Cassano · Fabio Deelan Cunden ·
Matteo Gallone · Marilena Ligabò ·
Alessandro Michelangeli
Editors

Singularities, Asymptotics, and Limiting Models

Springer

Editors
Biagio Cassano
Department of Mathematics and Physics
University of Campania "Luigi Vanvitelli"
Caserta, Italy

Matteo Gallone
Department of Mathematics
SISSA Trieste
Trieste, Italy

Alessandro Michelangeli
Department of Mathematics and Science
AUBG American University Bulgaria
Blagoevgrad, Bulgaria

Hausdorff Center for Mathematics
Bonn, Germany

TQT Trieste Institute for Theoretical
Quantum Technologies
Trieste, Italy

Fabio Deelan Cunden
Department of Mathematics
University of Bari
Bari, Italy

INFN Istituto Nazionale di Fisica Nucleare
Bari, Italy

Marilena Ligabò
Department of Mathematics
University of Bari
Bari, Italy

ISSN 2281-518X ISSN 2281-5198 (electronic)
Springer INdAM Series
ISBN 978-981-96-3586-3 ISBN 978-981-96-3584-9 (eBook)
https://doi.org/10.1007/978-981-96-3584-9

This Springer imprint is published by the registered company Springer Nature Singapore Pte Ltd.
The registered company address is: 152 Beach Road, #21-01/04 Gateway East, Singapore 189721, Singapore

Preface

The intensive programme "*Puglia Summer Trimester 2023*" took place in Bari, Italy, from April to July 2023.

The programme was structured around a series of main meetings, including a general conference and a summer school, supplemented by the local presence and activities of an amount of visiting scientists. Additionally, efforts were made to disseminate and popularise mathematics among schools and the general public, with the aim of extending the programme's impact beyond the immediate academic sphere.

The present volume collects a distinguished selection of contributions by scholars who participated in the programme as speakers or as visiting scientists, as well as further scholars who are experts in related fields.

Such contributions, in the form of retrospective reviews, overviews on recent developments, announcements and comments of new results, as well as outlooks on future perspectives, represent some of the main scientific instances of the trimester in Bari.

The trimester was actually focussed on a spectrum of mathematical problems, directly stemming or inspired from a variety of physical domains, involve *singular modelling, asymptotic and emergent phenomena, singular interactions, non-trivial limit effects.*

Natural backgrounds are quantum physics, cold atom physics, soft matter physics, with methods and tools, suitably adapted to such singular settings, spanning across operator and spectral theory, functional analysis, probability, differential geometry, partial differential equations, and numerical analysis.

This opportunity of beneficially connecting different mathematical areas is reflected in the present volume too: the reader would find a stimulating perspective on the analysis of *singular, asymptotic, limiting models,* which was also the running title of the trimester in Bari, spanning across one body quantum models with singular perturbations and their operator-theoretic apparatus, to several many-body problem settings, perturbations of integral quantum systems, singular differential boundary value problems of various sort, and so on and so forth.

As a by-product, young researchers and students may find in the very pedagogically written chapters of this volume a convenient way to learn an amount of technical tools directly in application to concrete, state-of-the-art investigations.

For the completion of this volume of the Springer-INdAM series, a sincere thank from our side is due both to the generous funding received from the INdAM for last year's spring trimester and its follow-up initiatives like the present one, and to all the authors and anonymous peer reviewers who carefully prepared each chapter.

We trust that these very much updated materials may inspire new prolific research over the next future.

Caserta, Italy	Biagio Cassano
Bari, Italy	Fabio Deelan Cunden
Trieste, Italy	Matteo Gallone
Bari, Italy	Marilena Ligabò
Blagoevgrad, Bulgaria	Alessandro Michelangeli
July 2024	

Contents

Globally Integrable Quantum Systems and Their Perturbations

Dario Bambusi and Beatrice Langella

Abstract In this paper we present the notion of globally integrable quantum system that we introduced in [18]: we motivate it using the spectral theory of pseudodifferential operators and then we give some results on linear and nonlinear perturbations of a globally integrable quantum system. In particular, we give a spectral result ensuring stability of most of its eigenvalues under relatively bounded perturbations and two results controlling the growth of Sobolev norms when it is subject either to a linear unbounded time dependent perturbation or to a small nonlinear Hamiltonian nonlinear perturbation.

Keywords Schrödinger operator · Normal form · Nekhoroshev theorem · Pseudo differential operators

MSC 2010 37K10 · 35Q55

1 Introduction

In this paper we review some results that we recently got on perturbations of "globally integrable quantum systems".

One of the key points is the definition of globally integrable quantum system. Indeed, as far as we know, there is not a generally accepted definition of integrable quantum system, thus in the first part of the paper (see Sects. 2, 3, 4 and 5) we first motivate and then give our definition of *globally integrable quantum system* (GIQS). Roughly speaking, a GIQS is a Hamiltonian operator which is a function of the quantum actions. In turn we define the quantum actions to be a set of commuting

D. Bambusi (✉)
Dipartimento di Matematica, Università degli Studi di Milano, Milano, Italy
e-mail: dario.bambusi@unimi.it

B. Langella
International School for Advanced Studies (SISSA), Trieste, Italy
e-mail: beatrice.langella@sissa.it

B. Cassano et al. (eds.), *Singularities, Asymptotics, and Limiting Models*, Springer INdAM Series 64, https://doi.org/10.1007/978-981-96-3584-9_1

self-adjoint operators with pure point spectrum contained in $\mathbb{Z} + \kappa$, where $\kappa \in \mathbb{R}$ is a real number.

The main feature of GIQS is that, on the one hand (as showed by the result presented in Sect. 6) they have very strong stability properties when perturbed, and on the other hand it turns out that they cover several nontrivial examples. That is the reason why we think they are of interest.

GQIS exhibit stability properties both from the spectral point of view and from a dynamical point of view, when the Schrödinger equation relative to a GIQS is perturbed by a relatively bounded family of time dependent pseudodifferential operators, or by a small nonlinear Hamiltonian perturbation.

Such results are reviewed respectively in the Sects. 6.1 and 6.2. Section 6.1 is split in a paragraph dealing with the spectral problem and a paragraph dealing with the dynamical problem. In Paragraph 6.1.1 we present the spectral result, according to which most of the eigenvalues of a perturbation of a GIQS admit an asymptotic expansion in inverse powers of their size. In Paragraph 6.1.2 we state the result on time dependent perturbations of the Schrödinger equation, according to which the Sobolev norms of the solution are bounded by $|t|^\epsilon$ for all $\epsilon > 0$.

Then, in Sect. 6.2 we study a nonlinear Hamiltonian PDE which is a perturbation of the Schrödinger equation with unperturbed Hamiltonian given by a GIQS. Here we give an almost global existence result, namely we prove that for initial data with Sobolev norm of order ϵ the Sobolev norm of the corresponding solution remains of order ϵ for times of order ϵ^{-N}, $\forall N$. We also give a couple of applications to concrete nonlinear PDEs.

The main tool for the proof of all the abstract results is a clustering property of the spectrum of a GIQS. Such a clustering is described in Sect. 7 and is constructed by quantizing the so called geometric part of the proof of classical Nekhoroshev's theorem. Such a clustering has also a property that was identified by Bourgain as a key property for the proof of KAM type results in PDEs. For this reason we call it Nekhoroshev-Bourgain partition.

In Sect. 7 we also give a qualitative description of the way the clustering is used in order to prove the results presented in Sect. 6.

We close this introduction by recalling that in the line of research of this paper one finds a very extensive literature. In particular we recall that this line of research on the quantization of classical perturbation theory was initiated many years ago, in particular we recall [8, 12, 30, 32, 33], and more recent developments were obtained in [2, 3, 23] for 1-d systems and [15, 16] for some higher dimensional systems. Then we investigated the case of a nontrivial unperturbed structure of the resonance relations in [18–21].

The case of nonlinear Hamiltonian with semilinear perturbations originated from the papers [1, 14] for 1-d systems, the case of the wave equation on Zoll manifolds was studied in [6] (see also [13, 38]). Recently such a theory was extended to the case of quasilinear perturbations [5, 24]. The case of general tori was dealt with in [10], which was also the starting point of our work [9]. We also recall the paper [31], which had a strong influence on our result [9].

2 Classical Integrable and Superintegrable Systems.

Let M be an n dimensional Riemannian manifold, and consider the symplectic manifold given by the cotangent boundle T^*M. Let $h \in C^\infty(T^*M)$ be a Hamiltonian function. Then to h one associates the corresponding Hamiltonian vector field X_h defined by $\omega_z(X_h(z), k) = dh(z)k, \forall k \in T_z(T^*M)$, where ω_z is the symplectic form at $z \in T^*M$. It is well known that in any canonical coordinate system of T^*M the Hamilton equations $\dot{z} = X_h(z)$ take the form

$$\dot{\xi} = -\partial_x h \ , \quad \dot{x} = \partial_\xi h \ .$$

In this case $n = dim(M)$ is called the number of degrees of freedom of the Hamiltonian system h.

One also defines the Poisson Brackets of two functions by

$$\{f_1; f_2\} := df_1 X_{f_2} \ .$$

2.1 *Integrable Systems*

Definition 2.1 Let $f_1, ..., f_k \in C^\infty(T^*M)$ be k smooth functions. They are said to be independent if the covectors

$$df_1(z), ..., df_k(z)$$

are independent for almost every z.

The functions are said to be in involution if

$$\left\{f_i; f_j\right\} = 0 \ , \forall i, j = 1, ..., k \ .$$

It is well known that the maximal number of functions which are independent and in involution in a manifold of dimension n is n.

Definition 2.2 A Classical Hamiltonian system is said to be integrable if it admits n prime integrals which are independent and in involution, namely if there exist $f_1, ..., f_n$ independent and in involution and s.t. $\left\{h; f_j\right\} \equiv 0$.

It is well known that the main result of classical integrable systems is the so called Arnold Liouville theorem, which can be stated as follows

Theorem 2.3 *Let h be an integrable system; let $z_0 \in T^*M$ be s.t. the differentials $df_j(z_0)$ are independent and denote $m_j = f_j(z_0)$. If the level surface*

$$S_0 := \left\{z \in T^*M \ : \ f_j(z) = m_j \quad \forall j = 1, \dots, n\right\}$$

is connected and compact, then S_0 is diffeomorphic to an n dimensional torus and there exists a neighborhood $\mathcal{U}$ of S_0 foliated in invariant n dimensional tori. Furthermore there exist $\mathcal{A} \subseteq \mathbb{R}^n$ and a symplectic coordinate system $(a, \alpha) : \mathcal{A} \times \mathbb{T}^n \to \mathcal{U}$, with the property that in such coordinates h is a function of a only.

The coordinates (a, α) are called Action Angle coordinates.

Remark 2.4 For the extension to quantum systems, the main property of the action coordinates is that the coordinate canonically conjugated to each a_j is an angle. This means that if one considers the Hamiltonian given by the action a_j, such a Hamiltonian generates a flow which is periodic of period 2π.

Our main goal is to define the quantum analogue of the action variable, which in principle is just the operator obtained by quantizing the classical action. However, quantization is an easy procedure only for globally defined functions: for this reason we are particularly interested in situations in which the action variables do not have singularites, or have only singularities which can be eliminated.

Remark 2.5 In this paper, we are interested in the behavior of functions at infinity. For this reason, the possibly harmful singularities are only those accumulating at infinity. A typical example are the circular orbits of the central motion problem, in which the effective radial Hamiltonian has critical points. It will turn out that such critical points can be regularized quite easily if they are elliptic, while the hyperbolic case is not covered by the theory developed here. We also remark that in more general cases one can meet also singularities of focus-focus kind, whose semiclassical theory has been developed by [48].

2.2 Superintegrable Systems

A further interesting case where the set of action angle variables is singular is the case of superintegrable Hamiltonian system, however in such a case one has that some of the action variables are typically globally defined and furthermore the Hamiltonian only depends on these globally defined action variables, a phenomenon which is also related to Gordon's theorem [39, 45]. As a consequence our theory will turn out to be applicable also to several superintegrable systems.

We recall that in the case of superintegrable systems the geometry of the phase space is described "semiglobally" by a classical theorem by Nekhoroshev [42] (see also [34] and literature cited therein). We will not state the corresponding theorem, instead we are going to describe the situation in the case of a free particle on the 2 dimensional sphere.

So, consider a particle on the two dimensional sphere, which is a system with two degrees of freedom. Here there exist three independent integrals of motion which are the three components of the angular momentum. From these variables one can extract two integrals of motion which are independent and in involution, namely the z component $\mathtt{M}_z$ of the angular momentum and the square $\mathtt{M}^2$ of the total angular momentum. A system of action coordinates is given by $\mathtt{M}$ and by $\mathtt{M}_z$. The corresponding angles can then be constructed by the standard procedure obtaining a system of action angle variables.

Such a system of coordinates has a singularity when (1) the total angular momentum is in the direction of the z axis, since here the two actions are no more independent and (2) when the total angular momentum vanishes. However, it is clear that the first singularity is just related to the choice of the z axis and that the other action $\mathtt{M}$ is smooth at such a singularity, while at $\mathtt{M} = 0$, there is a singularity of the action variable, but the Hamiltonian, which is a function just of $\mathtt{M}^2$ is smooth also at $\mathtt{M} = 0$.

This is the typical semiglobal situation described by Nekhoroshev's theorem [42]: one of the actions, say a, is defined intrinsecally: its level surfaces form a foliation, each leaf of which is identified by a value of the intrinsic action. For $a \in \mathcal{A}$ denote by $\mathcal{M}_a$ such a level surface. On $\mathcal{M}_a$, $\mathbb{T}^1$ acts through the flow generated by the vector field X_a of the Hamiltonian a. This gives $\mathcal{M}_a$ the stucture of a fibre bundle, whose fibre is isomorphic to $\mathbb{T}^1$ and whose basis is a manifold $\mathcal{B}$. In the case of the free motion on the sphere, such $\mathcal{B}$ is a sphere. The fact that it has a non trivial topological structure is the reason why there do not exist standard action angle coordinates in a whole neighborhood of $\mathcal{M}_a$.

A careful analysis allows to prove that a neighborhood $\mathcal{U}$ of the level surface has the structure of a bi-fibration $\mathcal{U} \to^1 \mathcal{B} \to^2 \mathcal{A} \subset \mathbb{R}$, where $\mathcal{A}$ is the domain of the true action on which the Hamiltonian depends, the fibers of the first map 1 are one dimensional tori, while the fibers of the second fibration 2 are typically compact surfaces.

2.3 Conclusion

As a conclusion of this section, we just summarize that classical integrable systems are systems in which locally the Hamiltonian is a function of the action variables. In turn, the action variables are functions on the phase space which, when used as Hamiltonian, generate a flow which is periodic with period 2π.

Typically the action variables are only defined locally in the phase space, but there are situations in which they are defined globally.

A particular situation is that of superintegrable systems, in which the Hamiltonian only depends on a number of actions smaller than the number of degrees of freedom, and in some interesting situations such actions are globally defined.

This is a first set of considerations that we use as a basis for the definition of quantum actions and quantum integrable system that we will give in Sect. 5.1.

3 Quantizing Hamiltonians with Periodic Flow

As observed in Remark 2.4 and Sect. 2.3, the actions of an integrable system generate a periodic Hamiltonian flow. With this in mind, in this section we discuss quantum operators obtained by quantizing a Hamiltonian whose flow is periodic. We will carry on in a parallel way the theory on $\mathbb{R}^n$ and on compact manifolds.

To start with consider the case of $\mathbb{R}^n$, where the phase space is $T^*\mathbb{R}^n \simeq \mathbb{R}^n \times \mathbb{R}^n$. We follow the theory developed by Helffer-Robert in [41], which is adapted to the study of anharmonic oscillators in $\mathbb{R}^n$, namely the systems with classical Hamiltonian

$$h_{An}(x,\xi) := \frac{\|\xi\|^2}{2} + \frac{\|x\|^{2\ell}}{2\ell}\,, \quad \ell \in \mathbb{N}^*\,. \tag{3.1}$$

In order to precisely state the result in [41], we need to introduce some setting. First we introduce a scale of Hilbert spaces $\{\mathcal{H}^s_{An}\}_{s\geq 0}$ adapted to this situation by defining the operator $K^{(0)}_{An} := (-\Delta + \|x\|^{2\ell})^{\frac{\ell+1}{2\ell}}$ and putting, for $s \geq 0$ $\mathcal{H}^s_{An} := D([K^{(s)}_0]^s)$ endowed by the graph norm. For $s < 0$ we also define $\mathcal{H}^s_{An}$ to be the dual to $\mathcal{H}^{-s}_{An}$ with respect to the L^2 scalar product.

We also need to introduce the concept of a quasihomogeneous function.

Definition 3.1 Fix $\ell \in \mathbb{N}$, then a function $a \in C^\infty(\mathbb{R}^{2n} \setminus \{0\})$ is said to be quasihomogeneous of degree $m \in \mathbb{R}$ if

$$a\left(\lambda^{\frac{1}{\ell+1}}x, \lambda^{\frac{\ell}{\ell+1}}\xi\right) = \lambda^m a(x,\xi) \quad \forall\lambda \in \mathbb{R}^+\,, \quad \forall(x,\xi) \in \mathbb{R}^{2n}\,. \tag{3.2}$$

Then we define also a class of symbols adapted to this situation: denote

$$\mathrm{k}_0(x,\xi) := (1 + \|x\|^{2\ell} + \|\xi\|^2)^{\frac{1+\ell}{2\ell}}\,.$$

Definition 3.2 A function f will be called a symbol of order $m \in \mathbb{R}$ if $f \in C^\infty(\mathbb{R}^n_x \times \mathbb{R}^n_\xi)$ and $\forall\alpha, \beta \in \mathbb{N}^n$, there exists $C_{\alpha,\beta} > 0$ s.t.

$$|\partial_x^\alpha \partial_\xi^\beta f(x,\xi)| \leq C_{\alpha,\beta}\, \mathrm{k}_0(x,\xi)^{m - \frac{\ell|\beta|+|\alpha|}{1+\ell}}\,. \tag{3.3}$$

We will write $f \in S^m_{An}$.

To a symbol $f \in S^m_{An}$ we associate the operator $F \equiv Op^W(f)$ which is obtained by standard Weyl quantization, namely

$$(Fu)(x) := \frac{1}{(2\pi)^n} \int_{\mathbb{R}^n}\int_{\mathbb{R}^n} e^{\mathrm{i}\langle x-\eta,\xi\rangle} f\left(\frac{x+\eta}{2},\xi\right) u(\eta)\, d\eta\, d\xi\,. \tag{3.4}$$

Definition 3.3 We say that F is a pseudodifferential operator of order M and we write $F \in \mathrm{OPS}^m_{An}$ if there exist $f \in S^m_{An}$ and a smoothing operator S such that $F =$

$Op^W(f) + S$. Here by smoothing operator we mean an operator S mapping $\mathcal{H}^s_{An}$ into $\mathcal{H}^{s+N}_{An}$ for any s and any $N > 0$.

In this framework we have the following theorem:

Theorem 3.4 *(Helffer-Robert, [41]) Let $a \in S^1_{An}$ be quasihomogeneous of degree 1 and assume that all the solutions of the corresponding Hamilton equations are periodic with period 2π. Assume also that the corresponding Weyl operator A is positive definite, then there exist an operator $Q \in \mathrm{OPS}^{-1}_{An}$ commuting with A and $\kappa \in \mathbb{R}$ s.t.*

$$\sigma(A + Q) \subset \mathbb{Z} + \kappa \ . \tag{3.5}$$

By $\sigma(A)$ we are denoting the spectrum of the operator A.

Idea of the proof. The idea is first to prove that

$$\exp\left(-2\pi \mathrm{i}\,(A - \kappa)\right) = \mathbf{1} + C\,, \tag{3.6}$$

with some $C \in \mathrm{OPS}^{-1}_{An}$ and then to pass to the logarithm. The intuition is that, by Egorov's theorem, one has that, for any pseudodifferential operator B with symbol b, the principal symbol of $\exp(-2\pi \mathrm{i} A)\, B \exp(2\pi \mathrm{i} A)$ is $b \circ \phi^{2\pi}_a = b$, by the periodicity of the flow of an action variable. Explicitly one has

$$\exp(-2\pi \mathrm{i} A)\, B \exp(2\pi \mathrm{i} A) = B + l.o.t. \tag{3.7}$$

so that, one expects that, up to lower order terms, $\exp(-2\pi \mathrm{i} A)$ is a purely imaginary phase.

The actual proof is slightly different: we are now going to summarize how it works. Consider again the operator $U(2\pi) := \exp(-\mathrm{i}2\pi A)$. The first step consists in exploiting Egorov theorem (3.7) in order to show that, since the classical flow of a is periodic, $U(2\pi)$ satisfies Beals' characterization [7] and therefore it is a pseudodifferential operator in a suitable class. Then one has to study the principal symbol $u_{2\pi}$ of $U(2\pi)$. First one gets that it is quasihomogeneous of degree 0. Then one has to show that it is constant. To this end one defines a quasihomogeneous function

$$M(x, \xi) := \lim_{\lambda \to \infty} u_{2\pi}(\lambda^2 x; \lambda^{2\ell} \xi)\,,$$

which is easily seen to exist everywhere and finally one proves that such a function is actually independent of (x, ξ). This last step is done by writing $u_{2\pi}(\lambda^2 x, \lambda^{2\ell}\xi)$ as the integral of its differential along the curve

$$t \mapsto ((\lambda t)^2 x, (\lambda t)^{2\ell} \xi)\,, \quad t \in [0, 1]\,, \tag{3.8}$$

and then by studying the integral of the differential of u_T on the closed curve γ constructed as in Fig. 1. In particular, given ε and $\lambda > 0$, one takes two points (x, ξ), (x', ξ') on the ellipse $x^2 + \xi^{\frac{2}{2\ell}} = \varepsilon^2$. Then the support of γ is built as

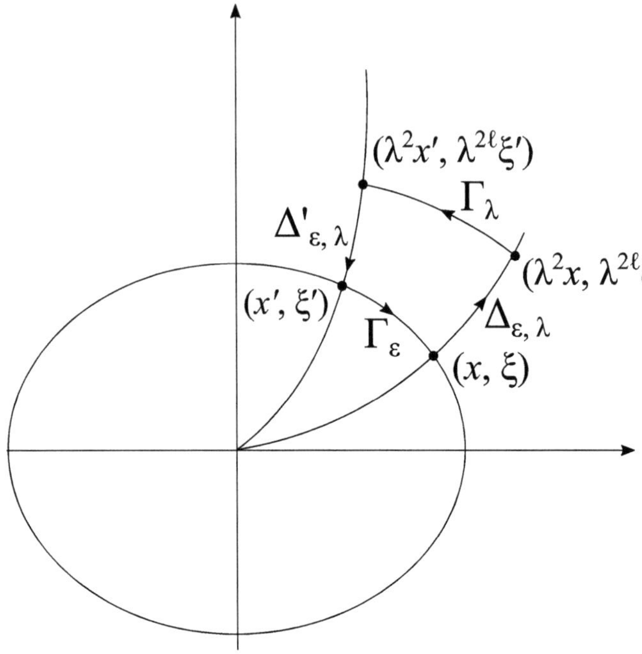

Fig. 1 Representation of the path γ

$$\gamma = \Gamma_\varepsilon \cup \Delta_{\varepsilon,\lambda} \cup \Gamma_\lambda \cup \Delta'_{\varepsilon,\lambda} ,$$

where

- Γ_ε is the shortest arc of the ellipse joining (x, ξ) and (x', ξ')
- $\Delta_{\varepsilon,\lambda}$ is the arc of the curve (3.8) joining (x, ξ) with $(\lambda^2 x, \lambda^{2\ell}\xi)$
- $\Delta'_{\varepsilon,\lambda}$ is constructed as $\Delta_{\varepsilon,\lambda}$ but replacing (x, ξ) with (x', ξ')
- Γ_λ is the arc of ellipse joining $(\lambda^2 x, \lambda^{2\ell}\xi)$ with $(\lambda^2 x', \lambda^{2\ell}\xi')$.

Then one proves that, as $\varepsilon \to 0$ and $\lambda \to \infty$, the integrals on Γ_ε and on Γ_λ tend to 0. Thus, one gets that the two integrals on $\Delta_{\varepsilon,\lambda}$ and on $\Delta_{\varepsilon',\lambda'}$ coincide in the limit $\varepsilon \to 0$ $\lambda \to \infty$, namely $M(x, \xi) = M(x', \xi')$. Finally, since U is unitary, one must have $|M| = 1$, and this allows to conclude the proof. □

The above theorem clarifies the situation in the case of operators on $\mathbb{R}^n$. We come now to the case of compact manifolds. So, let M be a compact Riemannian manifold and denote by $\Psi^m(M)$ the space of pseudodifferential operators of order m *à la Hörmander* (see [40], Chap. 18 for the precise definition). First we recall that a function $a \in C^\infty(T^*M \setminus \{0\})$ is homogeneous of degree m, if in any canonical coordinate system on T^*M it has the property that

$$a(x, \lambda\xi) = \lambda^m a(x, \xi) , \ \forall \lambda > 0 .$$

The analogue of Theorem 3.4 for the case of compact manifolds is the following theorem:

Theorem 3.5 (Theorem 1.1 of [27]) *Let $a \in C^\infty(T^*M \setminus \{0\})$ be homogeneous of degree 1 and denote by A its Weyl quantization; assume that all the orbits of the Hamiltonian vector field of a are periodic with period 2π, then there exists a pseudodifferential operator $Q \in \Psi^{-1}(M)$ of order -1, which commutes with A and a real number κ s.t.*

$$\sigma(A+Q) \subset (\mathbb{Z}+\kappa) . \tag{3.9}$$

Namely, up to a correction smoothing of order -1 the quantization of a Hamiltonian with periodic flow has spectrum contained in a translation of $\mathbb{Z}$.

3.1 Conclusion

Both in the case of $\mathbb{R}^n$ and in the case of compact manifolds, the quantization of a globally defined action is an operator A with the property that there exists an operator Q of order -1, which commutes with A and has spectrum contained in a translation of $\mathbb{Z}$.

Thus the idea that one can consider is to neglect the regularizing perturbation and to define directly a quantum action to be a pseudodifferential operator of order 1 with spectrum contained in a translation of $\mathbb{Z}$. As shown in the next section, this has the advantage that in some situations one can avoid to pass through the classical system.

4 Two Purely Quantum Examples

In this section we give two examples of quantum models where the Hamiltonian can be described as a function of a few number of operators $A_1, \ldots, A_{\mathrm{d}}$, each of which has spectrum lying on $\mathbb{Z}+\kappa_j$ for a suitable real κ_j.

The first simple example is that of a free quantum particle on the d dimensional sphere, whose Hamiltonian is the Laplace-Beltrami operator on the sphere. In this case it is well known that the spectrum of the Laplacian is given by $j(j+(n-1))$, and the corresponding eigenspace is spanned by the spherical Harmonics $Y_{j,l}$. Thus in particular one can write $-\Delta_g = (A+Q)^2$, with A defined by

$$AY_{j,l} := \left(j + \frac{n-1}{2}\right) Y_{j,l} ,$$

and Q defined as a spectral multiplier of order -1.

So, in this case one can define a quantum action to be the operator A and the Laplace-Beltrami operator turns out to be a bounded perturbation of A^2.

The situation is similar on Zoll manifolds, on which one still has $-\Delta_g = (A+Q)^2$ and Q is a pseudodifferential operator of order -1.

A second quantum example is that of a free quantum particle on a simply connected compact Lie group G. We remark that in the case of $M = SO(3)$ this is a quantum rigid body. Endow M by the bi-invariant metric, which in the case of the rigid body corresponds to the case of a spherically symmetric rigid body. The corresponding (quantum) Hamiltonian is the Laplace-Beltrami operator $-\Delta_g$. As we will show in a while, in this case the Laplacian turns out to be the function of a few number of operators with spectrum lying on $\mathbb{Z}+\kappa$, for some $\kappa \in \mathbb{R}$.

To describe this point we have to rapidly introduce the intrinsic Fourier series on Lie groups. The key point is that the Fourier coefficients of a function on a Lie group are labeled by the irreducible unitary representations of G. More precisely, if ξ is an irreducible unitary representation, then the corresponding Fourier coefficient of a function ψ is an element of the representation space of such a representation (see for instance [47]). Furthermore one has that such representations are in 1-1 correspondence with the elements of the cone $\Lambda^+(G)$ of the dominant weights, defined by

$$\Lambda^+(G) = \left\{ \mathtt{w} \in \mathbb{R}^d \;\middle|\; \mathtt{w} = \mathtt{w}^1 \mathtt{f}_1 + \cdots + \mathtt{w}^d \mathtt{f}_d \,, \quad \mathtt{w}^j \in \mathbb{N} \,,\, \forall j = 1, \ldots, d \right\} , \tag{4.1}$$

where $\mathtt{f}_1, \ldots, \mathtt{f}_d \in \mathbb{R}^d$ are the *fundamental weights* of G. In this language the Laplacian is a Fourier multiplier. Precisely, given a dominant weight $\mathtt{w}$, consider its decomposition on the basis $\mathtt{f}_j$, namely $\mathtt{w} = \sum_{j=1}^d \mathtt{w}^j \mathtt{f}_j$, then the Laplacian is the Fourier multiplier by

$$\sum_{j,i} (\mathtt{w}^j + 1)(\mathtt{w}^i + 1) \mathtt{f}_i \cdot \mathtt{f}_j - \sum_{j,i} \mathtt{f}_i \cdot \mathtt{f}_j \,,$$

which is a homogeneous quadratic function of the basic Fourier multipliers

$$A_j := \text{multiplier by } (\mathtt{w}^j + 1) \,. \tag{4.2}$$

Remark 4.1 Since $\mathtt{w}^j$ are integers, the spectrum of such operators are just contained in $\mathbb{Z}$. Furthermore, exploiting the results of [35], in [18] we proved that such operators are actually pseudodifferential operators in the sense of Hörmander.

Actually it can be seen that the situation is absolutely similar in homogeneous spaces. We also remark that the sphere is also a homogeneous space, and this suggests that there should be a connection between the two examples of this section. However, we also remark that Zoll manifolds are not, in general, homogeneous spaces.

5 Globally Integrable Quantum Systems

The theorems and the examples of the previous sections give a strong hint on a possible definition of quantum action and quantum integrable system.

We will treat in a unified way both the situation of a compact manifold or that of $\mathbb{R}^n$. In particular on $\mathbb{R}^n$ we are interested to the anharmonic oscillator with quantum Hamiltonian

$$H_{An} := -\frac{\Delta}{2} + \frac{\|x\|^{2\ell}}{2\ell} , \tag{5.1}$$

with some fixed $\ell \in \mathbb{N}$, $\ell \geq 1$. We define

$$K_0 := \begin{cases} (\mathbf{1} - \Delta_g)^{1/2} & \text{if } M \text{ is compact} \\ (\mathbf{1} + H_{An})^{(\ell+1)/2\ell} & \text{if } M \quad \text{is } \mathbb{R}^n \end{cases} . \tag{5.2}$$

Then we consider the scale of Hilbert spaces $\mathcal{H}^s := D(K_0^s)$, if $s > 0$ and $\mathcal{H}^s$ the dual of $\mathcal{H}^{-s}$ with respect to the L^2 scalar product, if $s < 0$.

Concerning pseudodifferential operators, we define

$$\mathcal{A}^m := \begin{cases} \Psi^m(M) \text{ if } M \text{ is compact} \\ \mathrm{OPS}^m_{An} \text{ if } M \quad \text{is } \mathbb{R}^n \end{cases} . \tag{5.3}$$

5.1 *Definitions*

As anticipated above, the actions will be a set of commuting self-adjoint operators, each of which with spectrum contained in a translation of $\mathbb{Z}$.

Before giving the precise definition we recall that, given a set of self-adjoint operators $A_1, ..., A_d$ with pure point spectrum, their joint spectrum is defined as follows:

Definition 5.1 The joint spectrum Λ of the operators A_j is the set of the $a = (a^1, \dots, a^d) \in \mathbb{R}^d$ s.t. there exists $\psi \in \mathcal{H}$ with

$$A_j\psi = a^j\psi , \quad \forall j = 1, ..., d . \tag{5.4}$$

Definition 5.2 (*Set of quantum actions*) A set of self-adjoint pseudodifferential operators $(A_1, ..., A_d)$, with $A_j \in \mathcal{A}^1$ are said to be a set of quantum actions if

i. They are pairwise commuting
ii. $\exists c_1 > 0$ s.t. $c_1 K_0^2 < \mathbf{1} + \sum_{j=0}^{d} A_j^2$.
iii. There exist a convex closed cone $\mathcal{C} \subseteq \mathbb{R}^d$ and a vector $\kappa = (\kappa_1, ..., \kappa_d) \in \mathbb{R}^d$, such that the joint spectrum Λ of the A_j's fulfills

$$\Lambda \subset (\mathbb{Z}^d + \kappa) \cap \mathcal{C} . \tag{5.5}$$

Definition 5.3 A self-adjoint operator H_L is said to be *globally integrable* if there exists a function $h_L : \mathbb{R}^d \to \mathbb{R}$ with the property that

$$H_L = h_L(A_1, ..., A_d) \ , \tag{5.6}$$

with the operator function spectrally defined.

We remark that Definition 5.3 is given in order to cover also the case of the quantization of superintegrable systems, in which the total number of globally defined action variables is less than the number of degrees of freedom. In particular with this definition one renounces to any assumption ensuring "completeness" of the set of quantum actions. This is recovered in the case where the multiplicity of the common eigenvalues of the actions is 1. Indeed one has the following very simple remark

Remark 5.4 Assume that the multiplicity of the common eigenfunctions of the action is 1, namely that $\forall a \in \Lambda$ there exists a **unique** $\psi_a \in \mathcal{H}$ s.t. $\|\psi\| = 1$ and (5.4) holds, then if $F \in \mathcal{A}^m$ for some m is a self-adjoint operator on $\mathcal{H}$ with the property that

$$[A_j, F] = 0 \ , \quad \forall j = 1, ..., d$$

then there exists a function $f : \Lambda \to \mathbb{R}$ s.t. $F = f(A)$. Indeed one can just define $f(a) := \langle \psi_a; F\psi_a \rangle_{\mathcal{H}}$.

The Definitions 5.2 and 5.3 of system of quantum actions and of integrable quantum system we gave have a strong limitation. This is hidden in the statements of Theorems 3.4 and 3.5. Indeed such theorems give the spectrum of an operator obtained by quantizing a function a which is *globally defined* on the phase space. This is the reason why we added the word *globally* to the word integrable in the above definition. Furthermore, in view of this limitation it is important to present some nontrivial examples of *globally integrable quantum systems*.

5.2 Examples of Globally Integrable Quantum Systems

As proven in [18], some examples of globally integrable quantum systems are:

1. $(M, g) = (\mathbb{T}^d, g)$, with g an arbitrary flat metric, and $H_L = -\Delta_g$. Here the quantum actions are given by $A_j := \mathrm{i}\partial_j$, and one easily sees that $-\Delta_g = H_L = \sum_{i,j=1}^d g^{ij} A_i A_j$, where g^{ij} is the inverse of the matrix of the metric.
2. (M, g) a Zoll manifold. This is the situation studied in [27], and is the same one described for spheres in Sect. 4. As for spheres, one has $-\Delta_g = (A + Q)^2$, for some commuting operators $Q \in \Psi^{-1}$ and $A \in \Psi^1$. The quantum action is A, and one again has $H_L = A^2 = -\Delta_g +$ lower order terms.
3. (M, g) a rotation invariant surface as follows: M is the level surface $f(x, y, z) = 1$ of a C^∞ function $f : \mathbb{R}^3 \to \mathbb{R}$ which is a submersion at $f(x, y, z) = 1$ and

invariant under rotation around the z axis. We endowed M with metric g induced by the embedding. If $\theta \in [0, L]$ is the curvilinear abscissa along the geodesic γ as in Fig. 2, and we introduce cylindrical coordinates (r, θ, ϕ), then in such coordinates one has $M = \{(r(\theta), \theta, \phi) \mid \theta \in [0, L], \quad \phi \in \mathbb{T}\}$. We assume that the function $r(\theta)$ has a unique critical point in $(0, L)$. This case was covered by [28], where it was shown that there exists a GIQS H_L which coincides with $-\Delta_g$ up to lower order corrections. Here the quantum actions are a lower order perturbation of of the quantization of the actions of the principal symbol of H_L.

4. (M, g) a simply connected, compact Lie Group, with H_L a bounded perturbation of $-\Delta_g$. This is the situation described in Sect. 4; further details will be discussed in Sect. 8.
5. $(M, g) = (\mathbb{R}^2, g)$ with g the Euclidean metric and H_L a correction of H_{An}, cf. Eq. (5.1). This model will be discussed in detail in Sect. 8

Remark 5.5 We point out that, in the examples 1,3 and 5, any a in the joint spectrum Λ has multiplicity 1 as a joint eigenvalue for $A_1, \ldots, A_d$, so we are in the situation of Remark 5.4.

On the contrary, in Examples 2 and 4 one deals with superintegrable quantum systems.

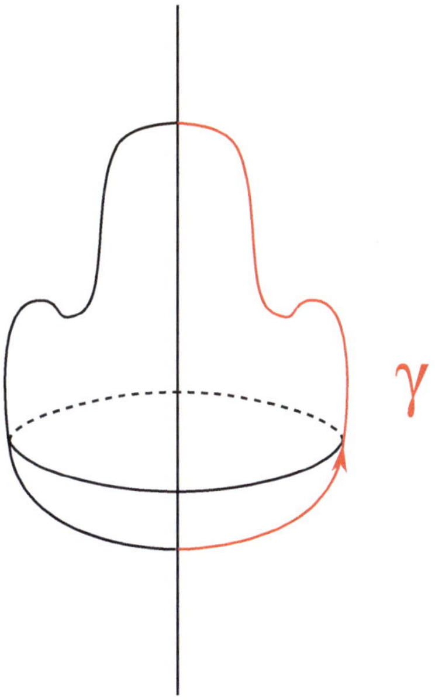

Fig. 2 An admissible rotation invariant surface M. In red, the geodesic γ parametrized by $\theta \in [0, L]$

5.3 Semiclassical Integrable Systems

We end this section by recalling the definition of a semiclassical integrable system given by Colin de Verdière in his book [26]. First we recall that Colin de Verdière deals with *semiclassical Hamiltonian*[1] operators on compact manifolds and studies the behavior of the various objects as $\hbar \to 0$.

Definition 5.6 A semiclassical Hamiltonian H on an n dimensional manifold is said to be *semiclassically completely integrable* if there exist n selfadjoint semiclassical operators $F_1, ..., F_n$ of order 0 which are pairwise commuting s.t.

i. The principal symbols $f_1, ..., f_n$ are in involution and almost everywhere independent;
ii. There exists a function $h_L \in C^\infty(\mathbb{R}^n : \mathbb{R})$ s.t.

$$H = h_L(F_1, ..., F_n) \ .$$

We remark that this definition does not make any reference to the action angle variables, but just to the existence of a sufficient number of "prime integrals" of the system.

In [26] one can also find a semiclassical analogue of Arnold Liouville theorem, showing that microlocally close to an invariant nondegenerate torus of the classical system, one can conjugate the Hamiltonian to a function of the operators $-i\hbar\partial_j$ (see also Charbonnel [29]) on $\mathbb{T}^d$. We emphasize that the semiclassically integrable systems dealt with in such studies are more general than those covered by our definition, but on the contrary these semiclassical studies only give local results (see also [48] for some semiglobal results). We remark that we do not know if the results of Sect. 6 can be extended to semiclassical integrable systems.

6 Perturbations of Globally Integrable Quantum Systems

In this section we will give the statement of two kind of results on perturbations of globally integrable quantum systems, the first one deals with linear perturbations, while the second deals with small nonlinear Hamiltonian perturbations.

The main assumption on the GIQS that we perturb is that h_L is homogeneous and steep. The main tool used in the proof is a result of clustering of the joint spectrum Λ that is given in the forthcoming Sect. 7 (see Theorem 7.4).

[1] for a precise definition, we refer to [26].

We recall that homogeneous functions are typically singular at the origin, but, in the context of pseudodifferential operators, the behavior of functions in a neighborhood of the origin is not important. In particular, in order to get rid of singularities at the origin remaining in the class of symbols, we give the following:

Definition 6.1 A function $f \in C^\infty(\mathbb{R}^d)$ is said to be homogeneous of degree m at infinity if there exists $R > 0$ such that

$$f(\lambda a) = \lambda^m f(a) \quad \forall \lambda > 1 , \quad \forall |a| \geq R .$$

We recall, from [36], the definition of steepness:

Definition 6.2 (Steepness) Let $\mathcal{U} \subset \mathbb{R}^d$ be a bounded connected open set with nonempty interior. A function $h_0 \in C^1(\mathcal{U})$, is said to be steep in $\mathcal{U}$ with steepness radius $\mathtt{r}$, steepness indices $\alpha_1, \dots, \alpha_{d-1}$ and (strictly positive) steepness coefficients $\mathtt{B}_1, ..., \mathtt{B}_{d-1}$, if its gradient $w_i(a) := \frac{\partial h_0}{\partial a_i}(a)$ fulfills: $\inf_{a \in \mathcal{U}} \|w(a)\| > 0$ and for any $a \in \mathcal{U}$ and for any s dimensional linear subspace $M \subset \mathbb{R}^d$ orthogonal to $w(a)$, one has

$$\max_{0 \leq \eta \leq \xi} \min_{u \in M : \|u\| = 1} \|\Pi_M w(a + \eta u)\| \geq \mathtt{B}_s \xi^{\alpha_s} \quad \forall \xi \in (0, \mathtt{r}] , \tag{6.1}$$

where Π_M is the orthogonal projector on M; the quantities u and η are also subject to the limitation $a + \eta u \in \mathcal{U}$.

It is well known that steepness is generic. We also recall that all convex or quasiconvex functions are also steep.

A very useful sufficient condition for steepness is contained in the following theorem by Niederman [46]. This is the theorem that we used in order to verify steepness in our concrete applications.

Theorem 6.3 (Niederman) *Let h_L be a function real analytic in an open set $\mathcal{U} \subset \mathbb{R}^d$. Assume that h_L has no critical points in $\mathcal{U}$, and its restriction $h_L\big|_M$ to any affine subspace $M \subset \mathbb{R}^d$ admits only isolated critical points, then h_L is steep on any compact subset of $\mathcal{U}$.*

So, we concentrate on perturbations of linear systems with Hamiltonian H_L which is globally integrable quantum system, namely $H_L = h_L(A)$ and make the following assumption

Assumption *(L)* h_L is homogeneous of degree $\mathtt{d} > 1$ at infinity and steep in an open set $\mathcal{U} \subset B_1 \setminus B_{1/2}$ with the property that

$$\Lambda \cap \mathcal{U} = \Lambda \cap (B_1 \setminus B_{1/2}) .$$

6.1 Linear Perturbations: Spectral Stability and Growth of Sobolev Norms

6.1.1 Stability of the Majority of Eigenvalues

The first result we are going to present is a stability result of the majority of eigenvalues of H_L when one adds a perturbation which is relatively bounded with respect to H_L this is a generalization of Theorems 2.8 of [19] and 2.8 of [22].

First we consider the general case where the common eigenvalue possible have some multiplicity.

In this case it is possible to compute the first correction to the eigenvalues bifurcating from the majority of eigenvalues of H_L. This requires to define the average of the perturbation V, namely the operator

$$\langle V\rangle := \frac{1}{(2\pi)^d}\int_{\mathbb{T}^d} e^{-i\varphi\cdot A} V e^{i\varphi\cdot A} d\varphi\,, \tag{6.2}$$

which commutes with all the actions. Denote also

$$\Pi_a := \Pi_{a^1}...\Pi_{a^d} \tag{6.3}$$

the projector on the common eigenspace of the actions corresponding to the common eigenvalue $a = (a^1, ..., a^d)$, and denote by n_a the dimension of $\Pi_a\mathcal{H}$, namely the multiplicity of the common eigenvalues, then the operator $\Pi_a\langle V\rangle\Pi_a$, gives the first variation of the eigenvalues bifurcating from λ_a. Precisely, denote by $\mu_{a,j}$ the eigenvalues of $\Pi_a\langle V\rangle\Pi_a$, then we have the following:

Theorem 6.4 *Consider the operator*

$$H := H_L + V\,, \tag{6.4}$$

assume that H_L fulfills assumption L and $V \in \mathcal{A}^{\mathtt{b}}$ with $\mathtt{b} < \mathtt{d}$. Then there exist a set $\Omega \subset \mathbb{R}^d$ and number $m_2 < \mathtt{b}$ such that the following holds. $\Omega \cap \Lambda$ has density one at infinity, namely there exists $\rho > 0$ such that

$$1 - \frac{\sharp(\Omega\cap\Lambda\cap B_R)}{\sharp(B_R\cap\Lambda)} = \mathcal{O}(R^{-\rho})\,, \tag{6.5}$$

and $\forall a \in \Omega$ and any $j = 1, ..., n_a$ there exists an eigenvalue $\lambda_{a,j}$ of (6.4) fulfilling

$$|\lambda_{a,j} - \lambda_a - \mu_{a,j}| \lesssim \langle a\rangle^{m_2}\,.$$

In the case in which the common eigenvalues of the actions have no multiplicity one can give a more precise result, to this end we need first the following:

Definition 6.5 Given a function $f : \mathcal{V} \subseteq \mathbb{R}^d \to \mathbb{R}$, we say that f admits an asymptotic expansion,

$$f \sim \sum_j f_j \,, \tag{6.6}$$

if there exist a real sequence $\{m_j\}_{j\in\mathbb{N}}$ and a sequence $\{f_j\}_{j\in\mathbb{N}}$ of smooth functions $f_j : \mathcal{V} \to \mathbb{R}$ such that

$$m_{j+1} < m_j \quad \forall j \,, \quad m_j \to -\infty \quad \text{as} \quad j \to \infty \,, \tag{6.7}$$

$$|f_j(a)| \le C_j \langle a \rangle^{m_j} \quad \forall a \,, \tag{6.8}$$

$$\left| f(a) - \sum_{j=0}^{N} f_j(a) \right| \le C_N \langle a \rangle^{m_{N+1}} \quad \forall a, \quad \forall N \in \mathbb{N} \,. \tag{6.9}$$

Theorem 6.6 *Under the same assumptions of Theorem 6.4, assume also that the multiplicity of each common eigenvalues of the actions is 1 (as in Remark 5.4). Then there exist a sequence of smooth functions $z_j : \mathcal{C} \to \mathbb{R}$, with $\mathcal{C}$ as in Definition 5.2, such that the functions z_j depend on a only, and $\forall a \in \Omega \cap \Lambda$ there exists an eigenvalue λ_a of (6.4) which admits the asymptotic expansion*

$$\lambda_a \sim h_L(a) + \sum_{j \ge 0} z_j(a) \,. \tag{6.10}$$

Such a theorem ensures that in the case of multiplicity one, the majority of the eigenvalues of the perturbed problem are stable, namely they move only by asymptotically small quantities and furthermore admit an asymptotic expansion.

It is also possible to give a description of all the eigenvalues outside Ω, following the presentation of [21], but the corresponding precise statement is quite complicated, so we avoid to give it here.

6.1.2 Time Dependent Perturbations

The second kind of results deals with the problem of growth of Sobolev norms, when the perturbation is time dependent. This is contained in the following Theorem.

Theorem 6.7 (Main Theorem of [18]) *Let $H = H(t)$ be of the form*

$$H(t) := H_L + V(t) \tag{6.11}$$

with H_L fulfilling assumption L. Assume also that $V(\cdot) \in C_b^\infty\left(\mathbb{R}; \mathcal{A}^{\mathtt{b}}\right)$ is a family of self-adjoint pseudodifferential operators with $\mathtt{b} < \mathtt{d}$; then for any $s \ge 0$ and for any *initial datum* $\psi \in \mathcal{H}^s$ *there exists a unique global solution* $\psi(t) := \mathcal{U}(t,\tau)\psi \in \mathcal{H}^s$ *of the initial value problem*

$$\mathrm{i}\partial_t \psi(t) = H(t)\psi(t)\,, \quad \psi(\tau) = \psi\,, \tag{6.12}$$

furthermore, for any $s > 0$ *and* $\varepsilon > 0$ *there exists a positive constant* $K_{s,\varepsilon}$ *such that for any* $\psi \in \mathcal{H}^s$

$$\|\mathcal{U}(t,\tau)\psi\|_s \leq K_{s,\varepsilon}\langle t-\tau\rangle^{\varepsilon}\|\psi\|_s\,, \quad \forall t,\tau \in \mathbb{R}\,. \tag{6.13}$$

In particular Theorem 6.7 applies to the following equations:

1. Quantum particle on $\mathbb{T}^n$ subjected to electro-magnetic potential:

$$\mathrm{i}\partial_t \psi = \sum_{j=1}^{n}(\mathrm{i}\partial_j + B_j(t,x))^2\psi + W(t,x)\psi\,, \quad x \in \mathbb{T}^n\,, \tag{6.14}$$

with $B_1,\dots,B_n, W \in C_b^\infty(\mathbb{R};\mathbb{T}^n)$. This case was also studied in [20], with techniques specific to the torus.

2. Perturbed two-dimensional quantum anharmonic oscillator:

$$\mathrm{i}\partial_t \psi = -\frac{1}{2}\Delta\psi + \frac{1}{2\ell}\|x\|^{2\ell}\psi + \mathrm{i}\sum_{i,j=1,2} v_{ij}(t)x_i\partial_{x_j}\psi\,, \quad x \in \mathbb{R}^2 \tag{6.15}$$

with $v_{i,j}(t) \in C_b^\infty(\mathbb{R})$.

3. Quantum particle on any one of the manifolds of Sect. 5.2 with time dependent potential:

$$\mathrm{i}\partial_t \psi = -\Delta_g\psi + V(t)\psi\,, \tag{6.16}$$

with $V \in C_b^\infty(\mathbb{R};M)$.

Remark 6.8 The classical counterpart to the problem of bounding the growth of Sobolev norms for (6.14) was treated in [4] for a particle on the torus; there the following result was proven: consider the classical Hamiltonian

$$h(x,\xi) = \frac{1}{2}\sum_{j=1}^{n}(\xi_j - B_j(t,x))^2 + W(t,x)\,, \quad (x,\xi) \in \mathbb{T}^n \times \mathbb{R}^n\,, \tag{6.17}$$

with $B_1,\dots,B_n, W \in C_b^\infty(\mathbb{R}\times\mathbb{T}^n)$. Then for any $\varepsilon > 0$ there exists $C_\varepsilon > 0$ such that all solutions of the Hamilton equations generated by h satisfy

$$\|\xi(t)\| \leq C_\varepsilon\langle t\rangle^{\varepsilon}(1+\|\xi(0)\|)\,, \quad \forall t \in \mathbb{R}\,.$$

6.2 Nonlinear Hamiltonian Perturbations

Concerning nonlinear perturbations, we have to restrict to the case of operators on compact manifolds so we assume that M is a compact manifold, and consider a nonlinear PDE of the form

$$\mathrm{i}\psi_t = H_L\psi + P(\psi, \bar{\psi}, x) , \tag{6.18}$$

where P is a Hamiltonian vector field whose properties will be specified in a while. Furthermore, besides assumption L, we are going to make a nonresonance assumption on the eigenvalues of H_L; this is typically achieved by considering the case where h_L depends on some external parameters that are used to tune the frequencies. This is the way we proceed in applications.

We start by giving the precise assumption on P.

Assumption (*NL*) There exists a function $F \in C^\infty(\mathcal{V} \times \mathcal{V} \times M; \mathbb{C})$ with $\mathcal{V} \subset \mathbb{C}$ a neighborhood of the origin with the following properties

i. $P = \partial_{\bar{\psi}} F$,
ii. there exists C s.t. for any $z \in \mathcal{V}$ one has $F(z, \bar{z}, x) \in \mathbb{R}$, and

$$|F(z, \bar{z}, x)| \leq C|z|^3 .$$

Actually a slightly more general assumption is enough and this is needed for the application to the nonlinear stability of the ground state in NLS, but we will not give here its abstract form.

We are now going to state the nonresonance assumption on the eigenvalues of H_L. As we will discuss in a while, this is a generalization of "0-th Melnikov condition", whose standard form is

$$\begin{gathered}\omega_{a^{(1)}} \pm \omega_{a^{(2)}} \pm \ldots \pm \omega_{a^{(r)}} \neq 0 \quad \Longrightarrow \\ |\omega_{a^{(1)}} \pm \omega_{a^{(2)}} \pm \ldots \pm \omega_{a^{(r)}}| \geq \frac{\gamma_r}{(\max |a^{(j)}|)^{\tau_r}} .\end{gathered} \tag{6.19}$$

To give the precise statement denote by

$$\omega_a := h_L(a) , \; a \in \Lambda$$

the eigenvalues of H_L and remark that by the assumption on the homogeneity of h_L one has a bound on the cardinality of the set of eigenvalues in a ball. From such a bound it is easy to deduce the following clustering property

Lemma 6.9 *There exist two sequences α_n, β_n $n \in \mathbb{N}$, $n \geq 0$ with $\alpha_n < \beta_n < \alpha_{n+1}$ and such that*

- $h_L(\Lambda) \subset \bigcup_n [\alpha_n, \beta_n]$;
- $\beta_n - \alpha_n \leq 2$;
- $\alpha_{n+1} - \beta_n \geq 2n^{-d/\mathtt{d}}$.

The nonresonance condition that we are going to assume ensures that one can use Birkhoff normal form procedure in order to eliminate from the nonlinearity terms enforcing exchanges of energy among modes labeled by indexes belonging to different intervals. This requires to characterize the linear combination of frequencies which correspond to monomials enforcing such exchange of energy. So, we give the following definition

Definition 6.10 Given $r \in \mathbb{N}$ and $j \in \{0, ...r\}$, a set of indexes $(a^{(1)}, ..., a^{(r)}) \in \Lambda^r$ will be said to be *dangerous* if r is even and for any $l = 1, ..., r/2$ there exists an interval $[\alpha_l, \beta_l]$ with the property that

$$a^{(l)} \text{ and } a^{(l+r/2)} \in [\alpha_l, \beta_l] \;, \forall l = 1, ..., r/2 \;. \tag{6.20}$$

A sequence which is not dangerous is called *nondangerous*.

Assumption (*NR*) For any $r \geq 3$, there are constants $\gamma_r, \tau_r > 0$ such that $\forall j = 1, ..., r$ one has

$$\left| \sum_{l=1}^{j} \omega_{a^{(l)}} - \sum_{l=j+1}^{r} \omega_{a^{(l)}} \right| \geq \frac{\gamma_r}{\left(\max |a^{(l)}|\right)^{\tau_r}} \;. \tag{6.21}$$

for any non dangerous sequence $(a^{(1)}, ..., a^{(r)})$. For dangerous sequences we require (6.21) to hold only for $j \neq r/2$.

The following theorem is the main abstract result of [9].

Theorem 6.11 *Consider the Hamiltonian system (6.18). Assume Hypotheses NL and NR, then for any integer $r \geq 3$, there exists $s_r \in \mathbb{N}$ such that, for any $s \geq s_r$, there are constants $\epsilon_0 > 0$, $c > 0$ and C for which the following holds: if the initial datum $\psi_0 \in H^s(M, \mathbb{C})$ fulfills*

$$\epsilon := \|\psi_0\|_s < \epsilon_0 \,,$$

then the Cauchy problem has a unique solution $u \in C^0\left((-T_\epsilon, T_\epsilon), H^s(M, \mathbb{C})\right)$ with $T_\epsilon > c\epsilon^{-r}$. Moreover, one has

$$\|\psi(t)\|_s \leq C\epsilon, \quad \forall t \in (-T_\epsilon, T_\epsilon) \;. \tag{6.22}$$

In particular, in [9] we show that Theorem 6.11 applies to the following models:

1. Schrödinger equation with convolution potential on one of the manifolds M of the examples 1, 2, 3, 4, of Sect. 5.2

$$\mathrm{i}\partial_t \psi = -\Delta_g \psi + V * \psi + f(x, |\psi|^2)\psi\,, \quad x \in M\,, \tag{6.23}$$

with $f(x, \cdot)$ a smooth function having a zero of order at least 2 at 0, and V belonging to a full measure subset of

$$\mathcal{V}_m := \left\{ V = \{V_k\}_{k \in \mathbb{Z}^d} \;\middle|\; \langle k \rangle^m |V_k| \in \left[-\frac{1}{2}, \frac{1}{2} \right] \right\} \quad \text{for some } m > 0\,.$$

The case of $M = \mathbb{T}^d$ with a flat metric was already dealt with in [10].

2. Beam equation on one of the manifolds M of the examples 1, 2, 3, 4, of Sect. 5.2

$$(\partial_{tt} + \Delta_g^2 + m)u = -\partial_u F(x, u)\,, \tag{6.24}$$

with $F \in \mathcal{C}^\infty(M \times \mathcal{V}; \mathbb{R})$, where $\mathcal{V}$ is a neighborhood of the origin, having a zero of order at least 3 at $u = 0$, and m belonging to a full measure subset of $\mathbb{R}^+$.

3. Stability of the ground state in Nonlinear Schrödinger equation on one of the above manifolds

$$\mathrm{i}\partial_t \psi = -\Delta_g \psi + f(|\psi|^2)\psi\,, \tag{6.25}$$

with $f \in \mathcal{C}^\infty(\mathcal{V}; \mathbb{R})$ having a zero of order at least 1 at the origin. Then, for any $p_0 \in \mathcal{V} \cap \mathbb{R}^+$ the plane wave

$$\psi_\star(t) = \sqrt{p_0} e^{\mathrm{i}\nu t}\,, \quad \nu = f(p_0)\,, \tag{6.26}$$

is a solution of (6.25). In this case, from the abstract Theorem 6.11 we deduce the following:

Theorem 6.12 *Let $\bar{\lambda}$ be the minimum nonvanishing eigenvalue of $-\Delta_g$ and assume there exists $\bar{p}_0 > 0$ such that $\bar{\lambda} + 2f(p_0) > 0$ for any $p_0 \in (0, \bar{p}_0]$. Then there exists a zero measure set $\mathcal{P} \subset (0, \bar{p}_0]$ such that if $p_0 \in (0, \bar{p}_0] \setminus \mathcal{P}$ then for any $r \in \mathbb{N}$ there exists s_r for which the following holds. For any $s \geq s_r$, there exist constants ϵ_0 and C such that if the initial datum ψ_0 fulfills*

$$\|\psi_0\|_0^2 = p_0, \qquad \inf_{\alpha \in \mathbb{T}} \|\psi_0 - \sqrt{p_0} e^{-i\alpha}\|_s = \epsilon \leq \epsilon_0\,,$$

then the corresponding solution fulfills

$$\inf_{\alpha \in \mathbb{T}} \|\psi(t) - \sqrt{p_0} e^{-i\alpha}\|_s \leq C\epsilon \qquad \forall\, |t| \leq C\epsilon^{-r}\,.$$

7 Nekhoroshev-Bourgain Partition of the Joint Spectrum

In this section we give a partition of the joint spectrum Λ, which is the quantum analogue of the partition of the action space of a classical system in resonant blocks introduced by Nekhoroshev in order to prove his celebrated theorem on exponential stability of quasi integrable systems. It turns out that such a partition also has a remarkable property that was introduced by Bourgain in order to prove KAM type results for nonlinear wave and Schrödinger equation on $\mathbb{T}^d$.

As a result, this partition is the main tool for proving all the results of the previous section.

7.1 Classical Nekhoroshev Theorem and Classical Nekhoroshev's Partition

We start with recalling the classical Nekhoroshev Theorem:

Theorem 7.1 *Let $h : \mathcal{A} \times \mathbb{T}^n \to \mathbb{R}$ be a function of the form*

$$h(a,\alpha) = h_L(a) + \varepsilon v(a,\alpha)\,,$$

where h_L and v are analytic on $\mathcal{A}$ and h_L is steep on $\mathcal{A}$. Let $\mathcal{U} \subset \mathcal{A}$ be an open set such that $\overline{\mathcal{U}} \subset \mathcal{A}$. Then there exist $\varepsilon_0 > 0$, and $\mathtt{a}, \mathtt{b} > 0$ such that, if $|\varepsilon| < \varepsilon_0$, then, for all initial data in $\mathcal{U} \times \mathbb{T}^n$ one has

$$|a(t) - a(0)| \leq \varepsilon^{\mathtt{b}} \quad \forall |t| \leq \varepsilon^{-1} e^{\varepsilon^{-\mathtt{a}}}\,. \tag{7.1}$$

The classical proof as formulated by Nekhoroshev in [43, 44] (see also [4, 17, 36, 37]) is based on two steps: an analytic part, in which the Hamiltonian h is put in local normal form, removing all non resonant contributions, and a geometric part, based on the analysis of resonances in the space of actions. We now are going to describe the latter geometric analysis of resonances and its role in the proof of Nekhoroshev Theorem, in order to enlighten the similarities with the procedure we adopt in the quantum case.

Actually, here we present the scheme of the proof of [17, 25], which is the closest to the quantum case and deals with long time stability in the case of smooth Hamiltonians:

Theorem 7.2 ($\mathcal{C}^\infty$ Nekhoroshev Theorem) *Let $h : \mathcal{A} \times \mathbb{T}^n \to \mathbb{R}$ be of the form*

$$h(a,\alpha) = h_L(a) + \varepsilon v(a,\alpha)\,,$$

with h_L steep on $\mathcal{A}$ and $h_L, v \in C_b^\infty(\mathcal{A} \times \mathbb{T}^d; \mathbb{R})$. Let $\mathcal{U} \subset \mathcal{A}$ be an open set such that $\overline{\mathcal{U}} \subset \mathcal{A}$. For any $b < \frac{1}{2}$ and $N > 0$ there exist $\varepsilon_N > 0$ and $C_N > 0$ such that, for $0 < \varepsilon < \varepsilon_N$, and any initial datum in $\mathcal{U} \times \mathbb{T}^n$ one has

$$|a(t) - a(0)| \leq \varepsilon^b \quad \forall |t| \leq C_N \varepsilon^{-N} . \tag{7.2}$$

The first step in the proof of Theorem 7.2 consists in performing a normal form procedure, which for any N enables to conjugated the Hamiltonian h to a new one of the form

$$h_N = h_L(a) + \varepsilon z_N(a, \alpha) + \varepsilon^N r_N(a, \alpha) ,$$

with

$$z_N(a, \alpha) = \sum_{\substack{k \in \mathbb{Z}^n \\ |k| \leq K}} \hat{z}_k(a) e^{\mathrm{i}k \cdot \alpha} , \tag{7.3}$$

$K = \varepsilon^{-\frac{1}{\zeta}}$ for some $\zeta > 0$. Furthermore, there exists $\delta > 0$ such that, for each k,

$$\operatorname{supp} \hat{z}_k(a) \subseteq \{a \mid |w(a) \cdot k| \leq \varepsilon^\delta\} , \tag{7.4}$$

where we recall that $w(\cdot) := \partial_a h_L(\cdot)$. Then the geometric part of the proof consists in analyzing the dynamics generated by $h_L + \varepsilon z_N$.

To this aim we first recall that a module of dimension $s = 0, \ldots, n$, is a subset $M \subseteq \mathbb{Z}^n$ such that $\operatorname{span}_{\mathbb{R}}(M) \cap \mathbb{Z}^n = M$. In this section we will denote by M a submodulus of $\mathbb{Z}^n$.

Now we start by defining the *resonant zones*

$$\begin{aligned} \mathcal{Z}_M^{(s)} := \{a \mid \exists\, k_1, \ldots, k_s \quad & \text{linearly indep. in } M \\ & \text{s.t. } |w(a) \cdot k_j| \leq \varepsilon^{\delta_s} \quad \forall j = 1, \ldots, s\} , \end{aligned} \tag{7.5}$$

with a suitable sequences of increasing $\{\delta_s\}_{s=1}^n$ with $\delta_1 = \delta$. For any M, the set $\mathcal{Z}_M^{(s)}$ collects all points a such that $w(a)$ is resonant with some vectors in M. Its role is that, due to the form of the truncated Hamiltonian $h_L + \varepsilon z_N$ (see (7.3)–(7.4)), points in $\mathcal{Z}_M^{(s)}$ only move along directions parallel to M. Then, as long as the point $a(t)$ belongs to some resonant zone, its motion is easy to describe. A key step in the geometric part of the proof is then to understand how the resonant zones intersect one each other, and whether or not is possible to pass from one resonant zone from the other, visiting sooner or later all the action space (this is the well known phenomenon of *overlapping of resonances*).

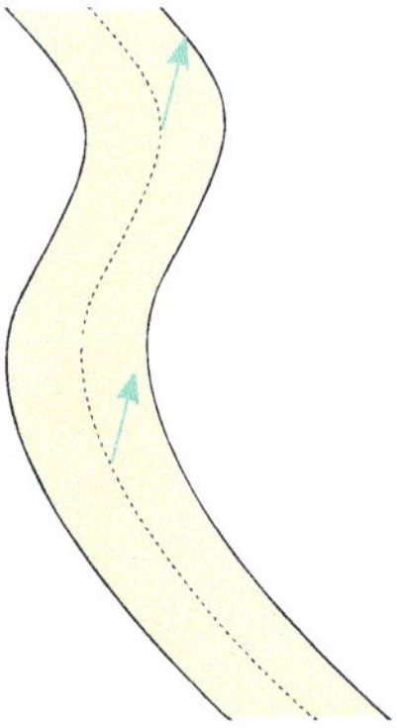

(a) In yellow, cartoon of a resonant zone $\mathcal{Z}_M^{(s)}$, with $M = \text{span}\{k\}$ for a given k. The dashed line corresponds to the curve $w(a) \cdot k = 0$. Inside the zone, the motion generated by $h_L + \varepsilon z_N$ is parallel to k, pictured in turquoise.

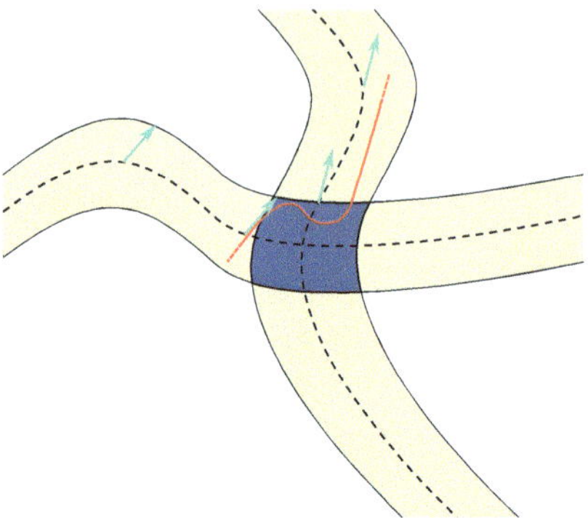

(b) In yellow, two resonant zones associated to different modules. Their intersection is pictured in blue. In orange, a possible dangerous trajectory, passing from one zone to the other (overlapping of resonances, which we want to exclude).

In order to show that the latter phenomenon does not occur, one needs to define a more refined decomposition of the actions space $\mathbb{R}^d$ with respect to the one given by the resonant zones $\{\mathcal{Z}_M^{(s)}\}_{s,M}$. In particular, one proves the existence of a partition

$$\mathbb{R}^d = \bigcup_{s=0}^{n} \bigcup_{\substack{M \subseteq \mathbb{Z}^n \\ \dim M = s \\ M \text{module}}} \bigcup_{j \in \mathbb{N}} \mathcal{E}_{M,j}^{(s)} \tag{7.6}$$

in sets $\mathcal{E}_{M,j}^{(s)}$ such that

1. Each $\mathcal{E}_{M,j}^{(s)}$ has small diameter:

$$\text{diam}\left(\mathcal{E}_{M,j}^{(s)}\right) \leq \varepsilon^{\mathtt{C}_2} \quad \forall s, M, j\,. \tag{7.7}$$

2. Each $\mathcal{E}_{M,j}^{(s)}$ is contained in $\mathcal{Z}_M^{(s)}$, and is far away from having resonances with vectors not in M: in particular, if $k \in \mathbb{Z}^n$ satisfies

$$|w(a) \cdot k| < \varepsilon^{\delta_1}\,, \quad |k| \leq K\,, \tag{7.8}$$

then $k \in M$ and $a + \eta k \in \mathcal{E}_{M,j}^{(s)}$, provided $|\eta| \ll 1$.
3. $\mathcal{E}_{M,j}^{(s)}$ is invariant for the dynamics of $h_L + \varepsilon z_N$.

Note that, once this decomposition has been obtained, the statement of Theorem 7.2 essentially follows combining Items 1 and 3.

7.2 *Invariant Partition in the Quantum Case*

Let us start with defining what we consider *resonant* in the case of a quasi-integrable quantum system:

Definition 7.3 Let $\delta < \mathtt{d}-1$, $\mu > 0$ and $\mathtt{R} > 0$. We say $a \in \Lambda$ *is resonant with* $k \in \mathbb{Z}^d$ if

$$|w(a)\cdot k| < \|a\|^{\delta}\|k\|\,, \quad \|k\| \le \|a\|^{\mu}\,, \quad \|a\| \ge \mathtt{R}\,, \tag{7.9}$$

where $w(a) := \partial_a h_L(a)$

A key point in proving Theorems 6.6, 6.4 and 6.7 is the following result, proven in [18] which gives the partition constituting the main step for the proof of the theorems of Sect. 6:

Theorem 7.4 *Under the assumptions of Theorem 6.7, there exist* $0 < \delta_0 < \mathtt{d}-1$, $\mu_0 > 0$ *and* $\mathtt{R}_0 > 0$ *such that, if* $\delta_0 < \delta < \mathtt{d}-1$, $0 < \mu < \mu_0$, *and* $\mathtt{R} > \mathtt{R}_0$, *the lattice* Λ *admits a partition*

$$\Lambda = \bigcup_{s=0}^{n} \bigcup_{\substack{M\subseteq\mathbb{Z}^d\\ \dim M = s\\ M \text{module}}} \bigcup_{j\in\mathbb{N}} E^{(s)}_{M,j} \tag{7.10}$$

with the following properties:

(i) Each block $E^{(s)}_{M,j}$ *is dyadic, in the sense that there exists* $C > 0$ *such that*

$$\max\{\|a\| \mid a \in E^{(s)}_{M,j}\} \le C \min\{\|a\| \mid a \in E^{(s)}_{M,j}\}\,. \tag{7.11}$$

(ii) If $a \in E^{(s)}_{M,j}$, *there exists a vector* $k \in \mathbb{Z}^d$ *such that* a *is resonant with* k. *Conversely, if* a is resonant with $k \in \mathbb{Z}^d$ *in the sense of Definition 7.3, then* $a + k \in E^{(s)}_{M,j}$.

(iii) There exists $K > 0$ *such that, if* $a \in E^{(s)}_{M,j}$ *and* $b \in E^{(s')}_{M',j'}$ *with* $(s, M, j) \neq (s', M', j')$, *then*

$$\|a-b\| + |\omega_a - \omega_b| \ge K(\|a\|^{\mu} + \|b\|^{\mu})\,. \tag{7.12}$$

We start by remarking the analogies with the properties of the covering $\{\mathcal{E}^{(s)}_{M,j}\}$ exhibited in the classical case in Sect. 7.1: Item (i) gives an upper bound on the diameter of the $E^{(s)}_{M,j}$, just as Property 1 of the sets $\mathcal{E}^{(s)}_{M,j}$. Furthermore, Item (ii) states that, if a is resonant with k, and $\|k\|$ is not too large, then $a+k$ belongs to the same block $E^{(s)}_{M,j}$, which is analogous to Property 2 of the sets $\mathcal{E}^{(s)}_{M,j}$.

We now describe how Theorem 7.4 is used to prove the different theorems.

7.3 Linear Perturbations

The first step of the proof consists in proving a "quantum normal form theorem" for Hamiltonians of the form (6.4) and (6.11).

This is contained in the following:

Proposition 7.5 *For any $N \in \mathbb{N}$ there exists a map $U_N \in C^\infty_b(\mathbb{R}; \mathcal{B}(\mathcal{H}^s, \mathcal{H}^s))$, such that $U_N(t)$ is unitary in $\mathcal{H}$ for any t and ψ solves (6.12) if and only if $\phi = U_N^{-1}(t)\psi$ solves*

$$
\mathrm{i}\partial_t\phi = (H_L + Z_N(t) + R_N(t))\phi\,, \tag{7.13}
$$

with Z_N and R_N self-adjoint operators as follows:

1. (Smoothing remainder) $R_N \in C^\infty_b(\mathbb{R}; \mathcal{B}(\mathcal{H}^s; \mathcal{H}^{s+N}))$ *for any s*
2. (Resonant normal form) $Z_N \in C^\infty_b(\mathbb{R}; \mathcal{B}(\mathcal{H}^s; \mathcal{H}^{s-\mathtt{b}}))$ *for any s. Moreover, for any $a \in \Lambda$ let Π_a be the projection operator defined in (6.3), then one has*

$$
\Pi_a Z_N(t)\Pi_{a+k} \neq 0 \tag{7.14}
$$

   only if a is resonant with k, or $a+k$ is resonant with k.
3. (Block diagonal structure) *As a consequence, Z_N is block diagonal along the blocks $E^{(s)}_{M,j}$ of Theorem 7.4, in the sense that for any t*

$$
\Pi_a Z_N(t)\Pi_b \neq 0 \quad \Rightarrow a, b \in E^{(s)}_{M,j} \tag{7.15}
$$

   for some s, M, j.

Proposition 7.5 is based on an iterative procedure along which, for any n, at the $n-$th step one is able to eliminate from the perturbation $V_n(t)$ all terms

$$
\Pi_a V_n(t)\Pi_{a+k}
$$

such that both a and $a+k$ are *not resonant* with k. Then all the remaining resonant contributions are collected in the normal form Z_N. This is the heart to the proof of Item 2. Note also that Item 3. follows directly combining Item 2. with property (ii) of Theorem 7.4.

We now discuss how Proposition 7.5 enables to Prove Theorems 6.6, 6.4 and 6.7.

Proof (Sketch of the proof of Theorems 6.4 and 6.6) We apply Proposition 7.5 in the time independent case. The main point is to define Ω as the set

$$\Omega := \bigcup_j E^{(0)}_{j,M}$$

where $M = \{0\}$ is the trivial modulus. By Item (ii) of Theorem 7.4, it follows that such a set contains all the points a with the property that $w(a)$ is non- resonant, namely the only vector $k \in \mathbb{Z}^n$ such that (7.9) is satisfied is $k = 0$. As a consequence, by Item 2 of Proposition 7.5 at these points the normal form Z_N turns out to leave invariant the common eigenspaces of the actions, namely one has

$$Z_N \Pi_a = \Pi_a Z_N \Pi_a \,, \quad \forall a \in \Omega \,.$$

Furthermore the algorithm of the proof of Proposition 7.5 also allows to show that $Z_N = \langle V \rangle$+lower order operators, with $\langle V \rangle$ defined by (6.2). This is why the spectrum of the restriction of $\langle V \rangle$ to $\Pi_a \mathcal{H}$ controls the first correction to the unperturbed eigenvalues. Then one should prove that the set Ω has density one at infinity. This can be obtained by following some ideas of the proof of the corresponding theorems of [19, 22], but exploiting the steepness of h_L instead than the arguments in [49].

The proof of Theorem 6.6 then simply follows by the remark that in this case the pseudo-differential operators commuting with all the actions are just functions of the actions.

Proof (Sketch of the proof of Theorem 6.7) By Item 3. of Proposition 7.5, neglecting the remainder $R_N(t)$, one can study the dynamics generated by $H_L + Z_N(t)$: by the fact that (1) Z_N is diagonal with respect to the blocks, (2) Z_N is self-adjoint so that the L^2 norm is conserved along the dynamics it generates, (3) since the blocks are dyadic, in each block the L^2 norm is equivalent to the $\mathcal{H}^s$ norm, one gets that, along the dynamics of $H_L + Z_N(t)$ one has

$$\|\varphi(t)\|_{\mathcal{H}^s} \lesssim_s \|\varphi(0)\|_{\mathcal{H}^s} \,.$$

Then the contribution of the smoothing remainder $R_N(t)$ is dealt with using Duhamel principle. Precisely, denote by $\mathcal{U}$ the flow map of the complete Hamiltonian (7.13) and by $\mathcal{U}_{app}$ the flow map of $H_L + Z_N(t)$, then, by Duhamel formula one has

$$\mathcal{U}(t,\tau)\psi = \mathcal{U}_{app}(t,\tau)\psi + \int_\tau^t \mathcal{U}_{app}(t,\tau')(-\mathrm{i}R^{(N)}(\tau'))\mathcal{U}(\tau',\tau)\psi \, d\tau' \,, \qquad (7.16)$$

Since $R^{(N)} \in C_b^\infty\left(\mathbb{R}; \mathcal{B}(\mathcal{H}^0; \mathcal{H}^N)\right)$ and since $H^{(N)}$ is self-adjoint in L^2, one gets

$$\|\mathcal{U}(t,\tau)\psi\|_N \lesssim \|\psi\|_N + \int_\tau^t \left\|R^{(N)}\right\|_{\mathcal{B}(\mathcal{H}^0;\mathcal{H}^N)} \|\mathcal{U}_{app}(\tau',\tau)\psi\|_0 \, d\tau'$$
$$\lesssim \|\psi\|_N + \int_\tau^t \left\|R^{(N)}\right\|_{\mathcal{B}(\mathcal{H}^0;\mathcal{H}^N)} \|\psi\|_0 \, d\tau' \lesssim \|\psi\|_N + \langle t-\tau\rangle \|\psi\|_0 ,$$
$$\lesssim \langle t-\tau\rangle \|\psi\|_N .$$

Then using interpolation one immediately gets the thesis.

7.4 Nonlinear Perturbations

Proof (Sketch of the proof of Theorem 6.11) Theorem 6.11 is the consequence of a Birkhoff normal form theorem conjugating (6.18) to its linear part plus a "normal form" plus a remainder which is of order ϵ^N, with arbitrary N. Now, the main point is that, as it is well known, in order to perform such a normal form one has to consider small divisor corresponding to the terms one wants to eliminate from the Hamiltonian. The kind of normal form one can obtain is thus directly related to the kind of nonresonant conditions which is satisfied by the frequencies ω_a. Using just the nonresonant condition (6.19) it is difficult to obtain a useful normal form in the general case. Actually the kind of nonresonant condition which is known to be very useful is

$$\omega_{a^{(1)}} \pm \omega_{a^{(2)}} \pm \ldots \pm \omega_{a^{(r)}} \neq 0 \implies$$
$$|\omega_{a^{(1)}} \pm \omega_{a^{(2)}} \pm \ldots \pm \omega_{a^{(r)}}| \geq \frac{\gamma_r}{(\max_3 |a^{(j)}|)^{\tau_r}} , \tag{7.17}$$

where $\max_3$ is the third largest index among the $a^{(j)}$'s.

The idea here is that if one only tries to eliminate from the nonlinearities terms coupling different blocks $E_{M,j}$, then it turns out that for the corresponding small denominators one has a selection rule, which entails the fact that condition (6.19) implies condition (7.17) for all the small denominators that really matter.

The actual implementation of such a scheme is however quite complicated technically and we refer to [9] for more details (see also [10]).

8 Two Examples of Quantum Integrable Systems

In this section we analyze in detail two models and we show that they satisfy the assumptions of globally integrable quantum systems.

8.1 Anharmonic Oscillator

Let us consider the operator

$$H_{An} = -\frac{\Delta}{2} + \frac{\|x\|^{2\ell}}{2\ell}, \quad x \in \mathbb{R}^2, \tag{8.1}$$

with $\ell \in \mathbb{N}$, $\ell > 1$. In this case, the class $\mathcal{A}^m$ of pseudodifferential operators that we consider is the one defined in the second line of (5.3), and we are going to show that there exist a function h_L and two action operators $A_1, A_2 \in \mathcal{A}^1$ such that

$$H_{An} = h_L(A_1, A_2) + R,$$

where h_L coincides outside a neighborhood of the origin with the classical Hamiltonian written in action variables and R is a lower order correction. It was proved in [18] that it is homogeneous of degree $\mathtt{d} = \frac{2\ell}{\ell+1}$; the operator R is a lower order operator in $\mathcal{A}^{\mathtt{b}}$ with $\mathtt{b} < \mathtt{d}$. Remark that lower order terms are not relevant in order to satisfy the assumptions of Theorem 6.7, since the term R can be included in the perturbation $V(t)$.

To prove this fact, following the strategy of Sect. 3: we start by remarking that H_L is the quantization of the classical Hamiltonian

$$h_{An}(x, \xi) = \frac{\|\xi\|^2}{2} + \frac{\|x\|^{2\ell}}{2\ell}, \tag{8.2}$$

and start by constructing suitable action angle variables for it. The action variables that we choose for h_{An} are the angular momentum

$$a_2(x, \xi) := x_1\xi_2 - x_2\xi_1 \tag{8.3}$$

and the radial action a_r, namely the action of the effective Hamiltonian

$$h^*_{An}(r, p_r, \mathtt{M}) := \frac{p_r^2}{2} + V^*_{\mathtt{M}}(r), \quad V^*_{\mathtt{M}}(r) := \frac{\mathtt{M}^2}{2r^2} + \frac{r^{2\ell}}{2\ell}, \tag{8.4}$$

where $r = \|x\|$, p_r is the conjugated momentum and $\mathtt{M}$ is a particular value of the angular momentum, in the sense that we put ourselves on a level surface $\{(x, \xi) \mid a_2(x, \xi) = \mathtt{M}\}$. In particular, one has

$$a_r = a_r(E, \mathtt{M}) := \frac{\sqrt{2}}{\pi} \int_{r_m}^{r_M} \sqrt{E - V^*_{\mathtt{M}}(r)} dr, \tag{8.5}$$

where $0 < r_m < r_M$ are the two solutions of the equation

$$E - V^*_{\mathtt{M}}(r) = 0.$$

However, the actions (a_2, a_r) are *not* good candidates to be quantized for our construction: indeed, while a_2 is a polynomial, a_r turns out to have a singularity at $\mathtt{M} = 0$. Anyway, this singularity can be removed, by performing a suitable unimodular transformation. This is the content of the following lemma, proved in [22]:

Lemma 8.1 (Lemma 4.5 of [22]) *The function*

$$a_1(E, \mathtt{M}) := \begin{cases} a_r(E, \mathtt{M}) & for\ \mathtt{M} > 0 \\ a_r(E, \mathtt{M}) - \mathtt{M} & for\ \mathtt{M} < 0 \end{cases} \tag{8.6}$$

has the following properties:

(1) it extends to a complex analytic function of $\mathtt{M}$ *and* E *in the region*

$$|\mathtt{M}| < \left(\frac{2\ell}{\ell+1}E\right)^{\frac{\ell+1}{2\ell}}, \quad E > 0\,. \tag{8.7}$$

(2) Define the cone $\mathcal{C}$ *by*

$$\mathcal{C} := \left\{a \in \mathbb{R}^2\,;\ a_1 \geq 0\ \text{if}\, a_2 \geq 0,\quad a_1 \geq |a_2|\ \text{if}\ a_2 < 0\right\}\,. \tag{8.8}$$

Then h_{An} *is an analytic function of* (a_1, a_2) *in the interior of* $\mathcal{C}$. *Furthermore it is homogeneous of degree* $\frac{2\ell}{\ell+1}$ *as a function of* (a_1, a_2).

(3) Let $a = (a_1, a_2)$, *there exist positive constants* C_1, C_2 *s.t.*

$$C_1\langle a\rangle \leq \mathtt{k}_0 \leq C_2\langle a\rangle\,.$$

Still, both h_{An} and a_1 have a singularity at the origin. But, essentially due to the fact that the behavior of functions in a finite neighborhood of the origin does not play any fundamental role in symbolic calculus, this is easily fixed: we take $K \gg 1$ and we consider the decomposition

$$h_{An}(x, \xi) = h_{An}(x, \xi)\chi(Kh_{An}(x, \xi)) + h_{An}(x, \xi)(1 - \chi(Kh_{An}(x, \xi)))\,. \tag{8.9}$$

Now the first term is smoothing and, denoting

$$h_L(x, \xi) := h_{An}(x, \xi)(1 - \chi(Kh_{An}(x, \xi)))\,, \tag{8.10}$$

actually h_L is the expression of the function (8.10) in action variables a_1', a_2, so that it coincide with h_{An} (when written in terms of action variables) outside a nighbourhood of the origin. Moreover, since by Item (2) of Theorem 8.1 h_{An} is a homogeneous function of (a_1, a_2), h_L is homogeneous at infinity.

Turning back to the quantum case, Theorems 3.1 and 3.2 of [28] guarantee the following:

Theorem 8.2 *There exist two pseudodifferential operators $R_j \in \mathcal{A}^\rho$, with $\rho < 1$, and two commuting pseudodifferential operators $A_j \in \mathcal{A}^1$ such that $A_j = Op^W(a_j) + R_j$, and A_1, A_2 are quantum actions according to Definition 5.2. In particular there exists a symbol $h \in S_{AN}^{\frac{2\ell}{\ell+1}}$ s.t.*

$$H_{An} = h(A) \, . \tag{8.11}$$

Furthermore one has an asymptotic expansion

$$h = h_L + \text{lower order terms.}$$

with h_L as in (8.10).

It remains to prove that h_L satisfies Assumption L. This is done in the following steps: first one proves that there exists an open cone $\mathcal{C}_e$ containing $\overline{\mathcal{C}} \setminus \{0\}$ such that h_L admits an homogeneous at infinity and analytic extension on $\mathcal{C}_e$. Then one checks that h_L is steep on the cone $\mathcal{C}_e$; this is done in [18] by applying the criterion given in Theorem 6.3. This is verified following the methods of [11], which consist in analyzing the expansion of the Hamiltonian at the circular orbits. In particular, we prove that the restriction of h_L on any affine line only admits isolated critical points.

8.2 Lie Groups

Consider the Laplace Beltrami operator $-\Delta_g$ in (M, g), where M is a simply connected Lie group endowed with the bi-invariant metric g. Here the class of pseudo-differential operators $\mathcal{A}^m$ is given by the first line of (5.3), namely Hörmander class. The discussion in Sect. 4 shows that, if $A_1, \ldots, A_d$ are operators defined in (4.2) and $h_L : \mathbb{R}^d \to \mathbb{R}$ is the function defined by

$$h_L(a) := \sum_{i,j=1}^{d} a_i a_j \mathtt{f}_i \cdot \mathtt{f}_j \, ,$$

with $\mathtt{f}_1, \ldots, \mathtt{f}_d$ the basis of fundamental weights of M (see (4.1)), then one has

$$-\Delta_g = h_L(a) - F \, , \quad F := \sum_{i,j=1}^{d} \mathtt{f}_i \cdot \mathtt{f}_j \, . \tag{8.12}$$

Now, as pointed out in Remark 4.1, the results of [35] prove that $A_1, \ldots, A_d$ are pseudo-differential operators in $\mathcal{A}^1$. Moreover, since they act as Fourier multipliers on the basis of the fundamental weights $\{f_j\}_j$, they mutually commute and their joint spectrum is given by elements $\Lambda = \Lambda_+(M)$ in the cone of dominant weights of M. Since one also has

$$-\Delta_g \leq -\Delta_g + |F| \leq |h_L(A)| + 2|F| \leq \sum_{j=1}^{d} A_j^2 \|\mathtt{f}_j\|^2 + 2|F| \lesssim \mathbf{1} + \sum_{j=1}^{d} A_j^2 ,$$

we deduce that $\mathbf{1} + \sum_{j=1}^{d} A_j^2 \lesssim -\Delta_g = K_0^2$. Hence we can conclude that $A_1, \ldots, A_d$ are a set of quantum actions according to Definition 5.2. We then have that $H_L = h_L(A) = -\Delta_g + F$ is globally integrable. Indeed, the function h_L is homogeneous of degree two and convex, thus in particular it is steep and satisfies Assumption L.

Acknowledgements The authors would like to thank Francesco Fassò for interesting explanations on superintegrable systems and Giorgio Gubbiotti for a discussion on quantum integrable systems. This project is supported by INDAM and by PRIN 2020 (2020XB3EFL001) "Hamiltonian and dispersive PDEs".

References

1. Bambusi, D.: Birkhoff normal form for some nonlinear PDEs. Commun. Math. Phys. **234**(2), 253–285 (2003). MR 1962462
2. Bambusi, D.: Reducibility of 1-d Schrödinger equation with time quasiperiodic unbounded perturbations, II. Commun. Math. Phys. **353**(1), 353–378 (2017). MR 3638317
3. Bambusi, D.: Reducibility of 1-d Schrödinger equation with time quasiperiodic unbounded perturbations, I. Trans. Am. Math. Soc. **370**(3), 1823–1865 (2018)
4. Bambusi, D.: Bounds on the growth of energy for particles on the torus with unbounded time dependent perturbations. J. Math. Phys. **65**(8), Paper No. 082701 (2024)
5. Berti, M., Delort, J.M.: Almost Global Solutions of Capillary-gravity Water Waves Equations on the Circle. Springer, UMI Lecture Notes (2018)
6. Bambusi, D., Delort, J.M., Grébert, B., Szeftel, J.: Almost global existence for Hamiltonian semilinear Klein-Gordon equations with small Cauchy data on Zoll manifolds. Commun. Pure Appl. Math. **60** (2005)
7. Beals, R.: Characterization of pseudodifferential operators and applications. Duke Math. J. **44**(1), 45–57 (1977). MR 435933
8. Bellissard, J.: Stability and instability in quantum mechanics, Trends and developments in the eighties (Bielefeld, 1982/1983), World Sci. Publishing, Singapore, 1985, pp. 1–106 (1985). MR 853743
9. Bambusi, D., Feola, R., Langella B., Monzani, F.: Almost global existence for some Hamiltonian PDEs on manifolds with globally integrable geodesic flow. Nonlinearity **38**, 055013 (2025)
10. Bambusi, D., Feola, R., Montalto, R.: Almost global existence for some Hamiltonian PDEs with small Cauchy data on general tori. Commun. Math. Phys. **405**(15), 253–285 (2024)
11. Bambusi, D., Fusè, A., Sansottera, M.: Exponential stability in the perturbed central force problem. Regul. Chaotic Dyn. **23**(7–8), 821–841 (2018). MR 3910168
12. Bambusi, D., Graffi, S.: Time quasi-periodic unbounded perturbations of Schrödinger operators and KAM methods. Commun. Math. Phys. **219**(2), 465–480 (2001). MR 1833810
13. Bambusi, D., Grébert, B.: Forme normale pour NLS en dimension quelconque. CR Math Acad Sci Paris **337**(6), 409–414 (2003). MR 2015085

14. Bambusi, D., Grébert, B.: Birkhoff normal form for partial differential equations with tame modulus. Duke Math J **135**(3), 507–567 (2006)
15. Bambusi, D., Grebert, B., Maspero, A., Robert, D.: Reducibility of the quantum Harmonic oscillator in d-dimensions with polynomial time dependent perturbation. Anal PDEs **11**(3), 775–799 (2018)
16. Bambusi D, Grébert B, Maspero, A., Robert, D.: Growth of Sobolev norms for abstract linear Schrödinger equations. J. Eur. Math. Soc. (JEMS) **23**(2):557–583 (2021). MR 4195741
17. Bambusi, D., Langella, B.: A C^∞ Nekhoroshev theorem. Math. Eng. **3**(2), 117 (2020)
18. Bambusi, D., Langella, B.: Growth of Sobolev norms in quasi integrable quantum systems. Annales Scientifiques de l École Normale Supérieure Serie, t. **58**, 997–1035 (2025)
19. Bambusi, D., Langella, B., Montalto, R.: On the spectrum of the Schrödinger operator on T^d: a normal form approach. Commun. Part. Diff. Equ. **45**, 1–18 (2020)
20. Bambusi, D., Langella, B., Montalto, R.: Growth of Sobolev norms for unbounded perturbations of the Schrödinger equation on flat tori. J. Diff. Equ. **318**, 344–358 (2022). MR 4387286
21. Bambusi, D., Langella, B., Montalto, R.: Spectral asymptotics of all the eigenvalues of Schrödinger operators on flat tori. Nonlinear Anal. **216**, 112679 (2022)
22. Bambusi, D., Langella, B., Rouveyrol, M.: On the stable eigenvalues of perturbed anharmonic oscillators in dimension two. Commun. Math. Phys. **390**(1), 309–348 (2022)
23. Bambusi, D., Montalto, R.: Reducibility of 1-d Schrödinger equation with time quasiperiodic unbounded perturbations, III. J. Math. Phys. **59**(12), 122702 (2018)
24. Berti, M., Maspero, A., Murgante, F.: Hamiltonian birkhoff normal form for gravity-capillary water waves with constant vorticity: almost global existence (2022). arXiv:2212.12255
25. Caimi, S.: The C^∞ Nekhoroshev theorem, Master thesis, Universita' degli Studi di Milano (2021)
26. Colin De Verdière, Y.: Méthodes semi-classiques et théorie spectrale. https://www-fourier.univ-grenoble-alpes.fr/~ycolver/All-Articles/93b.pdf
27. Colin De Verdière, Y.: Sur le spectre des opérateurs elliptiques à bicaractéristiques toutes périodiques, Comment. Math. Helv. **54**(3), 508–522 (1979). MR 543346
28. Colin de Verdière, Y.: Spectre conjoint d'opérateurs pseudo-différentiels qui commutent. II. Le cas intégrable. Mathematische Zeitschrift **171**, 51–74 (1980)
29. Charbonnel, A.-M.: Comportement semi-classique du spectre conjoint d'opérateurs pseudodifférentiels qui commutent. Asympt. Anal. **1**(3), 227–261 (1988). MR 962310
30. Combescure, M.: The quantum stability problem for some class of time-dependent Hamiltonians. Ann. Phys. **185**(1), 86–110 (1988). MR 954669
31. Delort, J., Imekraz, R.: Long-time existence for the semilinear Klein-Gordon equation on a compact boundary-less Riemannian manifold. Commun. Partial Diff. Equ. **42**(3), 388–416 (2017). MR 3620892
32. Duclos, P., Lev, O., Stovíček, P., Vittot, M.: Weakly regular Floquet Hamiltonians with pure point spectrum. Rev. Math. Phys. **14**(6), 531–568 (2002). MR 1915516
33. Duclos, P., Stovíček, P.: Floquet Hamiltonians with pure point spectrum. Commun. Math. Phys. **177**(2), 327–347 (1996). MR 1384138
34. Fassò, F.: Superintegrable Hamiltonian systems: geometry and perturbations. Acta Appl. Math. **87**(1–3), 93–121 (2005). MR 2151125
35. Fischer, V.: Intrinsic pseudo-differential calculi on any compact lie group. J. Funct. Anal. **268**(11), 3404–3477 (2015)
36. Guzzo, M., Chierchia, L., Benettin, G.: The steep Nekhoroshev's theorem. Commun. Math. Phys. **342**, 569–601 (2016)
37. Giorgilli, A.: Notes on exponential stability of Hamiltonian systems, Pubblicazioni della Classe di Scienze, Scuola Normale Superiore, Pisa. Centro di Ricerca Matematica "Ennio De Giorgi" (2003)
38. Grébert, B., Imekraz, R., Paturel, E.: Normal forms for semilinear quantum harmonic oscillators. Commun. Math. Phys. **291**(3), 763–798 (2009). MR 2534791
39. Gordon, W.B.: On the relation between period and energy in periodic dynamical systems. J. Math. Mech. **19**, 111–114 (1969/70). MR 245930

40. Hörmander, L.: The analysis of linear partial differential operators I–III. Springer, Berlin (1985)
41. Helffer, B., Robert, D.: Propriétés asymptotiques du spectre d'opérateurs pseudodifférentiels sur R^n. Commun. Partial Diff. Equ. **7**(7), 795–882 (1982). MR 662451
42. Nekhorošev, N.N.: Action-angle variables, and their generalizations. Trudy Moskov. Mat. Obšč. **26**, 181–198 (1972). MR 365629
43. Nekhoroshev, N.N.: Exponential estimate of the stability of near integrable Hamiltonian systems. Russ. Math. Surv. **32**(6), 1–65 (1977)
44. Nekhoroshev, N.N.: An exponential estimate of the time of stability of nearly integrable Hamiltonian systems. II, Trudy Sem. Petrovsk. (5), 5–50 (1979). MR 549621
45. Nekhoroshev, N.N.: Types of integrability on a submanifold and generalizations of Gordon's theorem. Tr. Mosk. Mat. Obs. **66**, 184–262 (2005). MR 2193433
46. Niederman, L.: Hamiltonian stability and subanalytic geometry. Ann. Inst. Fourier (Grenoble) **56**(3), 795–813 (2006). MR 2244230
47. Procesi, C.: Lie groups, Universitext. Springer, New York (2007). An approach through invariants and representations. MR 2265844
48. Pelayo, A., Vũ Ngoc, S.: Semiclassical inverse spectral theory for singularities of focus–focus type. Commun. Math. Phys. **329**(2), 809–820 (en) (2014)
49. Rüssmann, H.: Invariant tori in non-degenerate nearly integrable Hamiltonian systems. Regul. Chaotic Dyn. **6**(2), 119–204 (2001). MR 1843664

On Two-dimensional Dirac Operators with δ-Shell Interactions Supported on Unbounded Curves with Straight Ends

Jussi Behrndt, Pavel Exner, Markus Holzmann, and Matěj Tušek

Abstract In this paper we study the self-adjointness and spectral properties of two-dimensional Dirac operators with electrostatic, Lorentz scalar, and anomalous magnetic δ-shell interactions with constant weights that are supported on a smooth unbounded curve that is straight outside a compact set and whose ends are rays that are not parallel to each other. For all possible combinations of interaction strengths we describe the self-adjoint realizations and compute their essential spectra. Moreover, we prove in different situations the existence of geometrically induced discrete eigenvalues.

Keywords Two-dimensional Dirac operators · Delta-shell interactions · Self-adjoint realizations · Geometrically induced discrete spectrum

J. Behrndt
Institut für Angewandte Mathematik, Technische Universität Graz, Graz, Austria
e-mail: behrndt@tugraz.at

P. Exner
Doppler Institute for Mathematical Physics and Applied Mathematics, Czech Technical University in Prague, Prague, Czech Republic
e-mail: exner@ujf.cas.cz

Department of Theoretical Physics, Nuclear Physics Institute, Czech Academy of Sciences, Řež, Czech Republic

M. Holzmann (✉)
Institut für Angewandte Mathematik, Technische Universität Graz, Graz, Austria
e-mail: holzmann@math.tugraz.at

M. Tušek
Department of Mathematics, Faculty of Nuclear Sciences and Physical Engineering, Czech Technical University in Prague, Prague, Czech Republic
e-mail: matej.tusek@fjfi.cvut.cz

B. Cassano et al. (eds.), *Singularities, Asymptotics, and Limiting Models*, Springer INdAM Series 64, https://doi.org/10.1007/978-981-96-3584-9_2

1 Introduction

Operators with contact interactions are frequently used as quantum mechanical Hamiltonians. True, as models of real physical systems they are idealized but often they provide a description which preserves the salient features of the system one wants to describe while being mathematically more accessible than their 'regular' counterparts. In nonrelativistic quantum mechanics there is wealth of such models; for a survey and bibliography we refer to the monograph [2]. The motivation to investigate Dirac operators with contact interactions was weaker for a long time; the only system of that type appeared in connection with the so-called MIT bag model [23]. This changed with the advent of graphene and similar materials, in which the effective dynamics is described by the Dirac equation, see, e.g., [5], and the systems of interest may have many geometric features depending on their experimental design.

The goal in this paper is to study the self-adjointness and the spectrum of a Dirac operator in $\mathbb{R}^2$ with a combination of electrostatic, Lorentz scalar, and anomalous magnetic δ-shell potentials supported on an unbounded curve Γ that is straight outside a compact set such that its ends are not parallel to each other. Such an operator is associated with the formal differential expression

$$\mathcal{A}^{m,c}_{\eta,\tau,\lambda} := -ic(\sigma_1\partial_1 + \sigma_2\partial_2) + mc^2\sigma_3 + (\eta\sigma_0 + \tau\sigma_3 + \lambda i(\sigma_1\nu_1 + \sigma_2\nu_2)\sigma_3)\delta_\Gamma, \tag{1}$$

where $\sigma_1, \sigma_2, \sigma_3 \in \mathbb{C}^{2\times 2}$ are the Pauli spin matrices defined in (2) below, σ_0 is the 2×2-identity matrix, $c > 0$ is the speed of light, $m \geq 0$ is the mass of the particle (note that we will later consider $m \in \mathbb{R}$), δ_Γ stands for the δ- or single layer distribution supported on Γ and $\nu = (\nu_1, \nu_2)$ is a unit normal vector field along Γ; finally, $\eta, \tau, \lambda \in \mathbb{R}$ are coupling constants for the electrostatic, Lorentz scalar, and anomalous magnetic interaction, respectively. It is known that with these three parameters one can define a general symmetric δ-interaction with constant coefficients for the two-dimensional Dirac operator; a fourth possible parameter in the 2×2-matrix-valued coefficient of δ_Γ can be gauged away by a unitary transformation, see [16, 22].

Operators of the form $\mathcal{A}^{m,c}_{\eta,\tau,\lambda}$ were first studied in the one-dimensional case [32], in which the singular interaction is supported on a point in $\mathbb{R}$. By using this analysis and a decomposition to spherical harmonics, the three-dimensional counterpart of the operator in (1) was investigated at the end of the 1980s with the interaction support being a sphere [27]. It took then 25 years, until more general bounded C^2-surfaces in $\mathbb{R}^3$ were considered as supports of purely electrostatic interactions in [3]. After that the interest on the operators induced by (1) grew, the more general interactions including Lorentz scalar and anomalous magnetic terms were introduced, and the self-adjointness, the spectral and scattering properties were studied for bounded interaction supports in dimensions two and three, see [4, 7, 8, 11, 22], to mention just a few references, and the recent review [13] for more references. The case of unbounded interaction supports was not studied that extensively, despite the fact that for the case of the straight line in $\mathbb{R}^2$ interesting spectral transitions were observed in [15, 16]. We also mention the paper [18], where the self-adjointness and spectral properties for combinations of electrostatic and Lorentz scalar interactions supported on surfaces in $\mathbb{R}^3$, which are compact perturbations of the plane, were investigated, the works [21, 31, 40, 43, 44] about the self-adjointness of the operator in (1) under different assumptions on the interaction support including non-smooth broken lines, and the recent paper [30], where the spectrum of the operator $\mathcal{A}^{m,c}_{0,\tau,0}$ was studied, when Γ is a broken line.

It is the main goal in this paper to treat the two-dimensional Dirac operator in (1) with a singular potential that is supported on a smooth unbounded curve in $\mathbb{R}^2$ that is straight outside a compact set in a systematic way. The operator associated with $\mathcal{A}^{m,c}_{\eta,\tau,\lambda}$ lacks semi-boundedness and hence, form methods are not suitable to introduce it as a self-adjoint operator. Therefore, a different technique from extension theory of symmetric operators will be applied. We follow closely the approach in [13], where two- and three-dimensional Dirac operators with singular potentials supported on bounded curves and surfaces were studied with the help of boundary triples and their Weyl functions. More precisely, we construct a suitable ordinary boundary triple, which allows us to introduce $\mathcal{A}^{m,c}_{\eta,\tau,\lambda}$ as a self-adjoint operator for all combinations of the interaction strengths $\eta, \tau, \lambda \in \mathbb{R}$ and to identify the essential spectrum. In a similar way as it is known for bounded interaction supports, it turns out that there are special combinations of the interaction strengths η, τ, λ for which there is a lack of Sobolev regularity in the operator domain. Moreover, as in the case of the straight line [15, 16], unexpected transitions in the essential spectra appear in several situations.

Finally, we are going to show for particular geometries of the interaction support and/or parameter values the existence of isolated eigenvalues. Generally speaking, relations between spectral properties of differential operators and the geometry belong to the most traditional questions, and one of the intriguing examples is the fact that geometrically induced bound states may exist even in regions which are spatially infinitely extended. A classical example is the Dirichlet Laplacian in strips around a bent reference curve, for which we refer to the monograph [29] and the bibliography therein. The same effect was observed for two-dimensional Schrödinger operators with attractive δ-potentials supported on unbounded curves [28, 39], and recently also for Dirac operators with Lorentz scalar δ-shell interactions on broken lines [30]. We will prove that also for smooth unbounded interaction supports Γ that are straight outside a compact set geometrically induced eigenvalues exist in certain situations. We do this first in the case, where the electrostatic and the Lorentz scalar interaction strengths have the same negative value, no magnetic term is present, and the speed of light c is large by using the nonrelativistic limit of the Dirac operator in combination with the known result about the corresponding Schrödinger operator from [28]. Secondly, we prove the existence of geometrically induced bound states for attractive purely Lorentz scalar interactions that are supported on a compact perturbation of a broken line with a sufficiently small opening angle by using the claim proved in the paper [30]. While such existence results are of obvious interest, they lack the universal character of their nonrelativistic counterparts mentioned above, and their extension to a wider class of operators represents an important future challenge.

The paper is organized as follows: In Sect. 2 we state and discuss the main results of this paper in detail. Then, in Sect. 3 we provide some preliminary material that is necessary to prove the main results in Sects. 4 and 5. Finally, in an appendix in Sect. 6 we recall some useful abstract results about ordinary boundary triples, and in a second appendix in Sect. 7 we prove a technical statement about compactness properties of integral operators.

Notations

Let

$$\sigma_1 := \begin{pmatrix} 0 & 1 \\ 1 & 0 \end{pmatrix}, \quad \sigma_2 := \begin{pmatrix} 0 & -i \\ i & 0 \end{pmatrix}, \quad \sigma_3 := \begin{pmatrix} 1 & 0 \\ 0 & -1 \end{pmatrix}, \tag{2}$$

be the Pauli spin matrices and denote by σ_0 the 2×2-identity matrix. Note that the Pauli matrices satisfy

$$\sigma_j \sigma_k + \sigma_k \sigma_j = 2\delta_{jk}\sigma_0, \quad j, k \in \{1, 2, 3\}. \tag{3}$$

For $x = (x_1, x_2)$ we will often write $\sigma \cdot x = \sigma_1 x_1 + \sigma_2 x_2$ and $\sigma \cdot \nabla = \sigma_1 \partial_1 + \sigma_2 \partial_2$.

Let Ω be an open set. The L^2-based Sobolev spaces of k times weakly differentiable functions are $H^k(\Omega; \mathbb{C})$ and we write $H^k(\Omega; \mathbb{C}^2) := H^k(\Omega; \mathbb{C}) \otimes \mathbb{C}^2$. Moreover, if Γ is an infinite curve that is straight outside a compact set as in (6) with arc-length parametrization $\gamma : \mathbb{R} \to \mathbb{R}^2$, then we define the mapping

$$U_\gamma : \mathcal{S}'(\Gamma; \mathbb{C}) \to \mathcal{S}'(\mathbb{R}; \mathbb{C}), \quad (U_\gamma f)(\varphi) = f(\varphi(\gamma^{-1}(\cdot))), \tag{4}$$

where $\mathcal{S}'(\Gamma; \mathbb{C})$ and $\mathcal{S}'(\mathbb{R}; \mathbb{C})$ are the spaces of tempered distributions on Γ and $\mathbb{R}$, respectively. Clearly, if $f \in L^1(\Gamma; \mathbb{C})$, then the distribution $U_\gamma f \in L^1(\mathbb{R}; \mathbb{C})$ is generated by $f \circ \gamma$. With the help of U_γ we introduce the Sobolev space of order $s \in \mathbb{R}$ on Γ by

$$H^s(\Gamma) := \{ f \in \mathcal{S}'(\Gamma; \mathbb{C}) : U_\gamma f \in H^s(\mathbb{R}; \mathbb{C}) \}.$$

In the main part of the paper we will often use the mapping Λ defined by

$$(\mathcal{F} U_\gamma \Lambda f)(p) = \sqrt[4]{1 + p^2}\, (\mathcal{F} U_\gamma f)(p), \quad p \in \mathbb{R}, \tag{5}$$

where $\mathcal{F}$ is the Fourier transform in $\mathcal{S}'(\mathbb{R}; \mathbb{C})$. Evidently, for any $s \in \mathbb{R}$ the operator $\Lambda : H^s(\Gamma; \mathbb{C}) \to H^{s-1/2}(\Gamma; \mathbb{C})$ is well defined and bijective and for $s = \frac{1}{2}$ it can be viewed as an unbounded self-adjoint operator in $L^2(\Gamma; \mathbb{C})$ with dense domain $H^{1/2}(\Gamma; \mathbb{C})$.

If $\mathcal{H}$ is a Hilbert space and A is a closed operator in $\mathcal{H}$, then we denote its domain of definition, its range, and its kernel by $\operatorname{dom} A$, $\operatorname{ran} A$, and $\ker A$, respectively. Furthermore, we use the symbols $\sigma(A)$, $\sigma_\mathrm{p}(A)$, and $\rho(A)$ for the spectrum, the point spectrum, and the resolvent set of A, respectively. If A is self-adjoint, then $\sigma_{\text{disc}}(A)$ and $\sigma_{\text{ess}}(A)$ denote the discrete and the essential spectrum of A, respectively.

2 Main Results

2.1 Definition of $A^{m,c}_{\eta,\tau,\lambda}$ and Confinement

Throughout the paper, we assume that the support of the singular interaction is a smooth curve that does not intersect with itself, that is straight outside a compact set, and whose ends are rays that are not parallel to each other. To make this mathematically more precise, we make the following assumption:

Hypothesis 2.1 *Assume that $\gamma : \mathbb{R} \to \mathbb{R}^2$ is a C^∞-smooth and injective map that satisfies $|\dot\gamma(s)| = 1$ for all $s \in \mathbb{R}$ and*

$$\gamma(s) = \begin{cases} (a_- s + b_-, c_- s + d_-) & \text{for } s < -M, \\ (a_+ s + b_+, c_+ s + d_+) & \text{for } s > M, \end{cases} \tag{6}$$

for suitable constants $M > 0$ and $a_\pm, b_\pm, c_\pm, d_\pm \in \mathbb{R}$ with $a_\pm^2 + c_\pm^2 = 1$ and $(a_-, c_-) \neq -(a_+, c_+)$. We set $\Gamma := \operatorname{ran} \gamma$ and denote by $\nu(x)$ the unit normal vector at $x = \gamma(s) \in \Gamma$, $s \in \mathbb{R}$, that is given by $\nu(\gamma(s)) = (\dot\gamma_2(s), -\dot\gamma_1(s))$. The set Ω_+ is the domain with boundary Γ such that ν is pointing outwards of Ω_+ and we define $\Omega_- := \mathbb{R}^2 \setminus \overline{\Omega_+}$.

If not mentioned differently, we always assume that Γ satisfies the above hypothesis. The condition $(a_-, c_-) \neq -(a_+, c_+)$ in Hypothesis 2.1 means that the two asymptotes of Γ are not parallel rays (of the same orientation) as $t \to \pm\infty$. Note that for γ in Hypothesis 2.1 there exists $C_1 > 0$ such that for all $s, t \in \mathbb{R}$

$$C_1 |s - t| \leq |\gamma(s) - \gamma(t)| \leq |s - t| \tag{7}$$

holds. This will be proved, for the sake of completeness, at the beginning of Sect. 7.

In the following we use the notation $f_\pm := f \upharpoonright \Omega_\pm$ for a function f that is defined on $\mathbb{R}^2$. Recall that $\sigma_1, \sigma_2, \sigma_3$ are the $\mathbb{C}^{2\times 2}$-valued Pauli spin matrices defined in (2) and σ_0 is the 2×2 identity matrix. We define the spaces

$$H(\sigma, \Omega_\pm) := \left\{ f_\pm \in L^2(\Omega_\pm; \mathbb{C}^2) : (\sigma \cdot \nabla) f_\pm \in L^2(\Omega_\pm; \mathbb{C}^2) \right\}$$

and note that for $f_\pm \in H(\sigma; \Omega_\pm)$ the trace $T_\pm^D f_\pm$ exists in the Sobolev space $H^{-1/2}(\Gamma; \mathbb{C}^2)$; cf. Sect. 3.1 for details. This allows us to define for $m, \eta, \tau, \lambda \in \mathbb{R}$ and $c > 0$ in $L^2(\mathbb{R}^2; \mathbb{C}^2)$ the operator

$$
\begin{aligned}
A^{m,c}_{\eta,\tau,\lambda} f &:= \big(-ic(\sigma\cdot\nabla)+mc^2\sigma_3\big)f_+ \oplus \big(-ic(\sigma\cdot\nabla)+mc^2\sigma_3\big)f_-,\\
\operatorname{dom} A^{m,c}_{\eta,\tau,\lambda} &:= \Big\{ f = f_+ \oplus f_- \in H(\sigma,\Omega_+)\oplus H(\sigma,\Omega_-):\\
&\qquad -ic(\sigma\cdot\nu)(T^D_+ f_+ - T^D_- f_-)\\
&\qquad = \frac{1}{2}(\eta\sigma_0+\tau\sigma_3+\lambda i(\sigma\cdot\nu)\sigma_3)(T^D_+ f_+ + T^D_- f_-)\Big\}.
\end{aligned}
\tag{8}
$$

Using integration by parts, it can be seen that the operator in (8) is the rigorous mathematical definition of the formal expression in (1), see, e.g., [37].

A remarkable property of $A^{m,c}_{\eta,\tau,\lambda}$ is that it decomposes for special combinations of η, τ, λ to the orthogonal sum of two Dirac operators in $L^2(\Omega_\pm;\mathbb{C}^2)$ with boundary conditions. This phenomenon, which was observed originally in the three-dimensional setting and later in the current two-dimensional case (but always with bounded interaction supports) in [4, 8, 11, 13, 22, 27], is known as confinement and means that the δ-potential is, in this case, impenetrable for the quantum particle. The mentioned boundary conditions are known as quantum dot boundary conditions and play an important role in the mathematical analysis of graphene [17]. Note that in all studied cases the confinement occurs if and only if the parameter

$$
d := \eta^2 - \tau^2 - \lambda^2 \tag{9}
$$

satisfies $d = -4c^2$ and the same is also true in the current setting.

Proposition 2.2 *Let Γ be as in Hypothesis 2.1, $m, \eta, \tau, \lambda \in \mathbb{R}$ and $c > 0$. Then, the following is true:*

(i) *If $d \neq -4c^2$, then there exists a map $Q_{\eta,\tau,\lambda}: \Gamma \to \mathbb{C}^{2\times 2}$ depending on the parameters η, τ, λ such that $Q_{\eta,\tau,\lambda}$ is pointwise invertible and $f \in \operatorname{dom} A^{m,c}_{\eta,\tau,\lambda}$ if and only if $f \in H(\sigma;\Omega_+)\oplus H(\sigma;\Omega_-)$ and*

$$
T^D_+ f_+ = Q_{\eta,\tau,\lambda} T^D_- f_-.
$$

(ii) *If $d = -4c^2$, then $A^{m,c}_{\eta,\tau,\lambda} = A^+_{\eta,\tau,\lambda} \oplus A^-_{\eta,\tau,\lambda}$, where $A^\pm_{\eta,\tau,\lambda}$ are the operators acting in $L^2(\Omega_\pm;\mathbb{C}^2)$ given by*

$$
\begin{aligned}
A^\pm_{\eta,\tau,\lambda} f &= \big(-ic(\sigma\cdot\nabla)+mc^2\sigma_3\big)f,\\
\operatorname{dom} A^\pm_{\eta,\tau,\lambda} &= \big\{ f \in H(\sigma;\Omega_\pm):\\
&\qquad \big(2c\sigma_0 \mp i(\sigma\cdot\nu)(\eta\sigma_0+\tau\sigma_3+\lambda i(\sigma\cdot\nu)\sigma_3)\big)T^D_\pm f = 0\big\}.
\end{aligned}
$$

This result can be proved in the same way as, e.g., [4, Theorem 5.5], [11, Lemma 4.1], [13, Lemma 5.11], or [22, Theorem 2.3], and hence, we omit it in this paper.

2.2 Self-Adjointness and Essential Spectrum of $A^{m,c}_{\eta,\tau,\lambda}$

In the following theorem the self-adjointness and the essential spectrum of $A^{m,c}_{\eta,\tau,\lambda}$ are discussed. Note that, in a similar way as it was observed in [10, 11, 13, 22, 38] for compact interaction supports, there is a loss of Sobolev regularity in the operator domain in the case of critical interaction strengths $(\frac{d}{4} - c^2)^2 - \lambda^2 c^2 = 0$. The essential spectrum of $A^{m,c}_{\eta,\tau,\lambda}$ in the particular case of Γ being the straight line was computed originally in [16]. By Theorem 2.3 the essential spectrum is preserved under the geometric perturbations as long as Γ is straight outside a compact set. We point out that to the best of our knowledge item (i) of the following theorem on the Sobolev regularity in dom $A^{m,c}_{\eta,\tau,\lambda}$ was not known in this generality before even for the straight line.

Theorem 2.3 *Let $m, \eta, \tau, \lambda \in \mathbb{R}$, $c > 0$, d be given by (9), and Γ be as in Hypothesis 2.1. Then, the operator $A^{m,c}_{\eta,\tau,\lambda}$ is self-adjoint in $L^2(\mathbb{R}^2; \mathbb{C}^2)$ and the following holds:*

(i) *For the domain of $A^{m,c}_{\eta,\tau,\lambda}$ one has:*

  (a) *If $(\frac{d}{4} - c^2)^2 - \lambda^2 c^2 \neq 0$, then* dom $A^{m,c}_{\eta,\tau,\lambda} \subset H^1(\mathbb{R}^2 \setminus \Gamma; \mathbb{C}^2)$.

  (b) *If $(\frac{d}{4} - c^2)^2 - \lambda^2 c^2 = 0$, then* dom $A^{m,c}_{\eta,\tau,\lambda} \not\subset H^s(\mathbb{R}^2 \setminus \Gamma; \mathbb{C}^2)$ *for all $s > 0$.*

(ii) *For the essential spectrum of $A^{m,c}_{\eta,\tau,\lambda}$ one has:*

  (a) *If $d = 4c^2$ and $\lambda \neq 0$, then*

$$\sigma_{\text{ess}}(A^{m,c}_{\eta,\tau,\lambda}) = \sigma(A^{m,c}_{\eta,\tau,\lambda}) = \mathbb{R}.$$

  (b) *If $d = 4c^2$ and $\lambda = 0$, then*

$$\sigma_{\text{ess}}(A^{m,c}_{\eta,\tau,\lambda}) = (-\infty, -|m|c^2] \cup \left\{-\frac{\tau}{\eta} mc^2\right\} \cup [|m|c^2, +\infty).$$

  (c) *If $d \neq 4c^2$, then*

$$(-\infty, -|m|c^2] \cup [|m|c^2, +\infty) \subset \sigma_{\text{ess}}(A^{m,c}_{\eta,\tau,\lambda})$$

  and $\sigma_{\text{ess}}(A^{m,c}_{\eta,\tau,\lambda}) \cap (-|m|c^2, |m|c^2)$ coincides with the set

$$\overline{\left\{z_\pm(k) : (d - 4c^2)\left(\frac{\eta}{c^2} z_\pm(k) + \lambda k + \tau m\right) > 0\right\}} \cap (-|m|c^2, |m|c^2),$$

where $z_\pm(k)$, $k \in \mathbb{R}$, is given by

$$z_\pm(k) = \frac{1}{\frac{\eta^2}{c^2} + \left(\frac{d}{4c^2} - 1\right)^2} \Bigg(- \eta(\lambda k + \tau m) \pm \left|\frac{d}{4c^2} - 1\right| \sqrt{\left(\tau^2 c^2 + \left(\tfrac{d}{4} + c^2\right)^2\right)k^2 - 2\lambda\tau m k c^2 + \left(\lambda^2 c^2 + \left(\tfrac{d}{4} + c^2\right)^2\right)m^2} \Bigg).$$

Moreover, if Γ is the straight line, then $\sigma(A^{m,c}_{\eta,\tau,\lambda}) = \sigma_{\mathrm{ess}}(A^{m,c}_{\eta,\tau,\lambda})$, i.e. the spectrum of $A^{m,c}_{\eta,\tau,\lambda}$ is fully characterized by (ii) in this case.

We point out that an interesting spectral transition appears when $d = 4c^2$: If $\lambda \neq 0$, then always $\sigma(A^{m,c}_{\eta,\tau,\lambda}) = \mathbb{R}$, but for $\lambda = 0$ the essential spectrum within the interval $(-|m|c^2, |m|c^2)$ breaks down to the single point $-\frac{\tau}{\eta} mc^2$.

Theorem 2.3 will be proved in Sect. 4 with the help of a suitable boundary triple. In Sect. 4.4 we will also show a variant of the Birman-Schwinger principle, a Krein type resolvent formula, and an isospectral relation that are useful in the analysis of $A^{m,c}_{\eta,\tau,\lambda}$.

2.3 Special Cases

The next results follow immediately from Theorem 2.3 (and thus, no proofs of these corollaries are given), but we prefer to state them here as they are more explicit than Theorem 2.3 and of particular interest in applications. First, we discuss the case of purely electrostatic interactions, i.e. when $\eta \in \mathbb{R}$ and $\tau = \lambda = 0$. Note that, as in the case of the straight line [15], there appears a spectral transition at $\eta = \pm 2c$, where an interval of continuous spectrum of $A^{m,c}_{\eta,0,0}$ collapses to a single point in the essential spectrum.

Corollary 2.4 *Let $\eta \in \mathbb{R}$. Then, one has*

$$\sigma_{\mathrm{ess}}(A^{m,c}_{\eta,0,0}) = \begin{cases} (-\infty, \frac{4c^2-\eta^2}{4c^2+\eta^2}|m|c^2] \cup [|m|c^2, +\infty) & \text{for } \eta \in (-\infty, -2c), \\ (-\infty, -|m|c^2] \cup \{0\} \cup [|m|c^2, +\infty) & \text{for } \eta = -2c, \\ (-\infty, -|m|c^2] \cup [\frac{4c^2-\eta^2}{4c^2+\eta^2}|m|c^2, +\infty) & \text{for } \eta \in (-2c, 0), \\ (-\infty, -|m|c^2] \cup [|m|c^2, +\infty) & \text{for } \eta = 0, \\ (-\infty, \frac{\eta^2-4c^2}{\eta^2+4c^2}|m|c^2] \cup [|m|c^2, +\infty) & \text{for } \eta \in (0, 2c), \\ (-\infty, -|m|c^2] \cup \{0\} \cup [|m|c^2, +\infty) & \text{for } \eta = 2c, \\ (-\infty, -|m|c^2] \cup [\frac{\eta^2-4c^2}{\eta^2+4c^2}|m|c^2, +\infty) & \text{for } \eta \in (2c, +\infty). \end{cases}$$

For purely Lorentz scalar interactions, i.e. for $\tau \in \mathbb{R}$ and $\eta = \lambda = 0$, the following result holds. Note that for $\tau m \geq 0$ the essential spectrum of $A^{m,c}_{0,\tau,0}$ coincides with the spectrum of the free Dirac operator.

Corollary 2.5 *Let $\tau \in \mathbb{R}$. Then, one has*

$$\sigma_{\text{ess}}(A^{m,c}_{0,\tau,0}) = \begin{cases} (-\infty, -|m|c^2] \cup [|m|c^2, +\infty) & \textit{for } \tau m \geq 0, \\ (-\infty, -\frac{|4c^2-\tau^2|}{4c^2+\tau^2}|m|c^2] \cup [\frac{|4c^2-\tau^2|}{4c^2+\tau^2}|m|c^2, +\infty) & \textit{for } \tau m < 0. \end{cases}$$

Finally, we consider the case of purely anomalous magnetic interactions, i.e. when $\lambda \in \mathbb{R}$ and $\eta = \tau = 0$. Note that in this configuration the essential spectrum of $A^{m,c}_{0,0,\lambda}$ always coincides with the spectrum of the free Dirac operator.

Corollary 2.6 *Let $\lambda \in \mathbb{R}$. Then, one has*

$$\sigma_{\text{ess}}(A^{m,c}_{0,0,\lambda}) = (-\infty, -|m|c^2] \cup [|m|c^2, +\infty).$$

2.4 Geometrically Induced Bound States

In the present paper we also prove the existence of geometrically induced bound states in two situations: For $A^{m,c}_{\eta/2,\eta/2,0}$ for large values of c with the help of the nonrelativistic limit, and for $A^{m,c}_{0,\tau,0}$ by using the test functions from [30].

The nonrelativistic limit is one way to relate a relativistic Dirac operator $A^{m,c}$ with the Schrödinger operator H that describes the same physical system [46, Chapter 6]. To compute the nonrelativistic limit, one has to subtract the particle's rest energy mc^2, which is a purely relativistic quantity, from the total energy and compute the limit in the operator norm of the resolvent of the resulting operator $A^{m,c} - mc^2$ for $c \to +\infty$. The expected limit is then the resolvent of H times a projection onto the upper component. This result is interesting by itself, as it gives a physical interpretation of the model, but taking properties of convergence of the resolvents into account, this allows to relate the spectral properties of H to those of $A^{m,c}$ for large values of c.

In this paper we consider a simple case and compute the nonrelativistic limit of $A^{m,c}_{\eta/2,\eta/2,0}$, see also Remark 5.3 for a more general consideration. In the following, we assume that Γ is given as in Hypothesis 2.1 and that $m > 0$. First, we have to introduce the expected limit operator, a Schrödinger operator with a δ-potential supported on Γ. Define for $\eta \in \mathbb{R}$ in $L^2(\mathbb{R}^2; \mathbb{C})$ the sesquilinear form

$$\begin{aligned} \mathfrak{h}_\eta[f, g] &:= \frac{1}{2m}(\nabla f, \nabla g)_{L^2(\mathbb{R}^2;\mathbb{C})} + \eta(T^D f, T^D g)_{L^2(\Gamma;\mathbb{C})}, \\ \operatorname{dom} \mathfrak{h}_\eta &= H^1(\mathbb{R}^2; \mathbb{C}), \end{aligned} \tag{10}$$

where $T^D : H^1(\mathbb{R}; \mathbb{C}) \to H^{1/2}(\Gamma; \mathbb{C})$ is the Dirichlet trace operator. It is well-known that $\mathfrak{h}_\eta$ is densely defined, symmetric, and closed [19, 28]. Hence, by the first representation theorem [36, Chapter VI, Theorem 2.1] there exists a unique self-adjoint operator H_η that is associated with $\mathfrak{h}_\eta$. The operator H_η can be regarded as the rigorous way to interpret the formal expression $-\frac{1}{2m}\Delta + \eta\delta_\Gamma$. With this notation we can formulate the result about the nonrelativistic limit of $A^{m,c}_{\eta/2,\eta/2,0}$, which yields, combined with a result from [28], a statement about the existence of geometrically induced bound states:

Theorem 2.7 *Let $\eta \in \mathbb{R}$, $m > 0$, and Γ be as in Hypothesis 2.1. Then, there exists $z_0 < 0$ such that any fixed $z \in (-\infty, z_0)$ belongs to $\rho(A^{m,c}_{\eta/2,\eta/2,0} - mc^2) \cap \rho(H_\eta)$, whenever c is above a z-dependent threshold value c_z. Moreover, there exists a constant $K > 0$ such that*

$$\left\| \left(A^{m,c}_{\eta/2,\eta/2,0} - (z + mc^2)\right)^{-1} - (H_\eta - z)^{-1} \begin{pmatrix} 1 & 0 \\ 0 & 0 \end{pmatrix} \right\|_{L^2(\mathbb{R}^2;\mathbb{C}^2) \to L^2(\mathbb{R}^2;\mathbb{C}^2)} \leq \frac{K}{c}$$

for all $c > c_z$. In particular, if Γ is not the straight line, $\eta < 0$, and c is sufficiently large, then $\sigma_{\text{disc}}(A^{m,c}_{\eta/2,\eta/2,0}) \neq \emptyset$.

Theorem 2.7 will be proved in Sect. 5 with the help of the Krein type resolvent formula for $A^{m,c}_{\eta/2,\eta/2,0}$ provided in Sect. 4. In this resolvent formula, convergence of each term appearing there can be proved, when $c \to +\infty$, which yields the claim about the nonrelativistic limit. The statement about the discrete spectrum is then a simple consequence of the nonrelativistic limit, the non-emptiness of $\sigma_{\text{disc}}(H_\eta)$ under the stated assumptions shown in [28], and general results that are known for operators that converge in the norm resolvent sense.

Another possibility to show the existence of geometrically induced bound states is to transfer a recent result from [30]. Define for $\omega \in (0, \frac{\pi}{2})$ the broken line $\widetilde{\Gamma}_\omega$ of opening angle 2ω by

$$\widetilde{\Gamma}_\omega := \left\{(r\cos(\omega), r\sin(\omega)) : r > 0\right\} \cup \left\{(r\cos(\omega), -r\sin(\omega)) : r > 0\right\}. \quad (11)$$

It was shown in [30] that a Dirac operator with a Lorentz scalar δ-shell interaction supported on $\widetilde{\Gamma}_\omega$ of strength $\tau \in (-\infty, 0) \setminus \{-2\}$ always has non-empty discrete spectrum, if ω is sufficiently small. This can be transferred to obtain an associated result for $A^{m,c}_{0,\tau,0}$, if the interaction support Γ is a compact smooth perturbation of $\widetilde{\Gamma}_\omega$. We remark that any curve Γ satisfying Hypothesis 2.1, that is not a compact perturbation of the straight line, is, up to a translation and rotation, of this form and that the following result, in contrast to the one in Theorem 2.7, can be shown for arbitrary $c > 0$.

Theorem 2.8 *Let $\tau \in (-\infty, 0) \setminus \{-2\}$, Γ be as in Hypothesis 2.1 such that Γ coincides with $\widetilde{\Gamma}_\omega$ away from a compact set in $\mathbb{R}^2$, and $m, c > 0$. Then, for any $N \in \mathbb{N}$ there exists an angle $\omega_\star \in (0, \frac{\pi}{2})$ depending on N, τ such that for all $\omega \in (0, \omega_\star]$ the operator $A^{m,c}_{0,\tau,0}$ has at least N discrete eigenvalues with multiplicities taken into account.*

Theorem 2.8 is also shown in Sect. 5 with the help of the min-max principle applied to the non-negative and self-adjoint operator $(A^{m,c}_{0,\tau,0})^2$ and a family of test functions constructed in [30] that are supported in a region where Γ and $\widetilde{\Gamma}_\omega$ coincide, which allows to transfer the result from [30] to the situation studied in this paper.

3 Preliminaries

In this section we collect several results, that are needed to prove the main claims made on $A^{m,c}_{\eta,\tau,\lambda}$ in Sect. 2. First, in Sect. 3.1 we summarize some well-known properties of the free Dirac operator $A^{m,c}_0$ and in Sect. 3.2 we state some properties of integral operators that are associated with the integral kernel of the resolvent of $A^{m,c}_0$. In Sect. 3.3 we construct a boundary triple that is useful to study $A^{m,c}_{\eta,\tau,\lambda}$. Throughout this section, we always assume that $\Gamma \subset \mathbb{R}^2$ is as in Hypothesis 2.1.

3.1 The Free Dirac Operator and Function Spaces for Dirac Operators

In this section we recall the definition of the free Dirac operator in $\mathbb{R}^2$ and state several related results that are needed in the main part of the paper. Let $\sigma_1, \sigma_2, \sigma_3$ be the Pauli spin matrices defined in (2). For $m \in \mathbb{R}$ and $c > 0$ the free Dirac operator in $\mathbb{R}^2$ is

$$A^{m,c}_0 f := -ic(\sigma \cdot \nabla) f + mc^2 \sigma_3 f, \quad \mathrm{dom}\, A^{m,c}_0 = H^1(\mathbb{R}^2; \mathbb{C}^2). \tag{12}$$

We will sometimes make use of the relation $A^{m,c}_0 = cA^{mc,1}_0$. With the help of the Fourier transform it is not difficult to show that $A^{m,c}_0$ is self-adjoint in $L^2(\mathbb{R}^2; \mathbb{C}^2)$ and that

$$\sigma(A^{m,c}_0) = \sigma_{\mathrm{ess}}(A^{m,c}_0) = (-\infty, -|m|c^2] \cup [|m|c^2, +\infty).$$

Moreover, for $z \in \rho(A^{m,c}_0)$ the resolvent of $A^{m,c}_0$ is given by

$$(A^{m,c}_0 - z)^{-1} f(x) = \int_{\mathbb{R}^2} G^{m,c}_z(x-y) f(y) dy, \quad f \in L^2(\mathbb{R}^2; \mathbb{C}^2),\ x \in \mathbb{R}^2,$$

where

$$\begin{aligned} G^{m,c}_z(x) = {}& \frac{1}{2\pi c}\sqrt{\tfrac{z^2}{c^2} - (mc)^2}\, K_1\left(-i\sqrt{\tfrac{z^2}{c^2} - (mc)^2}\,|x|\right) \frac{(\sigma \cdot x)}{|x|} \\ &+ \frac{1}{2\pi c} K_0\left(-i\sqrt{\tfrac{z^2}{c^2} - (mc)^2}\,|x|\right)\left(\frac{z}{c}\sigma_0 + mc\sigma_3\right) \end{aligned} \tag{13}$$

and K_j is the modified Bessel function of the second kind and order j; cf. [1]. We remark that we always choose the complex square root such that for all $w \in \mathbb{C} \setminus [0, +\infty)$ one has $\mathrm{Im}(\sqrt{w}) > 0$ and hence $\overline{\sqrt{w}} = -\sqrt{\overline{w}}$ for $w \in \mathbb{C} \setminus [0, +\infty)$. Note also that $A^{m,c}_0 = cA^{mc,1}_0$ yields

$$(A_0^{m,c} - z)^{-1} = \frac{1}{c}\left(A_0^{mc,1} - \frac{z}{c}\right)^{-1} \quad \text{and} \quad G_z^{m,c} = \frac{1}{c} G_{z/c}^{mc,1}, \quad z \in \rho(A_0^{m,c}). \tag{14}$$

Recall that $\Omega_\pm \subset \mathbb{R}^2$ are the two open sets with joint boundary $\partial\Omega_\pm = \Gamma$, see Hypothesis 2.1, and that we denote elements in $L^2(\Omega_\pm; \mathbb{C}^2)$ with a subscript $\pm$. We consider the spaces

$$H(\sigma, \Omega_\pm) := \left\{ f_\pm \in L^2(\Omega_\pm; \mathbb{C}^2) : (\sigma \cdot \nabla) f_\pm \in L^2(\Omega_\pm; \mathbb{C}^2) \right\}, \tag{15}$$

which are, endowed with the norms

$$\| f_\pm \|^2_{H(\sigma,\Omega_\pm)} := \| f_\pm \|^2_{L^2(\Omega_\pm;\mathbb{C}^2)} + \| (\sigma \cdot \nabla) f_\pm \|^2_{L^2(\Omega_\pm;\mathbb{C}^2)},$$

Hilbert spaces; cf. [17, Lemma 2.1] for similar arguments in the case when Γ is bounded. An important property of $H(\sigma, \Omega_\pm)$ is the fact that the trace operator can be extended to these spaces. The proof of the next lemma for our geometric setting follows very similar ideas as the ones in [17, Lemma 2.3] or [40, Lemma 2.3]:

Lemma 3.1 *The map $T^D_{\pm,0} : H^1(\Omega_\pm; \mathbb{C}^2) \to H^{1/2}(\Gamma; \mathbb{C}^2)$, $T^D_{\pm,0} f = f|_\Gamma$, extends uniquely by continuity to a bounded linear operator*

$$T^D_\pm : H(\sigma, \Omega_\pm) \to H^{-1/2}(\Gamma; \mathbb{C}^2).$$

In the main part of the paper we will work with the densely defined closed symmetric restriction S of $A_0^{m,c}$ onto $H_0^1(\mathbb{R}^2 \setminus \Gamma; \mathbb{C}^2)$, that is,

$$Sf = -ic(\sigma \cdot \nabla) f + mc^2 \sigma_3 f, \quad \operatorname{dom} S = H_0^1(\mathbb{R}^2 \setminus \Gamma; \mathbb{C}^2), \tag{16}$$

and its adjoint

$$\begin{aligned} S^* f &= \big(-ic(\sigma \cdot \nabla) + mc^2 \sigma_3 \big) f_+ \oplus \big(-ic(\sigma \cdot \nabla) + mc^2 \sigma_3 \big) f_-, \\ \operatorname{dom} S^* &= H(\sigma, \Omega_+) \oplus H(\sigma, \Omega_-). \end{aligned} \tag{17}$$

3.2 Integral Operators for the Dirac Operator

In this section, we introduce two families of integral operators that will play an important role in the analysis of $A^{m,c}_{\eta,\tau,\lambda}$. For this, recall the fundamental solution $G_z^{m,c}$ in (13) and note that Γ as in Hypothesis 2.1 belongs to the class of curves considered in [12].

Following [12, Appendix C], we introduce for

$$z \in \rho(A_0^{m,c}) = \mathbb{C} \setminus \big((-\infty, -|m|c^2] \cup [|m|c^2, +\infty) \big)$$

the operator

$$(\Phi_z^{m,c})^* := T^D(A_0^{m,c} - \overline{z})^{-1} : L^2(\mathbb{R}^2; \mathbb{C}^2) \to H^{1/2}(\Gamma; \mathbb{C}^2), \tag{18}$$

which is well-defined and bounded due to the mapping properties of the Dirichlet trace $T^D : H^1(\mathbb{R}^2) \to H^{1/2}(\Gamma)$ and the resolvent of $A_0^{m,c}$. Hence, also its anti-dual

$$\Phi_z^{m,c} := \big((\Phi_z^{m,c})^*\big)' : H^{-1/2}(\Gamma; \mathbb{C}^2) \to L^2(\mathbb{R}^2; \mathbb{C}^2) \tag{19}$$

is well-defined and bounded. Using (18) and $G_z^{m,c}(x)^* = G_{\overline{z}}^{m,c}(-x)$ one finds that the action of $\Phi_z^{m,c}$ on $\varphi \in L^2(\Gamma; \mathbb{C}^2)$ is

$$\Phi_z^{m,c}\varphi(x) = \int_\Gamma G_z^{m,c}(x-y)\varphi(y)d\sigma(y), \quad x \in \mathbb{R}^2 \setminus \Gamma.$$

Moreover, as $\ker(\Phi_z^{m,c})^* = \operatorname{ran}(S - \overline{z})$, where S is the symmetric operator in (16), and $\operatorname{ran}(\Phi_z^{m,c})^*$ is closed, we get that

$$\operatorname{ran} \Phi_z^{m,c} = \ker(S^* - z) \subset H(\sigma, \Omega_+) \oplus H(\sigma, \Omega_-). \tag{20}$$

Since by (14) one has $\Phi_z^{m,c} = \frac{1}{c}\Phi_{z/c}^{mc,1}$ and $\Phi_{z/c}^{mc,1} = \Phi_{z/c}$ in the notation of [12] (when m is replaced by mc), we conclude from [12, Proposition 2.8] that $\Phi_z^{m,c}$ has the bounded restriction

$$\widetilde{\Phi}_z^{m,c} : H^{1/2}(\Gamma; \mathbb{C}^2) \to H^1(\mathbb{R}^2 \setminus \Gamma; \mathbb{C}^2). \tag{21}$$

The boundary integral operator formally given by

$$C_z^{m,c}\varphi(x) := \text{p.v.} \int_\Gamma G_z^{m,c}(x-y)\varphi(y)d\sigma(y), \quad x \in \Gamma,\ \varphi \in L^2(\Gamma; \mathbb{C}^2), \tag{22}$$

where $z \in \rho(A_0^{m,c})$ and the integral is understood as the principal value, will be used throughout this paper. Since $C_z^{m,c} = \frac{1}{c}C_{z/c}^{mc,1}$ holds by (14) and $C_{z/c}^{mc,1} = C_{z/c}$ in the notation of [12] (when m is replaced by mc), we get with [12, Proposition 2.9] that $C_z^{m,c}$ gives rise to a bounded operator in $H^r(\Gamma; \mathbb{C}^2)$ for $r \in [-\frac{1}{2}, \frac{1}{2}]$. Moreover, as $G_z^{m,c}(x)^* = G_{\overline{z}}^{m,c}(-x)$, $x \in \mathbb{R}^2 \setminus \{0\}$, it can be shown that the adjoint of $C_z^{m,c}$, viewed as an operator in $L^2(\Gamma; \mathbb{C}^2)$, is $(C_z^{m,c})^* = C_{\overline{z}}^{m,c}$.

Next, we explain that in a suitable sense $C_z^{m,c}$ can be regarded as a compact perturbation of the same mapping $C_z^{m,c,\Sigma}$ on the straight line $\Sigma = \mathbb{R} \times \{0\}$, i.e. of the operator that acts on $\varphi \in L^2(\Sigma; \mathbb{C}^2)$ as

$$C_z^{m,c,\Sigma}\varphi(s,0) := \text{p.v.} \int_{-\infty}^{\infty} G_z^{m,c}(s-t,0)\varphi(t,0)dt, \quad s \in \mathbb{R}. \tag{23}$$

In the formulation of the next proposition, the multiplication operator with the matrix-valued function

$$V = \begin{pmatrix} 1 & 0 \\ 0 & \bar{t} \end{pmatrix}, \tag{24}$$

where $t = t_1 + it_2$ with $(t_1, t_2) = \dot{\gamma} \circ \gamma^{-1}$ being the unit tangent vector at Γ, will play an important role. Since t is C^∞-smooth and $|t| = 1$, one clearly has that for any $r \in \mathbb{R}$ the multiplication with V gives rise to a bounded and bijective operator in $H^r(\Gamma; \mathbb{C}^2)$. Moreover, recall the definition of the map U_g from (4) for any arc-length parametrization g of an unbounded curve. The technical proof of Proposition 3.2 is given in an appendix in Sect. 7.

Proposition 3.2 *Let $z \in \rho(A_0^{m,c}) = \mathbb{C} \setminus ((-\infty, -|m|c^2] \cup [|m|c^2, +\infty))$, Γ be as in Hypothesis 2.1, and $x(t) = (t, 0)$ be a parametrization of the straight line Σ. Then, the operator*

$$U_\gamma V(\sigma \cdot \nu) C_z^{m,c} (\sigma \cdot \nu) V^* U_\gamma^{-1} - U_x \sigma_2 C_z^{m,c,\Sigma} \sigma_2 U_x^{-1} : H^{-1/2}(\mathbb{R}; \mathbb{C}^2) \to H^{1/2}(\mathbb{R}; \mathbb{C}^2)$$

is compact.

Remark 3.3 In fact, we will show in the proof of Proposition 3.2 that the operator $U_\gamma V(\sigma \cdot \nu) C_z^{m,c} (\sigma \cdot \nu) V^* U_\gamma^{-1} - U_x \sigma_2 C_z^{m,c,\Sigma} \sigma_2 U_x^{-1}$ can be extended, for any $r \in [-1, 0]$, to a compact operator from $H^r(\mathbb{R}; \mathbb{C}^2)$ to $H^{r+1}(\mathbb{R}; \mathbb{C}^2)$.

Finally, we state a useful variant of the Plemelj-Sokhotskii formula, which relates the trace of $\Phi_z^{m,c}\varphi$ to $C_z^{m,c}\varphi$; cf. [12, Proposition 2.9 (ii)]. For a direct proof of this formula one can also follow line by line the proof of [11, Proposition 3.4] using the classical Plemelj-Sokhotskii formula for the Cauchy transform on unbounded curves, see, e.g., [33, Proposition 4.6.1].

Lemma 3.4 *Let $z \in \rho(A_0^{m,c}) = \mathbb{C} \setminus ((-\infty, -|m|c^2] \cup [|m|c^2, +\infty))$. Then, one has for all $\varphi \in H^{-1/2}(\Gamma; \mathbb{C}^2)$*

$$T_\pm^D \Phi_z^{m,c} \varphi = \mp \frac{i}{2c} (\sigma \cdot \nu)\varphi + C_z^{m,c} \varphi.$$

Eventually, we note that Lemma 3.4 allows to show, in the same way as in [3, Lemma 3.3 (ii)], the identity

$$-4c^2 \big((\sigma \cdot \nu) C_z^{m,c}\big)^2 = -4c^2 \big(C_z^{m,c} (\sigma \cdot \nu)\big)^2 = I, \quad z \in \rho(A_0^{m,c}). \tag{25}$$

3.3 A Boundary Triple for $A_{\eta,\tau,\lambda}^{m,c}$

In this section we show how the operator $A_{\eta,\tau,\lambda}^{m,c}$ defined in (8) can be described with the help of a boundary triple. The general definition and some useful properties of boundary triples are recalled in an appendix in Sect. 6. As before, Γ, $\Omega_\pm$, and ν are as in Hypothesis 2.1, and we use for $f : \mathbb{R}^2 \to \mathbb{C}^2$ the notation $f_\pm := f \upharpoonright \Omega_\pm$. First, we introduce an auxiliary Dirac operator and discuss its properties. Recall that $A_0^{m,c}$ is the free Dirac operator given by (12), define the unitary operator

$$U : L^2(\mathbb{R}^2; \mathbb{C}^2) \to L^2(\mathbb{R}^2; \mathbb{C}^2), \quad U(f_+ \oplus f_-) = f_+ \oplus (-f_-),$$

and introduce in $L^2(\mathbb{R}^2; \mathbb{C}^2)$ the operator

$$B := U^* A_0^{m,c} U. \tag{26}$$

In a more explicit form the operator B is given by

$$\begin{aligned} Bf &= \big(-ic(\sigma \cdot \nabla) + mc^2\sigma_3 \big) f_+ \oplus \big(-ic(\sigma \cdot \nabla) + mc^2\sigma_3 \big) f_-, \\ \operatorname{dom} B &= \big\{ f = f_+ \oplus f_- \in H^1(\Omega_+; \mathbb{C}^2) \oplus H^1(\Omega_-; \mathbb{C}^2) : T_+^D f_+ + T_-^D f_- = 0 \big\}. \end{aligned}$$

Some basic properties of B are stated in the following lemma. Recall that $\widetilde{\Phi}_z^{m,c}$ and $C_z^{m,c}$ are the operators defined by (21) and (22), respectively.

Lemma 3.5 *The operator B is self-adjoint in $L^2(\mathbb{R}^2; \mathbb{C}^2)$, its spectrum is given by $\sigma(B) = \sigma_{\mathrm{ess}}(B) = \sigma(A_0^{m,c}) = (-\infty, -|m|c^2] \cup [|m|c^2, +\infty)$, and for $z \in \rho(B)$ one has*

$$(B - z)^{-1} = (A_0^{m,c} - z)^{-1} + 4c^2 \widetilde{\Phi}_z^{m,c} (\sigma \cdot \nu) C_z^{m,c} (\sigma \cdot \nu) (\Phi_{\overline{z}}^{m,c})^*.$$

Proof Since the operator U is unitary, it follows from the definition of B in (26) and the properties of $A_0^{m,c}$ in (12) that B is self-adjoint in $L^2(\mathbb{R}^2; \mathbb{C}^2)$ and that $\sigma(B) = \sigma(A_0^{m,c})$. It remains to show the claimed resolvent formula. For this, fix $z \in \rho(B) = \rho(A_0^{m,c})$ and $f \in L^2(\mathbb{R}^2; \mathbb{C}^2)$, and define

$$g := (A_0^{m,c} - z)^{-1} f + 4c^2 \widetilde{\Phi}_z^{m,c} (\sigma \cdot \nu) C_z^{m,c} (\sigma \cdot \nu) (\Phi_{\overline{z}}^{m,c})^* f.$$

Then, by the mapping properties of $(A_0^{m,c} - z)^{-1}$, $\widetilde{\Phi}_z^{m,c}$, $C_z^{m,c}$, and $(\Phi_{\overline{z}}^{m,c})^*$ in (21), (22), and (18), respectively, we get $g \in H^1(\Omega_+; \mathbb{C}^2) \oplus H^1(\Omega_-; \mathbb{C}^2)$. Moreover, using Lemma 3.4 and (25), we find

$$\begin{aligned} T_+^D g_+ + T_-^D g_- &= 2\, T^D (A_0^{m,c} - z)^{-1} f + 8c^2 C_z^{m,c} (\sigma \cdot \nu) C_z^{m,c} (\sigma \cdot \nu) (\Phi_{\overline{z}}^{m,c})^* f \\ &= 2\, T^D (A_0^{m,c} - z)^{-1} f - 2 (\Phi_{\overline{z}}^{m,c})^* f = 0 \end{aligned}$$

and hence, $g \in \operatorname{dom} B$. Finally, equation (20) yields

$$((B - z)g)_\pm = (-ic(\sigma \cdot \nabla) + mc^2\sigma_3 - z\sigma_0) \big((A_0^{m,c} - z)^{-1} f \big)_\pm = f_\pm.$$

Taking the injectivity of $B - z$ and the definition of g into account, we see that also the claimed resolvent formula is true. □

Next, we provide a boundary triple that is useful to study $A^{m,c}_{\eta,\tau,\lambda}$, see also [13, Theorem 4.8] and [11, 22] for similar constructions. In order to formulate the result, recall the definition of the map Λ in (5), of the free Dirac operator $A^{m,c}_0$ in (12), of its restriction S and extension S^* in (16) and (17), respectively, of the boundary integral operator $C^{m,c}_z$ in (22), and of the unitary map V in (24).

Proposition 3.6 *Let* $\zeta \in \rho(A^{m,c}_0)$. *Define the maps* $\Gamma_0, \Gamma_1 : \mathrm{dom}\, S^* \to L^2(\Gamma; \mathbb{C}^2)$ *acting on* $f = f_+ \oplus f_- \in \mathrm{dom}\, S^*$ *by*

$$\begin{aligned}\Gamma_0 f &:= \frac{1}{2}\Lambda^{-1} V\big(T^D_+ f_+ + T^D_- f_-\big),\\ \Gamma_1 f &:= -\Lambda\big(icV(\sigma\cdot\nu)(T^D_+ f_+ - T^D_- f_-)\\ &\qquad + c^2\, V(\sigma\cdot\nu)(C^{m,c}_\zeta + C^{m,c}_{\overline{\zeta}})(\sigma\cdot\nu)(T^D_+ f_+ + T^D_- f_-)\big).\end{aligned}$$

Then, $\{L^2(\Gamma;\mathbb{C}^2), \Gamma_0, \Gamma_1\}$ *is a boundary triple for* S^*. *Moreover, the associated* γ*-field and Weyl function are given by*

$$\rho(A^{m,c}_0) \ni z \mapsto \gamma(z) = -4c^2 \Phi^{m,c}_z (\sigma\cdot\nu) C^{m,c}_z (\sigma\cdot\nu) V^* \Lambda$$

and

$$\rho(A^{m,c}_0) \ni z \mapsto M(z) = 4c^2 \Lambda V(\sigma\cdot\nu)\left(C^{m,c}_z - \frac{1}{2}(C^{m,c}_\zeta + C^{m,c}_{\overline{\zeta}})\right)(\sigma\cdot\nu) V^* \Lambda,$$

respectively.

Proof We apply the construction from the end of Sect. 6 leading to Proposition 6.3 with the self-adjoint operator B in (26) and the mapping $\mathcal{T} : \mathrm{dom}\, B \to L^2(\Gamma;\mathbb{C}^2)$ given by

$$\mathcal{T} f = -ic\Lambda V(\sigma\cdot\nu)(T^D_+ f_+ - T^D_- f_-), \quad f \in \mathrm{dom}\, B. \tag{27}$$

Since $T^D_\pm : H^1(\Omega_\pm;\mathbb{C}^2) \to H^{1/2}(\Gamma;\mathbb{C}^2)$ is surjective, V and $\sigma\cdot\nu$ give rise to bijective maps in $H^{1/2}(\Gamma;\mathbb{C}^2)$ (these functions are pointwise invertible and C^∞-smooth), and $\Lambda : H^{1/2}(\Gamma;\mathbb{C}^2) \to L^2(\Gamma;\mathbb{C}^2)$ is bijective by construction in (5), the operator $\mathcal{T}$ is surjective. Moreover, as $\mathrm{dom}\, B$ is a closed subspace of $H^1(\mathbb{R}^2\setminus\Gamma;\mathbb{C}^2)$, the embedding of $\mathrm{dom}\, B$ equipped with the norm inherited from $H^1(\mathbb{R}^2\setminus\Gamma;\mathbb{C}^2)$ into $\mathrm{dom}\, B$ equipped with the graph norm is bounded and bijective. Therefore, the graph norm induced by B and the norm in $H^1(\mathbb{R}^2\setminus\Gamma;\mathbb{C}^2)$ are equivalent on $\mathrm{dom}\, B$. Hence, by the properties of $T^D_\pm$, $\sigma\cdot\nu$, V, and Λ the map $\mathcal{T}$ is continuous. Furthermore, $\ker\mathcal{T} = H^1_0(\mathbb{R}^2\setminus\Sigma;\mathbb{C}^2)$ is dense in $L^2(\mathbb{R}^2;\mathbb{C}^2)$. Next, with Lemma 3.4 and Lemma 3.5, one finds

$$\mathcal{G}_z = (\mathcal{T}(B-\overline{z})^{-1})^* = -4c^2\Phi^{m,c}_z(\sigma\cdot\nu)C^{m,c}_z(\sigma\cdot\nu)V^*\Lambda \tag{28}$$

for $z \in \rho(A^{m,c}_0) = \rho(B)$. Recall that $\zeta \in \rho(B)$. For $f = f_\zeta + \mathcal{G}_\zeta\xi$ with $f_\zeta \in \mathrm{dom}\, B$ and $\xi \in L^2(\Gamma;\mathbb{C}^2)$ it follows from (28), Lemma 3.4, and (25) that

$$
\begin{aligned}
&\frac{1}{2}\Lambda^{-1}V\big(T_+^D + T_-^D\big)(f_\zeta + \mathcal{G}_\zeta\xi) \\
&\qquad = -2c^2\Lambda^{-1}V\big(T_+^D + T_-^D\big)\Phi_\zeta^{m,c}(\sigma\cdot\nu)C_\zeta^{m,c}(\sigma\cdot\nu)V^*\Lambda\xi \\
&\qquad = -4c^2\Lambda^{-1}V\big(C_\zeta^{m,c}(\sigma\cdot\nu)\big)^2 V^*\Lambda\xi \\
&\qquad = \xi,
\end{aligned} \tag{29}
$$

and hence we obtain the representation of Γ_0 from Proposition 6.3. In order to obtain the formula for Γ_1 we note first that the mapping $\mathcal{T}$ in (27) can be extended onto $\operatorname{dom} S^*$. Using (27), (28), and Lemma 3.4 we then compute for $z \in \rho(A_0^{m,c}) = \rho(B)$ and $\xi \in L^2(\Gamma;\mathbb{C}^2)$

$$
\begin{aligned}
\mathcal{T}\mathcal{G}_z\xi &= 4ic^3\Lambda V(\sigma\cdot\nu)(T_+^D - T_-^D)\Phi_z^{m,c}(\sigma\cdot\nu)C_z^{m,c}(\sigma\cdot\nu)V^*\Lambda\xi \\
&= 4c^2\Lambda V(\sigma\cdot\nu)C_z^{m,c}(\sigma\cdot\nu)V^*\Lambda\xi.
\end{aligned} \tag{30}
$$

Now consider $f = f_\zeta + \mathcal{G}_\zeta\xi = f_{\bar\zeta} + \mathcal{G}_{\bar\zeta}\xi$ and observe that by Proposition 6.3

$$
\Gamma_1 f = \frac{1}{2}\mathcal{T}(f_\zeta + f_{\bar\zeta}) = \frac{1}{2}\mathcal{T}(f - \mathcal{G}_\zeta\xi + f - \mathcal{G}_{\bar\zeta}\xi) = \mathcal{T}f - \frac{1}{2}\mathcal{T}\big(\mathcal{G}_\zeta + \mathcal{G}_{\bar\zeta}\big)\xi. \tag{31}
$$

Inserting $\mathcal{T}f$ and the expressions $\mathcal{T}\mathcal{G}_\zeta\xi, \mathcal{T}\mathcal{G}_{\bar\zeta}\xi$ from (30) in (31), and making use of $\xi = \Gamma_0 f = \frac{1}{2}\Lambda^{-1}V(T_+^D f_+ + T_-^D f_-)$ we obtain the representation of Γ_1. Finally, the expressions for the γ-field and Weyl function follow immediately from $\gamma(z) = \mathcal{G}_z$ and $M(z) = \mathcal{T}(\mathcal{G}_z - \frac{1}{2}(\mathcal{G}_\zeta + \mathcal{G}_{\bar\zeta}))$ in Proposition 6.3. □

In the following proposition we describe the operator $A_{\eta,\tau,\lambda}^{m,c}$ with the help of the boundary triple from the previous proposition. For similar arguments see [13, Section 6] and also [22, Proposition 6.3] or [11, Proposition 4.3]. Recall that V is given by (24) and define for $\eta, \tau, \lambda \in \mathbb{R}$ the matrix

$$
F := \begin{pmatrix} \eta + \tau & \lambda \\ \lambda & \eta - \tau \end{pmatrix}, \tag{32}
$$

implying

$$
V^*FV = \eta\sigma_0 + \tau\sigma_3 + i\lambda(\sigma\cdot\nu)\sigma_3. \tag{33}
$$

Proposition 3.7 *Let $\eta, \tau, \lambda \in \mathbb{R}$ and define in $L^2(\Gamma; \mathbb{C}^2)$ the operator*

$$\begin{aligned} \Theta\varphi &:= \Lambda\big[F - 2c^2\, V(\sigma\cdot\nu)(C_\zeta^{m,c} + C_{\overline{\zeta}}^{m,c})(\sigma\cdot\nu)V^*\big]\Lambda\varphi, \\ \operatorname{dom}\Theta &:= \big\{\varphi \in L^2(\Gamma;\mathbb{C}^2) : \\ &\qquad \big[F - 2c^2\, V(\sigma\cdot\nu)(C_\zeta^{m,c} + C_{\overline{\zeta}}^{m,c})(\sigma\cdot\nu)V^*\big]\Lambda\varphi \in H^{1/2}(\Gamma;\mathbb{C}^2)\big\}. \end{aligned} \tag{34}$$

Then, one has $A_{\eta,\tau,\lambda}^{m,c} = S^* \restriction \ker(\Gamma_1 - \Theta\Gamma_0)$.

Proof Recall that $f \in \operatorname{dom} A_{\eta,\tau,\lambda}^{m,c}$ if and only if $f_\pm \in H(\sigma, \Omega_\pm)$ and

$$-ic(\sigma\cdot\nu)(T_+^D f_+ - T_-^D f_-) = \frac{1}{2}(\eta\sigma_0 + \tau\sigma_3 + i\lambda(\sigma\cdot\nu)\sigma_3)(T_+^D f_+ + T_-^D f_-).$$

Using (33) and

$$\begin{aligned} \frac{1}{2}(T_+^D f_+ + T_-^D f_-) &= V^*\Lambda\Gamma_0 f, \\ -ic(\sigma\cdot\nu)(T_+^D f_+ - T_-^D f_-) &= V^*\Lambda^{-1}\Gamma_1 f + 2c^2(\sigma\cdot\nu)(C_\zeta^{m,c} + C_{\overline{\zeta}}^{m,c})(\sigma\cdot\nu)V^*\Lambda\Gamma_0 f, \end{aligned}$$

one finds that $f \in \operatorname{dom} A_{\eta,\tau,\lambda}^{m,c}$ if and only if $f \in \operatorname{dom} S^*$ and $\Gamma_1 f - \Theta\Gamma_0 f = 0$. □

4 Proof of Theorem 2.3 and Further Properties of $A_{\eta,\tau,\lambda}^{m,c}$

The claims in Theorem 2.3 are shown in various steps: Sect. 4.1 is devoted to the self-adjointness of $A_{\eta,\tau,\lambda}^{m,c}$, the claim in Theorem 2.3 (i) about the Sobolev regularity of $\operatorname{dom} A_{\eta,\tau,\lambda}^{m,c}$ is then shown in Sect. 4.2. Afterwards, $\sigma_{\mathrm{ess}}(A_{\eta,\tau,\lambda}^{m,c})$ is computed in Sect. 4.3. Finally, in Sect. 4.4 we state some further properties of $A_{\eta,\tau,\lambda}^{m,c}$ like a useful Krein type resolvent formula and a Birman-Schwinger principle.

We will make use of the following two main ingredients: We use that $A_{\eta,\tau,\lambda}^{m,c}$ can be described with the boundary triple in Proposition 3.6 as explained in Proposition 3.7, and the abstract results in the appendix in Sect. 6. And we will use that the claimed properties can be obtained explicitly when the interaction support is the straight line Σ; cf. [16]. To distinguish the appearing objects on the straight line from the case of the more general interaction support Γ, we will denote them with a superindex Σ. Eventually, we will often make use of the arc-length parametrization

$$x : \mathbb{R} \to \mathbb{R}^2, \quad x(t) = (t, 0), \tag{35}$$

of the straight line Σ.

4.1 Self-Adjointness of $A^{m,c}_{\eta,\tau,\lambda}$

Let $\{L^2(\Gamma;\mathbb{C}^2),\Gamma_0,\Gamma_1\}$ be the boundary triple for S^* from Proposition 3.6 and Θ be defined by (34). Recall that

$$A^{m,c}_{\eta,\tau,\lambda} = S^* \upharpoonright \ker(\Gamma_1 - \Theta\Gamma_0).$$

Hence, by Theorem 6.2 the operator $A^{m,c}_{\eta,\tau,\lambda}$ is self-adjoint in $L^2(\mathbb{R}^2;\mathbb{C}^2)$ if and only if Θ is self-adjoint in $L^2(\Gamma;\mathbb{C}^2)$. We show that the latter is the case.

Denote by Λ^Σ the map in (5) on Σ, by $C^{m,c,\Sigma}_z$ the operator in (22) on Σ, and we write Θ^Σ for the parameter defined as in (34) on Σ. Since $\sigma\cdot\nu^\Sigma = -\sigma_2$ and $V^\Sigma = \sigma_0$ on Σ, one has the more explicit representation

$$\Theta^\Sigma = \Lambda^\Sigma\big[F - 2c^2\sigma_2(C^{m,c,\Sigma}_\zeta + C^{m,c,\Sigma}_{\overline{\zeta}})\sigma_2\big]\Lambda^\Sigma \tag{36}$$

and as in (34) this map is defined on the maximal domain. This operator is self-adjoint in $L^2(\Sigma;\mathbb{C}^2)$ by Theorem 6.2 and Proposition 3.7 (applied for $\Gamma=\Sigma$), as $A^{m,c,\Sigma}_{\eta,\tau,\lambda}$ is self-adjoint in $L^2(\mathbb{R}^2;\mathbb{C}^2)$; this follows for $d\neq -4c^2$ from [16, Section 6] noting that $A^{m,c,\Sigma}_{\eta,\tau,\lambda} = cA^{mc,1,\Sigma}_{\eta/c,\tau/c,\lambda/c}$ and $A^{mc,1,\Sigma}_{\eta/c,\tau/c,\lambda/c} = A_{\eta/c,\tau/c,\lambda/c}$ is the operator studied in [16], when m is replaced by mc, and for $d=-4c^2$ from [34]. Next, recall that γ is the arc-length parametrization of Γ given by (6) and that x is the arc-length parametrization of Σ in (35). By definition, one has $\Lambda^\Sigma = U_x^{-1}U_\gamma\Lambda U_\gamma^{-1}U_x$. Then, by Proposition 3.2

$$\begin{aligned}
&U_\gamma\Theta U_\gamma^{-1} - U_x\Theta^\Sigma U_x^{-1}\\
&= -2c^2U_\gamma\Lambda U_\gamma^{-1}\big[U_\gamma V(\sigma\cdot\nu)C^{m,c}_\zeta(\sigma\cdot\nu)V^*U_\gamma^{-1} - U_x\sigma_2C^{m,c,\Sigma}_\zeta\sigma_2U_x^{-1}\\
&+ U_\gamma V(\sigma\cdot\nu)C^{m,c}_{\overline{\zeta}}(\sigma\cdot\nu)V^*U_\gamma^{-1} - U_x\sigma_2C^{m,c,\Sigma}_{\overline{\zeta}}\sigma_2U_x^{-1}\big]U_\gamma\Lambda U_\gamma^{-1},
\end{aligned} \tag{37}$$

defined on its natural maximal domain, can be extended to a compact operator in $L^2(\mathbb{R};\mathbb{C}^2)$. Since the adjoint of $C^{m,c}_\zeta$ in $L^2(\Gamma;\mathbb{C}^2)$ is given by $C^{m,c}_{\overline{\zeta}}$, the right hand side of the latter equation is symmetric and can be extended to a self-adjoint operator. As Θ^Σ is self-adjoint in $L^2(\Sigma;\mathbb{C}^2)$ and $U_\gamma: L^2(\Gamma;\mathbb{C}^2)\to L^2(\mathbb{R};\mathbb{C}^2)$ and $U_x: L^2(\Sigma;\mathbb{C}^2)\to L^2(\mathbb{R};\mathbb{C}^2)$ are unitary, this implies that Θ is self-adjoint.

4.2 Sobolev Regularity in dom $A^{m,c}_{\eta,\tau,\lambda}$

In this subsection we show item (i) of Theorem 2.3. For this we prove an auxiliary lemma in which we analyze dom Θ. Afterwards the Krein type resolvent formula for $A^{m,c}_{\eta,\tau,\lambda}$ from Theorem 6.2 allows us to show the claims on dom $A^{m,c}_{\eta,\tau,\lambda}$. Since $mc^2\sigma_3$ is a bounded perturbation of the self-adjoint operator $A^{m,c}_{\eta,\tau,\lambda}$, it does not affect dom $A^{m,c}_{\eta,\tau,\lambda}$, and hence, it is no restriction to assume in this subsection that $m \neq 0$ and to choose $\zeta = 0$ in Proposition 3.6.

Lemma 4.1 *Assume that $m \neq 0$. Then, for the domain of the operator Θ defined in* (34) *with $\zeta = 0$ one has:*

(i) *If* $(\frac{d}{4} - c^2)^2 - \lambda^2 c^2 \neq 0$, *then* $\operatorname{dom}\Theta = H^1(\Gamma; \mathbb{C}^2)$.
(ii) *If* $(\frac{d}{4} - c^2)^2 - \lambda^2 c^2 = 0$, *then* $\operatorname{dom}\Theta \not\subset H^s(\Gamma; \mathbb{C}^2)$ *for all* $s > 0$.

Proof The proof of this lemma is split into two steps: We analyze $\operatorname{dom}\Theta^\Sigma$ first and with the help of Proposition 3.2 we transfer these results to the case of more general curves Γ.

Step 1. We use the Fourier transform $\mathcal{F}$ in $L^2(\mathbb{R}; \mathbb{C}^2) = U_x L^2(\Sigma; \mathbb{C}^2)$ and note first that $\mathcal{F} U_x C_z^{m,c,\Sigma} U_x^{-1} \mathcal{F}^{-1}$ is the maximal multiplication operator associated with the function

$$c_z(p) = \frac{1}{2\sqrt{p^2c^2 + (mc^2)^2 - z^2}} \begin{pmatrix} \frac{z}{c} + mc & p \\ p & \frac{z}{c} - mc \end{pmatrix};$$

to see this, one can apply [15, Lemma 2.1] and note with (14) that $C_z^{m,c,\Sigma} = \frac{1}{c} C_{z/c}^{mc,1,\Sigma}$ and $C_z = C_z^{mc,1,\Sigma}$ in the notation of [15]. Recall that we chose $\zeta = 0$ in the definition of Θ^Σ. Hence, with (36) one sees that the operator $\mathcal{F} U_x \Theta^\Sigma U_x^{-1} \mathcal{F}^{-1}$ is the maximal multiplication operator associated with the function

$$\theta(p) = \sqrt{p^2+1}\left[F - \sigma_2 \frac{2c^2}{\sqrt{p^2c^2 + (mc^2)^2}}\big(mc\sigma_3 + p\sigma_1\big)\sigma_2\right] = \theta_1(p) + \theta_2(p)$$

with

$$\theta_1(p) = \sqrt{p^2+1}\begin{pmatrix} \eta + \tau & \lambda + \frac{2pc}{\sqrt{p^2+(mc)^2}} \\ \lambda + \frac{2pc}{\sqrt{p^2+(mc)^2}} & \eta - \tau \end{pmatrix},$$

$$\theta_2(p) = \frac{2\sqrt{p^2+1}}{\sqrt{p^2+(mc)^2}} mc^2 \sigma_3.$$

Since θ_2 is a bounded function, its associated multiplication operator is bounded and everywhere defined in $L^2(\mathbb{R}; \mathbb{C}^2)$. The eigenvalues of $\theta_1(p)$ are given by

$$\begin{aligned} \mu_\pm(p) &= \sqrt{p^2+1}\left(\eta \pm \sqrt{\eta^2 - d + \frac{4p^2c^2}{p^2+(mc)^2} + \frac{4pc\lambda}{\sqrt{p^2+(mc)^2}}}\right) \\ &= \sqrt{p^2+1}\left(\eta \pm \sqrt{\eta^2 - d + 4c^2 + 4\lambda c \operatorname{sign}(p) + f(p)}\right), \end{aligned}$$

where d is given by (9) and f is a continuous function such that $f(p) = \mathcal{O}(\frac{1}{p^2})$ for $p \to \pm\infty$. Hence, if $(\frac{d}{4} - c^2)^2 - \lambda^2 c^2 \neq 0$, then $\mu_\pm(p) \sim \sqrt{p^2+1}$ and thus,

$\operatorname{dom}\Theta^\Sigma = H^1(\Sigma;\mathbb{C}^2)$. On the other hand, if $(\frac{d}{4}-c^2)^2-\lambda^2c^2=0$, then at least one of the functions $\mu_\pm$ is bounded on $\mathbb{R}_+$ or on $\mathbb{R}_-$ (or bounded everywhere on $\mathbb{R}$, if $0=\lambda=d-4c^2$), implying that $\operatorname{dom}\Theta^\Sigma \not\subset H^s(\Sigma;\mathbb{C}^2)$ for all $s>0$. Thus, the claim of the lemma is true for the straight line $\Gamma=\Sigma$.

Step 2. Recall that $U_\gamma\Theta U_\gamma^{-1} - U_x\Theta^\Sigma U_x^{-1}$ can be extended to a compact operator in $L^2(\mathbb{R};\mathbb{C})$, see Proposition 3.2 and (37). Hence, the maximal domain of Θ can be described in the form $\operatorname{dom}\Theta = \operatorname{dom}(\Theta^\Sigma U_x^{-1}U_\gamma)$ and now the claim follows from our observation in *Step 1*. □

Proof of Theorem 2.3 (i) As explained above we can assume that $m\neq 0$, and choose $\zeta=0$ in (34). To prove item (a) we use $\operatorname{dom}A^{m,c}_{\eta,\tau,\lambda} = \operatorname{ran}(A^{m,c}_{\eta,\tau,\lambda}-z)^{-1}$ for $z\in\mathbb{C}\setminus\mathbb{R}$ and the resolvent formula

$$(A^{m,c}_{\eta,\tau,\lambda}-z)^{-1} = (A^{m,c}_0-z)^{-1} + \gamma(z)\big(\Theta - M(z)\big)^{-1}\gamma(\overline{z})^* \tag{38}$$

from Theorem 6.2 (iv). If $(\frac{d}{4}-c^2)^2-\lambda^2c^2\neq 0$, then by Lemma 4.1 one has

$$\operatorname{dom}(\Theta - M(z)) = \operatorname{dom}\Theta = H^1(\Gamma;\mathbb{C}^2).$$

Hence, $\operatorname{ran}(A^{m,c}_0-z)^{-1} = \operatorname{dom}A^{m,c}_0 = H^1(\mathbb{R}^2;\mathbb{C}^2)$, the explicit form of the map $\gamma(z) = -4c^2\Phi^{m,c}_z(\sigma\cdot\nu)C^{m,c}_z(\sigma\cdot\nu)V^*\Lambda$ from Proposition 3.6, and the mapping properties in (5), (21), and (22) imply $\operatorname{dom}A^{m,c}_{\eta,\tau,\lambda}\subset H^1(\mathbb{R}^2\setminus\Gamma;\mathbb{C}^2)$, that is, the assertion in Theorem 2.3 (i) (a) holds.

Next, we show item (b), i.e. that $\operatorname{dom}A^{m,c}_{\eta,\tau,\lambda}\not\subset H^s(\mathbb{R}^2\setminus\Gamma;\mathbb{C}^2)$ for all $s>0$, if $(\frac{d}{4}-c^2)^2-\lambda^2c^2=0$. Note that, due to the surjectivity of (Γ_0,Γ_1), for all $\varphi\in\operatorname{dom}\Theta$ there exists $f\in\operatorname{dom}S^*$ such that $\Gamma_0 f=\varphi$ and $\Gamma_1 f=\Theta\varphi$. Hence, for each $\varphi\in\operatorname{dom}\Theta$ there exists $f\in\operatorname{dom}A^{m,c}_{\eta,\tau,\lambda}$ such that $\Gamma_0 f=\varphi$. If we assume now $\operatorname{dom}A^{m,c}_{\eta,\tau,\lambda}\subset H^s(\mathbb{R}^2\setminus\Gamma;\mathbb{C}^2)$ for an $s>0$, then by the definition of Γ_0 we would conclude that any $\varphi\in\operatorname{dom}\Theta$ also belongs to $H^s(\Gamma;\mathbb{C}^2)$. However, this contradicts Lemma 4.1, showing that also Theorem 2.3 (i) (b) is true. □

4.3 Essential Spectrum of $A^{m,c}_{\eta,\tau,\lambda}$

First, we claim that for all combinations $\eta,\tau,\lambda\in\mathbb{R}$ we have

$$(-\infty,-|m|c^2]\cup[|m|c^2,+\infty)\subset\sigma_{\text{ess}}(A^{m,c}_{\eta,\tau,\lambda}), \tag{39}$$

which can be proved via a suitable singular sequence. To show (39), we can assume that $\Gamma\subset[0,+\infty)\times\mathbb{R}$. In fact, if this is not the case, then via a suitable rotation and translation one can transform $A^{m,c}_{\eta,\tau,\lambda}$ to a unitary equivalent Dirac operator, which may have different Dirac matrices $\alpha_1,\alpha_2,\beta\in\mathbb{C}^{2\times 2}$ instead of $\sigma_1,\sigma_2,\sigma_3$, see, e.g.,

[35, p. 150] or [44, Proposition 4], with a singular potential that is supported on a curve contained in $[0, +\infty) \times \mathbb{R}$, so this assumption is not a restriction.

Let now $z \in (-\infty, -|m|c^2) \cup (|m|c^2, +\infty)$ be fixed and consider for $n \in \mathbb{N}$ the function

$$f_n(x_1, x_2) := \frac{1}{n}\chi\left(\frac{1}{n}|x - y_n|\right) e^{i\sqrt{z^2-m^2c^4}x_1/c}\left(\sqrt{z^2 - m^2c^4}\sigma_1 + mc^2\sigma_3 + z\sigma_0\right)\xi,$$

where $\chi : \mathbb{R} \to [0, 1]$ is a C^∞-function such that $\chi(t) = 1$ for $|t| \leq \frac{1}{2}$ and $\chi(t) = 0$ for $|t| \geq 1$, $\xi \in \mathbb{C}^2$ is chosen such that $(\sqrt{z^2 - m^2c^4}\sigma_1 + mc^2\sigma_3 + z\sigma_0)\xi \neq 0$, and we set $y_n := (0, -n^2)$, $n \in \mathbb{N}$. Then $f_n \in \operatorname{dom} A^{m,c}_{\eta,\tau,\lambda}$ and one can show as in [10, Theorem 5.7 (i)] that f_n converges weakly to zero and $\|(A^{m,c}_{\eta,\tau,\lambda} - z)f_n\|/\|f_n\| \to 0$ for $n \to \infty$. Since $z \in (-\infty, -|m|c^2) \cup (|m|c^2, +\infty)$ was arbitrary, the inclusion (39) follows.

For $m = 0$ the inclusion (39) implies $\sigma_{\mathrm{ess}}(A^{0,c}_{\eta,\tau,\lambda}) = \mathbb{R}$, so there is nothing left to show. Hence we can assume $m \neq 0$ in the following. To analyze $\sigma_{\mathrm{ess}}(A^{m,c}_{\eta,\tau,\lambda}) \cap (-|m|c^2, |m|c^2)$, we use that by Proposition 3.7 and Theorem 6.2 (iii)

$$z \in \sigma_{\mathrm{ess}}(A^{m,c}_{\eta,\tau,\lambda}) \cap (-|m|c^2, |m|c^2) \quad \text{if and only if} \quad 0 \in \sigma_{\mathrm{ess}}(\Theta - M(z)), \tag{40}$$

where M is the Weyl function of the boundary triple in Proposition 3.6. Using the definition of Θ in (34) one verifies for $z \in (-|m|c^2, |m|c^2)$

$$\Theta - M(z) = \Lambda[F - 4c^2\, V(\sigma \cdot \nu)C^{m,c}_z(\sigma \cdot \nu)V^*]\Lambda$$

and in the same way for the straight line

$$\Theta^\Sigma - M^\Sigma(z) = \Lambda^\Sigma[F - 4c^2\sigma_2 C^{m,c,\Sigma}_z\sigma_2)]\Lambda^\Sigma.$$

Since $\Lambda^\Sigma = U_x^{-1}U_\gamma\Lambda U_\gamma^{-1}U_x$ we obtain

$$\begin{aligned} &U_\gamma(\Theta - M(z))U_\gamma^{-1} - U_x(\Theta^\Sigma - M^\Sigma(z))U_x^{-1} \\ &\quad = -4c^2 U_\gamma\Lambda U_\gamma^{-1}\big[U_\gamma V(\sigma\cdot\nu)C^{m,c}_z(\sigma\cdot\nu)V^*U_\gamma^{-1} - U_x\sigma_2 C^{m,c,\Sigma}_z\sigma_2 U_x^{-1}\big]U_\gamma\Lambda U_\gamma^{-1} \end{aligned}$$

and it follows from Proposition 3.2 that this operator can be extended to a compact operator in $L^2(\mathbb{R}; \mathbb{C}^2)$. Since U_γ and U_x are unitary, we conclude that

$$\sigma_{\mathrm{ess}}(\Theta - M(z)) = \sigma_{\mathrm{ess}}(\Theta^\Sigma - M^\Sigma(z)),$$

and hence, by (40),

$$z \in \sigma_{\mathrm{ess}}(A^{m,c}_{\eta,\tau,\lambda}) \cap (-|m|c^2, |m|c^2) \quad \text{if and only if} \quad 0 \in \sigma_{\mathrm{ess}}(\Theta^\Sigma - M^\Sigma(z)).$$

Applying Theorem 6.2 (iii) and Proposition 3.7 to the Dirac operator $A^{m,c,\Sigma}_{\eta,\tau,\lambda}$ with the singular interaction supported on the straight line Σ, we find $\sigma_{\mathrm{ess}}(A^{m,c}_{\eta,\tau,\lambda}) = \sigma_{\mathrm{ess}}(A^{m,c,\Sigma}_{\eta,\tau,\lambda})$, which finally yields the statement in Theorem 2.3 (ii); this follows in the case $\eta^2 - \tau^2 - \lambda^2 \neq -4c^2$ from [16, Theorem 6.2] noting that $A^{m,c,\Sigma}_{\eta,\tau,\lambda} = cA^{mc,1,\Sigma}_{\eta/c,\tau/c,\lambda/c}$ and $A^{mc,1,\Sigma}_{\eta/c,\tau/c,\lambda/c} = A_{\eta/c,\tau/c,\lambda/c}$ is the operator studied in [16], when m is replaced by mc; to obtain the result, it is convenient to replace the term k in [16, Theorem 6.2] by kc. The claim is also true for $\eta^2 - \tau^2 - \lambda^2 = -4c^2$, which is excluded in [16], see [34].

Finally, the claim that $\sigma(A^{m,c}_{\eta,\tau,\lambda}) = \sigma_{\mathrm{ess}}(A^{m,c}_{\eta,\tau,\lambda})$, if $\Gamma = \Sigma$ is the straight line, is a part of the results in [16, 34].

4.4 Krein Type Resolvent Formula and Further Spectral Properties of $A^{m,c}_{\eta,\tau,\lambda}$

In the next proposition we formulate a Birman-Schwinger principle to characterize discrete eigenvalues of $A^{m,c}_{\eta,\tau,\lambda}$. The proof of the following result is the same as the proof of [13, Proposition 6.5 (ii)] and will be omitted here.

Proposition 4.2 *Let $\eta, \tau, \lambda \in \mathbb{R}$ and $z \notin \sigma_{\mathrm{ess}}(A^{m,c}_{\eta,\tau,\lambda})$. Then,*

$$z \in \sigma_{\mathrm{disc}}(A^{m,c}_{\eta,\tau,\lambda}) \quad \textit{if and only if} \quad -1 \in \sigma_{\mathrm{p}}\big((\eta\sigma_0 + \tau\sigma_3 + i\lambda(\sigma\cdot\nu)\sigma_3)C^{m,c}_z\big),$$

where $(\eta\sigma_0 + \tau\sigma_3 + i\lambda(\sigma\cdot\nu)\sigma_3)C^{m,c}_z$ is viewed as an operator in $H^{-1/2}(\Gamma;\mathbb{C}^2)$.

Next, we state a Krein type resolvent formula for $A^{m,c}_{\eta,\tau,\lambda}$. Recall that the operators $A^{m,c}_0$ and $\Phi^{m,c}_z$ are given by (12) and (19), respectively.

Proposition 4.3 *Let $\eta, \tau, \lambda \in \mathbb{R}$. Then, $z \in \rho(A^{m,c}_{\eta,\tau,\lambda})$ if and only if the operator $\sigma_0 + (\eta\sigma_0 + \tau\sigma_3 + i\lambda(\sigma\cdot\nu)\sigma_3)C^{m,c}_z$ admits a bounded inverse from $H^{1/2}(\Gamma;\mathbb{C}^2)$ to $H^{-1/2}(\Gamma;\mathbb{C}^2)$ and in this case the resolvent formula*

$$\begin{aligned}(A^{m,c}_{\eta,\tau,\lambda} - z)^{-1} &= (A^{m,c}_0 - z)^{-1}\\ &- \Phi^{m,c}_z\big(\sigma_0 + (\eta\sigma_0 + \tau\sigma_3 + i\lambda(\sigma\cdot\nu)\sigma_3)C^{m,c}_z\big)^{-1}(\eta\sigma_0 + \tau\sigma_3 + i\lambda(\sigma\cdot\nu)\sigma_3)(\Phi^{m,c}_{\bar{z}})^*\end{aligned}$$

holds.

Proof First, one shows with Lemma 3.5 in exactly the same way as in [13, Proposition 6.5 (i)] that for $z \in \rho(A^{m,c}_{\eta,\tau,\lambda})$ the map $\sigma_0 + (\eta\sigma_0 + \tau\sigma_3 + i\lambda(\sigma\cdot\nu)\sigma_3)C^{m,c}_z$ admits a bounded inverse from $H^{1/2}(\Gamma;\mathbb{C}^2)$ to $H^{-1/2}(\Gamma;\mathbb{C}^2)$ and that the claimed resolvent formula is true.

Conversely, assume that $\sigma_0 + (\eta\sigma_0 + \tau\sigma_3 + i\lambda(\sigma\cdot\nu)\sigma_3)C^{m,c}_z$ admits a bounded inverse from $H^{1/2}(\Gamma;\mathbb{C}^2)$ to $H^{-1/2}(\Gamma;\mathbb{C}^2)$. Then, by Proposition 4.2 the operator

$A^{m,c}_{\eta,\tau,\lambda} - z$ is injective. Furthermore, by the mapping properties of $\Phi^{m,c}_z$ and $(\Phi^{m,c}_{\bar z})^*$ in (19), (20), and (18), respectively, the map

$$\begin{aligned} R_z &:= (A^{m,c}_0 - z)^{-1} \\ &- \Phi^{m,c}_z\big(\sigma_0 + (\eta\sigma_0 + \tau\sigma_3 + i\lambda(\sigma\cdot\nu)\sigma_3)C^{m,c}_z\big)^{-1}(\eta\sigma_0 + \tau\sigma_3 + i\lambda(\sigma\cdot\nu)\sigma_3)(\Phi^{m,c}_{\bar z})^* \end{aligned}$$

is bounded in $L^2(\mathbb{R}^2;\mathbb{C}^2)$ and ran $R_z \subset H(\sigma, \mathbb{R}^2 \setminus \Gamma)$. Moreover, with the help of Lemma 3.1 one finds that for any $f \in L^2(\mathbb{R}^2;\mathbb{C}^2)$ the function $R_z f$ satisfies the transmission condition in dom $A^{m,c}_{\eta,\tau,\lambda}$ and thus, $R_z f \in$ dom $A^{m,c}_{\eta,\tau,\lambda}$. Finally, one obtains with (20) that

$$((A^{m,c}_{\eta,\tau,\lambda} - z)R_z f)_\pm = (-ic(\sigma\cdot\nabla) + mc^2\sigma_3 - z\sigma_0)\big((A^{m,c}_0 - z)^{-1} f\big)_\pm = f_\pm.$$

Since $A^{m,c}_{\eta,\tau,\lambda} - z$ is injective, we see that also the claimed resolvent formula is true. □

Eventually, two symmetry relations about the eigenvalues of $A^{m,c}_{\eta,\tau,\lambda}$, that are sometimes referred to as isospectral relations, hold. For the proof of the following result, one can follow the one of [22, Proposition 4.2].

Proposition 4.4 *Let $\eta, \tau, \lambda \in \mathbb{R}$ and d be given by* (9). *Then, the following holds:*

(i) *If $d \neq 0$, then $z \in \sigma_p(A^{m,c}_{\eta,\tau,\lambda})$ if and only if $z \in \sigma_p(A^{m,c}_{-4\eta/d,-4\tau/d,-4\lambda/d})$.*
(ii) *$z \in \sigma_p(A^{m,c}_{\eta,\tau,\lambda})$ if and only if $-z \in \sigma_p(A^{m,c}_{-\eta,\tau,-\lambda})$.*

5 Proof of Theorems 2.7 and 2.8

Throughout this section we assume that $m > 0$. First, we follow ideas from [7, Section 5] and show Theorem 2.7 about the nonrelativistic limit of $A^{m,c}_{\eta/2,\eta/2,0}$ and the geometrically induced bound states for large c, for which some preparations are necessary. Recall that H_η, $\eta \in \mathbb{R}$, is the self-adjoint operator associated with the quadratic form $\mathfrak{h}_\eta$ defined by (10). To state the resolvent formula for H_η, define for $z \in \mathbb{C} \setminus [0, +\infty)$ the single layer potential $SL(z) : L^2(\Gamma;\mathbb{C}) \to L^2(\mathbb{R}^2;\mathbb{C})$ acting as

$$SL(z)\varphi(x) = \frac{m}{\pi}\int_\Gamma K_0\left(-i\sqrt{2mz}|x-y|\right)\varphi(y)d\sigma(y), \quad x \in \mathbb{R}^2 \setminus \Gamma,\ \varphi \in L^2(\Gamma;\mathbb{C}), \tag{41}$$

and the single layer boundary integral operator $\mathcal{S}(z) : L^2(\Gamma;\mathbb{C}) \to L^2(\Gamma;\mathbb{C})$ given by

$$\mathcal{S}(z)\varphi(x) = \frac{m}{\pi}\int_\Gamma K_0\left(-i\sqrt{2mz}|x-y|\right)\varphi(y)d\sigma(y), \quad x \in \Gamma,\ \varphi \in L^2(\Gamma;\mathbb{C}). \tag{42}$$

It is well-known that both operators $SL(z)$ and $\mathcal{S}(z)$ are well-defined and bounded, see, e.g., [6, Theorem 2.8 (a)] applied for $d = 2$ for a reference under the present assumptions. In the following, we denote by $-\Delta$ the free Laplacian defined on $H^2(\mathbb{R}^2; \mathbb{C})$. Then, one has the following relation of $SL(z)$ and $\mathcal{S}(z)$ and the resolvent of H_η, see [19, Lemma 2.3] or [6, Theorem 2.8]:

Lemma 5.1 *Let $\eta \in \mathbb{R}$ and H_η be the self-adjoint operator associated with the quadratic form $\mathfrak{h}_\eta$ in (10). If $z \in \mathbb{C} \setminus [0, \infty)$ and $0 \in \rho(I + \eta\mathcal{S}(z))$, then $z \in \rho(H_\eta)$ and*

$$(H_\eta - z)^{-1} = \left(-\frac{1}{2m}\Delta - z\right)^{-1} - SL(z)\big(I + \eta\mathcal{S}(z)\big)^{-1}\eta SL(\overline{z})^*.$$

In particular, there exists a number $z_0 < 0$ such that for any $z \in (-\infty, z_0)$ one has $0 \in \rho(I + \eta\mathcal{S}(z))$.

In the following proposition we state some preliminary estimates for the convergence of the operators $\Phi^{m,c}_{z+mc^2}$ and $C^{m,c}_{z+mc^2}$ given by (19) and (22), as $c \to \infty$, which are necessary to compute the desired nonrelativistic limit. In the following we make use of

$$P_+ := \begin{pmatrix} 1 & 0 \\ 0 & 0 \end{pmatrix} \in \mathbb{C}^{2\times 2} \quad \text{and} \quad e := \begin{pmatrix} 1 \\ 0 \end{pmatrix} \in \mathbb{C}^2. \tag{43}$$

Note that $P_+ = ee^\top$.

Proposition 5.2 *Let $z \in \mathbb{C} \setminus [0, \infty)$ and $c > \sqrt{\frac{|z|}{|m|}}$. Then, there exists a constant K depending on m, z, Γ such that*

$$\left\| \left(A_0^{m,c} - (z + mc^2)\right)^{-1} - \left(-\frac{1}{2m}\Delta - z\right)^{-1} P_+ \right\|_{L^2(\mathbb{R}^2;\mathbb{C}^2)\to L^2(\mathbb{R}^2;\mathbb{C}^2)} \leq \frac{K}{c}, \tag{44a}$$

$$\left\| \Phi^{m,c}_{z+mc^2} e - SL(z)e \right\|_{L^2(\Gamma;\mathbb{C})\to L^2(\mathbb{R}^2;\mathbb{C}^2)} \leq \frac{K}{c}, \tag{44b}$$

$$\left\| e^\top (\Phi^{m,c}_{z+mc^2})^* - e^\top SL(z)^* \right\|_{L^2(\mathbb{R}^2;\mathbb{C}^2)\to L^2(\Gamma;\mathbb{C})} \leq \frac{K}{c}, \tag{44c}$$

$$\left\| e^\top C^{m,c}_{z+mc^2} e - \mathcal{S}(z) \right\|_{L^2(\Gamma;\mathbb{C})\to L^2(\Gamma;\mathbb{C})} \leq \frac{K}{c}. \tag{44d}$$

Proof First, the estimate in (44a) is well-known, see, e.g., [14, Lemma 3.1] for a proof in the case $m = \frac{1}{2}$. To prove (44b), we note first that $\Phi^{m,c}_{z+mc^2} e - SL(z)e$ acts on $\varphi \in L^2(\Gamma; \mathbb{C})$ as

$$\big(\Phi^{m,c}_{z+mc^2} e - SL(z)e\big)\varphi(x) = \int_\Gamma \big(k_1(x-y) + k_2(x-y)\big)\varphi(y)d\sigma(y), \quad x \in \mathbb{R}^2,$$

with

$$k_1(x) := \frac{1}{2\pi}\left(\left(2m + \frac{z}{c^2}\right) K_0\left(-i\sqrt{\tfrac{z^2}{c^2} + 2mz}|x|\right) - 2mK_0\left(-i\sqrt{2mz}|x|\right)\right) e,$$
$$k_2(x) := \frac{1}{2\pi c}\sqrt{\tfrac{z^2}{c^2} + 2mz}K_1\left(-i\sqrt{\tfrac{z^2}{c^2} + 2mz}|x|\right)\frac{x_1 + ix_2}{|x|}\begin{pmatrix}0\\1\end{pmatrix}. \tag{45}$$

In the same way as in the proof of [14, Lemma 3.1] (see the considerations on the functions t_1, t_2, t_3 there) one finds that there exists constant $M_1, \kappa_1, R > 0$ such that

$$\left|k_1(x) + k_2(x)\right|^2 \leq \frac{M_1^2}{c^2}k_3(x)k_4(x), \tag{46}$$

where

$$k_3(x) := \begin{cases} |x|^{-3/2}, & \text{if } |x| \leq R, \\ e^{-\kappa_1|x|}, & \text{if } |x| > R, \end{cases}$$

and

$$k_4(x) := \begin{cases} |x|^{-1/2}, & \text{if } |x| \leq R, \\ e^{-\kappa_1|x|}, & \text{if } |x| > R. \end{cases}$$

Clearly, by the translation invariance of the Lebesgue measure we have for any $y \in \Gamma$ that

$$\int_{\mathbb{R}^2} |k_3(x-y)|dx = \int_{\mathbb{R}^2} |k_3(x)|dx$$

and thus,

$$M_2 := \sup_{y\in\Gamma}\int_{\mathbb{R}^2} |k_3(x-y)|dx = \int_{\mathbb{R}^2} |k_3(x)|dx < \infty. \tag{47}$$

Next, recall that γ is the arc-length parametrization of Γ defined in (6) and choose for a fixed $x \in \mathbb{R}^2$ a number $t = t(x) \in \mathbb{R}$ such that $|x - \gamma(t)| = \min_{s\in\mathbb{R}} |x - \gamma(s)|$. Then, taking (7) into account, we see that

$$\frac{C_1}{2}|t(x) - s| \leq \frac{1}{2}|\gamma(t(x)) - \gamma(s)| \leq |x - \gamma(s)|.$$

This implies $B_R(x) \cap \Gamma \subset \{\gamma(s) : |t(x) - s| \leq \frac{2R}{C_1}\}$, where $B_R(x)$ is the ball of radius R around x, and

$$
\begin{aligned}
M_3 &:= \sup_{x\in\mathbb{R}^2} \int_\Gamma |k_4(x-y)| d\sigma(y) = \sup_{x\in\mathbb{R}^2} \int_\mathbb{R} |k_4(x-\gamma(s))| ds \\
&= \sup_{x\in\mathbb{R}^2} \left(\int_{\gamma^{-1}(B_R(x)\cap\Gamma)} |x-\gamma(s)|^{-1/2} ds + \int_{\gamma^{-1}(\Gamma\setminus B_R(x))} e^{-\kappa_1|x-\gamma(s)|} ds \right) \\
&\leq \sup_{x\in\mathbb{R}^2} \left(\int_{t(x)-2R/C_1}^{t(x)+2R/C_1} |x-\gamma(s)|^{-1/2} ds + \int_\mathbb{R} e^{-\kappa_1|x-\gamma(s)|} ds \right) \\
&\leq \sup_{x\in\mathbb{R}^2} M_4 \left(\int_{t(x)-2R/C_1}^{t(x)+2R/C_1} |t(x)-s|^{-1/2} ds + \int_\mathbb{R} e^{-\kappa_2|t(x)-s|} ds \right) < +\infty,
\end{aligned}
\tag{48}
$$

where $M_4, \kappa_2 > 0$ are suitable constants. Using (46)–(48) in the Schur test, see, e.g., [48, Theorem 6.24] or [36, Example III 2.4], we get

$$
\left\| \Phi^{m,c}_{z+mc^2} e - SL(z)e \right\|_{L^2(\Gamma;\mathbb{C})\to L^2(\mathbb{R}^2;\mathbb{C}^2)} \leq \frac{M_1}{c}\sqrt{M_2 M_3},
$$

which is exactly (44b). By taking adjoints this also implies (44c).

It remains to prove (44d). Note that for $\varphi \in L^2(\Gamma;\mathbb{C})$ one has

$$
\left(e^\top C^{m,c}_{z+mc^2} e - \mathcal{S}(z)\right)\varphi(x) = \int_\Gamma e^\top k_1(x-y)\varphi(y) d\sigma(y), \quad x \in \Gamma,
$$

where k_1 is given by (45). The function $e^\top k_1$ can be further decomposed in the sum $e^\top k_1 = k_5 + k_6$ with

$$
\begin{aligned}
k_5(x) &:= \frac{1}{2\pi}\left(2m + \frac{z}{c^2}\right)\left(K_0\left(-i\sqrt{\tfrac{z^2}{c^2} + 2mz}|x|\right) - K_0\left(-i\sqrt{2mz}|x|\right)\right), \\
k_6(x) &:= \frac{z}{2\pi c^2} K_0\left(-i\sqrt{2mz}|x|\right).
\end{aligned}
$$

With the same argument as in [14, equation (3.11)] one sees that

$$
|k_5(x)| \leq \frac{M_5}{c} \begin{cases} 1, & \text{if } |x| \leq R, \\ e^{-\kappa_3|x|}, & \text{if } |x| > R, \end{cases}
$$

while for k_6 one gets with the asymptotics for the Bessel function K_0 in [1, Sections 9.6 and 9.7] that

$$
|k_6(x)| \leq \frac{M_5}{c} \begin{cases} |\log|x|| + 1, & \text{if } |x| \leq R, \\ e^{-\kappa_3|x|}, & \text{if } |x| > R, \end{cases}
$$

where $M_5, \kappa_3 > 0$ are suitable constants. Note that for a fixed $y = \gamma(s) \in \Gamma$, $s \in \mathbb{R}$, we have $B_R(\gamma(s)) \cap \Gamma \subset \{\gamma(t) : |t-s| \leq \frac{R}{C_1}\}$ by (7). Hence, one obtains as in (48) with constants $M_6, M_7, M_8, \kappa_4 > 0$ that

$$
\begin{aligned}
\sup_{x\in\Gamma}\int_\Gamma |k_1(x-y)|d\sigma(y) &= \sup_{y\in\Gamma}\int_\Gamma |k_1(x-y)|d\sigma(x) = \sup_{y\in\Gamma}\int_{\mathbb{R}} |k_1(\gamma(s)-y)|ds \\
&\le \sup_{y\in\Gamma}\frac{M_5}{c}\left(\int_{\gamma^{-1}(B_R(x)\cap\Gamma)}\big(\big|\log|\gamma(s)-y|\big|+1\big)ds + \int_{\gamma^{-1}(\Gamma\setminus B_R(x))} e^{-\kappa_3|\gamma(s)-y|}ds\right) \\
&\le \sup_{y=\gamma(t)\in\Gamma}\frac{M_6}{c}\left(\int_{t-R/C_1}^{t+R/C_1}\big(|\gamma(s)-\gamma(t)|^{-1/2}+1\big)ds + \int_{\mathbb{R}} e^{-\kappa_3|\gamma(s)-\gamma(t)|}ds\right) \\
&\le \sup_{y=\gamma(t)\in\Gamma}\frac{M_7}{c}\left(\int_{t-R/C_1}^{t+R/C_1}\big(|s-t|^{-1/2}+1\big)ds + \int_{\mathbb{R}} e^{-\kappa_4|s-t|}ds\right) \\
&\le \frac{M_8}{c} < +\infty.
\end{aligned}
$$

This yields, together with the Schur test, the claim in (44d). □

Using the result from Proposition 5.2 it is not difficult to prove that the nonrelativistic limit of $A^{m,c}_{\eta/2,\eta/2,0}$ is H_η. Recall that the matrix P_+ is defined in (43).

Proof of Theorem 2.7 First, we prove the result about the nonrelativistic limit. Let z_0 be as in Lemma 5.1, $z < z_0$, and c be sufficiently large so that $z + mc^2 \in \rho(A_0^{m,c})$. Note that (44d) implies

$$
\big\|(I+\eta\mathcal{S}(z)) - (I+\eta e^\top C^{m,c}_{z+mc^2}e)\big\|_{L^2(\Gamma;\mathbb{C})\to L^2(\Gamma;\mathbb{C})} \le \frac{K}{c},
$$

and hence, for sufficiently large $c > 0$ the operator $I + \eta e^\top C^{m,c}_{z+mc^2}e$ is boundedly invertible in $L^2(\Gamma;\mathbb{C})$, as $I + \eta\mathcal{S}(z)$ has this property by Lemma 5.1. Applying [36, Theorem IV 1.16] and (44d) yields

$$
\big\|(I+\eta\mathcal{S}(z))^{-1} - (I+\eta e^\top C^{m,c}_{z+mc^2}e)^{-1}\big\|_{L^2(\Gamma;\mathbb{C})\to L^2(\Gamma;\mathbb{C})} \le \frac{K}{c}. \tag{49}
$$

Since $I + \eta e^\top C^{m,c}_{z+mc^2}e$ is boundedly invertible, also

$$
\sigma_0 + \eta e e^\top C^{m,c}_{z+mc^2} = \sigma_0 + \left(\frac{\eta}{2}\sigma_0 + \frac{\eta}{2}\sigma_3\right)C^{m,c}_{z+mc^2}
$$

is boundedly invertible in $L^2(\Gamma;\mathbb{C}^2)$. Thus, it follows from Proposition 4.3 that $z \in \rho(A^{m,c}_{\eta/2,\eta/2,0})$ and

$$
\begin{aligned}
\big(A^{m,c}_{\eta/2,\eta/2,0} - (z+mc^2)\big)^{-1} &= \big(A_0^{m,c} - (z+mc^2)\big)^{-1} \\
&\quad - \Phi^{m,c}_{z+mc^2}\big(\sigma_0 + \eta e e^\top C^{m,c}_{z+mc^2}\big)^{-1}\eta e e^\top(\Phi^{m,c}_{\overline{z}+mc^2})^* \\
&= \big(A_0^{m,c} - (z+mc^2)\big)^{-1} \\
&\quad - \Phi^{m,c}_{z+mc^2}e\big(I + \eta e^\top C^{m,c}_{z+mc^2}e\big)^{-1}\eta e^\top(\Phi^{m,c}_{\overline{z}+mc^2})^*.
\end{aligned}
$$

Using Proposition 5.2 and (49) in this equation, one finds

$$\lim_{c\to+\infty}\big(A^{m,c}_{\eta/2,\eta/2,0}-(z+mc^2)\big)^{-1}=\left(-\frac{1}{2m}\Delta-z\right)^{-1}P_+ \\ -SL(z)e\big(I+\eta\mathcal{S}(z)\big)^{-1}\eta e^\top SL(\overline{z})^*$$

and that the order of convergence is $\frac{1}{c}$. Since $ee^\top=P_+$, we see with Lemma 5.1 that the last expression is $(H_\eta-z)^{-1}P_+$.

To show $\sigma_{\text{disc}}(A^{m,c}_{\eta/2,\eta/2,0})\neq\emptyset$ in the case $\eta<0$ and Γ not being the straight line, we follow closely the proof of [7, Proposition 5.5] and transfer the result from [28] about the spectrum of H_η to $A^{m,c}_{\eta/2,\eta/2,0}$. Note first that, due to the assumption $\eta<0$ and Γ not being the straight line, $\sigma_{\text{disc}}(H_\eta P_+)\neq\emptyset$, see [28, Theorem 5.2]. Now fix some $z<z_0$. We remark that $(H_\eta-z)^{-1}P_+$ is the resolvent of a self-adjoint relation (multivalued operator) in $L^2(\mathbb{R}^2;\mathbb{C}^2)$, whose operator part is $H_\eta P_+$. Since $(A^{m,c}_{\eta/2,\eta/2,0}-(z-+mc^2))^{-1}$ converges in the operator norm to $(H_\eta-z)^{-1}P_+$ for $c\to+\infty$, it follows in the same way as in [45, Theorem VIII.23 (b)] or [47, Satz 9.24 b)] that the spectral measures of $A^{m,c}_{\eta/2,\eta/2,0}-mc^2$ converge to the spectral measures of $H_\eta P_+$. Thus, one finds with [47, Satz 2.58 a)] that $\sigma_{\text{disc}}(A^{m,c}_{\eta/2,\eta/2,0})\neq\emptyset$. □

Remark 5.3 One can show, in a similar way as in Theorem 2.7, that $A^{m,c}_{\eta,\tau,\lambda}$ converges in the nonrelativistic limit to $H_{\eta+\tau}$. Hence, as long as Γ is not the straight line and $\eta+\tau<0$, one concludes as above that $\sigma_{\text{disc}}(A^{m,c}_{\eta,\tau,\lambda})\neq\emptyset$ in this situation. However, since the coefficients of the Lorentz scalar and the anomalous magnetic interactions should be scaled differently to obtain a nonrelativistic limit modelling the same physics, we prefer to state the simpler result on $A^{m,c}_{\eta/2,\eta/2,0}$ in this section.

Next, we prove Theorem 2.8. Recall that $\widetilde{\Gamma}_\omega$ is the broken line defined in (11) and that the essential spectrum of $A^{m,c}_{0,\tau,0}$ was identified in Corollary 2.5.

Proof of Theorem 2.8 In the following we denote by $\widetilde{\Omega}_\pm$ the two disjoint domains with joint boundary $\widetilde{\Gamma}_\omega$, which are chosen such that $(1,0)\in\widetilde{\Omega}_+$, and we denote by $\widetilde{\nu}$ the unit normal vector field at $\widetilde{\Gamma}_\omega$ that is pointing outwards of $\widetilde{\Omega}_+$. We will make use of the operator

$$\widetilde{S}^{m,c}_\tau f:=\big(-ic(\sigma\cdot\nabla)+mc^2\sigma_3\big)f_+\oplus\big(-ic(\sigma\cdot\nabla)+mc^2\sigma_3\big)f_-,$$

$$\operatorname{dom}\widetilde{S}^{m,c}_\tau:=\Big\{f=f_+\oplus f_-\in H^1(\widetilde{\Omega}_+;\mathbb{C}^2)\oplus H(\widetilde{\Omega}_-;\mathbb{C}^2): \\ -ic(\sigma\cdot\widetilde{\nu})\big(f_+|_{\widetilde{\Gamma}_\omega}-f_-|_{\widetilde{\Gamma}_\omega}\big)=\frac{\tau}{2}\sigma_3\big(f_+|_{\widetilde{\Gamma}_\omega}+f_-|_{\widetilde{\Gamma}_\omega}\big)\Big\}.$$

Note that $\widetilde{S}_{\tau}^{m,c} = c\widetilde{S}_{\tau/c}^{mc,1} = cS_{\tau/c}$, where S_τ is the operator considered in [30], when m is replaced by mc. Hence, by [30, Section 2] the operator $\widetilde{S}_{\tau}^{m,c}$ is closed and symmetric, but it has deficiency indices $\dim\ker((\widetilde{S}_{\tau}^{m,c})^* \mp i) = 1$, i.e. $\widetilde{S}_{\tau}^{m,c}$ is symmetric, but not self-adjoint.

Next, fix $L_0 > 0$ such that $\Gamma \cap ([L_0, +\infty) \times \mathbb{R}) = \widetilde{\Gamma}_\omega \cap ([L_0, +\infty) \times \mathbb{R})$, i.e. Γ and $\widetilde{\Gamma}_\omega$ coincide for $x \geq L_0$. Consider for $L > L_0$ and $c_1, \dots, c_N \in \mathbb{C}$ the function

$$f(x, y) := \sum_{n=1}^{N} c_n u_n(x) v(y) w(x, y)$$

with

$$u_n(x) := \sin\left(\frac{2n\pi}{L}x\right) \chi_{[L,2L]}(x),$$

where $\chi_{[L,2L]}$ denotes the characteristic function for $[L, 2L]$,

$$v(y) := \begin{cases} 1, & |y| \leq 2r, \\ e^{-\gamma(|y|-2r)}, & |y| > 2r, \end{cases}$$

where $r := L\tan(\omega)$ and $\gamma := -\frac{4mc^2\tau}{4c^2+\tau^2} > 0$, and

$$w(x, y) := \begin{cases} \begin{pmatrix} 1 \\ 0 \end{pmatrix}, & (x, y) \in \Omega_+, \\ \begin{pmatrix} \frac{4c^2+\tau^2}{4c^2-\tau^2} \\ \frac{4c\tau}{4c^2-\tau^2} e^{i\omega} \end{pmatrix}, & (x, y) \in \Omega_- \ \wedge \ y > 0, \\ \begin{pmatrix} \frac{4c^2+\tau^2}{4c^2-\tau^2} \\ -\frac{4c\tau}{4c^2-\tau^2} e^{-i\omega} \end{pmatrix}, & (x, y) \in \Omega_- \ \wedge \ y < 0. \end{cases}$$

As in the proof of [30, Proposition 6.3] one sees that, when varying the coefficients $c_1, \dots, c_N$, the functions f span an N dimensional space and since each f is supported in $[L, 2L] \times \mathbb{R}$, we find that $f \in \operatorname{dom} \widetilde{S}_{\tau}^{m,c} \cap \operatorname{dom} A_{0,\tau,0}^{m,c}$ and $\widetilde{S}_{\tau}^{m,c} f = A_{0,\tau,0}^{m,c} f$. Recall that by the scaling property $\widetilde{S}_{\tau}^{m,c} = cS_{\tau/c}$ one has to replace m and τ in the formulas in [30] by mc and τ/c, respectively. Hence, one gets with [30, equation (6.3)]

$$\begin{aligned}
&\|A^{m,c}_{0,\tau,0} f\|^2_{L^2(\mathbb{R}^2;\mathbb{C}^2)} - \left(mc^2\frac{\tau^2-4c^2}{\tau^2+4c^2}\right)^2 \|f\|^2_{L^2(\mathbb{R}^2;\mathbb{C}^2)} \\
&= \|\widetilde{S}^{m,c}_\tau f\|^2_{L^2(\mathbb{R}^2;\mathbb{C}^2)} - \left(mc^2\frac{\tau^2-4c^2}{\tau^2+4c^2}\right)^2 \|f\|^2_{L^2(\mathbb{R}^2;\mathbb{C}^2)} \\
&= c^2\left(\|S_{\tau/c} f\|^2_{L^2(\mathbb{R}^2;\mathbb{C}^2)} - \left(mc\frac{\tau^2-4c^2}{\tau^2+4c^2}\right)^2 \|f\|^2_{L^2(\mathbb{R}^2;\mathbb{C}^2)}\right) \\
&< c^2 \sum_{n=1}^N |c_n|^2 \Bigg\{ \tan(\omega)\left[3+\frac{(4c^2+\tau^2)^2+16c^2\tau^2}{(4c^2-\tau^2)^2}\right](2N^2\pi^2+m^2c^2L^2) \\
&\qquad + \frac{4mc^2L\tau}{4c^2+\tau^2} - \frac{N^2\pi^2[(4c^2+\tau^2)^2+16\tau^2c^2](4c^2+\tau^2)}{2mc^2L\tau(4c^2-\tau^2)^2}\Bigg\}.
\end{aligned}$$

Since we consider $m > 0$ and $\tau < 0$, the sum of the second and the third term in the curly brackets is negative, when L is sufficiently large. For these values of L, one can choose $\omega_\star = \omega_\star(L)$ such that the right hand side of the last displayed formula is negative for any $\omega < \omega_\star$. As the space of all f's has dimension N, we find with the min-max principle that $(A^{m,c}_{0,\tau,0})^2$ has at least N discrete eigenvalues below the bottom of its essential spectrum $\left(mc^2\frac{\tau^2-4c^2}{\tau^2+4c^2}\right)^2$; cf. Corollary 2.5. Therefore, by the spectral theorem, $A^{m,c}_{0,\tau,0}$ has at least N discrete eigenvalues in the gap of its essential spectrum. This finishes the proof of this proposition. □

Acknowledgements This research was funded in part by the Austrian Science Fund (FWF) 10.55776/P 33568-N. P. Exner and M. Tušek were supported by the grant No. 21-07129S of the Czech Science Foundation (GAČR). J. Behrndt, M. Holzmann, and M. Tušek thank the Ministry of Education, Youth and Sports of the Czech Republic, the BMBWF, and the Austrian Agency for International Cooperation in Education and Research (OeAD) for support within the project 8J23AT025 and CZ 16/2023, respectively. This publication is based upon work from COST Action CA 18232 MAT-DYN-NET, supported by COST (European Cooperation in Science and Technology), www.cost.eu. The authors thank the referee for helpful comments and remarks that led to an improvement of the manuscript.

6 Appendix: Boundary Triples

In this section we briefly recall some notions related to the theory of (ordinary) boundary triples. Boundary triples are an abstract tool in the extension and spectral theory of symmetric and self-adjoint operators and they are particularly useful to handle boundary value and transmission problems. Here, we only state the definitions and results needed in the main part of the paper, for proofs and further results we refer to [9, 20, 25, 26].

Throughout this section $\mathcal{H}$ is a complex Hilbert space and S is a densely defined closed symmetric operator in $\mathcal{H}$.

Definition 6.1 Let $\mathcal{G}$ be a complex Hilbert space and $\Gamma_0, \Gamma_1 : \operatorname{dom} S^* \to \mathcal{G}$ be linear mappings. Then $\{\mathcal{G}, \Gamma_0, \Gamma_1\}$ is called a boundary triple for S^*, if the following conditions are fulfilled:

(i) For all $f, g \in \operatorname{dom} S^*$ the abstract Green's identity

$$(S^* f, g)_{\mathcal{H}} - (f, S^* g)_{\mathcal{H}} = (\Gamma_1 f, \Gamma_0 g)_{\mathcal{G}} - (\Gamma_0 f, \Gamma_1 g)_{\mathcal{G}}$$

holds.

(ii) The map $(\Gamma_0, \Gamma_1) : \operatorname{dom} S^* \to \mathcal{G} \times \mathcal{G}$ is surjective.

We note that a boundary triple for S^* exists if and only if S admits self-adjoint extensions. In the following we always assume that this is the case and that $\{\mathcal{G}, \Gamma_0, \Gamma_1\}$ is a boundary triple associated with S^*. Then, the extension $B_0 := S^* \upharpoonright \ker \Gamma_0$ of S is self-adjoint and one has the direct sum decomposition

$$\operatorname{dom} S^* = \operatorname{dom} B_0 \dot{+} \ker(S^* - z) = \ker \Gamma_0 \dot{+} \ker(S^* - z), \quad z \in \rho(B_0).$$

Hence, the following mappings are well defined: The *γ-field*

$$\rho(B_0) \ni z \mapsto \gamma(z) := \big(\Gamma_0 \upharpoonright \ker(S^* - z)\big)^{-1},$$

whose values $\gamma(z)$, $z \in \rho(B_0)$, are bounded and everywhere defined operators from $\mathcal{G}$ to $\mathcal{H}$, and the *Weyl function*

$$\rho(B_0) \ni z \mapsto M(z) := \Gamma_1\big(\Gamma_0 \upharpoonright \ker(S^* - z)\big)^{-1},$$

whose values $M(z)$, $z \in \rho(B_0)$, are bounded and everywhere defined operators in $\mathcal{G}$. We remark that

$$\gamma(z)^* = \Gamma_1(B_0 - \bar{z})^{-1}, \quad z \in \rho(B_0). \tag{50}$$

In the following we discuss how the self-adjointness of certain extensions of S can be shown and how their spectral properties can be analyzed with the help of a boundary triple. Let Θ be a linear operator in $\mathcal{G}$. Then we define in $\mathcal{H}$ the linear operator B_Θ as the restriction of S^* onto

$$\operatorname{dom} B_\Theta := \big\{f \in \operatorname{dom} S^* : \Gamma_1 f = \Theta \Gamma_0 f\big\}. \tag{51}$$

Note that in the above definition the equation $\Gamma_1 f = \Theta \Gamma_0 f$ contains the condition $\Gamma_0 f \in \operatorname{dom} \Theta$. We remark that for the description of all closed extensions of S via (51) it is necessary to extend the class of parameters Θ to the class of closed linear relations in $\mathcal{G}$; cf. [9, Chapter 2.2]. The following theorem shows that several properties of B_Θ are encoded in the parameter Θ.

Theorem 6.2 *Let $\{\mathcal{G}, \Gamma_0, \Gamma_1\}$ be a boundary triple for S^* with associated γ-field γ and Weyl function M. Let Θ be a linear operator in $\mathcal{G}$ and let B_Θ be defined by* (51).

Then, B_Θ is self-adjoint in $\mathcal{H}$ if and only if Θ is self-adjoint in $\mathcal{G}$. If Θ is self-adjoint, then the following holds for all $z \in \rho(B_0)$:

(i) *$z \in \sigma(B_\Theta)$ if and only if $0 \in \sigma(\Theta - M(z))$.*
(ii) *$z \in \sigma_p(B_\Theta)$ if and only if $0 \in \sigma_p(\Theta - M(z))$.*
(iii) *$z \in \sigma_{ess}(B_\Theta)$ if and only if $0 \in \sigma_{ess}(\Theta - M(z))$.*
(iv) *For $z \in \rho(B_\Theta)$ one has*

$$(B_\Theta - z)^{-1} = (B_0 - z)^{-1} + \gamma(z)\big(\Theta - M(z)\big)^{-1}\gamma(\overline{z})^*.$$

Finally, we recall an efficient way how a boundary triple can be constructed, see [41, 42] or also [20, Section 1.4.2]. Let B be a self-adjoint operator in a Hilbert space $\mathcal{H}$, $\mathcal{G}$ be another Hilbert space, and assume that $\mathcal{T} : \operatorname{dom} B \to \mathcal{G}$ is surjective, continuous (when $\operatorname{dom} B$ is endowed with the graph norm of B), and that $\ker \mathcal{T}$ is dense in $\mathcal{H}$. Then, $S := B \upharpoonright \ker \mathcal{T}$ is a densely defined closed symmetric operator in $\mathcal{H}$.

To proceed, define for $z \in \rho(B)$ the operator

$$\mathcal{G}_z := \big(\mathcal{T}(B - \overline{z})^{-1}\big)^*. \tag{52}$$

It can be shown that $\mathcal{G}_z$ is bounded and injective. Moreover, one has $\operatorname{ran} \mathcal{G}_z = \ker(S^* - z)$ and that for any $z \in \rho(B)$ the direct sum decomposition

$$\operatorname{dom} S^* = \operatorname{dom} B \dot{+} \ker(S^* - z) = \operatorname{dom} B \dot{+} \operatorname{ran} \mathcal{G}_z$$

holds. Hence, for any $f \in \operatorname{dom} S^*$ there exist an element $f_z \in \operatorname{dom} B$ and $\xi \in \mathcal{G}$ such that $f = f_z + \mathcal{G}_z\xi$; it can be shown that ξ is independent of the choice of $z \in \rho(B)$. Then, the following result holds.

Proposition 6.3 *Let $\zeta \in \rho(B)$ be fixed and define the maps $\Gamma_0, \Gamma_1 : \operatorname{dom} S^* \to \mathcal{G}$ acting on $f = f_\zeta + \mathcal{G}_\zeta\xi = f_{\overline{\zeta}} + \mathcal{G}_{\overline{\zeta}}\xi$ with $f_\zeta, f_{\overline{\zeta}} \in \operatorname{dom} B$, $\xi \in \mathcal{G}$, as*

$$\Gamma_0 f = \xi \quad \text{and} \quad \Gamma_1 f = \frac{1}{2}\mathcal{T}(f_\zeta + f_{\overline{\zeta}}).$$

Then, $\{\mathcal{G}, \Gamma_0, \Gamma_1\}$ is a boundary triple for S^*. Moreover, the values of the γ-field and Weyl function are given by $\gamma(z) = \mathcal{G}_z$ and $M(z) = \mathcal{T}(\mathcal{G}_z - \frac{1}{2}(\mathcal{G}_\zeta + \mathcal{G}_{\overline{\zeta}}))$, $z \in \rho(B)$, respectively.

7 Appendix: Proof of Proposition 3.2

First, we prove (7), which will be used in the proof of Proposition 3.2, i.e. we show that for γ in Hypothesis 2.1 there exists $C_1 > 0$ such that for all $s, t \in \mathbb{R}$

$$C_1|s-t| \leq |\gamma(s) - \gamma(t)| \leq |s-t|$$

holds. The second inequality is clear. To see the first one consider the function

$$\mathbb{R}^2 \setminus \{(u, v) \in \mathbb{R}^2 : u \neq v\} \ni (s, t) \mapsto r(s, t) := \frac{|\gamma(s) - \gamma(t)|}{|s-t|},$$

which is positive and continuous, being extended to $r(s, s) = 1$ as Γ is smooth by assumption. In the first and the third quadrant of the (s, t) plane we have $r(s, t) = 1$ for all s, t large enough. Moreover, for any $a \in [0, +\infty)$,

$$\lim_{s \to \pm\infty} \inf_{\pm t \in [0,a]} r(s, t) = \lim_{s \to \pm\infty} \frac{|\gamma(s)|}{|s|} = 1,$$

and similarly with the roles of s and t interchanged. In the other two quadrants we have $r(s, t) \leq 1$ and, e.g. for $s > M, t < -M$, with $t = -\mu s$ for some $\mu = \mu(t) > 0$, it is not difficult to check that

$$\begin{aligned} r(s, -\mu s) &\geq \frac{|(a_+, c_+) + \mu(a_-, c_-)|}{1+\mu} - \frac{|(b_+, d_+) - (b_-, d_-)|}{|s|(1+\mu)} \\ &\geq \frac{\sqrt{1+\mu^2 - 2\mu\cos\theta}}{1+\mu} - \frac{|(b_+, d_+) - (b_-, d_-)|}{|s|} \end{aligned}$$

holds, where θ is the angle between the asymptotes; the first term on the right-hand side reaches its minimum $\sin\frac{\theta}{2}$ at the axis of the quadrants, $\mu = 1$. Again, the same is true with the roles of s and t interchanged. We conclude that there is a compact subset $\mathcal{M}$ of the plane outside of which we have $r(s, t) > \frac{1}{2}\sin\frac{\theta}{2}$. In view of the compactness, the minimum is attained within $\mathcal{M}$ and $C_2 := \min_{\mathcal{M}} r(s, t) > 0$, because the curve does not intersect itself; it is then enough to choose $C_1 = \min\{C_2, \frac{1}{2}\sin\frac{\theta}{2}\}$.

Next, we provide an abstract result about the compactness of integral operators under suitable assumptions on the integral kernel that will be used several times in the proof of Proposition 3.2.

Lemma 7.1 *Let $k : \mathbb{R}^2 \to \mathbb{C}$ be continuous in $\mathbb{R}^2$ and differentiable on the set $\{(s, t) \in \mathbb{R}^2 : s \neq t\}$. Assume that there exist constants $R, C, \kappa > 0$ and $\alpha \in [0, \frac{1}{2})$ such that for $p \in \{0, 1\}$ and all $(s, t) \in \mathbb{R}^2$ with $s \neq t$*

$$\max\left\{\left|\frac{\partial^p}{\partial s^p}k(s,t)\right|, \left|\frac{\partial^p}{\partial s^p}\overline{k(t,s)}\right|\right\} \leq C \begin{cases} |s-t|^{-\alpha}, & |s|,|t| \leq 3R, \\ e^{-\kappa|s-t|}, & |s-t| > 2R, \\ 0, & s,t > R \text{ or } s,t < -R, \end{cases} \tag{53}$$

holds. Then, for any $r \in [-1, 0]$ *the mapping*

$$\mathcal{K}u(s) = \int_{\mathbb{R}} k(s,t)u(t)dt, \quad u \in C_0^\infty(\mathbb{R};\mathbb{C}),\ s \in \mathbb{R},$$

can be extended to a compact operator $\mathcal{K} : H^r(\mathbb{R};\mathbb{C}) \to H^{r+1}(\mathbb{R};\mathbb{C})$.

Proof Define for $p \in \{0, 1\}$ the function $k_p(s,t) := \frac{\partial^p}{\partial s^p}k(s,t)$, $s \neq t$, and the integral operator $\mathcal{K}_p$

$$\mathcal{K}_p : L^2(\mathbb{R};\mathbb{C}) \to L^2(\mathbb{R};\mathbb{C}), \qquad \mathcal{K}_p u(s) := \int_{\mathbb{R}} k_p(s,t)u(t)dt.$$

The proof is split into several steps: In *Step 1* we show that $k_p \in L^2(\mathbb{R}^2;\mathbb{C})$, which implies that $\mathcal{K}_p$ is a Hilbert-Schmidt operator in $L^2(\mathbb{R};\mathbb{C})$ and hence, it is well-defined and compact. In *Step 2* we verify that for any $u \in L^2(\mathbb{R};\mathbb{C})$ the function $\mathcal{K}u = \mathcal{K}_0 u$ is weakly differentiable and that $\frac{d}{ds}\mathcal{K}_0 u = \mathcal{K}_1 u$. Using this, it is proved in *Step 3* that $\mathcal{K}$ gives rise to a compact operator from $L^2(\mathbb{R};\mathbb{C})$ to $H^1(\mathbb{R};\mathbb{C})$. Then, in *Step 4* it is argued that $\mathcal{K}$ admits a compact extension from $H^{-1}(\mathbb{R};\mathbb{C})$ to $L^2(\mathbb{R};\mathbb{C})$. Finally, in *Step 5* an interpolation argument is used to deduce that also for any $r \in (-1, 0)$ the map $\mathcal{K}$ has a compact extension from $H^r(\mathbb{R};\mathbb{C})$ to $H^{r+1}(\mathbb{R};\mathbb{C})$.

Step 1. We claim that $k_p \in L^2(\mathbb{R}^2;\mathbb{C})$. For two intervals I_1 and I_2 set

$$\mathcal{I}(I_1, I_2) := \int_{I_1 \times I_2} |k_p(s,t)|^2 ds dt.$$

Then

$$\|k_p\|^2_{L^2(\mathbb{R}^2;\mathbb{C})} = \mathcal{I}((-\infty, -R], \mathbb{R}) + \mathcal{I}((-R, R), \mathbb{R}) + \mathcal{I}([R, +\infty), \mathbb{R}). \tag{54}$$

We further decompose

$$\begin{aligned}\mathcal{I}((-\infty, -R], \mathbb{R}) = {} & \mathcal{I}((-\infty, -R], (-\infty, -R]) + \mathcal{I}((-\infty, -3R], (-R, R)) \\ & + \mathcal{I}((-3R, -R], (-R, R)) + \mathcal{I}((-\infty, -R], [R, +\infty)).\end{aligned} \tag{55}$$

By (53) we have

$$\mathcal{I}((-\infty, -R], (-\infty, -R]) = 0 \tag{56}$$

and also

$$\mathcal{I}((-\infty,-R],[R,+\infty)) \le C^2 \int_{-\infty}^{-R}\int_{R}^{+\infty} e^{2\kappa(t-s)}\,ds\,dt < +\infty. \tag{57}$$

In the same way one finds that $\mathcal{I}((-\infty,-3R],(-R,R)) < +\infty$. Eventually, taking the properties of k from (53) for $|s|,|t| \le 3R$ into account, one gets for bounded intervals $I_1, I_2 \subset [-3R, 3R]$ that

$$\mathcal{I}(I_1, I_2) < +\infty. \tag{58}$$

Using (56)–(58) in (55) one concludes that $\mathcal{I}((-\infty,-R],\mathbb{R}) < +\infty$. In exactly the same way one gets that $\mathcal{I}([R,+\infty),\mathbb{R}) < +\infty$. Finally, employing again similar estimates as in (57)–(58) one finds that

$$\begin{aligned}\mathcal{I}((-R,R),\mathbb{R}) = \mathcal{I}((-R,R),(-\infty,-3R]) + \mathcal{I}((-R,R),(-3R,3R))\\ + \mathcal{I}((-R,R),[3R,+\infty)) < +\infty.\end{aligned}$$

Hence, we get from (54) that $k_p \in L^2(\mathbb{R}^2;\mathbb{C})$ showing that $\mathcal{K}_p$ is a Hilbert-Schmidt operator and thus, compact in $L^2(\mathbb{R};\mathbb{C})$.

Step 2. Let $u \in C_0^\infty(\mathbb{R};\mathbb{C})$ be fixed. We claim that $\mathcal{K}_0 u \in H^1(\mathbb{R};\mathbb{C})$ and $\frac{d}{ds}\mathcal{K}_0 u = \mathcal{K}_1 u$. Taking the result from *Step 1* into account, it suffices to prove the latter identity for the weak derivative. Let $\varphi \in C_0^\infty(\mathbb{R};\mathbb{C})$. By (53) one has for any fixed $t \in \mathbb{R}$ that $\varphi' k(\cdot,t) \in L^1(\mathbb{R};\mathbb{C})$. Hence, we can apply Fubini's theorem and the dominated convergence theorem and get

$$\begin{aligned}\int_{\mathbb{R}} \varphi'(s)\mathcal{K}_0 u(s)ds &= \int_{\mathbb{R}}\int_{\mathbb{R}} \varphi'(s)k(s,t)ds\, u(t)dt\\ &= \int_{\mathbb{R}} \lim_{\varepsilon\searrow 0} \int_{\mathbb{R}\setminus(t-\varepsilon,t+\varepsilon)} \varphi'(s)k(s,t)ds\, u(t)dt.\end{aligned}$$

Next, we use integration by parts, which is allowed, as for a fixed t the function $s \mapsto k(s,t)$ is differentiable in $\mathbb{R}\setminus(t-\varepsilon,t+\varepsilon)$. Using the continuity of k and φ, the dominated convergence theorem, which is applicable, since for any fixed $t \in \mathbb{R}$ one has $\varphi\frac{\partial}{\partial s}k(\cdot,t) \in L^1(\mathbb{R};\mathbb{C})$ due to (53), and Fubini's theorem we find that

$$\begin{aligned}\int_{\mathbb{R}} \varphi'(s)\mathcal{K}_0 u(s)ds &= \int_{\mathbb{R}} \lim_{\varepsilon\searrow 0}\left[-\varphi(s)k(s,t)\Big|_{t-\varepsilon}^{t+\varepsilon} - \int_{\mathbb{R}\setminus(t-\varepsilon,t+\varepsilon)} \varphi(s)\frac{\partial}{\partial s}k(s,t)ds\right]u(t)dt\\ &= -\int_{\mathbb{R}}\int_{\mathbb{R}} \varphi(s)\frac{\partial}{\partial s}k(s,t)ds\, u(t)dt = -\int_{\mathbb{R}} \varphi(s)\mathcal{K}_1 u(s)ds.\end{aligned}$$

This shows the claim about the weak derivative of $\mathcal{K}_0 u$. Note that by density and continuity $\frac{d}{ds}\mathcal{K}_0 u = \mathcal{K}_1 u$ also holds for $u \in L^2(\mathbb{R};\mathbb{C})$.

Step 3. We show the claim of the lemma for $r = 0$, i.e. that the mapping

$$\mathcal{K} : L^2(\mathbb{R}; \mathbb{C}) \to H^1(\mathbb{R}; \mathbb{C}), \qquad \mathcal{K}u(s) := \int_{\mathbb{R}} k(s,t)u(t)dt,$$

is compact. First, note that $\mathcal{K}$ is well-defined by the results in *Step 1 & 2*. Next, consider the map

$$\mathcal{D} : H^1(\mathbb{R}; \mathbb{C}) \to L^2(\mathbb{R}; \mathbb{C}), \qquad \mathcal{D}f = f + f'.$$

With the help of the Fourier transform it is not difficult to see that $\mathcal{D}$ is bijective. Using the results from *Step 1 & 2* we find that $\mathcal{D}\mathcal{K} = \mathcal{K}_0 + \mathcal{K}_1$ and the right hand side defines a compact operator in $L^2(\mathbb{R}; \mathbb{C})$. Therefore, $\mathcal{K} = \mathcal{D}^{-1}(\mathcal{K}_0 + \mathcal{K}_1)$ is compact as an operator from $L^2(\mathbb{R}; \mathbb{C})$ to $H^1(\mathbb{R}; \mathbb{C})$.

Step 4. We claim that $\mathcal{K}$ admits a compact extension $\widetilde{\mathcal{K}} : H^{-1}(\mathbb{R}; \mathbb{C}) \to L^2(\mathbb{R}; \mathbb{C})$. Due to the assumptions on the function $\overline{k(t,s)}$ in (53) it follows as in *Step 1–3* that the map

$$\mathcal{L} : L^2(\mathbb{R}; \mathbb{C}) \to H^1(\mathbb{R}; \mathbb{C}), \qquad \mathcal{L}u(s) := \int_{\mathbb{R}} \overline{k(t,s)}u(t)dt,$$

is compact. Hence, also the anti-dual operator

$$\widetilde{\mathcal{K}} := \mathcal{L}' : H^{-1}(\mathbb{R}; \mathbb{C}) \to L^2(\mathbb{R}; \mathbb{C})$$

is compact. We claim that $\widetilde{\mathcal{K}}$ is an extension of $\mathcal{K}$. To see this, we note that $\mathcal{L}u = \mathcal{K}_0^* u$ holds for any $u \in L^2(\mathbb{R}; \mathbb{C})$, and compute for $u, v \in L^2(\mathbb{R}; \mathbb{C})$

$$\begin{aligned}(\widetilde{\mathcal{K}}u, v)_{L^2(\mathbb{R};\mathbb{C})} = (u, \mathcal{L}v)_{H^{-1}(\mathbb{R};\mathbb{C})\times H^1(\mathbb{R};\mathbb{C})} &= (u, \mathcal{L}v)_{L^2(\mathbb{R};\mathbb{C})} \\ &= (\mathcal{K}_0 u, v)_{L^2(\mathbb{R};\mathbb{C})} = (\mathcal{K}u, v)_{L^2(\mathbb{R};\mathbb{C})};\end{aligned}$$

in the latter computation we used the symbol $(\cdot,\cdot)_{H^{-1}(\mathbb{R};\mathbb{C})\times H^1(\mathbb{R};\mathbb{C})}$ for the sesquilinear duality product in $H^{-1}(\mathbb{R}; \mathbb{C}) \times H^1(\mathbb{R}; \mathbb{C})$. Since this is true for all $v \in L^2(\mathbb{R}; \mathbb{C})$, we conclude that $\widetilde{\mathcal{K}}u = \mathcal{K}u$, which shows that $\widetilde{\mathcal{K}}$ is a compact extension of $\mathcal{K}$.

Step 5. By the results in *Step 3 & 4* the claim of the lemma is true for $r = -1$ and $r = 0$. By interpolation, one can apply, e.g., [24, Theorem 10] with $X_0 = H^{-1}(\mathbb{R}; \mathbb{C})$, $X_1 = L^2(\mathbb{R}; \mathbb{C})$, and $W = H^{-2}(\mathbb{R}; \mathbb{C})$, it follows that also for $r \in (-1, 0)$ the map $\mathcal{K}$ gives rise to a compact operator from $H^r(\mathbb{R}; \mathbb{C})$ to $H^{r+1}(\mathbb{R}; \mathbb{C})$. □

Proof of Proposition 3.2 Define the operator

$$\mathcal{K} := U_\gamma V(\sigma\cdot\nu)C_z^{m,c}(\sigma\cdot\nu)V^* U_\gamma^{-1} - U_x \sigma_2 C_z^{m,c,\Sigma}\sigma_2 U_x^{-1}.$$

The map $\mathcal{K}$ is an integral operator and we will do a fine analysis of the integral kernel of $\mathcal{K}$, so that we can apply Lemma 7.1 for $r = -\frac{1}{2}$ to each of the entries of the 2×2 block operator matrix $\mathcal{K}$ to conclude the claimed result.

Recall that γ is given by (6). We use for $y = (y_1, y_2) \in \mathbb{R}^2$ the notation $\boldsymbol{y} = y_1 + iy_2$, implying that $|y|_{\mathbb{R}^2} = |\boldsymbol{y}|_{\mathbb{C}}$. Furthermore, for $s \in \mathbb{R}$ one has $\boldsymbol{t}(\gamma(s)) = \dot{\boldsymbol{\gamma}}(s)$ and $\boldsymbol{\nu}(\gamma(s)) = \dot{\gamma}_2(s) - i\dot{\gamma}_1(s) = -i\dot{\boldsymbol{\gamma}}(s)$, and thus

$$V(\gamma(s)) = \begin{pmatrix} 1 & 0 \\ 0 & \overline{\dot{\boldsymbol{\gamma}}(s)} \end{pmatrix} \quad \text{and} \quad \sigma \cdot \nu(\gamma(s)) = \begin{pmatrix} 0 & i\overline{\dot{\boldsymbol{\gamma}}(s)} \\ -i\dot{\boldsymbol{\gamma}}(s) & 0 \end{pmatrix}.$$

Therefore, by using the definition of U_γ, U_x, and $C_z^{m,c}$ we see for $u \in L^2(\mathbb{R}; \mathbb{C}^2)$ that

$$\mathcal{K}u(s) = \lim_{\varepsilon \searrow 0} \int_{\mathbb{R}\setminus B(s,\varepsilon)} k(s,t)u(t)dt, \tag{59}$$

where

$$k(s,t) := \begin{pmatrix} 0 & i\overline{\dot{\boldsymbol{\gamma}}(s)} \\ -i & 0 \end{pmatrix} G_z^{m,c}\big(\gamma(s) - \gamma(t)\big) \begin{pmatrix} 0 & i \\ -i\dot{\boldsymbol{\gamma}}(t) & 0 \end{pmatrix} - \sigma_2 G_z^{m,c}(s-t, 0)\sigma_2.$$

Define the number $\zeta(z) := -i\sqrt{\frac{z^2}{c^2} - (mc)^2}$ and note that

$$\begin{aligned} \frac{\sigma \cdot (\gamma(s) - \gamma(t))}{|\gamma(s) - \gamma(t)|} &= \frac{1}{|\gamma(s) - \gamma(t)|} \begin{pmatrix} 0 & \overline{\boldsymbol{\gamma}(s)} - \overline{\boldsymbol{\gamma}(t)} \\ \boldsymbol{\gamma}(s) - \boldsymbol{\gamma}(t) & 0 \end{pmatrix} \\ &= |\gamma(s) - \gamma(t)| \begin{pmatrix} 0 & \frac{1}{\boldsymbol{\gamma}(s) - \boldsymbol{\gamma}(t)} \\ \frac{1}{\overline{\boldsymbol{\gamma}(s)} - \overline{\boldsymbol{\gamma}(t)}} & 0 \end{pmatrix}. \end{aligned}$$

Then, by the definition of $G_z^{m,c}$ in (13), the function k can be further decomposed as $k(s,t) = k_1(s,t) + k_2(s,t) + k_3(s,t)$ with

$$\begin{aligned} k_1(s,t) &:= \frac{1}{2\pi c} \overline{\dot{\boldsymbol{\gamma}}(s)}\big(\dot{\boldsymbol{\gamma}}(t) - \dot{\boldsymbol{\gamma}}(s)\big) \begin{pmatrix} \frac{z}{c} - mc & 0 \\ 0 & 0 \end{pmatrix} K_0\left(\zeta(z)|\gamma(s) - \gamma(t)|\right), \\ k_2(s,t) &:= \frac{1}{2\pi c}\left(\frac{z}{c}\sigma_0 - mc\sigma_3\right)\big(K_0\left(\zeta(z)|\gamma(s) - \gamma(t)|\right) - K_0\left(\zeta(z)|s-t|\right)\big), \\ k_3(s,t) &:= \frac{i\zeta(z)}{2\pi c}\Bigg(K_1\left(\zeta(z)|s-t|\right)|s-t| \begin{pmatrix} 0 & \frac{1}{s-t} \\ \frac{1}{s-t} & 0 \end{pmatrix} \\ &\qquad - K_1\left(\zeta(z)|\gamma(s) - \gamma(t)|\right)|\gamma(s) - \gamma(t)| \begin{pmatrix} 0 & \frac{\overline{\dot{\boldsymbol{\gamma}}(s)}}{\overline{\boldsymbol{\gamma}(s)} - \overline{\boldsymbol{\gamma}(t)}} \\ \frac{\dot{\boldsymbol{\gamma}}(t)}{\boldsymbol{\gamma}(s) - \boldsymbol{\gamma}(t)} & 0 \end{pmatrix}\Bigg). \end{aligned}$$

Denote the corresponding integral operators by $\mathcal{K}_j$, $j \in \{1, 2, 3\}$, i.e.

$$\mathcal{K}_j\varphi(s) = \lim_{\varepsilon \searrow 0} \int_{\mathbb{R}\setminus B(s,\varepsilon)} k_j(s,t)\varphi(t)dt, \quad s \in \mathbb{R}, \ \varphi \in C_0^\infty(\mathbb{R}; \mathbb{C}^2).$$

We show that the kernels k_j satisfy the assumptions in Lemma 7.1 implying none of the operators is strongly singular (i.e. the limit for $\varepsilon \searrow 0$ can be removed) and has the claimed mapping properties.

Let $M > 0$ be the number specified in the definition of γ in (6). First, we note that, due to the asymptotic properties of the modified Bessel functions and their derivatives for large arguments [1, Sections 9.6 and 9.7] and (7), there exist constants $M_1, \kappa_1 > 0$ such that for $j \in \{1, 2, 3\}$ and $p \in \{0, 1\}$

$$\max\left\{\left|\frac{\partial^p k_j(s,t)}{\partial s^p}\right|, \left|\frac{\partial^p \overline{k_j(t,s)}}{\partial s^p}\right|\right\} \leq M_1 e^{-\kappa_1|s-t|}, \quad |s-t| \geq M. \tag{60}$$

Next, taking the power series representation of K_0 from [1, equation 9.6.13] into account, there exist a constant M_2 and holomorphic functions g_1, g_2 such that

$$K_0(\xi) = -\log\xi + M_2 + g_1(\xi^2) + \xi^2 g_2(\xi^2)\log\xi = -\log\xi + g_3(\xi), \quad \xi \in \mathbb{C}\setminus\{0\}, \tag{61}$$

with the C^1-smooth function $g_3(\xi) := M_2 + g_1(\xi^2) + \xi^2 g_2(\xi^2)\log\xi$; cf. the proof of [11, Lemma 3.2] for a similar consideration. We will often use that the function

$$\log\big(\zeta(z)|\gamma(s)-\gamma(t)|\big) - \log\big(\zeta(z)|s-t|\big) = \log\left|\frac{\boldsymbol{\gamma}(s)-\boldsymbol{\gamma}(t)}{s-t}\right| \tag{62}$$

is C^∞-smooth. For $s \neq t$, this follows from the injectivity and the smoothness of γ, and for $s = t$ from the Taylor series expansion of $\boldsymbol{\gamma}$ and (7).

In the following, the properties of $\mathcal{K}_1$ and its kernel k_1 are further analyzed. With (61) and (62) it is not difficult to see that k_1 can be written as

$$k_1(s,t) = \frac{1}{2\pi c}\overline{\dot{\boldsymbol{\gamma}}(s)}\big(\dot{\boldsymbol{\gamma}}(t) - \dot{\boldsymbol{\gamma}}(s)\big)\begin{pmatrix} \frac{z}{c} - mc & 0 \\ 0 & 0 \end{pmatrix}$$
$$\cdot\left(-\log\big(\zeta(z)|s-t|\big) - \log\left|\frac{\boldsymbol{\gamma}(s)-\boldsymbol{\gamma}(t)}{s-t}\right| + g_3\big(\zeta(z)|\gamma(s)-\gamma(t)|\big)\right).$$

Since $\boldsymbol{\gamma}$ is smooth, the function $(\dot{\boldsymbol{\gamma}}(t) - \dot{\boldsymbol{\gamma}}(s))\log(\zeta(z)|s-t|)$ is continuous in $\mathbb{R}^2$, differentiable for $s \neq t$, and its first derivatives are bounded by $\widetilde{M}_1\big|\log|s-t|\big| + \widetilde{M}_2$ for some constants $\widetilde{M}_1, \widetilde{M}_2 > 0$. Hence, k_1 is continuous in $\mathbb{R}^2$ and differentiable for $s \neq t$, for $p \in \{0, 1\}$ one has

$$\max\left\{\left|\frac{\partial^p k_1(s,t)}{\partial s^p}\right|,\left|\frac{\partial^p \overline{k_1(t,s)}}{\partial s^p}\right|\right\}\le M_3\big|\log|s-t|\big|+M_4,\quad |s|,|t|\le 2M, \tag{63}$$

where M_3, M_4 are positive constants, and $k_1(s,t)=0$, if $s,t<-M$ or $s,t>M$, as in that case $\dot{\boldsymbol{\gamma}}(s)=\dot{\boldsymbol{\gamma}}(t)$. Since k_1 fulfils (60), this function satisfies the assumptions in Lemma 7.1 and thus, $\mathcal{K}_1: H^{-1/2}(\mathbb{R};\mathbb{C}^2)\to H^{1/2}(\mathbb{R};\mathbb{C}^2)$ is compact.

Next, $\mathcal{K}_2$ and its kernel k_2 are considered. With (61) and (62) one finds that $k_2\in C^1(\mathbb{R}^2;\mathbb{C}^2)$. In particular, this shows for a constant $M_5>0$ that

$$\max\left\{\left|\frac{\partial^p k_2(s,t)}{\partial s^p}\right|,\left|\frac{\partial^p \overline{k_2(t,s)}}{\partial s^p}\right|\right\}\le M_5,\quad |s|,|t|\le 2M.$$

Finally, since for $s,t\ge M$

$$|\gamma(s)-\gamma(t)|=\left|\begin{pmatrix}a_+\\c_+\end{pmatrix}(s-t)\right|=|s-t| \tag{64}$$

holds, we get that $k_2(s,t)=0$, if $s,t\ge M$. A similar consideration can be done also for $s,t\le -M$. With (60) one sees that k_2 satisfies (53) and hence, by Lemma 7.1 the map $\mathcal{K}_2: H^{-1/2}(\mathbb{R};\mathbb{C}^2)\to H^{1/2}(\mathbb{R};\mathbb{C}^2)$ is compact.

It remains to analyze $\mathcal{K}_3$ and its kernel k_3. By the power series representation of K_1 in [1, equation 9.6.11] there exist holomorphic functions g_4, g_5 such that

$$K_1(\xi)=\frac{1}{\xi}+\xi g_4(\xi^2)\log\xi+\xi g_5(\xi^2),\quad \xi\in\mathbb{C}\setminus\{0\}; \tag{65}$$

cf. the proof of [11, Lemma 3.2] for a similar consideration. Hence, we can further decompose $k_3(s,t)$ as $k_3(s,t)=k_4(s,t)+k_5(s,t)+k_6(s,t)$ with

$$\begin{aligned}
k_4(s,t)&:=\frac{i}{2\pi c}\begin{pmatrix}0&\frac{1}{s-t}\\ \frac{1}{s-t}&0\end{pmatrix}-\frac{i}{2\pi c}\begin{pmatrix}0&\frac{\overline{\dot{\boldsymbol{\gamma}}(s)}}{\overline{\boldsymbol{\gamma}(s)-\boldsymbol{\gamma}(t)}}\\ \frac{\dot{\boldsymbol{\gamma}}(t)}{\boldsymbol{\gamma}(s)-\boldsymbol{\gamma}(t)}&0\end{pmatrix},\\
k_5(s,t)&:=\frac{i\zeta(z)^2}{2\pi c}\Bigg(g_4(\zeta(z)^2|s-t|^2)\log(\zeta(z)|s-t|)\begin{pmatrix}0&s-t\\ s-t&0\end{pmatrix}\\
&\qquad -g_4(\zeta(z)^2|\gamma(s)-\gamma(t)|^2)\log(\zeta(z)|\gamma(s)-\gamma(t)|)\\
&\qquad\cdot\begin{pmatrix}0&\overline{\dot{\boldsymbol{\gamma}}(s)}(\boldsymbol{\gamma}(s)-\boldsymbol{\gamma}(t))\\ \dot{\boldsymbol{\gamma}}(t)\overline{(\boldsymbol{\gamma}(s)-\boldsymbol{\gamma}(t))}&0\end{pmatrix}\Bigg),\\
k_6(s,t)&:=\frac{i\zeta(z)^2}{2\pi c}\Bigg(g_5(\zeta(z)^2|s-t|^2)\begin{pmatrix}0&s-t\\ s-t&0\end{pmatrix}\\
&\qquad -g_5(\zeta(z)^2|\gamma(s)-\gamma(t)|^2)\begin{pmatrix}0&\overline{\dot{\boldsymbol{\gamma}}(s)}(\boldsymbol{\gamma}(s)-\boldsymbol{\gamma}(t))\\ \dot{\boldsymbol{\gamma}}(t)\overline{(\boldsymbol{\gamma}(s)-\boldsymbol{\gamma}(t))}&0\end{pmatrix}\Bigg).
\end{aligned}$$

Since g_5 is analytic, we observe that k_6 can be extended to a C^∞-smooth function in $\mathbb{R}^2$. The same is also true for k_4. Indeed, employing the fundamental theorem of calculus twice, we get

$$\begin{aligned} \frac{1}{s-t} - \frac{\dot{\gamma}(t)}{\gamma(s)-\gamma(t)} &= \frac{\gamma(s)-\gamma(t)-\dot{\gamma}(t)(s-t)}{(s-t)(\gamma(s)-\gamma(t))} \\ &= \frac{\int_0^1 \dot{\gamma}(t+\xi(s-t))(s-t)d\xi - \dot{\gamma}(t)(s-t)}{(s-t)(\gamma(s)-\gamma(t))} \\ &= \int_0^1 \int_0^1 \ddot{\gamma}(t+\zeta\xi(s-t))\xi d\zeta d\xi \cdot \frac{s-t}{\gamma(s)-\gamma(t)}. \end{aligned}$$

The first factor in the last line is a smooth function everywhere in $\mathbb{R}^2$, and due to (7) the same is true also for the second factor. Therefore, k_4 can be extended to a C^∞-function at $s = t$.

Next, we analyze the properties of $k_5(s,t)$ for s, t being close to each other. In a very similar way as in the consideration of k_1 one gets that k_5 is continuous everywhere in $\mathbb{R}^2$ and that there exist constants $M_6, M_7 > 0$ such that for $p \in \{0, 1\}$

$$\max\left\{\left|\frac{\partial^p k_5(s,t)}{\partial s^p}\right|, \left|\frac{\partial^p \overline{k_5(t,s)}}{\partial s^p}\right|\right\} \le M_6 \big|\log|s-t|\big| + M_7, \quad |s|, |t| \le 2M,$$

holds; cf. (63). Since k_4 and k_6 are smooth functions and $k_3 = k_4 + k_5 + k_6$, we conclude that there exist constants $M_8, M_9 > 0$ such that for $p \in \{0, 1\}$

$$\max\left\{\left|\frac{\partial^p k_3(s,t)}{\partial s^p}\right|, \left|\frac{\partial^p \overline{k_3(t,s)}}{\partial s^p}\right|\right\} \le M_8 \big|\log|s-t|\big| + M_9, \quad |s|, |t| \le 2M, \tag{66}$$

is satisfied.

Finally, we show that $k_3(s,t) = 0$ for $s, t > M$ or $s, t < -M$. For $s, t \ge M$ we have (64) and

$$\frac{\dot{\gamma}(t)}{\gamma(s)-\gamma(t)} = \frac{a_+ + ic_+}{a_+(s-t) + ic_+(s-t)} = \frac{1}{s-t}.$$

Similar identities are also true for $s, t \le -M$. Hence, we conclude from the definition of k_3 that $k_3(s,t) = 0$ for $s, t > M$ or $s, t < -M$. Combining this with (60) and (66), it follows from Lemma 7.1 that also $\mathcal{K}_3 : H^{-1/2}(\mathbb{R}; \mathbb{C}^2) \to H^{1/2}(\mathbb{R}; \mathbb{C}^2)$ is compact. Together with the mapping properties of $\mathcal{K}_1$ and $\mathcal{K}_2$ shown above this yields that

$$\mathcal{K} = \mathcal{K}_1 + \mathcal{K}_2 + \mathcal{K}_3 : H^{-1/2}(\mathbb{R}; \mathbb{C}^2) \to H^{1/2}(\mathbb{R}; \mathbb{C}^2)$$

is a compact operator. The proof of Proposition 3.2 is complete. □

Competing Interests The authors have no conflicts of interest to declare that are relevant to the content of this chapter.

References

1. Abramowitz, M., Stegun, I.: Handbook of Mathematical Functions with Formulas, Graphs, and Mathematical Tables. Wiley, Washington, D.C. (1984)
2. Albeverio, S., Gesztesy, F., Høegh-Krohn, R., Holden, H.: Solvable Models in Quantum Mechanics, 2nd edn. AMS Chelsea Publishing, Providence, R.I. (2005)
3. Arrizabalaga, N., Mas, A., Vega, L.: Shell interactions for Dirac operators. J. Math. Pures Appl. **102**, 617–639 (2014)
4. Arrizabalaga, N., Mas, A., Vega, L.: Shell interactions for Dirac operators: on the point spectrum and the confinement. SIAM J. Math. Anal. **47**, 1044–1069 (2015)
5. Beenakker, C.: Colloquium: Andreev reflection and Klein tunneling in graphene. Rev. Mod. Phys. **80**, 1337–1354 (2008)
6. Behrndt, J., Exner, P., Holzmann, M., Lotoreichik, V.: Approximation of Schrödinger operators with δ-interactions supported on hypersurfaces. Math. Nachr. **290**, 1215–1248 (2017)
7. Behrndt, J., Exner, P., Holzmann, M., Lotoreichik, V.: On the spectral properties of Dirac operators with electrostatic δ-shell interactions. J. Math. Pures Appl. **111**, 47–78 (2018)
8. Behrndt, J., Exner, P., Holzmann, M., Lotoreichik, V.: On Dirac operators in $\mathbb{R}^3$ with electrostatic and Lorentz scalar δ-shell interactions. Quantum Stud. Math. Found. **6**, 295–314 (2019)
9. Behrndt, J., Hassi, S., de Snoo, H.S.V.: Boundary value problems, Weyl Functions, and differential operators. Monographs in Mathematics, vol. 108, Birkhäuser/Springer, Cham (2020)
10. Behrndt, J., Holzmann, M.: On Dirac operators with electrostatic δ-shell interactions of critical strength. J. Spectr. Theory **10**, 147–184 (2020)
11. Behrndt, J., Holzmann, M., Ourmières-Bonafos, T., Pankrashkin, K.: Two-dimensional Dirac operators with singular interactions supported on closed curves. J. Funct. Anal. **279**, 108700 (47 pages) (2020)
12. Behrndt, J., Holzmann, M., Stelzer-Landauer, C.: Approximation of Dirac operators with δ-shell potentials in the norm resolvent sense (2023). arXiv:2308.13344
13. Behrndt, J., Holzmann, M., Stelzer-Landauer, C., Stenzel, G.: Boundary triples and Weyl functions for Dirac operators with singular interactions. Rev. Math. Phys. **36**, 2350036 (65 pages) (2024)
14. Behrndt, J., Holzmann, M., Stenzel, G.: Schrödinger operators with oblique transmission conditions in $\mathbb{R}^2$. Commun. Math. Phys. **401**, 3149–3167 (2023)
15. Behrndt, J., Holzmann, M., Tušek, M.: Spectral transition for Dirac operators with electrostatic δ-shell potentials supported on the straight line. Integr. Equ. Oper. Theory **94**, 33 (13 pages) (2022)
16. Behrndt, J., Holzmann, M., Tušek, M.: Two-dimensional Dirac operators with general δ-shell interactions supported on a straight line. J. Phys. A **56**, 045201 (29 pages) (2023)
17. Benguria, R.D., Fournais, S., Stockmeyer, E., Van Den Bosch, H.: Self-adjointness of two-dimensional Dirac operators on domains. Ann. Henri Poincaré **18**, 1371–1383 (2017)
18. Benhellal, B.: Spectral properties of the Dirac operator coupled with δ-shell interactions. Lett. Math. Phys. **112**, 52 (52 pages) (2022)
19. Brasche, J., Exner, P., Kuperin, Y., Šeba, P.: Schrödinger operators with singular interactions. J. Math. Anal. Appl. **184**, 112–139 (1994)
20. Brüning, J., Geyler, V., Pankrashkin, K.: Spectra of self-adjoint extensions and applications to solvable Schrödinger operators. Rev. Math. Phys. **20**, 1–70 (2008)

21. Camarasa, M., Pizzichillo, F.: Dirac operators with infinite mass boundary conditions on unbounded domains with infinite corners. Integr. Equ. Oper. Theory **97**, 20 (2025)
22. Cassano, B., Lotoreichik, V., Mas, A., Tušek, M.: General δ-shell interactions for the two-dimensional Dirac operator: self-adjointness and approximation. Rev. Mat. Iberoam. **39**, 1443–1492 (2023)
23. Chodos, A., Jaffe, R.L., Johnson, K., Thorn, C.B., Weisskopf, V.F.: New extended model of hadrons. Phys. Rev. D **9**, 3471–3495 (1974)
24. Cwikel, M., Kalton, N.J.: Interpolation of compact operators by the methods of Calderón and Gustavsson-Peetre. Proc. Edinburgh Math. Soc. **38**, 261–276 (1995)
25. Derkach, V.A., Malamud, M.M.: Generalized resolvents and the boundary value problems for Hermitian operators with gaps. J. Funct. Anal. **95**, 1–95 (1991)
26. Derkach, V.A., Malamud, M.M.: The extension theory of Hermitian operators and the moment problem. J. Math. Sci. **73**, 141–242 (1995)
27. Dittrich, J., Exner, P., Šeba, P.: Dirac operators with a spherically symmetric δ-shell interaction. J. Math. Phys. **30**, 2875–2882 (1989)
28. Exner, P., Ichinose, T.: Geometrically induced spectrum in curved leaky wires. J. Phys. A **34**, 1439–1450 (2001)
29. Exner, P., Kovařík, H.: Quantum Waveguides. Theoretical and Mathematical Physics, Springer, Cham (2015)
30. Frymark, D., Holzmann, M., Lotoreichik, V.: Spectral analysis of the Dirac operator with a singular interaction on a broken line. J. Math. Phys. **65**, 083514 (24 pages) (2024)
31. Frymark, D., Lotoreichik, V.: Self-adjointness of the 2D Dirac operator with singular interactions supported on star graphs. Ann. Henri Poincaré **24**, 179–221 (2023)
32. Gesztesy, F., Šeba, P.: New analytically solvable models of relativistic point interactions. Lett. Math. Phys. **13**, 345–358 (1987)
33. Grafakos, L.: Modern Fourier Analysis. Graduate Texts in Mathematics, vol. 250. Springer, New York (2014)
34. Gruber, M.J., Leitner, M.: Spontaneous edge currents for the Dirac equation in two space dimensions. Lett. Math. Phys. **75**, 25–37 (2006)
35. Jost, J.: Riemannian Geometry and Geometric Analysis. Universitext, Springer, Berlin (2005)
36. Kato, T.: Perturbation Theory for Linear Operators. Classics in Mathematics. Springer, Berlin (1995). Reprint of the 1980 edition
37. Ourmières-Bonafos, T., Pizzichillo, F.: Dirac operators and shell interactions: a survey. In: Michelangeli, A. (eds.), Mathematical Challenges of Zero-Range Physics, Springer INdAM Series, vol 42, Springer (2021)
38. Ourmières-Bonafos, T., Vega, L.: A strategy for self-adjointness of Dirac operators: application to the MIT bag model and δ-shell interactions. Publ. Mat. **62**, 397–437 (2018)
39. Pankrashkin, K.: Variational proof of the existence of eigenvalues for star graphs. In: Dittrich, J., Kovařík, H., Laptev, A. (eds.), Functional Analysis and Operator Theory for Quantum Physics. The Pavel Exner Anniversary Volume, EMS Series of Congress Reports, pp. 447–458. EMS Publishing House, Berlin (2018)
40. Pizzichillo, F., Van Den Bosch, H.: Self-adjointness of two-dimensional Dirac operators on corner domains. J. Spectr. Theory **11**, 1043–1079 (2021)
41. Posilicano, A.: Boundary triples and Weyl functions for singular perturbations of self-adjoint operators. Meth. Funct. Anal. Topol. **10**, 57–63 (2004)
42. Posilicano, A.: Self-adjoint extensions of restrictions. Oper. Matr. **2**, 483–506 (2008)
43. Rabinovich, V.: Two-dimensional Dirac operators with interactions on unbounded smooth curves. Russ. J. Math. Phys. **28**, 524–542 (2021)
44. Rabinovich, V.: Dirac operators with delta-interactions on smooth hypersurfaces in $\mathbb{R}^n$. J. Fourier Anal. Appl. **28**, 20 (26 pages) (2022)
45. Reed, M., Simon, B.: Methods of Modern Mathematical Physics I. Functional Analysis, Academic Press (1972)
46. Thaller, B.: The Dirac Equation. Texts and Monographs in Physics, Springer, Berlin (1992)
47. Weidmann, J.: Lineare Operatoren in Hilberträumen. Teil 1. Teubner, Stuttgart (2000)
48. Weidmann, J.: Linear Operators in Hilbert Spaces. Springer, New York (1980)

Attractor Subspace and Decoherence-Free Algebra of Quantum Dynamics

Daniele Amato, Paolo Facchi, and Arturo Konderak

Abstract In this review we discuss some results on the asymptotic dynamics of finite-dimensional open quantum systems in the Heisenberg picture. Both the spectral and algebraic approaches to this topic are addressed, with particular emphasis on their relationship. The analysis is conducted in both the discrete-time and the continuous-time Markovian settings. In the final part of the work, some issues emerging in the infinite-dimensional case are also discussed. In particular, we provide an example of a Markovian evolution whose decoherence-free algebra is a type III von Neumann algebra.

Keywords Decoherence-free algebra · Asymptotic evolution · Open quantum systems · von Neumann algebras

1 Motivation

The study of the asymptotic evolution of a system has drawn the attention of the scientific community already in classical mechanics since the seminal work by Poincaré [44], as solving the problem at all times is quite a formidable task. Not surprisingly, the importance of the analysis at large times appears in the quantum realm too. In realistic situations a quantum system is always open, i.e. it interacts with

---

D. Amato (✉) · P. Facchi · A. Konderak
Dipartimento di Fisica, Università di Bari, Bari, Italy
e-mail: daniele.amato@ba.infn.it

P. Facchi
e-mail: paolo.facchi@ba.infn.it

A. Konderak
e-mail: akonderak@cft.edu.pl

Istituto Nazionale di Fisica Nucleare, Bari, Italy

A. Konderak
Center for Theoretical Physics, Polish Academy of Sciences, Warsaw, Poland

B. Cassano et al. (eds.), *Singularities, Asymptotics, and Limiting Models*, Springer INdAM Series 64, https://doi.org/10.1007/978-981-96-3584-9_3

its surroundings, usually referred to as the environment or the bath, so that its evolution turns out to be non-unitary. Consequently, the general open-system dynamics is quite more complicated than the closed-system one, and much research has been devoted to understanding the asymptotic behavior of open systems, which resembles the evolution of an isolated system. Moreover, the investigation of what happens only at large times is a crucial question in physics, e.g. in statistical mechanics [31]. In particular, in recent years the large-time evolution of open quantum systems has turned out to be of upmost importance for applications in quantum computation [35].

The asymptotic evolution of open quantum systems has been intensively studied by both physicists and mathematicians since the seventies, see e.g. [23, 27, 52, 53]. These works focused on Markovian continuous-time evolutions, called quantum dynamical (or quantum Markov) semigroups and characterized by Gorini, Kossakowski, Lindblad, and Sudarshan (GKLS) [28, 36]. In this direction, see also the more recent works [1, 24, 26, 49], as well as [13, 14] in the more general context of unital completely positive (UCP) maps. The asymptotic evolution may also be studied in terms of quantum channels in the dynamically equivalent Schrödinger picture [2–7, 42, 43, 47, 48, 56].

It is not hard to recognize two possible approaches to the study of open-quantum-system asymptotics: a *spectral* approach based on the study of the unimodular eigenvalues and eigenoperators of the UCP map (or of the quantum dynamical semigroup) describing the evolution, and an *algebraic* approach in which the asymptotic dynamics is identified with a $*$-homomorphism (or with a semigroup of $*$-homomorphisms). Historically, the first approach may be traced back to the works by Jacobs [32], de Leeuw and Glicksberg [20, 21]. In particular, in the context of semigroups on Banach spaces, they identified the emergence of a direct-sum decomposition into a reversible algebra related to the peripheral eigenvalues of the semigroup and a stable subspace of asymptotically vanishing vectors.

The second approach is based on the concept of decoherence-free algebra, introduced by Evans [23] and whose interest by the quantum probability community [12, 13, 26, 27, 49] was renewed by the seminal paper of Blanchard and Olkiewicz [10]. In particular, in [10] various physical models were proposed, in which the algebra of observables decomposes into a decoherence-free algebra, on which the semigroup acts homomorphically, and a transient subspace, whose elements vanish in the long-time limit. Most works on open-system asymptotics tend to emphasize one of the two schemes, rather than relating them.

The purpose of this Article is to review the connection between the asymptotic subspaces and evolutions considered in the two approaches. The analysis will be mostly performed in a finite-dimensional setting. However, in the last section we will also briefly discuss the issues that arise in the infinite-dimensional case.

The paper is organized as follows: we recall some basic notions on open-system asymptotics in Sect. 2, both in the spectral (Subsect. 2.1) and algebraic (Subsect. 2.2) approaches. Their connection is then discussed in Sects. 3 and 4. In particular, Sect. 3 focuses on the discrete-time setting, where the evolution is given by a UCP map, while in Sect. 4 the continuous framework, governed by a GKLS generator, is studied. Finally, Sect. 5 dwells on the study of the asymptotic dynamics in the Markovian regime on infinite-dimensional Hilbert spaces. Here, we also give an explicit example of a quantum dynamical semigroup whose decoherence-free algebra is described by a type III von Neumann algebra, which is the type of algebras appearing in quantum field theory.

2 Preliminary Notions

In this paper we will investigate the dynamics in the Heisenberg picture, namely we look at the evolution of the observables. This allows us to properly compare the spectral and algebraic schemes of the asymptotic dynamics.

The dynamics in the unit time is given by a unital completely positive map Φ on $\mathcal{B}(\mathcal{H})$ [15], the space of linear operators on a finite-dimensional Hilbert space $\mathcal{H}$. Explicitly, the evolved observable $A(n)$ at time $t = n \in \mathbb{N}$ will be the n-fold composition Φ^n of the map Φ on the observable A at the initial time $t = 0$, i.e. $A(n) = \Phi^n(A)$. Similarly, in the dynamically equivalent Schrödinger picture, the unit-time evolution is described by $\Phi^\dagger$, the adjoint of Φ with respect to the *Hilbert-Schmidt scalar product*

$$\langle X \rangle \, Y_{\mathrm{HS}} = \mathrm{tr}(\mathrm{X}^* \mathrm{Y}) \ , \quad \mathrm{X}, \mathrm{Y} \in \mathcal{B}(\mathcal{H}) \ . \tag{1}$$

It is easy to see that $\Phi^\dagger$ is a completely positive trace-preserving (CPTP) map on $\mathcal{B}(\mathcal{H})$, called a *quantum channel*, and it describes the evolution of the state of an open quantum system given by a density operator ϱ, viz. a positive-semidefinite operator on $\mathcal{H}$ with unit trace.

In the next two subsections, we will present the basic notions behind the spectral and algebraic approaches to open-system asymptotics, whose motivation was already discussed in Sect. 1.

2.1 *Spectral Approach to the Asymptotics*

In this subsection we will discuss the basics of the spectral approach to the asymptotic evolution. Given a UCP map Φ, we will denote by $\mathrm{spect}(\Phi)$ its spectrum, viz. the set of its eigenvalues. It follows from the definition of Φ that all its eigenvalues have modulus not larger than one, a property which holds also for the larger class of unital positive maps [9, 55]. Moreover, the spectrum of any UCP map Φ satisfies the following properties

a) $1 \in \mathrm{spect}(\Phi)$,
b) $\lambda \in \mathrm{spect}(\Phi) \Rightarrow \bar{\lambda} \in \mathrm{spect}(\Phi)$,
c) $\mathrm{spect}(\Phi) \subseteq \{\lambda \in \mathbb{C} \,|\, |\lambda| \leqslant 1\}$.

We can decompose the spectrum of Φ as

$$\mathrm{spect}(\Phi) = \mathrm{spect}_{\mathrm{P}}(\Phi) \cup \mathrm{spect}_{\mathrm{B}}(\Phi)\ , \tag{2}$$

where

$$\mathrm{spect}_{\mathrm{P}}(\Phi) \equiv \{\lambda \in \mathrm{spect}(\Phi) \mid |\lambda| = 1\}\ , \tag{3}$$

$$\mathrm{spect}_{\mathrm{B}}(\Phi) \equiv \{\lambda \in \mathrm{spect}(\Phi) \mid |\lambda| < 1\}\ , \tag{4}$$

are called the *peripheral* and *bulk* spectra of Φ, respectively. Note that $\mathrm{spect}_{\mathrm{P}}(\Phi) \neq \emptyset$ because of property a), while $\mathrm{spect}_{\mathrm{B}}(\Phi) = \emptyset$ if and only if the UCP map is unitary, i.e. of the form $\Phi(X) = U^* X U$, with $X \in \mathcal{B}(\mathcal{H})$ and U unitary on $\mathcal{H}$.

The asymptotic dynamics resulting from the infinite n-limit of the evolution takes place inside the asymptotic, peripheral or *attractor* subspace of Φ, defined as

$$\mathrm{Attr}(\Phi) \equiv \mathrm{span}\{X \in \mathcal{B}(\mathcal{H}) \mid \Phi(X) = \lambda X \text{ for some } \lambda \in \mathrm{spect}_{\mathrm{P}}(\Phi)\}\ . \tag{5}$$

As an immediate consequence of definition (5), the attractor subspace $\mathrm{Attr}(\Phi)$ is invariant under Φ, indeed

$$\Phi(\mathrm{Attr}(\Phi)) = \mathrm{Attr}(\Phi)\ . \tag{6}$$

Importantly, the Jordan decomposition [34] of Φ implies a direct-sum decomposition of the algebra of observables $\mathcal{B}(\mathcal{H})$ into asymptotic and transient components of the evolution, which is related to decomposition (2) of the spectrum. In particular, we have [47, Theorem 3.1]

$$\mathcal{B}(\mathcal{H}) = \mathrm{Attr}(\Phi) \oplus \mathcal{T}(\Phi)\ , \tag{7}$$

where

$$\mathcal{T}(\Phi) = \left\{X \in \mathcal{B}(\mathcal{H}) \,\middle|\, \lim_{n\to\infty} \Phi^n(X) = 0\right\} \tag{8}$$

is the *transient part* of the dynamics, involving the generalized eigenoperators corresponding to bulk eigenvalues. Equation (7) is an example of Jacobs-de Leeuw-Glicksberg (JdLG) decomposition [20, 21, 32]. Notice that the rate of convergence to equilibrium is related to the *spectral gap* of the UCP map Φ defined as

$$g = 1 - \max_{\lambda \in \mathrm{spect}_{\mathrm{B}}(\Phi)} |\lambda|. \tag{9}$$

The *peripheral projection* $\mathcal{P}_P$ of Φ is defined as

$$\mathcal{P}_{\mathrm{P}} = \sum_{\lambda_k \in \mathrm{spect}_{\mathrm{P}}(\Phi)} \mathcal{P}_k\ , \tag{10}$$

where $\mathcal{P}_k$ denote the spectral projection of Φ corresponding to the kth peripheral eigenvalue $\lambda_k \in \mathrm{spect}_{\mathrm{P}}(\Phi)$. It is possible to relate $\mathcal{P}_{\mathrm{P}}$ with the UCP map Φ via the following formula

$$\mathcal{P}_{\mathrm{P}} = \lim_{i\to\infty} \Phi^{n_i} , \tag{11}$$

for some increasing sequence $(n_i)_{i\in\mathbb{N}}$. This ensures that $\mathcal{P}_{\mathrm{P}}$ is a UCP map as well. Clearly, the range of the projection $\mathcal{P}_{\mathrm{P}}$ is the attractor subspace $\mathrm{Attr}(\Phi)$ of Φ. An important subspace of $\mathrm{Attr}(\Phi)$ is the *fixed-point subspace* of Φ

$$\mathrm{Fix}(\Phi) = \{X \in \mathcal{B}(\mathcal{H}) \mid \Phi(X) = X\} , \tag{12}$$

i.e. the space of invariant observables of the system. Once the attractor subspace has been defined, the asymptotic dynamics is described in our setting by the *asymptotic map* of Φ

$$\hat{\Phi}_{\mathrm{P}} = \Phi|_{\mathrm{Attr}(\Phi)} . \tag{13}$$

By (5) the attractor subspace is a vector space. Now, in order to compare it with the decoherence-free algebra of Φ, which we will introduce in Subsect. 2.2, it is important to understand whether $\mathrm{Attr}(\Phi)$ is an algebra with respect to the composition product of $\mathcal{B}(\mathcal{H})$. We introduce the following notion, which is equivalent to the notion introduced in [47, Definition 1.1].

Definition 1 Let Φ be a UCP map. Then Φ is called peripherally automorphic if and only if $\mathrm{Attr}(\Phi)$ is closed under the composition product, i.e.

$$X , Y \in \mathrm{Attr}(\Phi) \Rightarrow XY \in \mathrm{Attr}(\Phi) . \tag{14}$$

Remark 1 If Φ is peripherally automorphic, then $\mathrm{Attr}(\Phi)$ is a C^*-subalgebra of $\mathcal{B}(\mathcal{H})$.

The name given to this class of maps may be easily justified thanks to the following proposition [47, Theorem 2.10].

Proposition 1 *Let Φ be a UCP map with asymptotic map $\hat{\Phi}_P$. The following conditions are equivalent:*

a) *Φ is peripherally automorphic;*
b) $\hat{\Phi}_{\mathrm{P}}(XY) = \hat{\Phi}_{\mathrm{P}}(X)\hat{\Phi}_{\mathrm{P}}(Y)$ *for all* $X, Y \in \mathrm{Attr}(\Phi)$.

We now temporarily move to the Schrödinger picture, where we can define the attractor subspace $\mathrm{Attr}(\Phi^\dagger)$ as well as the fixed-point space $\mathrm{Fix}(\Phi^\dagger)$ of the quantum channel $\Phi^\dagger$ using (5) and (12) respectively. The assumption of *faithfulness*, introduced in the following definition, is a property concerning $\mathrm{Fix}(\Phi^\dagger)$ which has profound consequences on the asymptotics in both pictures, as we shall see below.

Definition 2 A UCP map Φ is said to be faithful if there exists an invertible stationary state ϱ for the CPTP map $\Phi^\dagger$, the Hilbert-Schmidt adjoint of Φ, i.e.

$$\Phi^\dagger(\varrho) = \varrho > 0 . \tag{15}$$

Importantly, peripheral automorphism is a weaker condition than faithfulness [47, Corollary 2.8].

Proposition 2 *Let Φ be a faithful UCP map. Then it is peripherally automorphic.*

Remark 2 A peripherally automorphic map Φ may not be faithful, see [47, Example 2.4].

2.2 Algebraic Approach to the Asymptotics

In this subsection, we will discuss an alternative approach to study the long-time evolution of an open quantum system. This approach addresses the algebraic properties of the asymptotics. Let us recall the definition and some basic properties of the decoherence-free algebra of Φ, i.e. the asymptotic subspace of the dynamics in this approach. The starting point of the analysis is the following: the discrete-time semigroup $(\Phi_n)_{n\in\mathbb{N}}$ describing the evolution of a closed quantum system is made up of $*$-automorphisms of $\mathcal{B}(\mathcal{H})$, and since we expect open-system asymptotic dynamics to mimic a closed-system evolution, the asymptotic semigroup should involve $*$-automorphisms of the asymptotic algebra.

The *decoherence-free algebra* of a UCP map Φ is defined as

$$\mathcal{N} \equiv \{X \in \mathcal{B}(\mathcal{H}) \mid \Phi^n(YX) = \Phi^n(Y)\Phi^n(X) , \\ \Phi^n(XY) = \Phi^n(X)\Phi^n(Y) , \ \forall Y \in \mathcal{B}(\mathcal{H}) , \ \forall n \in \mathbb{N}\} . \quad (16)$$

It is well-known that $\mathcal{N}$ is a C^*-subalgebra of $\mathcal{B}(\mathcal{H})$ invariant under the action of Φ,

$$\Phi(\mathcal{N}) \subseteq \mathcal{N} , \quad (17)$$

see e.g. [13]. The operator Schwarz inequality

$$\Phi(X^*X) \geqslant \Phi(X)^*\Phi(X) , \quad X \in \mathcal{B}(\mathcal{H}) , \quad (18)$$

which holds for any UCP map, implies the following simple characterization of the space $\mathcal{N}$, see [55, Theorem 5.4].

Proposition 3 *Let Φ be a UCP map. Then*

$$\mathcal{N} = \{X \in \mathcal{B}(\mathcal{H}) \mid \Phi^n(X^*X) = \Phi^n(X)^*\Phi^n(X) , \\ \Phi^n(XX^*) = \Phi^n(X)\Phi^n(X)^* , \ \forall n \in \mathbb{N}\} . \quad (19)$$

In order to describe the asymptotic evolution in this setting, let us introduce the *decoherence-free map* of Φ

$$\Phi_{\mathcal{N}} = \Phi|_{\mathcal{N}} , \quad (20)$$

and the *decoherence-free semigroup*

$$\Phi_{\mathcal{N},n} := \Phi^n|_{\mathcal{N}}\,, \quad n \in \mathbb{N}\,. \tag{21}$$

The following corollary is an immediate consequence of the previous results.

Corollary 1 *Let Φ be a UCP map. Then its decoherence-free algebra is the largest C^*-subalgebra $\mathcal{M}$ of $\mathcal{B}(\mathcal{H})$ invariant under Φ for which the semigroup $(\Phi_{\mathcal{M},n})_{n\in\mathbb{N}}$, defined in* (21)*, is made up of $*$-homomorphisms.*

Proof First, note that the invariance of $\mathcal{M}$ under Φ implies that $\Phi_{\mathcal{M},n} = \Phi|^n_{\mathcal{M}}$ is a semigroup. Take $X \in \mathcal{M}$, implying $X^* \in \mathcal{M}$ too, so that for all $n \in \mathbb{N}$:

$$\Phi^n(X^*X) = \Phi_{\mathcal{M},n}(X^*X) = \Phi_{\mathcal{M},n}(X)^*\Phi_{\mathcal{M},n}(X) = \Phi^n(X)^*\Phi^n(X)\,, \tag{22}$$

$$\Phi^n(XX^*) = \Phi_{\mathcal{M},n}(XX^*) = \Phi_{\mathcal{M},n}(X)\Phi_{\mathcal{M},n}(X)^* = \Phi^n(X)\Phi^n(X)^*\,. \tag{23}$$

Thus, from Proposition 3, $X \in \mathcal{N}$ and the assertion follows. □

Remark 3 As already highlighted in [13, Remark 1], $\Phi_{\mathcal{N},n}$ is not generally made up of $*$-automorphisms.

3 Interplay Between the Spectral and the Algebraic Framework

In this section we are going to analyze the connection between the spectral and the algebraic approach. In particular, we aim at investigating the relation between the attractor subspace and the decoherence-free algebra, i.e. the asymptotic subspaces in the two schemes. Before stating the first main result of the section, we need a preparatory lemma [30, Lemma 2.4].

Lemma 1 *Let Φ be an idempotent UCP map, namely $\Phi^2 = \Phi$. Then*

$$\Phi(\Phi(X)\Phi(Y)) = \Phi(\Phi(X)Y) = \Phi(X\Phi(Y))\,, \quad X\,,\, Y \in \mathcal{B}(\mathcal{H})\,. \tag{24}$$

Under the faithfulness assumption the attractor subspace reduces to the decoherence-free algebra, and we can say that the spectral and algebraic approaches to the asymptotics become equivalent.

Theorem 1 *Let Φ be a faithful UCP map. Then*

$$\mathrm{Attr}(\Phi) = \mathcal{N}\,. \tag{25}$$

Proof Given $X \in \mathrm{Attr}(\Phi)$, by combining Proposition 1 and Proposition 2 we have

$$\Phi(X^*X) = \Phi(X)^*\Phi(X)\,. \tag{26}$$

Since $\mathrm{Attr}(\Phi)$ is invariant under Φ by (6), we have

$$\Phi^n(X^*X) = \Phi^n(X)^*\Phi^n(X) , \quad n \in \mathbb{N} . \tag{27}$$

By exchanging $X \leftrightarrow X^*$, we get $\mathrm{Attr}(\Phi) \subseteq \mathcal{N}$. For the other inclusion, given $X \in \mathcal{N}$ and using (11)

$$\Phi^n(X^*X) = \Phi^n(X)^*\Phi^n(X) \Rightarrow \mathcal{P}_\mathrm{P}(X^*X) = \mathcal{P}_\mathrm{P}(X)^*\mathcal{P}_\mathrm{P}(X) . \tag{28}$$

Thus, by applying Lemma 1 to $\mathcal{P}_\mathrm{P}$ we obtain

$$\mathcal{P}_\mathrm{P}((X - \mathcal{P}_\mathrm{P}(X))^*(X - \mathcal{P}_\mathrm{P}(X))) = 0 . \tag{29}$$

Call $Y = X - \mathcal{P}_\mathrm{P}(X)$, and let $\varrho = \Phi^\dagger(\varrho) > 0$ be an invertible invariant state for $\Phi^\dagger$. It will be invariant for $\mathcal{P}_\mathrm{P}^\dagger$ too by (11), and we have

$$\mathrm{tr}(\mathcal{P}_\mathrm{P}(\mathrm{Y}^*\mathrm{Y})\varrho) = \mathrm{tr}(\mathrm{Y}^*\mathrm{Y}\mathcal{P}_\mathrm{P}^\dagger(\varrho)) = \mathrm{tr}(\mathrm{Y}^*\mathrm{Y}\varrho) = 0 , \tag{30}$$

which implies $Y^*Y = 0$, and so $Y = 0$. Thus, $X = \mathcal{P}_\mathrm{P}(X)$, and $\mathcal{N} \subseteq \mathrm{Attr}(\Phi)$. □

Corollary 2 *Let Φ be a faithful UCP map. Then $\Phi|_\mathcal{N}$ is a $*$-automorphism.*

Proof The assertion follows by combining Theorem 1, claim *b)* of Proposition 1 and the invertibility of $\hat{\Phi}_\mathrm{P}$. □

The relation between the attractor subspace and the decoherence-free algebra may be derived for the larger class of peripherally automorphic UCP maps introduced in Definition 1, see [47, Theorem 2.6] and [26, Remark 10].

Theorem 2 *Let Φ be a UCP map. Then, the following conditions are equivalent.*

a) *Φ peripherally automorphic;*
b) $\mathrm{Attr}(\Phi) \subseteq \mathcal{N}$.

Corollary 3 *Let Φ be a peripherally automorphic UCP map. Then* $\mathrm{Fix}(\Phi) \subseteq \mathcal{N}$.

Therefore, if Φ is not peripherally automorphic, *b)* does not hold and, indeed, it can happen that $\mathrm{Fix}(\Phi) \nsubseteq \mathcal{N}$, see [47, Example 3.6].

4 Markovian Evolutions

As discussed in Sect. 1, Markovian continuous-time evolutions were the first ones addressed in the theory of open quantum systems, given their significance in areas such as quantum optics [11]. In this section we will first discuss the basic properties of continuous-time evolutions, with particular focus on their asymptotics. Afterwards, we will revisit the connection between the spectral and the algebraic approach to the asymptotic evolution in the continuous setting.

First, recall that a quantum dynamical semigroup (QDS) $(\Phi_t)_{t\in\mathbb{R}^+}$ is a continuous one-parameter semigroup of UCP maps on $\mathcal{B}(\mathcal{H})$. The semigroup property and continuity guarantee the existence of the *generator* $\mathcal{L}$ of the QDS [22, Theorem I.3.7], that is

$$\Phi_t = e^{t\mathcal{L}}, \quad t \in \mathbb{R}^+ . \tag{31}$$

Moreover, the generator $\mathcal{L}$ of a QDS may be always written in the *GKLS form* [28, 36]

$$\mathcal{L}(X) = \mathrm{i}[H, X] + \sum_{k=1}^{d^2-1} (L_k^* X L_k - \frac{1}{2}\{L_k^* L_k, X\}) , \tag{32}$$

with $H = H^*$ and $d = \dim \mathcal{H}$. Thus, it turns out that $\mathcal{L}$ is the sum of two contributions. The first one

$$\mathcal{L}_{\mathrm{H}}(X) = \mathrm{i}[H, X] , \quad X \in \mathcal{B}(\mathcal{H}) \tag{33}$$

is called the *Hamiltonian part* of $\mathcal{L}$, since it is related to a self-adjoint operator H that can be interpreted as an effective Hamiltonian of the system. The second one reads

$$\mathcal{L}_{\mathrm{D}}(X) = \sum_{k=1}^{d^2-1} (L_k^* X L_k - \frac{1}{2}\{L_k^* L_k, X\}) , \tag{34}$$

and it is called the *dissipative part* of $\mathcal{L}$. It involves the operators $L_k \in \mathcal{B}(\mathcal{H})$ called noise or *jump operators* and, physically, it describes phenomena arising from the interaction of the system with the environment, like dissipation and decoherence. It vanishes for isolated systems, for which the generator $\mathcal{L} = \mathcal{L}_{\mathrm{H}}$ becomes Hamiltonian. Observe that the GKLS form of $\mathcal{L}$ and, in particular, the decomposition of $\mathcal{L}$ into Hamiltonian and dissipative parts is not unique [11].

From a spectral perspective, the asymptotics of QDSs is governed by the unimodular eigenvalues of the dynamics, as underlined in the general discrete-time setting in Subsect. 2.1. Specifically, given a QDS $(\Phi_t)_{t\in\mathbb{R}^+}$ with generator $\mathcal{L}$, the spectrum of $\mathcal{L}$ satisfies

$$\mathrm{spect}(\mathcal{L}) \subseteq \{\lambda \in \mathbb{C} \mid \Re(\lambda) \leqslant 0\} , \tag{35}$$

i.e. it is contained in the left half-plane. Moreover, it is closed under complex conjugation, as is the spectrum of a UCP map, and $0 \in \mathrm{spect}(\mathcal{L})$. The asymptotic dynamics of the semigroup $(\Phi_t)_{t\in\mathbb{R}^+}$ takes place in the attractor subspace, which in this case is defined as

$$\mathrm{Attr}(\Phi_t) \equiv \{X \in \mathcal{B}(\mathcal{H}) \mid \mathcal{L}(X) = \lambda X \text{ for some } \lambda \in \mathrm{spect}_{\mathrm{P}}(\mathcal{L})\} = \mathrm{Attr}(\Phi) . \tag{36}$$

Here $\Phi \equiv \Phi_1 = e^{\mathcal{L}}$ is the unit-time dynamics, and

$$\mathrm{spect}_P(\mathcal{L}) = \{\lambda \in \mathrm{spect}(\mathcal{L}) \mid \Re(\lambda) = 0\} \tag{37}$$

is the peripheral spectrum of $\mathcal{L}$. Also, the fixed-point subspace of the QDS $(\Phi_t)_{t\in\mathbb{R}^+}$ is defined as

$$\mathrm{Fix}(\Phi_t) \equiv \mathrm{Ker}(\mathcal{L}) = \{X \in \mathcal{B}(\mathcal{H}) \mid \mathcal{L}(X) = 0\} , \tag{38}$$

and it generally differs from the fixed-point subspace $\mathrm{Fix}(\Phi)$ of Φ. More precisely,

$$\mathrm{Fix}(\Phi_t) \subseteq \mathrm{Fix}(\Phi) , \tag{39}$$

where the inclusion may be strict. As an example, consider a Hamiltonian GKLS generator with a couple of non-zero peripheral eigenvalues being multiples of 2π.

In the Schrödinger picture, the dynamics is given by the (Hilbert-Schmidt) adjoint semigroup $(\Phi_t^\dagger)_{t\in\mathbb{R}^+}$ and, again, the study of the asymptotic evolution involves the analysis of the attractor and fixed-point subspaces of the semigroup defined as in (36) and (38). In particular, the latter reads

$$\mathrm{Fix}(\Phi_t^\dagger) = \mathrm{Ker}(\mathcal{L}^\dagger) . \tag{40}$$

Consistently with the discrete-time setting, we will keep calling stationary the states belonging to $\mathrm{Fix}(\Phi_t^\dagger)$. A notion of faithfulness may now be introduced for QDSs.

Definition 3 A QDS $(\Phi_t)_{t\in\mathbb{R}^+}$ is said to be faithful if there exists an invertible stationary state ϱ for the adjoint semigroup $(\Phi_t^\dagger)_{t\in\mathbb{R}^+}$, viz.

$$\mathcal{L}(\varrho) = 0 , \quad \varrho > 0 . \tag{41}$$

The asymptotic dynamics is given by the asymptotic semigroup

$$\Phi_{t,\mathrm{P}} = \Phi_t|_{\mathrm{Attr}(\Phi_t)} . \tag{42}$$

From the algebraic viewpoint, the asymptotic subspace is identified with the decoherence-free algebra of the semigroup, given by

$$\mathcal{N} = \{X \in \mathcal{B}(\mathcal{H}) \mid \Phi_t(YX) = \Phi_t(Y)\Phi_t(X) , \tag{43}$$
$$\Phi_t(XY) = \Phi_t(X)\Phi_t(Y) , \ \forall Y \in \mathcal{B}(\mathcal{H}) , \ \forall t \in \mathbb{R}^+\} . \tag{44}$$

Similarly to the discrete-time setting (21), the asymptotic dynamics is described by the decoherence-free semigroup

$$\Phi_{\mathcal{N},t} = \Phi_t|_{\mathcal{N}} , \quad t \in \mathbb{R}^+ . \tag{45}$$

The analogues of Proposition 3 and Corollary 1 clearly hold in the Markovian continuous-time case.

Remark 4 $\Phi_t|_{\mathcal{N}}$ is a $*$-automorphism for all $t \in \mathbb{R}^+$, since Φ_t is invertible by (31). As discussed in Remark 3, this is not generally true in the non-Markovian case.

Once the spectral and algebraic approaches to the Markovian asymptotic dynamics have been introduced, we would like to discuss their relation. First, let us note that the concept of peripheral automorphism introduced in Definition 1 can be naturally extended to the present setting by (36). Now, the Markovian continuous-time analogues of Theorems 1 and 2 are still valid, as stated in the following propositions, see [49, Theorem 22] and [26, Remark 10] for the proofs.

Theorem 3 *Let* $(\Phi_t)_{t\in\mathbb{R}^+}$ *be a faithful QDS. Then*

$$\mathrm{Attr}(\Phi_t) = \mathcal{N} \,. \tag{46}$$

Remark 5 It is possible to see that equation (46) is not a sufficient condition for faithfulness, see [7, Example 2].

Theorem 4 *Let* $(\Phi_t)_{t\in\mathbb{R}^+}$ *be a QDS. Then, the following conditions are equivalent:*

a) $(\Phi_t)_{t\in\mathbb{R}^+}$ *peripherally automorphic;*
b) $\mathrm{Attr}(\Phi_t) \subseteq \mathcal{N}$.

Again, Theorem 3 implies the equivalence of the two approaches to the asymptotics for faithful evolutions. We conclude the section with an important property for the decoherence-free algebra of Markovian evolutions [25, Remark 1].

Proposition 4 *Let* $(\Phi_t)_{t\in\mathbb{R}^+}$ *be a QDS with generator* $\mathcal{L}$ *having GKLS form* (32). *Then*

$$\mathcal{N} \subseteq \{X \in \mathcal{B}(\mathcal{H}) \mid \Phi_t(X) = \mathrm{e}^{\mathrm{i}tH} X \mathrm{e}^{-\mathrm{i}tH} \,, \ \forall t \in \mathbb{R}^+\} \,. \tag{47}$$

Remark 6 The proposition holds for any H in a GKLS representation of $\mathcal{L}$. Moreover, equality does not hold in general, see [25, Example 8].

The latter result ensures that for a Markovian continuous-time evolution, the decoherence-free algebra is a subspace in which the evolution of the system is unitary, i.e. asymptotically the quantum system behaves as if it is closed, in line with our physical intuition. In particular, this justifies the idea behind the algebraic approach, viz. finding the subspace in which the dynamics acts as a *group* of $*$-automorphisms.

However, if we relax the Markovianity assumption, the dynamics on the decoherence-free algebra is no longer unitary in general, even in the faithful case. Indeed, general structure theorems of the asymptotics, see [6], [5, Theorem 3.2], and [56, Theorem 8], reveal the occurrence of an additional (non-unitary) asymptotic evolution described by permutations. The latter disappears for Markovian continuous evolutions, see [5, Theorem 5.3].

5 Remarks on the Infinite-Dimensional Case

In this section we will briefly outline several results regarding open-system asymptotics in the infinite-dimensional Markovian case. Specifically, we will call a quantum dynamical semigroup a family $(\Phi_t)_{t\in\mathbb{R}^+}$ of norm-continuous one-parameter semigroup of UCP normal maps on $\mathcal{B}(\mathcal{H})$, the von Neumann algebra of bounded operators on a (generally infinite-dimensional, but separable) Hilbert space $\mathcal{H}$. Such a semigroup is generated by a GKLS operator in the form [36]

$$\mathcal{L}(X) = \mathrm{i}[H, X] + \sum_{k\in\mathbb{N}} (L_k^* X L_k - \frac{1}{2}\{L_k^* L_k, X\}) \,, \tag{48}$$

with X, $H = H^*$, $(L_k)_{k\in\mathbb{N}} \in \mathcal{B}(\mathcal{H})$, such that $\sum_{k\in\mathbb{N}} L_k^* L_k$ is strongly convergent.

Again, we can follow two strategies for the analysis of the asymptotics: the spectral one seeking a JdLG decomposition of the form (7) and the algebraic one looking at the decoherence-free algebra (16) of the dynamics. Regarding the first approach, the infinite-dimensional analogue of the attractor subspace of Φ_t is called the *reversible algebra*, and it reads

$$\mathfrak{M}_\mathrm{r} = \overline{\mathrm{span}}\{X \in \mathcal{B}(\mathcal{H}) \mid \mathcal{L}(X) = \lambda X \text{ for some } \lambda, \text{ with } |\lambda| = 1\} \,, \tag{49}$$

where the closure is taken with respect to the weak-$*$ topology. Importantly, $\mathfrak{M}_\mathrm{r} = \mathcal{N}$ in the faithful case, see [49, Theorem 22]. In this situation, $\mathfrak{M}_\mathrm{r}$ and, consequently, $\mathcal{N}$ are atomic von Neumann algebras (i.e. every nonzero projection has a minimal nonzero subprojection) and there exists a JdLG decomposition of $\mathcal{B}(\mathcal{H})$

$$\mathcal{B}(\mathcal{H}) = \mathfrak{M}_\mathrm{r} \oplus \mathfrak{M}_s \,, \tag{50}$$

where

$$\mathfrak{M}_\mathrm{s} = \{X \in \mathcal{B}(\mathcal{H}) \mid 0 \in \overline{\{\Phi_t(X)\}_{t\geqslant 0}}\} \,, \tag{51}$$

is called the *stable subspace* of the semigroup. Therefore, also for infinite-dimensional open systems, the spectral and algebraic approaches turn out to be equivalent for faithful Markovian evolutions.

For non-faithful Markovian evolutions interesting issues arise in the infinite-dimensional case: the decoherence-free algebra $\mathcal{N}$ can be non-atomic, and also of type different from I. In the following example we will construct a decoherence-free algebra which is a factor of type III, generalizing the construction discussed in [49, Sect. II.A]. These algebras are the most singular ones in the original classification by von Neumann [37]. In particular, all the projections on such factor are similar, and their extremely singular nature made them less attractive in von Neumann's seminal papers [38, 41]. Nevertheless, they turned out to be extremely important in the algebraic description of quantum field theory [8, 29, 33]. Interestingly, they can also be obtained in the alternative description of quantum mechanics based on the

Schwinger picture [16], where the building blocks of the theory are the transitions, rather than the states or the observables [17–19, 50].

Example 1 (*Type III decoherence-free algebra*) In this example, we will construct a quantum dynamical semigroup whose decoherence-free algebra is a factor of type III. In the standard approach, one considers the action of a countable group on an Abelian von Neumann algebra of operators, and then takes the von Neumann algebra generated by operators representing such action on a proper Hilbert space. To our purpose, we will simply identify a countable subfamily of the generators with the jump operators of the GKLS generator. Specifically, we will employ the construction discussed in Appendix 7 and hereafter summarized.

Let Ω_∞ be the set of binary sequences, and consider the group $\mathcal{G} = (\Omega, \oplus)$, where $\Omega \subseteq \Omega_\infty$ is the set of finite sequences [which are eventually zero, see (61)], and the composition law $\oplus$ is the entrywise sum modulo 2. Moreover, consider the probability measure μ_λ on the binary sequences which on the nth component x_n assigns a probability λ if $x_n = 0$, and $1 - \lambda$ if $x_n = 1$, with λ a fixed parameter in $(0, 1/2]$. Also, let $\mathcal{H}$ be the Hilbert space of complex-valued μ_λ-measurable square-integrable functions on $\Omega_\infty \times \Omega$, defined in (63).

Let $e_k = (\delta_{jk})_{j\in\mathbb{N}} \in \Omega$ and consider the jump operators

$$M_k = V_{e_k} , \tag{52}$$

with V_{e_k} being the unitary representative on $\mathcal{H}$ of the group composition with e_k, see (65). We also need to consider the following family of jump operators on $\mathcal{H}$ in the form

$$N_k = L_{\psi_k} , \tag{53}$$

with L_ϕ being the multiplication operator by ϕ, defined as in (64), and ψ_k of the form

$$\psi_k(x) = \begin{cases} +1 \text{ for } x_k = 0, \\ -1 \text{ for } x_k = 1. \end{cases} \tag{54}$$

We can now construct the GKLS generator as

$$\mathcal{L}_\lambda(X) = \sum_{k\in\mathbb{N}} m_k(M_k^* X M_k - X) + \sum_{k\in\mathbb{N}} n_k(N_k^* X N_k - X) , \tag{55}$$

with $(m_k)_{k\in\mathbb{N}}$ and $(n_k)_{k\in\mathbb{N}}$ two strictly positive sequences with $\sum_k m_k < +\infty$, $\sum_k n_k < +\infty$. We claim that the decoherence-free algebra of the QDS $e^{t\mathcal{L}_\lambda}$ is non-atomic. Indeed, we have the following result.

Proposition 5 *The decoherence-free algebra of the quantum dynamical semigroup generated by $\mathcal{L}_\lambda$ in (55) is a type III factor for all $\lambda \in (0, 1/2)$, and a type II_1 factor for $\lambda = 1/2$.*

Proof The decoherence-free algebra is given by

$$\mathcal{N} = \{M_k, N_k \,|\, k \in \mathbb{N}\}' , \tag{56}$$

as a consequence of [26, Proposition 2]. Here, note that M_k and N_k are both self-adjoint and unitary operators on $\mathcal{H}$.

Since the type of an algebra is the same as the type of its commutant [54, Corollary 2.24], we only need to show that the von Neumann algebra generated by the jump operators is exactly the Pukánszky-Powers factor. Call the latter algebra $\mathcal{M}$, so that $\mathcal{M} = \mathcal{N}'$. First, observe that every $x^o \in \Omega$ is a finite sum of some e_k, so every $V_{x^o} \in \mathcal{M}$, as V is a unitary representation of the group $\mathcal{G}$.

Furthermore, we can prove that all L_ϕ defined by (64) belong to $\mathcal{M}$. First, the cylinders defining the σ-algebra over Ω_∞ are in the form

$$\mathscr{C}(y_1, \ldots, y_n) = \{x \in \Omega_\infty \,|\, x_1 = y_1, \ldots, x_n = y_n\} = \bigcap_{k=1}^{n} \{x \in \Omega_\infty \,|\, x_k = y_k\} \equiv \bigcap_{k=1}^{n} \mathscr{C}_k , \tag{57}$$

where $(y_1, \ldots, y_n) \in \{0, 1\}^n$. The characteristic function over $\mathscr{C}_k$ in (57) is

$$\mathbf{1}_{\mathscr{C}_k}(x) = (1 - y_k)\psi_k(x) - y_k\psi_k(x) . \tag{58}$$

Therefore, the characteristic function $\mathbf{1}_{\mathscr{C}}$ over the cylinder $\mathscr{C} \equiv \mathscr{C}(y_1, \ldots, y_n)$ reads

$$\mathbf{1}_{\mathscr{C}} = \prod_{k=1}^{n} \mathbf{1}_{\mathscr{C}_k} . \tag{59}$$

As a result, the multiplication operators $L_{\mathbf{1}_{\mathscr{C}_k}}$ and $L_{\mathbf{1}_{\mathscr{C}}}$, belong to $\mathcal{M}$. As the span of the characteristic functions $\mathbf{1}_{\mathscr{C}}$ of all cylinders is dense in $\mathrm{L}^\infty(\Omega_\infty, \mu_\lambda)$ in the $*$-weak topology, we can conclude that $\mathcal{M}$ is indeed the Pukánszky-Powers algebra.

In particular, $\mathcal{M}$, as well as the decoherence-free algebra $\mathcal{N}$, is a type III factor whenever $0 < \lambda < 1/2$, and a type II_1 factor for $\lambda = 1/2$ [46]. □

6 Conclusions

In the present work, we provided an updated short overview of the results regarding open-system Heisenberg asymptotics in the finite-dimensional case. We highlighted the existence of two kinds of strategies to address this topic: a spectral approach related to the peripheral (unimodular) eigenvalues of the dynamics, and an algebraic one where the asymptotic evolution is characterized by a $*$-homomorphism.

After introducing the basic notions of the two strategies, we presented the main connections between the two approaches in literature. Specifically, we examined the relation between the attractor subspace and the decoherence-free algebra, which are the asymptotic subspaces corresponding to the spectral and algebraic approaches, respectively. We illustrated how faithfulness implies the equality of such spaces and, consequently, the equivalence of the two methods, unveiling the strength of this assumption. Additionally, the implications of the milder property of peripheral automorphism were discussed.

Finally, we commented on the infinite dimensional case, where new issues arise for both the algebraic and topological viewpoints. In particular, we were able to construct a quantum dynamical semigroup whose decoherence-free algebra is non-atomic and is given by a type III factor.

This short review aims to bridge the gap between the mathematical and physical communities, thereby enhancing the understanding of the asymptotic problem, both conceptually and technically.

Acknowledgements We acknowledge support from INFN through the project "QUANTUM", from the Italian National Group of Mathematical Physics (GNFM-INdAM), and from the Italian funding within the "Budget MUR – Dipartimenti di Eccellenza 2023–2027" – Quantum Sensing and Modelling for One-Health (QuaSiModO). PF acknowledges support from Italian PNRR MUR project CN00000013 -"Italian National Centre on HPC, Big Data and Quantum Computing". DA acknowledges support from PNRR MUR project PE0000023-NQSTI. AK acknowledges the support from the "Young Scientist" project at the Center for Theoretical Physics, Polish Academy of Sciences.

7 Pukánszky-Powers Factors

In this appendix, we summarize some aspects regarding the construction of type II and III factors. This approach was originally discussed by von Neumann [39], and subsequently by Pukánszky and Powers [45, 46], who used it to identify inequivalent classes of type III algebras. The construction is based on the ergodic and free actions of countable groups on Abelian algebras [54], but it can be also framed in the context of infinite products of Hilbert spaces [40]. By defining a proper convolution between elements, one defines a crossed product algebra, and the type of the algebra is completely characterized by the structure of the underlying Abelian algebra [54, Theorem V.7.12].

We will not give the general theorem, but we will limit to present the construction from Pukánszky [46], see also [16] for technical details. In particular, we consider the set of binary sequences

$$\Omega_\infty = \left\{(x_n)_{n\in\mathbb{N}} \mid x_n \in \{0, 1\}\right\}, \tag{60}$$

equipped with the product σ-algebra Σ generated by the cylinders defined in (57). Then the measurable space (Ω_∞, Σ) may be equipped with the following probability measure. On the nth component x_n, a probability λ is assigned if $x_n = 0$, and $1 - \lambda$ if $x_n = 1$, with λ a fixed parameter in $(0, 1/2]$. Consequently, the probability measure

μ_λ on Ω_∞ is constructed through Kolmogorov theorem, see [51, Theorem II.3.3]. The Abelian algebra underlying the factor is thus $L^\infty(\Omega_\infty, \mu_\lambda)$. Afterwards, a countable group is introduced as the set Ω of finite sequences of the form

$$\Omega = \{(x_k) \in \Omega_\infty \mid x_k = 0 \ , \ k > N \ , \ \text{for some } N \in \mathbb{N}\} \ . \tag{61}$$

Using von Neumann's notation [40], we will call x, y, z the elements of Ω_∞, and with x^o, y^o, z^o those of Ω. The composition rule on Ω_∞ and, consequently, Ω is given by the entry-wise sum modulo 2. More explicitly, given $x^o = (x_k)_{k\in\mathbb{N}}$, $y^o = (y_k)_{k\in\mathbb{N}} \in \Omega$, the composition law reads

$$x \oplus y = (x_k \oplus y_k)_{k\in\mathbb{N}} \ , \tag{62}$$

where, with a slight abuse of notation, $\oplus$ also denotes the sum modulo 2 in $\{0, 1\}$. The algebra $L^\infty(\Omega_\infty, \mu_\lambda)$ and the group $\mathcal{G} \equiv (\Omega, \oplus)$ can be represented on the Hilbert space

$$\mathcal{H} = \left\{ F : \Omega_\infty \times \Omega \to \mathbb{C} \,\middle|\, \sum_{x^o\in\Omega} \int_{\Omega_\infty} |F(x, x^o)|^2 d\mu_\lambda(x) < +\infty \right\} . \tag{63}$$

In formulas, given $\phi \in L^\infty(\Omega_\infty, \mu_\lambda)$ and $y^o \in \Omega$, let us define

$$(L_\phi F)(x, x^o) = \phi(x) F(x, x^o) \ , \tag{64}$$

$$(V_{y^o} F)(x, x^o) = \sqrt{\frac{d\mu_\lambda^{y^o}}{d\mu_\lambda}}(x) F(x \oplus y^o, x^o \oplus y^o) \ . \tag{65}$$

Here, $d\mu_\lambda^{y^o}/d\mu_\lambda$ is the Radon-Nikodym derivative of the translation $\mu_\lambda^{y^o}$ of the measure μ_λ by y^o with respect to μ_λ. In particular, V is a unitary representation of $\mathcal{G}$, and we can consider the von Neumann algebra $\mathcal{M}_P$ generated by the operators L_ϕ and V_{y^o}. Then it turns out that:

a) When $\lambda = 1/2$ the algebra $\mathcal{M}_P$ is the hyperfinite type II_1 factor;
b) When $0 < \lambda < 1/2$ we obtain type III factors. In particular, these algebras are inequivalent von Neumann algebras [45], and are indicated as $\mathcal{R}_{\tilde{\lambda}}$, with $\tilde{\lambda} = \lambda/(1-\lambda)$.

As a final remark, observe that the crossed product algebra of the group Ω over the algebra $L^\infty(\Omega_\infty, \mu_\lambda)$ with unitary action V is given by the commutant of $\mathcal{M}_P$ [54].

References

1. Agredo, J., Fagnola, F., Rebolledo, R.: Decoherence free subspaces of a quantum Markov semigroup. J. Math. Phys. **55**, 112201 (2014)
2. Albert, V.V.: Asymptotics of quantum channels: conserved quantities, an adiabatic limit, and matrix product states. Quantum **3**, 151 (2019)
3. Amato, D., Facchi, P.: Number of steady states of quantum evolutions. Sci. Rep. **14**, 14366 (2024)
4. Amato, D., Facchi, P., Konderak, A.: Asymptotic dynamics of open quantum systems and modular theory. In: Correggi, M., Falconi, M. (eds.), Quantum Mathematics II. INdAM 2022. Springer INdAM Series, vol. 58, pp. 169–181 (2023)
5. Amato, D., Facchi, P., Konderak, A.: Asymptotics of quantum channels. J. Phys. A: Math. Theor. **56**, 265304 (2023)
6. Amato, D., Facchi, P., Konderak, A.: Asymptotic dynamics in the Heisenberg picture: attractor subspace and Choi-Effros product. in preparation (2025)
7. Amato, D., Facchi, P., Konderak, A.: Decoherence-free algebras in quantum dynamics (2024). arXiv:2403.12926 [quant-ph]
8. Araki, H.: Type of von Neumann algebra associated with free field. Prog. Theor. Phys. **32**, 956–965 (1964)
9. Asorey, M., Kossakowski, A., Marmo, G., Sudarshan, E.C.G.: Unital positive maps and quantum states. Open Syst. Inf. Dyn. **15**, 123–134 (2008)
10. Blanchard, P., Olkiewicz, R.: Decoherence induced transition from quantum to classical dynamics. Rev. Math. Phys. **15**, 217–243 (2003)
11. Breuer, H.P., Petruccione, F.: The Theory of Open Quantum Systems, 1st edn. Oxford University Press, Oxford (2002)
12. Bulinski, A.V.: Some asymptotic properties of W*-dynamical systems. Funct. Anal. Appl. **29**, 123–126 (1995)
13. Carbone, R., Jenčová, A.: On period, cycles and fixed points of a quantum channel. Ann. Henri Poincaré **21**, 155–188 (2020)
14. Carbone, R., Pautrat, Y.: Irreducible decompositions and stationary states of quantum channels. Rep. Math. Phys. **77**, 293–313 (2016)
15. Chruściński, D.: Dynamical maps beyond Markovian regime. Phys. Rep. **992**, 1–85 (2022)
16. Ciaglia, F.M., Di Cosmo, F., Facchi, P., Ibort, A., Konderak, A., Marmo, G.: Groupoid and algebra of the infinite quantum spin chain. J. Geom. Phys. **191**, 104901 (2023)
17. Ciaglia, F.M., Di Cosmo, F., González-Bravo, L., Ibort, A., Marmo, G.: The categorical foundations of quantum information theory: categories and the cramer-rao inequality. Mod. Phys. Lett. A **38**, 2350085 (2023)
18. Ciaglia, F.M., Ibort, A., Marmo, G.: Schwinger's picture of quantum mechanics I: Groupoids. Int. J. Geom. Methods Mod. Phys. **16**, 1950119 (2019)
19. Ciaglia, F.M., Ibort, A., Marmo, G.: Schwinger's picture of quantum mechanics II: Algebras and Observables. Int. J. Geom. Methods Mod. Phys. **16**, 1950136 (2019)
20. DeLeeuw, K., Glicksberg, I.: Almost periodic compactifications. Bull. Am. Math. Soc. **65**, 134–139 (1959)
21. DeLeeuw, K., Glicksberg, I.: Applications of almost periodic compactifications. Acta Math. **105**, 63–97 (1961)
22. Engel, K.J., Nagel, R.: One-parameter Semigroups for Linear Evolution Equations, 1st edn. Springer, New York (2000)
23. Evans, D.E.: Irreducible quantum dynamical semigroups. Commun. Math. Phys. **54**, 293–297 (1977)
24. Fagnola, F., Rebolledo, R.: On the existence of stationary states for quantum dynamical semigroups. J. Math. Phys. **42**, 1296–1308 (2001)
25. Fagnola, F., Sasso, E., Umanità, V.: Structure of uniformly continuous quantum markov semigroups with atomic decoherence-free algebra. Open Syst. Inf. Dyn. **24**, 1740005 (2017)

26. Fagnola, F., Sasso, E., Umanità, V.: The role of the atomic decoherence-free subalgebra in the study of quantum Markov semigroups. J. Math. Phys. **60**, 072703 (2019)
27. Frigerio, A., Verri, M.: Long-Time Asymptotic Properties of Dynamical Semigroups on W*-algebras. Math Z. **180**, 275–286 (1982)
28. Gorini, V., Kossakowski, A., Sudarshan, E.: Completely positive dynamical semigroups of N-level systems. J. Math. Phys. **17**, 821–825 (1976)
29. Haag, R.: Local Quantum Physics, 2nd edn. Springer, Berlin, Heidelberg (1996)
30. Hamana, M.: Injective envelopes of C^*-algebras. J. Math. Soc. Jpn. **31**, 181–197 (1979)
31. Huang, K.: Statistical Mechanics, 2nd edn. Wiley, Hoboken (1987)
32. Jacobs, K.: Fastperiodizitätseigenschaften allgemeiner halbgruppen in banach-räumen. Math. Z. **67**, 83–92 (1957)
33. Kadison, R.V.: Remarks on the type of von Neumann algebras of local observables in quantum field theory. J. Math. Phys. **4**, 1511–1516 (1963)
34. Kato, T.: Perturbation Theory for Linear Operators, 2nd edn. Springer, Berlin, Heidelberg (2013)
35. Lidar, D.A., Chuang, I.L., Whaley, K.B.: Decoherence-Free Subspaces for Quantum Computation. Phys. Rev. Lett. **81**, 2594 (1998)
36. Lindblad, G.: On the generators of quantum dynamical semigroups. Commun. Math. Phys. **48**, 119–130 (1976)
37. Murray, F.J., von Neumann, J.: On rings of operators. Ann. Math. **37**, 116–229 (1936)
38. Murray, F.J., von Neumann, J.: On Rings of operators. II. Trans. Am. Math. Soc. **41**, 208–248 (1937)
39. Murray, F.J., von Neumann, J.: On rings of operators. IV. Ann. Math. **44**, 716–808 (1943)
40. von Neumann, J.: On infinite direct products. Compos. Math. **6**, 1–77 (1939)
41. von Neumann, J.: On rings of operators III. Ann. Math. **41**, 94–161 (1940)
42. Novotný, J., Alber, G., Jex, I.: Asymptotic properties of quantum Markov chains. J. Phys. A: Math. Theor. **45**, 485301 (2012)
43. Novotný, J., Maryska, J., Jex, I.: Quantum Markov processes: from attractor structure to explicit forms of asymptotic states. Eur. Phys. J. Plus **133**, 1–17 (2018)
44. Poincaré, H.: New Methods of Celestial Mechanics, 1st edn. American Institute of Physics, College Park (1992)
45. Powers, R.T.: Representations of uniformly hyperfinite algebras and their associated von Neumann rings. Bull. Am. Math. Soc. **73**, 572–575 (1967)
46. Pukánszky, L.: Some example of factors. Publ. Math. **4**, 135–156 (1956)
47. Bhat, B.V.R., Kar, S., Talwar, B.: Peripherally automorphic unital completely positive maps. Linear Algebra Appl. **678**, 191–205 (2023)
48. Bhat, B.V.R., Talwar, B., Kar, S.: Peripheral Poisson Boundary (2022). arXiv: 2209.07731 [math.OA]
49. Sasso, E., Umanità, V.: The general structure of the decoherence-free subalgebra for uniformly continuous quantum Markov semigroups. J. Math. Phys. **64**, 042703 (2023)
50. Schwinger, J.: Quantum Kinematics and Dynamics, 1st edn. CRC Press, Boca Raton (2000)
51. Shiryaev, A.N.: Probability, 3rd edn. Springer Nature, Cham (2019)
52. Spohn, H.: An algebraic condition for the approach to equilibrium of an open N-level system. Lett. Math. Phys. **2**, 33–38 (1977)
53. Spohn, H.: Kinetic equations from Hamiltonian dynamics: Markovian limits. Rev. Mod. Phys. **52**, 569 (1980)
54. Takesaki, M.: Theory of Operator Algebras I, 1st edn. Springer, Berlin, Heidelberg (2002)
55. Wolf, M.M.: Quantum channels and operations: guided tour (2012). https://www-m5.ma.tum.de/foswiki/pub/M5/Allgemeines/MichealWolf/QChannelLecture.pdf
56. Wolf, M.M., Perez-Garcia, D.: The inverse eigenvalue problem for quantum channels (2010). arXiv:1005.4545 [quant-ph]

Algebraic Localization of Generalized Wannier Bases Implies Roe Triviality in Any Dimension

Vincenzo Rossi and Gianluca Panati

Abstract With the aim of understanding the localization-topology correspondence for *non-periodic* gapped quantum systems, we investigate the relation between the existence of an algebraically well-localized generalized Wannier basis and the topological triviality of the corresponding projection operator. Inspired by the work of M. Ludewig and G.C. Thiang, we consider the triviality of a projection in the sense of "coarse geometry", i.e. as triviality in the K_0-theory of the Roe C^*-algebra of $\mathbb{R}^d$. We obtain in Theorem 2.8 a threshold, depending on the dimension, for the decay rate of the generalized Wannier functions which implies topological triviality in Roe's sense. This threshold reduces, for $d = 2$, to the almost-optimal threshold appearing in the Localization Dichotomy Conjecture.

Keywords Wannier bases · Chern insulators · Roe C*-algebra · Topological K-theory

1 Introduction

Wannier bases, and their generalization to non-periodic systems, have been for decades a popular tool in solid state physics, both for practical and conceptual reasons. On the first side, the existence of a well-localized (composite) Wannier basis (WB) allows computational algorithms whose cost scales only linearly with the system size [7]. On the conceptual side, well-localized WBs allow a theoretical understanding of several phenomena in crystalline solids as e.g. macroscopic polarization and orbital magnetization in insulators [5]. We refer to [19] for a comprehensive review about WBs for solid state systems.

V. Rossi (✉)
GSSI - Gran Sasso Science Institute, L'Aquila, Italy
e-mail: vincenzo.rossi@gssi.it

G. Panati
Dipartimento di Matematica, "La Sapienza" Università di Roma,Rome, Italy
e-mail: panati@mat.uniroma1.it

B. Cassano et al. (eds.), *Singularities, Asymptotics, and Limiting Models*, Springer INdAM Series 64, https://doi.org/10.1007/978-981-96-3584-9_4

More recently, with the discovery of topological insulators, WBs gained a new role as a tool to distinguish between ordinary and Chern insulators. A correspondence between the topological non-triviality of the vector bundle of occupied states (Bloch bundle) and the non-existence of a composite Wannier basis (CWB) with finite second moment of the position operator, has been noticed and proved in [20], for periodic systems in dimension $d = 2$ and $d = 3$. More precisely, for any gapped periodic Schrödinger operator, the following *Localization Dichotomy* holds true:

(i) either there exists a CWB which is exponentially localized, and, correspondingly, the Chern class of the Bloch bundle is zero; or
(ii) any possible choice of a CWB yields an infinite expectation value of the squared position operator. Explicitly, this means

$$\sum_{a=1}^{m} \int_{\mathbb{R}^d} \langle x - \gamma \rangle^2 |w_{\gamma,a}(x)|^2 \, dx = +\infty \quad \forall \gamma \in \Gamma,$$

where $w_{\gamma,a}(x) = w_a(x - \gamma)$ are the elements of a CWB, Γ is the periodicity lattice, and $a \in \{1, \ldots, m\}$ is an index corresponding to the number m of relevant Bloch bands.

After the appearance of [20], it became clear that such a profound relation between localization and topology cannot depend on periodicity in an essential way, despite the fact that both terms of the correspondence (the Bloch bundle and, respectively, the Wannier functions) are defined by exploiting periodicity.

The generalization of the Localization Dichotomy to non-periodic systems is grounded on the definition of Generalized Wannier Basis (GWB), which traces back to the pioneering work of Kohn and Onfroy [11] and Kivelson [10], and has been formulated in clear mathematical terms by Nenciu and Nenciu [24] (see Definition 2.6 in the next Section). In [18], the existence of a localized GWB with finite second moment for the position operator has been conjecturally related to the vanishing of the Chern marker [1, 2]. Moreover, the authors proved that the algebraic decay of a GWB implies the vanishing of the Chern marker, however for a degree of localization s far from the optimal threshold $s \geq 1$. (With this notation, as in Definition 2.6, $s = 1$ corresponds to a finite second moment for the position operator). An almost-optimal threshold has been later reached by Lu and Stubbs [14, 15], who proved a similar result for every $s > 1$ in dimension $d = 2$.

A more general concept of Wannier basis has been proposed by Prodan [27], and has been later associated to the name Ultra Generalized Wannier Basis (UGWB). While the latter bases might be useful from the computational viewpoint as they can be constructed for any magnetic Schrödinger operator, the price to pay for such a generality is that UGWBs are not able to distinguish between ordinary and Chern insulators, as recently shown in [22].

Following an independent line of thought, several researchers approached the same problem from the viewpoint of C^*-algebras and Hilbert modules [3, 16, 17]. In particular, the latter paper suggests that the topological triviality of a projection may be better investigated from the viewpoint of "coarse geometry", *i.e.* by considering the corresponding class in the K-theory of the Roe C^*-algebra of the position space. Indeed, the authors of [17] proved that the existence of a Wannier basis which is s-localized for all $s > 0$ in the sense of Definition 2.6, with the L^2-decay condition replaced by the L^∞-one, implies the triviality of the corresponding projection in the K_0-group of the Roe C^*-algebra, in the setting of Riemannian manifolds. In this general context, a more geometric criterion for the triviality in the Roe sense has been obtained in [12].

In this paper, we combine ideas and methods from both lines of thoughts. Inspired by [17], we prove that, when the position space is $\mathbb{R}^d$ for $d \geq 1$, the existence of a GWB with L^2-localization parameter $s > \frac{d}{2}$ implies the triviality of the corresponding projection in the K_0-group of the Roe C^*-algebra (Theorem 2.8). While our setting is much restricted in comparison with [17], where general Riemannian manifolds are considered, our hypotheses are weaker, as only s-localization for $s > \frac{d}{2}$ is assumed. In comparison with Lu and Stubbs [14, 15], we obtain the same almost-optimal threshold $s > 1$ in $\mathbb{R}^2$, but we have a result which is also valid in any dimension $d \geq 1$. Moreover, the structure of the proof emphasizes the role of the Roe C^*-algebra to describe the "large scale" geometric properties which are relevant to physics, as suggested by Ludewig and Thiang. Clearly, it remains to clarify under which conditions the triviality in the sense of Roe proved in this paper is equivalent to the Chern triviality considered in [15, 18, 21]. While we are confident that the two notions agree for $\mathbb{R}^2$, we postpone a general analysis of this equivalence problem to future investigations.

2 Preliminaries and Main Result

2.1 The Roe C*-Algebra: A Quick Review

Consider the Hilbert space $L^2(\mathbb{R}^d)$, $d \geq 1$, and let $C_c(\mathbb{R}^d)$ be the set of compactly supported continuous functions on $\mathbb{R}^d$. To each $f \in C_c(\mathbb{R}^d)$ one associates a bounded multiplication operator on $L^2(\mathbb{R}^d)$, namely

$$(f\varphi)(x) = f(x)\varphi(x), \quad \forall \varphi \in L^2(\mathbb{R}^d).$$

We also denote by $\mathcal{B}(L^2(\mathbb{R}^d)$ the set of bounded linear operators over $L^2(\mathbb{R}^d)$ and by d the distance between sets, which is given by

$$\mathrm{d}(A, B) := \inf_{x \in A, y \in B} \|x - y\|, \quad \forall A, B \subseteq \mathbb{R}^d.$$

In order to define the Roe C^*-algebra of $\mathbb{R}^d$ we begin with the next definition, following [6, 8, 17, 28].

Definition 2.1 Consider $T \in \mathcal{B}(L^2(\mathbb{R}^d))$. We say that

(i) T is locally compact if Tf, fT are compact operators on $L^2(\mathbb{R}^d)$, $\forall f \in C_c(\mathbb{R}^d)$;
(ii) T has finite propagation if there exists $R > 0$ such that $fTg = 0$, whenever $\mathrm{d}(\mathrm{supp}(f), \mathrm{supp}(g)) > R, \forall f, g \in C_c(\mathbb{R}^d)$.

We define $\mathbb{C}[\mathbb{R}^d] := \{T \in \mathcal{B}(L^2(\mathbb{R}^d)) : T \text{ is locally compact with finite propagation}\}$.

One may think of finite propagation operators as the continuum analogous of a matrix, with discrete indices replaced by $(x, y) \in \mathbb{R}^d \times \mathbb{R}^d$, with vanishing entries outside a neighborhood of the diagonal.

Example 2.2 (Convolution operator with integral kernel in $C_c(\mathbb{R}^d)$) For $k \in C_c(\mathbb{R}^d)$, define the operator $K \in \mathcal{B}(L^2(\mathbb{R}^d))$ acting on $L^2(\mathbb{R}^d)$ by convolution, *i.e.*

$$(K\varphi)(x) = \int_{\mathbb{R}^d} \mathrm{d}y\, k(x-y)\varphi(y), \quad \varphi \in L^2(\mathbb{R}^d).$$

The operator K is locally compact, because for every $f \in C_c(\mathbb{R}^d)$, Kf, fK are still integral operators with compactly supported, continuous integral kernels. To see that K has also finite propagation define $R := \sup\{\|x\| : k(x) \neq 0\}$. Then, if $f, g \in C_c(\mathbb{R}^d)$, $\mathrm{d}(\mathrm{supp}(f), \mathrm{supp}(g)) > R$, we have that

$$(fKg\varphi)(x) = \int_{\mathbb{R}^d} \mathrm{d}y\; f(x)k(x-y)g(y)\varphi(y) = 0, \quad \forall \varphi \in L^2(\mathbb{R}^d), x \in \mathbb{R}^d,$$

because if $x \in \mathrm{supp}(f)$, $y \in \mathrm{supp}(g)$, then $\|x - y\| \geq \mathrm{d}(\mathrm{supp}(f), \mathrm{supp}(g)) > R$ and thus $k(x - y) = 0$. In particular $K \in \mathbb{C}[\mathbb{R}^d]$. ◇

Lemma 2.3 *$\mathbb{C}[\mathbb{R}^d]$ is a $*$-subalgebra of $\mathcal{B}(L^2(\mathbb{R}^d))$.*

Proof The only non-trivial part to prove is that, if $T, S \in \mathbb{C}[\mathbb{R}^d]$ then TS is also in $\mathbb{C}[\mathbb{R}^d]$. Indeed, if T, S are locally compact their composition is too, because the compact operators form an ideal in $\mathcal{B}(L^2(\mathbb{R}^d))$. Now suppose T, S have finite propagation R_1, R_2, respectively. Fix $\varepsilon > 0$, let $\{x_n\}_{n\in\mathbb{N}} \subseteq \mathbb{R}^d$ be a dense subset and let $\{\psi_n\}_{n\in\mathbb{N}} \subseteq C_c(\mathbb{R}^d)$ be a partition of unity subordinated to $\{B_\varepsilon(x_n)\}_{n\in\mathbb{N}}$, where $B_\varepsilon(x_n) := \{x \in \mathbb{R}^d : \|x - x_n\| < \varepsilon\}$. Then, if $f, g \in C_c(\mathbb{R}^d)$, we have that

$$fTSg = fT\left(\sum_{n\in N_1\cap N_2} \psi_n\right)Sg,$$

with $N_1 := \{n : \mathrm{d}(\mathrm{supp}(\psi_n), \mathrm{supp}(f)) \leq R_1\}$, $N_2 := \{n : \mathrm{d}(\mathrm{supp}(\psi_n), \mathrm{supp}(g)) \leq R_2\}$. Suppose $fTSg \neq 0$. Then there exists $n \in N_1 \cap N_2$ and, by Weierstrass Theorem, there exist $z, z' \in \mathrm{supp}(\psi_n)$, $x \in \mathrm{supp}(f)$, $y \in \mathrm{supp}(g)$ such that

$$\mathrm{d}(\mathrm{supp}(f), \mathrm{supp}(\psi_n)) = \|x - z\| \leq R_1, \quad \mathrm{d}(\mathrm{supp}(g), \mathrm{supp}(\psi_n)) = \left\|y - z'\right\| \leq R_2.$$

Therefore, since $\operatorname{supp}(\psi_n) \subseteq B_\varepsilon(x_m)$ for some $m \in \mathbb{N}$, we obtain

$$\mathrm{d}(\operatorname{supp}(f), \operatorname{supp}(g)) \leq \|x - y\| \leq \|x - z\| + \|z - z'\| + \|z' - y\| \leq R_1 + R_2 + 2\varepsilon.$$

By the arbitrarity of ε, it follows that TS has finite propagation $R_1 + R_2$. □

Definition 2.4 The Roe C^*-algebra $C^*(\mathbb{R}^d)$ is the closure in $\mathcal{B}(L^2(\mathbb{R}^d))$ of $\mathbb{C}[\mathbb{R}^d]$.

Observe that $C^*(\mathbb{R}^d)$ is not a unital C^*-algebra. Indeed, suppose that the identity operator is the norm-limit of a sequence of locally compact operators $\{T_n\}_{n\in\mathbb{N}}$. Then, for all $f \in C_c(\mathbb{R}^d)$ we would have

$$\|f - T_n f\| \leq \|\mathbb{1} - T_n\| \, \|f\| \xrightarrow{n\to\infty} 0.$$

As a consequence the multiplication operator by f on $L^2(\mathbb{R}^d)$ would be compact for every $f \in C_c(\mathbb{R}^d)$, which is absurd.

2.2 *Generalized Wannier Bases and Roe Triviality*

In this section we introduce the concept of generalized Wannier basis. We first consider a discrete set with some uniformity property, which has the role of modelling a non-crystalline material.

Definition 2.5 Let $\mathfrak{D}$ be a subset of $\mathbb{R}^d$. We say that $\mathfrak{D}$ is r-uniformly discrete if there exists $r > 0$ such that the set $\{B_r(\gamma)\}_{\gamma\in\mathfrak{D}}$ is composed of mutually disjoint balls.

Definition 2.6 Let $\mathcal{H}$ be a closed subspace of $L^2(\mathbb{R}^d)$. Given $s > 0$ we say that $\mathcal{H}$ admits a s-localized generalized Wannier basis (GWB) if there exists a uniformly discrete set $\mathfrak{D}$ and a set $\{\psi_\gamma\}_{\gamma\in\mathfrak{D}} \subseteq L^2(\mathbb{R}^d)$ such that

(i) $\{\psi_\gamma\}_{\gamma\in\mathfrak{D}}$ is an orthonormal basis for $\mathcal{H}$;
(ii) the functions ψ_γ are uniformly s-localized around $\mathfrak{D}$, *i.e.* there exists $M > 0$ such that
$$\int_{\mathbb{R}^d} \mathrm{d}x \, \langle x - \gamma \rangle^{2s} |\psi_\gamma(x)|^2 \leq M, \quad \forall \gamma \in \mathfrak{D},$$

where we used the standard notation $\langle x \rangle = (1 + \|x\|^2)^{\frac{1}{2}}$.

We now consider the topological K-theory of the C^*-algebra $C^*(\mathbb{R}^d)$. In order to proceed, we first give the following definition of equivalence between projections.

Definition 2.7 Let $\mathcal{A}$ be a C^*-algebra. Two projections $P, Q \in \mathcal{A}$ are said to be Murray-von Neumann (MvN) equivalent if there exists $V \in \mathcal{A}$ such that $P = VV^*$ and $Q = V^*V$.

By [6, Corollary 2.20] the elements in $K_0(C^*(\mathbb{R}^d))$ can be represented as formal differences of MvN equivalence classes of projections in $C^*(\mathbb{R}^d)$. In particular, to each projection $P \in C^*(\mathbb{R}^d)$ we can associate an element $[P] \in K_0(C^*(\mathbb{R}^d))$ which is given by the formal difference of the MvN equivalence classes of P with the zero projection in $C^*(\mathbb{R}^d)$. If such a class defines a trivial element in $K_0(C^*(\mathbb{R}^d))$, we say that the projection P is Roe trivial.

2.3 Main Result

We are now in position to state our main result, namely that the existence of a s-localized GWB for $s > \frac{d}{2}$ implies the Roe triviality of the corresponding projection.

Theorem 2.8 *Let $\mathcal{H}$ be a closed subspace of $L^2(\mathbb{R}^d)$ admitting a s-localized GWB, $\{\psi_\gamma\}_{\gamma\in\mathfrak{D}}$, with $s > \frac{d}{2}$, where $\mathfrak{D}$ is a uniformly discrete subset of $\mathbb{R}^d$.*
Then the projection operator $P_{\mathcal{H}} = \sum_{\gamma\in\mathfrak{D}} |\psi_\gamma\rangle\langle\psi_\gamma|$ defines a trivial element in $K_0(C^*(\mathbb{R}^d))$, namely $[P_{\mathcal{H}}] = 0$ in $K_0(C^*(\mathbb{R}^d))$.

Observe that, as mentioned in the Introduction, the Theorem holds for any dimension. In particular, when $d = 2$ we almost reach the threshold value $s_* = 1$, under which the s-localization of a GWB is not expected to imply topological triviality of $P_{\mathcal{H}}$. The same threshold appears also in [15], where the authors proved instead the Chern triviality of the projection $P_{\mathcal{H}}$.

2.4 Application to a Physically Relevant Model

To show a possible application of our Theorem, we consider here a deformation "á la Gubanov" of the two-dimensional Kronig-Penney model. The latter is given by the following one-particle Schrödinger operator acting in $L^2(\mathbb{R}^2)$

$$H_0 = -\Delta + V_0,$$

where V_0 is a two-dimensional $\mathbb{Z}^2$-periodic well potential, *i.e.*

$$V_0(x) = v_0 \sum_{n\in\mathbb{Z}^2} \chi_{\mathcal{R}(n)}(x),$$

where $v_0 \in \mathbb{R}$ is the height of the barrier and $\mathcal{R}(n) := \mathcal{R}(0) + n = [n_1 - \frac{a}{2}, n_1 + \frac{a}{2}] \times [n_2 - \frac{a}{2}, n_2 + \frac{a}{2}]$, with $0 < a < 1$ a length parameter. It is known that, for suitable height and length parameters, the spectrum of H_0 admits gaps ([13], for numerical simulations see [26]). We denote by σ_0 one of the connected components of the spectrum (spectral island), for which it holds

$$\operatorname{dist}(\sigma(H_0), \sigma_0) > 0.$$

We also denote by $P_0 = \chi_{\sigma_0}(H_0)$ the spectral projection onto σ_0. Since the Hamiltonian H_0 is time-reversal symmetric, by [25, Theorem 1] and [20, Theorem 3.5], there exists an orthornormal (Wannier) basis $\{w_n\}_{n\in\mathbb{Z}^2} = \{w_0(\cdot - n)\}_{n\in\mathbb{Z}^2}$ of Ran P_0 which is exponentially localized, *i.e.* there exists $\alpha_0 > 0$ such that

$$\int_{\mathbb{R}^2} \mathrm{d}x\, e^{2\alpha_0\|x-n\|} |w_n(x)|^2 < \infty, \quad \forall n \in \mathbb{Z}^2,$$

where we assumed, for simplicity, no degeneracy of the localization centers $n \in \mathbb{Z}^2$. Clearly $\{w_n\}_{n\in\mathbb{Z}^2}$ is also a s-localized GWB around $\mathbb{Z}^2$, for each $s > 0$.

In order to describe possible deformations of the system, we consider a vector-valued function $g \in C^3(\mathbb{R}^2; \mathbb{R}^2)$ satisfying

$$\xi := \sup_{x\in\mathbb{R}^2} \max_{1\le i,j,k,\ell\le 2} \left\{ |\partial_i g_k(x)|, |\partial_i\partial_j g_k(x)|, |\partial_i\partial_j\partial_\ell g_k(x)| \right\} < \infty.$$

We represent the deformed system through the so called quasi-crystalline Gubanov model ([4, 18, 23]), given by the Hamiltonian

$$H_g = -\Delta + V_g,$$

where $V_g(x) = V_0(x + g(x))$. By the Hadamard-Caccioppoli Theorem the function $f(x) := x + g(x)$ is globally invertible for ξ small enough, and we denote its inverse by $h(x) := f^{-1}(x)$. For each $\varphi \in L^2(\mathbb{R}^2)$, we consider the following operators

$$(Y\varphi)(x) = J(x)^{\frac{1}{2}}\varphi(f(x)), \qquad (Y^{-1}\varphi)(x) = J(h(x))^{-\frac{1}{2}}\varphi(h(x)),$$

where $J(x) = |\det Df(x)| = |\det(\mathbb{1} + Dg(x))|$. By the change of variable formula, it's easy to see that $Y^* = Y^{-1}$, so Y is a unitary operator. By a direct computation one shows that, for ξ small enough

$$Y^{-1} H_g Y = H_0 + \xi D_\xi,$$

where D_ξ is a H_0-bounded operator, *i.e.* there exist $c > 0$, $d > 0$, independent of ξ, such that

$$\left\| D_\xi\varphi \right\| \le c \left\| H_0\varphi \right\| + d \left\| \varphi \right\|, \quad \forall \varphi \in \mathcal{D}(H_0).$$

Then, by [9, Theorem 6.5.12], we have that the spectrum of H_g has a spectral island σ_g that varies continuously w.r.t the Hausdorff distance, and we denote by $P_g = \chi_{\sigma_g}(H_g)$ its spectral projection onto σ_g. As discussed e. g. in [18, Example 2.8], thanks to the Combes-Thomas estimate and the Kato-Nagy formula, it is possible to unitarily map the GWB in Ran P_0 to a GWB in $\mathrm{Ran}(Y^{-1}P_gY)$, hereafter denoted by $\{\psi_n\}_{n\in\mathbb{Z}^2}$, without losing its exponential localization around $\mathbb{Z}^2$, *i. e.* there exist $M > 0, \alpha > 0$ such that

$$\int_{\mathbb{R}^2} \mathrm{d}x\, e^{2\alpha\|x-n\|} |\psi_n(x)|^2 \le M, \quad \forall n \in \mathbb{Z}^2.$$

The deformed lattice $\mathfrak{D} := h(\mathbb{Z}^2) = \left\{\gamma \in \mathbb{R}^2 : \gamma = h(n),\ n \in \mathbb{Z}^2\right\}$ is a r-uniformly discrete set, with $2r = (1+2\xi)^{-1}$. By setting

$$\varphi_\gamma := Y\psi_n, \quad \forall \gamma = h(n) \in \mathfrak{D},$$

the set $\left\{\varphi_\gamma\right\}_{\gamma\in\mathfrak{D}}$ is a GWB for Ran P_g, which is exponentially localized around $\mathfrak{D}$. Indeed, notice that for all $x, y \in \mathbb{R}^2$, for $\xi < \frac{1}{2}$, applying the mean value Theorem to each component of g, we have

$$\|f(x) - f(y)\| \ge (1-2\xi)\,\|x-y\| \implies \|h(x) - h(y)\| \le (1-2\xi)^{-1}\,\|x-y\|\,.$$

Therefore, for all $\gamma = h(n) \in \mathfrak{D}$, taking $\beta = \alpha(1-2\xi)$ and using the change of variable $x = h(y)$, we obtain

$$\begin{aligned}\int_{\mathbb{R}^2} \mathrm{d}x\, e^{2\beta\|x-\gamma\|} |\varphi_\gamma(x)|^2 &= \int_{\mathbb{R}^2} \mathrm{d}x\, e^{2\beta\|x-h(n)\|} J(x) |\psi_n(f(x))|^2 \\ &= \int_{\mathbb{R}^2} \mathrm{d}y\, e^{2\beta\|h(y)-h(n)\|} |\psi_n(y)|^2 \le \int_{\mathbb{R}^2} \mathrm{d}y\, e^{2\alpha\|y-n\|} |\psi_n(y)|^2 \le M,\end{aligned}$$

which proves exponential localization. In particular Theorem 2.8 applies to $\mathcal{H} =$ Ran P_g, since it admits a GWB $\left\{\varphi_\gamma\right\}_{\gamma\in\mathfrak{D}}$ which is also s-localized around a r-uniformly discrete set $\mathfrak{D}$, for all $s > 0$. As a consequence we get that $[P_g] = 0$ in $K_0(C^*(\mathbb{R}^2))$.

We stress out that both the Kronig-Penney model and its Gubanov deformation can be generalized to the $\mathbb{R}^d$ case, for any dimension $d \ge 1$. Theorem 2.8 applies to d-generalizations as well.

3 Proofs

3.1 Proof of the Main Theorem

In order to prove our main result, it will be useful to introduce the concept of extremely localized generalized Wannier basis.

Definition 3.1 Let $\mathcal{H}$ be a closed subspace of $L^2(\mathbb{R}^d)$. We say that $\mathcal{H}$ admits an extremely localized GWB if there exists a r-uniformly discrete set $\mathfrak{D}$ and a set

$\{\phi_\gamma\}_{\gamma\in\mathfrak{D}} \subseteq L^2(\mathbb{R}^d)$ which is an orthonormal basis for $\mathcal{H}$, such that $\mathrm{supp}(\phi_\gamma) \subseteq B_r(\gamma)$, for all $\gamma \in \mathfrak{D}$.

A projection operator whose range admits an extremely localized GWB is an element of $C^*(\mathbb{R}^d)$ and its K-theory class $[P]$ is trivial, as shown in the next proposition, whose proof is postponed to the next section.

Proposition 3.2 *Let $\mathcal{H}$ be a closed subspace of $L^2(\mathbb{R}^d)$. Suppose that $\mathcal{H}$ admits an extremely localized GWB $\{\phi_\gamma\}_{\gamma\in\mathfrak{D}}$, where $\mathfrak{D} \subseteq \mathbb{R}^d$ is a r-uniformly discrete set. Then*

(i) The associated projection operator $P_\mathcal{H} := \sum_{\gamma\in\mathfrak{D}} |\phi_\gamma\rangle\langle\phi_\gamma|$ belongs to $\mathbb{C}[\mathbb{R}^d]$;
(ii) $[P_\mathcal{H}] = 0$ in $K_0(C^*(\mathbb{R}^d))$.

In light of the previous proposition, we are able to explain the strategy of the proof of Theorem 2.8: it consists in showing that a projection whose range admits a sufficiently fast decaying GWB is MvN equivalent to a projection onto the span of some extremely localized GWB. In this way, their K_0-classes would define the same element in $K_0(C^*(\mathbb{R}^d))$, which is the trivial one by Proposition 3.2 (ii), thus proving the thesis of our main Theorem.

To implement this idea, we need the following lemma, which is a specialization to our setting of [17, Proposition 3.4]. Again, the proof is postponed to the next section.

Lemma 3.3 *Let $\mathfrak{D}$ be a r-uniformly discrete subset of $\mathbb{R}^d$. Then for every $s > \frac{d}{2}$ there exists $C > 0$ such that*

$$\sum_{\substack{\gamma\in\mathfrak{D}\\ \|x-\gamma\|\geq R}} \langle x-\gamma\rangle^{-2s} \leq C(1+R)^{d-2s}, \quad \forall x \in \mathbb{R}^d, R \geq 0.$$

Observe that the hypothesis $s > \frac{d}{2}$ is necessary to guarantee the convergence of the series.

Now we are going to prove the key result of the article, following the strategy we have just described.

Proposition 3.4 *Let $\mathcal{H}$ be a closed subspace of $L^2(\mathbb{R}^d)$ admitting a s-localized GWB, $\{\psi_\gamma\}_{\gamma\in\mathfrak{D}}$, with $s > \frac{d}{2}$, where $\mathfrak{D}$ is a r-uniformly discrete subset of $\mathbb{R}^d$. Then for every closed Hilbert subspace $\widetilde{\mathcal{H}}$ admitting an extremely localized GWB $\{\phi_\gamma\}_{\gamma\in\mathfrak{D}}$ there exists $V \in C^*(\mathbb{R}^d)$ such that*

$$P_\mathcal{H} := \sum_{\gamma\in\mathfrak{D}} |\psi_\gamma\rangle\langle\psi_\gamma| = V^*V, \qquad P_{\widetilde{\mathcal{H}}} := \sum_{\gamma\in\mathfrak{D}} |\phi_\gamma\rangle\langle\phi_\gamma| = VV^*.$$

As a consequence, we have that $P_\mathcal{H} \in C^*(\mathbb{R}^d)$ and that $P_\mathcal{H}$ is MvN equivalent to $P_{\tilde{\mathcal{H}}}$.

Proof Let $\widetilde{\mathcal{H}}$ be a closed Hilbert subspace of $L^2(\mathbb{R}^d)$ admitting an extremely localized GWB $\{\phi_\gamma\}_{\gamma\in\mathfrak{D}}$. Define the operator V by

$$V := \sum_{\gamma\in\mathfrak{D}} |\phi_\gamma\rangle\langle\psi_\gamma| .$$

It is clear from the definition that $P_{\mathcal{H}} = V^*V$, $P_{\widetilde{\mathcal{H}}} = VV^*$. Indeed

$$V^*V = \sum_{\gamma,\eta\in\mathfrak{D}} |\psi_\gamma\rangle\langle\phi_\gamma|\phi_\eta\rangle\langle\psi_\eta| = \sum_{\gamma,\eta\in\mathfrak{D}} |\psi_\gamma\rangle\langle\psi_\eta|\,\delta_{\gamma,\eta} = P_{\mathcal{H}},$$

where we used the orthonormality property of $\{\phi_\gamma\}_{\gamma\in\mathfrak{D}}$. Analogously, using the fact that $\{\psi_\gamma\}_{\gamma\in\mathfrak{D}}$ is an orthonormal basis, we have also that $VV^* = P_{\widetilde{\mathcal{H}}}$. In order to show that $V \in C^*(\mathbb{R}^d)$ we want to approximate it with operators $V^R \in \mathbb{C}[\mathbb{R}^d]$, with R tending to infinity. More precisely, for every $R \geq 0$, we define

$$V^R := \sum_{\gamma\in\mathfrak{D}} |\phi_\gamma\rangle\langle\psi_\gamma^R| ,$$

where $\psi_\gamma^R = \chi_{B_R(\gamma)}\psi_\gamma$ is the restriction of ψ_γ to the ball $B_R(\gamma)$. Clearly, $V^R \in \mathbb{C}[\mathbb{R}^d]$. Indeed, if $f \in C_c(\mathbb{R}^d)$, we have that

$$fV^R = \sum_{\gamma\in\mathfrak{D}_1} |f\phi_\gamma\rangle\langle\psi_\gamma^R| , \qquad V^R f = \sum_{\gamma\in\widetilde{\mathfrak{D}}_1} |\phi_\gamma\rangle\langle\psi_\gamma^R \bar{f}| ,$$

where $\mathfrak{D}_1 := \{\gamma : \mathrm{supp}(\phi_\gamma)\cap\mathrm{supp}(f) \neq \emptyset\}$, $\widetilde{\mathfrak{D}}_1 := \left\{\gamma : \mathrm{supp}(\psi_\gamma^R)\cap\mathrm{supp}(f) \neq \emptyset\right\}$. Since $\mathfrak{D}$ is r-uniformly discrete and because $\mathrm{supp}(\psi_\gamma^R) \subseteq B_R(\gamma)$, $\mathrm{supp}(\phi_\gamma) \subseteq B_r(\gamma)$, for all $\gamma\in\mathfrak{D}$, we have that both $\mathfrak{D}_1$, $\widetilde{\mathfrak{D}}_1$ have only a finite number of elements. Thus both $V^R f$, fV^R are finite rank operators and are thus compact. Considering now $f, g \in C_c(\mathbb{R}^d)$, it holds that

$$fV^R g = \sum_{\gamma\in\mathfrak{D}_1\cap\mathfrak{D}_2} |f\phi_\gamma\rangle\langle\psi_\gamma^R \bar{g}| ,$$

where $\mathfrak{D}_2 := \{\gamma : \mathrm{supp}(\psi_\gamma^R)\cap\mathrm{supp}(g) \neq \emptyset\}$. If $fV^R g \neq 0$ then there exists $\gamma \in \mathfrak{D}_1\cap\mathfrak{D}_2$. In particular there exist $x \in B_r(\gamma)\cap\mathrm{supp}(f)$, $y \in B_R(\gamma)\cap\mathrm{supp}(g)$. Thus we obtain that

$$\mathrm{d}(\mathrm{supp}(f), \mathrm{supp}(g)) \leq \|x - y\| \leq \|x-\gamma\| + \|y-\gamma\| \leq r + R.$$

So V^R has finite propagation $r + R$. Therefore we only need to show that

$$\lim_{R\to\infty} \left\| V - V^R \right\| = 0, \tag{3.1}$$

which implies that $V \in C^*(\mathbb{R}^d)$. In order to prove (3.1), we consider the operator

$$V - V^R = \sum_{\gamma\in\mathfrak{D}} |\phi_\gamma\rangle\langle\psi_\gamma - \psi_\gamma^R| ,$$

which is an integral operator with integral kernel given by

$$K^R(x, y) = \sum_{\gamma\in\mathfrak{D}} \phi_\gamma(x)\overline{\big(\psi_\gamma(y) - \psi_\gamma^R(y)\big)}, \quad x, y \in \mathbb{R}^d. \tag{3.2}$$

It is useful to notice that the sum in (3.2) is finite, for fixed $x \in \mathbb{R}^d$. Indeed, if $x \in B_r(\eta)$, $\eta \in \mathfrak{D}$, then the summands corresponding to $\gamma \neq \eta$ give no contribution to the sum, because the supports of ϕ_γ are all disjoint. On the other hand, if $x \notin B_r(\gamma)$, for any $\gamma \in \mathfrak{D}$, then $K^R(x, y) = 0$, for all $y \in \mathbb{R}^d$. Now, for any $\varphi \in L^2(\mathbb{R}^d)$, we have that

$$\begin{aligned}
\left\|(V - V^R)\varphi\right\|^2 &= \int_{\mathbb{R}^d} \mathrm{d}x \left| \int_{\mathbb{R}^d} \mathrm{d}y\, K^R(x, y)\varphi(y)\right|^2 \\
&= \int_{\mathbb{R}^d} \mathrm{d}x \left| \int_{\mathbb{R}^d} \mathrm{d}y \left[\sum_{\gamma\in\mathfrak{D}} \phi_\gamma(x)\overline{\big(\psi_\gamma(y) - \psi_\gamma^R(y)\big)}\right]\varphi(y)\right|^2 \\
&= \sum_{\eta\in\mathfrak{D}} \int_{B_r(\eta)} \mathrm{d}x \left| \int_{\mathbb{R}^d} \mathrm{d}y \left[\sum_{\gamma\in\mathfrak{D}} \phi_\gamma(x)\overline{\big(\psi_\gamma(y) - \psi_\gamma^R(y)\big)}\right]\varphi(y)\right|^2 \\
&= \sum_{\gamma\in\mathfrak{D}} \int_{B_r(\gamma)} \mathrm{d}x \left| \int_{\mathbb{R}^d} \mathrm{d}y\, \phi_\gamma(x)\overline{\big(\psi_\gamma(y) - \psi_\gamma^R(y)\big)}\varphi(y)\right|^2 \\
&= \sum_{\gamma\in\mathfrak{D}} \left| \int_{\mathbb{R}^d} \mathrm{d}y\, \overline{\big(\psi_\gamma(y) - \psi_\gamma^R(y)\big)}\varphi(y)\right|^2 ,
\end{aligned} \tag{3.3}$$

where in the third equality we used that the function $x \mapsto K^R(x, y)$ is supported in $\cup_{\eta\in\mathfrak{D}} B_r(\eta)$, in the second-to-last one we used the fact that, if $x \in B_r(\eta)$, then $\phi_\gamma(x) = 0$, for all $\gamma \neq \eta$, and in the last one we used that $\left\|\phi_\gamma\right\| = 1$, $\forall \gamma \in \mathfrak{D}$. Then, multiplying and dividing by $\langle y - \gamma\rangle^s$ inside the integral and using Cauchy-Schwarz inequality and the property of uniform localization of $\left\{\psi_\gamma\right\}_{\gamma\in\mathfrak{D}}$, we obtain

$$\left\|(V - V^R)\varphi\right\|^2 = \sum_{\gamma\in\mathfrak{D}} \left| \int_{\mathbb{R}^d} \mathrm{d}y\, \overline{\psi_\gamma(y)}\left(1 - \chi_{B_R(\gamma)}(y)\right) \frac{\langle y - \gamma\rangle^s}{\langle y - \gamma\rangle^s}\varphi(y)\right|^2$$

$$
\begin{aligned}
&\leq \sum_{\gamma\in\mathfrak{D}} \left(\int_{\mathbb{R}^d} \mathrm{d}y\, |\psi_\gamma(y)|^2 \langle y-\gamma\rangle^{2s} \right) \left(\int_{\mathbb{R}^d} \mathrm{d}y\, \frac{|\varphi(y)|^2}{\langle y-\gamma\rangle^{2s}} \left(1-\chi_{B_R(\gamma)}(y)\right) \right) \\
&\leq M \sum_{\gamma\in\mathfrak{D}} \int_{\mathbb{R}^d} \mathrm{d}y\, |\varphi(y)|^2 \langle y-\gamma\rangle^{-2s} \left(1-\chi_{B_R(\gamma)}(y)\right) \\
&= M \int_{\mathbb{R}^d} \mathrm{d}y\, |\varphi(y)|^2 \sum_{\gamma\in\mathfrak{D}} \langle y-\gamma\rangle^{-2s} \left(1-\chi_{B_R(y)}(\gamma)\right) \\
&= M \int_{\mathbb{R}^d} \mathrm{d}y\, |\varphi(y)|^2 \sum_{\substack{\gamma\in\mathfrak{D} \\ \|y-\gamma\|\geq R}} \langle y-\gamma\rangle^{-2s} \leq C(1+R)^{d-2s} \|\varphi\|^2 ,
\end{aligned}
$$

where in the second-to-last line we used Tonelli's Theorem to exchange the series with the integral, and in the last inequality we used the result of Proposition 3.3. In conclusion we obtained

$$
\left\| V - V^R \right\|^2 \leq C(1+R)^{d-2s} \xrightarrow{R\to\infty} 0,
$$

because $s > \frac{d}{2}$, and this implies that $V \in C^*(\mathbb{R}^d)$. □

The previous argument is similar to the proof of [17, Proposition 3.4], where the auxiliary operator V is replaced with its adjoint, interchanging the role of bra and ket. Although this exchange has no impact on the MvN equivalence, it makes a difference when estimating $\left\|(V - V^R)\varphi\right\|$, since both the test function and the s-localized generalized Wannier functions are now calculated in the same integration variable, namely

$$
\left((V - V^R)\varphi\right)(x) = \int_{\mathbb{R}^d} \mathrm{d}y \left[\sum_{\gamma\in\mathfrak{D}} \phi_\gamma(x) \overline{\left(\psi_\gamma(y) - \psi^R_\gamma(y)\right)} \varphi(y) \right].
$$

This little but crucial modification, allowed us to bring the series appearing in the expression of K^R outside of the integral in (3.3), thanks to the orthonormality property of $\left\{\phi_\gamma\right\}_{\gamma\in\mathfrak{D}}$, making the subsequent estimates easier. In view of that, we were able to reach an almost-optimal threshold, *i. e.* $s > \frac{d}{2}$.

Proof of Theorem 2.8 By Proposition 3.4 we know that $P_{\mathcal{H}}$ is MvN equivalent to a projection operator $P_{\widetilde{\mathcal{H}}}$ whose range admits an extremely localized GWB. As a consequence, they have the same class in $K_0(C^*(\mathbb{R}^d))$. Thus, by Proposition 3.2(ii), we conclude that

$$
[P_{\mathcal{H}}] = [P_{\widetilde{\mathcal{H}}}] = 0, \quad \text{in } K_0(C^*(\mathbb{R}^d)).
$$

□

3.2 Proofs of Technical Results

Proof of Proposition 3.2 (i). Let $f \in C_c(\mathbb{R}^d)$, then $f P_{\mathcal{H}}$, $P_{\mathcal{H}} f$ are compact operators. Indeed

$$f P_{\mathcal{H}} = \sum_{\gamma \in \mathfrak{D}_1} |f \phi_\gamma\rangle\langle\phi_\gamma|, \qquad P_{\mathcal{H}} f = \sum_{\gamma \in \mathfrak{D}_1} |\phi_\gamma\rangle\langle\phi_\gamma \bar{f}|,$$

where $\mathfrak{D}_1 := \{\gamma \in \mathfrak{D} : \operatorname{supp}(\phi_\gamma) \cap \operatorname{supp}(f) \neq \emptyset\}$. Since $\mathfrak{D}$ is r-uniformly discrete and $\operatorname{supp}(\phi_\gamma) \subseteq B_r(\gamma)$, for all $\gamma \in \mathfrak{D}$, we have that $\mathfrak{D}_1$ has only a finite number of elements. This implies that both $f P_{\mathcal{H}}$, $P_{\mathcal{H}} f$ are finite-rank operators and are thus compact. To show that $P_{\mathcal{H}}$ has also finite propagation consider $f, g \in C_c(\mathbb{R}^d)$. Then we have

$$f P_{\mathcal{H}} g = \sum_{\gamma \in \mathfrak{D}_1 \cap \mathfrak{D}_2} |f \phi_\gamma\rangle\langle\phi_\gamma \bar{g}|,$$

where $\mathfrak{D}_2 := \{\gamma \in \mathfrak{D} : \operatorname{supp}(\phi_\gamma) \cap \operatorname{supp}(g) \neq \emptyset\}$. If $f P_{\mathcal{H}} g \neq 0$ then there exists $\gamma \in \mathfrak{D}_1 \cap \mathfrak{D}_2$. Specifically there exist $x \in B_r(\gamma) \cap \operatorname{supp}(f)$, $y \in B_r(\gamma) \cap \operatorname{supp}(g)$. Thus it holds that

$$\mathrm{d}(\operatorname{supp}(f), \operatorname{supp}(g)) \leq \|x - y\| \leq \|x - \gamma\| + \|\gamma - y\| \leq 2r.$$

This implies that $P_{\mathcal{H}}$ has finite propagation $2r$.

(ii). Consider the decomposition $\mathbb{R}^d = \mathbb{R}^d_+ \cup \mathbb{R}^d_-$, where $\mathbb{R}^d_\pm$ are the upper and lower half euclidean spaces. Then, by [28, Proposition 9.4], we have that $K_0(\mathbb{R}^d_+) = K_0(\mathbb{R}^d_-) = 0$, therefore we obtain the thesis using [17, Proposition 2.5]. □

Proof of Lemma 3.3 Let $x \in \mathbb{R}^d$, $R > 0$ and $0 < \varepsilon < \min\{1, r, R\}$. Then, for every $\gamma \in \mathfrak{D}$ such that $\|x - \gamma\| \geq R$ and for every $y \in B_\varepsilon(\gamma)$ we have that

$$\|x - y\| \geq \|x - \gamma\| - \|\gamma - y\| > R - \varepsilon, \quad 1 + \|x - \gamma\| \geq 1 + \|x - y\| - \varepsilon > 0. \tag{3.4}$$

Thus we can estimate the series in the following way

$$\begin{aligned} \sum_{\substack{\gamma \in \mathfrak{D} \\ \|x-\gamma\| \geq R}} \left(1 + \|x - \gamma\|^2\right)^{-s} &= \sum_{\substack{\gamma \in \mathfrak{D} \\ \|x-\gamma\| \geq R}} \frac{1}{|B_\varepsilon(\gamma)|} \int_{B_\varepsilon(\gamma)} \mathrm{d}y \left(1 + \|x - \gamma\|^2\right)^{-s} \\ &\leq \sum_{\substack{\gamma \in \mathfrak{D} \\ \|x-\gamma\| \geq R}} \frac{2^s}{|B_\varepsilon(\gamma)|} \int_{B_\varepsilon(\gamma)} \mathrm{d}y \, (1 + \|x - \gamma\|)^{-2s} \\ &\leq \frac{2^s}{c\,\varepsilon^d} \sum_{\substack{\gamma \in \mathfrak{D} \\ \|x-\gamma\| \geq R}} \int_{B_\varepsilon(\gamma)} \mathrm{d}y \, (1 + \|x - y\| - \varepsilon)^{-2s}, \end{aligned} \tag{3.5}$$

where in the first inequality we used $a^2 + b^2 \geq \frac{1}{2}(a+b)^2$ and in the last one we used the second inequality in (3.4). Now, since the set $\{B_r(\gamma)\}_{\gamma\in\mathfrak{D}}$ is composed of disjoint balls and because $\varepsilon < r$, we can estimate the series of integrals in (3.5) with the integral over all the space. Furthermore, because of the first inequality in (3.4), we can restrict the integration to the complement of the ball $B_{R-\varepsilon}(x)$. Thus we obtain

$$
\begin{aligned}
\sum_{\substack{\gamma\in\mathfrak{D}\\ \|x-\gamma\|\geq R}} \left(1+\|x-\gamma\|^2\right)^{-s} &\leq \frac{2^s}{c\,\varepsilon^d}\int_{\mathbb{R}^d\setminus B_{R-\varepsilon}(x)} \mathrm{d}y\,(1+\|x-y\|-\varepsilon)^{-2s} \\
&= \frac{2^s}{c\,\varepsilon^d}\int_{\mathbb{R}^d\setminus B_{R-\varepsilon}(x)} \mathrm{d}y\,(1+\|x-y\|)^{-2s}\left(1-\frac{\varepsilon}{1+\|x-y\|}\right)^{-2s} \\
&\leq \frac{2^s(1-\varepsilon)^{-2s}}{c\,\varepsilon^d}\int_{\mathbb{R}^d\setminus B_{R-\varepsilon}(x)} \mathrm{d}y\,(1+\|x-y\|)^{-2s} \\
&= \frac{2^s(1-\varepsilon)^{-2s}}{\hat{c}\,\varepsilon^d}\int_{R-\varepsilon}^{\infty} \mathrm{d}\rho\,\frac{\rho^{d-1}}{(1+\rho)^{2s}} \\
&\leq \frac{2^s(1-\varepsilon)^{-2s}}{\hat{c}\,\varepsilon^d(d-2s)}(1+R-\varepsilon)^{d-2s} = C(1+R)^{d-2s},
\end{aligned}
$$

where $C = \dfrac{2^s(1-\varepsilon)^{-2s}}{\hat{c}\,\varepsilon^d(d-2s)}\left(1-\dfrac{\varepsilon}{1+R}\right)^{d-2s}$. The proof of the lemma is concluded.□

Acknowledgements We are grateful to G. Marcelli and D. Monaco for interesting discussions on related topics. We gratefully acknowledge financial support from MUR Italian Ministry of University and Research and Next Generation EU (PRIN project 2022AKRC5P "Interacting Quantum Systems: Topological Phenomena and Effective Theories") and from the National Quantum Science and Technology Institute (PNRR MUR project PE0000023-NQSTI).

References

1. Avron, J.E., Seiler, R., Simon, B.: Charge deficiency, charge transport and comparison of dimensions. Comm. Math. Phys. **159**, 399–422 (1994)
2. Bellissard, J.: Ordinary quantum Hall effect and non-commutative cohomology. In: Weller, W., Zieche, P. (eds.), Localization in Disordered Systems. Teubner, Leipzig (1986)
3. Bourne, C., Mesland, B.: Localised module frames and Wannier bases from groupoid Morita equivalences. J. Fourier. Anal. Appl. **27**, 69 (2021). Revised version of: gabor frames and Wannier bases from groupoid Morita equivalence (2020). arXiv:2009.13806
4. Boutet De Monvel-Berthier, A., Nenciu, A., Nenciu, G.: Perturbed periodic Hamiltonians: essential spectrum and exponential decay of eigenfunctions. Lett. Maths. Phys. **34**, 119 (1995)
5. Ceresoli, D., Thonhauser, T., Vanderbilt, D., Resta, R.: Orbital magnetization in crystalline solids: multi-band insulators, Chern insulators, and metals. Phys. Rev. B **74**, 024408 (2006)

6. Ewert, E.E., Meyer, R.: Coarse geometry and topological phases. Commun. Math. Phys. **366**, 1069–1098 (2019)
7. Goedecker, S.: Linear scaling electronic structure methods. Rev. Mod. Phys. **71**, 1085–1111 (1999)
8. Higson, N., Roe, J., Yu, G.: A coarse Mayer-Vietoris principle. Math. Proc. Camb. Philos. Soc. **114**, 85–97 (1993)
9. Kato, T.: Perturbation Theory for Linear Operators, 2nd edn. Springer, Berlin (1995)
10. Kivelson, S.: Wannier functions in one-dimensional disordered systems: application to fractionally charged solitons. Phys. Rev. B **26**, 4269 (1982)
11. Kohn, W., Onffroy, J.R.: Wannier functions in a simple nonperiodic system. Phys. Rev. B **8**, 2485 (1973)
12. Kordyukov, Yu., Manuilov, V.: On topological obstructions to the existence of non-periodic Wannier bases. J. Math. Phys. **65**, 033506 (2024)
13. Kronig, R.. De., L., Penney, W.G.: Quantum mechanics of electrons in crystal lattices. Proc. R. Soc. Lond. A **130**, 499–513 (1931)
14. Lu, J., Stubbs, K.: Algebraic localization implies exponential localization in nonperiodic insulators (2021). arXiv:2101.02626
15. Lu, J., Stubbs, K.: Algebraic localization of Wannier functions implies Chern triviality in non-periodic insulators (2021). arXiv:2107.10699
16. Ludewig, L., Thiang, G.C.: Good Wannier bases in Hilbert modules associated to topological insulators. J. Math. Phys. **61**, 061902 (2020)
17. Ludewig, L., Thiang, G.C.: Large-scale geometry obstructs localization. J. Math. Phys. **63**, 091902 (2022)
18. Marcelli, G., Moscolari, M., Panati, G.: Localization of generalized Wannier bases implies Chern triviality in non-periodic insulators. Ann. Henri Poincaré **24**, 895- 930 (2023). Revised version of: Localization implies Chern triviality in non-periodic insulators (2020). arXiv:2012.14407
19. Marzari, N., Mostofi, A.A., Yates, J.R., Souza, I., Vanderbilt, D.: Maximally localized Wannier functions: theory and applications. Rev. Mod. Phys. **84**, 1419 (2012)
20. Monaco, D., Panati, G., Pisante, A., Teufel, S.: Optimal decay of Wannier functions in Chern and quantum Hall insulators. Commun. Math. Phys. **359**, 61 (2018)
21. Moscolari, M., Panati, G.: Symmetry and localization for magnetic Schrödinger operators: Landau levels, Gabor frames, and all that. Acta Appl. Math. **162**(1) (2019)
22. Moscolari, M., Panati, G.: Ultra-generalized Wannier bases: are they relevant to topological transport? J. Math. Phys. **64**, 071901 (2023)
23. Nenciu, A., Nenciu, G.: The existence of energy gaps for three-dimensional systems without long-range order. J. Phys. A. **14**, L109–L112 (1981)
24. Nenciu, A., Nenciu, G.: Existence of exponentially localized Wannier functions for nonperiodic systems. Commun. Math. Phys. **190**, 541–548 (1998)
25. Panati, G.: Triviality of Bloch and Bloch-Dirac bundles. Ann. Henri Poincaré **8**, 995–1011 (2007)
26. Pavelich, R.L., Marsiglio, F.: Calculation of 2D electronic band structure using matrix mechanics. Am. J. Phys. **84**, 924–935 (2016)
27. Prodan, E.: On the generalized Wannier functions. J. Math. Phys. **56**(11), 113511 (2015)
28. Roe, J.: Index theory, coarse geometry, and topology of manifolds. CBMS Regional Conf. Ser. Math. **90** (1996)

Hearing the Boundary Conditions of the One-Dimensional Dirac Operator

Giuliano Angelone

Abstract We study the isospectrality problem for a relativistic free quantum particle, described by the Dirac Hamiltonian, confined in a one-dimensional ring with a junction. We analyze all the self-adjoint extensions of the Hamiltonian in terms of the boundary conditions at the junction, characterizing the energy spectrum by means of a spectral function. By determining the symmetries of the latter, we are able to divide the self-adjoint extensions in two classes, identifying all the families of isospectral Hamiltonians, and thus completely characterizing the isospectrality problem.

Keywords Quantum boundary conditions · Self-adjoint extensions · Dirac operator · Isospectrality

1 Introduction

As is well known, in any bounded region $\Omega \subset \mathbb{R}^2$ the Laplacian with Dirichlet boundary conditions (BCs) has a discrete spectrum, say $\{\lambda_n\}_{n\in\mathbb{N}}$, which can in principle *always* be determined by solving the spectral problem

$$-\Delta u_n(x, y) = \lambda_n u_n(x, y), \qquad u_n(x, y)|_{\partial\Omega} = 0, \tag{1}$$

assuming that Ω and its boundary $\partial\Omega$ are sufficiently regular. This setting emerges as an idealized model of a drum, described as a two-dimensional vibrating membrane Ω whose boundary $\partial\Omega$ is kept fixed by a frame, so that the λ_n, the eigenvalues of the Laplacian, represent its normal frequencies. In a famous paper of 1966, Mark Kac popularized the related *inverse* spectral problem, which can be formulated as follows [1]: can we *uniquely* determine the region Ω just by knowing the full set

G. Angelone (✉)
Dipartimento di Fisica, Università di Bari, Bari, Italy
e-mail: giuliano.angelone@ba.infn.it

INFN, Sezione di Bari, Bari, Italy

B. Cassano et al. (eds.), *Singularities, Asymptotics, and Limiting Models*, Springer INdAM Series 64, https://doi.org/10.1007/978-981-96-3584-9_5

of the corresponding Laplace eigenvalues λ_n? Or, conversely, can we construct two *isospectral* and non-isometric regions Ω_1 and Ω_2, i.e. having different shapes but the same spectrum?

Remarkably, even after more than fifty years, this problem has been only partially solved [2, 3]. A first negative answer has been given in 1992 by C. Gordon et al. [4], who constructed an isospectral pair of two-dimensional polygons, depicted in Fig. 1. In the case of two-dimensional domains with a smooth boundary, a positive result can be recovered by requiring some additional symmetries, see e.g. the work of S. Zelditch [5], but the general problem is still unsolved. Besides, many related questions (which can generally fall under the umbrella of "isospectrality", or of inverse problems) are still unsolved, and object of research. Here we briefly mention some interesting topics, involving the isospectrality problem for regions with a fractal boundary [6], relativistic billiards [7], quantum graphs [8–12], photonic systems [13], and also the experimental construction of some isospectral regions [14, 15].

In this paper, we are interested in the following variation of the Kac's problem, formulated for a "quantum drum", that is *quantum billiard* consisting of a free quantum particle moving in a region of space Ω: which boundary conditions, for a given domain Ω, are associated with the same energy eigenvalues? In other words we fix the shape of the drum and we ask whether it is possible to distinguish different quantum boundary conditions just by looking at the energy spectrum of the particle. In [16], in particular, the authors studied the case of a non-relativistic particle (whose kinetic-energy is described by the free Schrödinger operator, i.e. the Laplacian) moving in a one-dimensional ring with a junction. This one-dimensional system could be implemented in a SQUID, a superconducting ring where different BCs can be obtained by a flux-loop tunable junction [17–19]. Here, we extend the analysis of [16] by considering a spin-$\frac{1}{2}$ *relativistic* particle, described by the free Dirac operator, moving on the same domain. Note that from a physical perspective this inverse spectral problem might have applications in quantum metrology with SQUIDs [20], atoms in cavities [21], and ions and atoms in magnetic traps [22].

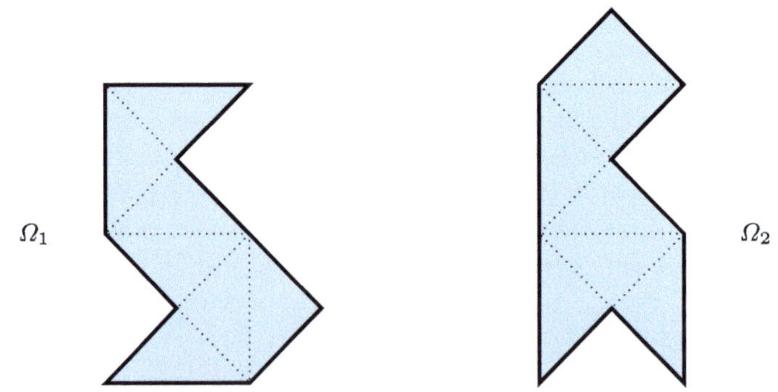

Fig. 1 A pair of isospectral polygons, constructed with seven half-square tiles differently arranged (the dotted lines are a guide for the eye), see [4]

The paper is organized as follows. In Sect. 2 we review the isospectrality problem for a non-relativistic particle, which can be analyzed by determining a certain spectral function. In Sect. 3 we then extend the analysis to a relativistic particle, drawing our conclusions in Sect. 4.

2 Hearing the Boundary Conditions of a Quantum Billiard

Let us consider a free non-relativistic quantum particle of mass m in a ring with a junction, as the one depicted in Fig. 2, idealizing the junction as a (generic) point interaction. This system, which is essentially one-dimensional, is described by the kinetic-energy operator

$$H = -\frac{\hbar^2}{2m}\frac{\mathrm{d}^2}{\mathrm{d}x^2} \tag{2}$$

that is the free Schrödinger operator, acting on the Hilbert space of square integrable functions $\mathcal{L}^2(-L/2, L/2)$, where $L > 0$ denotes the length of the ring. Equation (2) describes the action of H in the ring *away* from the junction. In order to generate a well-defined dynamics, the Hamiltonian H should be equipped with suitable BCs, which specify the behavior of the particle at the junction. In quantum mechanics the possible behaviors at the boundary, encoded in the domain $\mathfrak{D}(H)$ of H, cannot be arbitrary, but are constrained by a basic principle: H must be *self-adjoint*, i.e. $\mathfrak{D}(H) = \mathfrak{D}(H^\dagger)$ and $H = H^\dagger$. Indeed, self-adjointness is a necessary and sufficient condition for the operator to generate a unitary dynamics, and ensures a purely real spectrum.

Different domains correspond to different behaviors of the particle at the junction and thus give rise to different dynamics. All the self-adjoint realizations of the Hamiltonian (2) are known to be in one-to-one correspondence with the set of 2×2 unitary matrices $U \in \mathrm{U}(2)$ [23, 24]. Each of these realizations, which we henceforth denote by H_U, is defined on the domain

$$\mathfrak{D}(H_U) = \left\{\psi \in \mathcal{H}^2\left(-\tfrac{L}{2}, \tfrac{L}{2}\right) : \Psi_- = U\Psi_+\right\}, \tag{3}$$

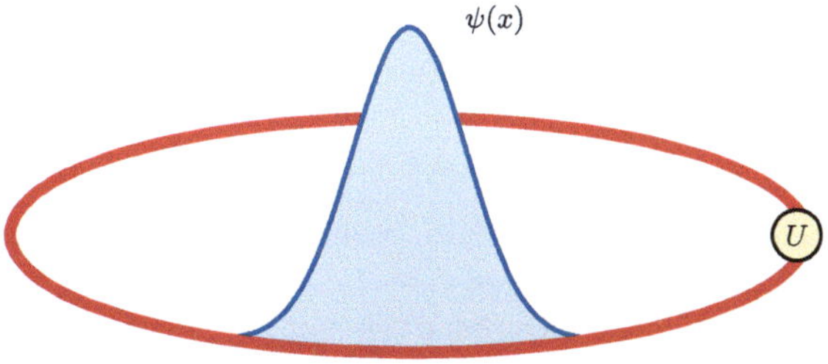

Fig. 2 A quantum particle in a one-dimensional ring with a junction: schematic diagram, where U parametrizes the boundary conditions at the junction, according to Eq. (5)

where $\mathcal{H}^2(\Omega)$ denotes the space of wavefunctions with square-integrable first and second derivative (i.e. the second Sobolev space) having support on Ω and

$$\Psi_{\pm} = \begin{pmatrix} -L_0\psi'(-\frac{L}{2}) \pm \psi(-\frac{L}{2}) \\ +L_0\psi'(+\frac{L}{2}) \pm \mathrm{i}\psi(+\frac{L}{2}) \end{pmatrix} \tag{4}$$

are the relevant boundary values of $\psi(x)$, with L_0 an arbitrary scale length which we henceforth take as $L_0 = L$ for convenience. In other words, to each unitary matrix U there corresponds the quantum BC

$$\Psi_- = U\Psi_+ , \qquad U \in \mathrm{U}(2) , \tag{5}$$

which in turn prescribes a linear relation between the vectors of boundary data Ψ_+ and Ψ_-. We mention that, at least in one-dimensional settings, there is a one-to-one mapping between self-adjoint BCs and self-adjoint (delta-like) point interactions, see e.g. [25]. The expression (5) admits many interesting BCs as particular cases, such as the Robin conditions

$$-\psi'(-\tfrac{L}{2}) = -\tfrac{1}{L}\cot(\tfrac{\alpha}{2})\psi(-\tfrac{L}{2}) , \qquad \psi'(\tfrac{L}{2}) = -\tfrac{1}{L}\cot(\tfrac{\alpha}{2})\psi(\tfrac{L}{2}) , \tag{6}$$

and the pseudo-periodic conditions

$$\psi(\tfrac{L}{2}) = \mathrm{e}^{\mathrm{i}\alpha}\psi(-\tfrac{L}{2}) , \qquad \psi'(\tfrac{L}{2}) = \mathrm{e}^{\mathrm{i}\alpha}\psi'(-\tfrac{L}{2}) , \tag{7}$$

obtained respectively with the unitary matrices

$$U_{\mathrm{R}}(\alpha) = \begin{pmatrix} \mathrm{e}^{\mathrm{i}\alpha} & 0 \\ 0 & \mathrm{e}^{\mathrm{i}\alpha} \end{pmatrix} \quad \text{and} \quad U_{\mathrm{pp}}(\alpha) = \begin{pmatrix} 0 & -\mathrm{e}^{-\mathrm{i}\alpha} \\ -\mathrm{e}^{\mathrm{i}\alpha} & 0 \end{pmatrix} \tag{8}$$

for each $\alpha \in [0, 2\pi)$. Before proceeding, let us observe that although the "shape" of the system is fixed, its topology is not, being determined by the BCs: while non-local BCs effectively describe a particle in a ring, local conditions (associated with a diagonal matrix U) model instead a particle confined in a interval, i.e. a one-dimensional box.

2.1 Isospectrality

The isospectrality problem for this one-dimensional system can be tackled by computing the *spectral function* $F_U(E)$, whose real zeroes, for each choice of U, characterize the spectrum $\sigma(H_U)$ of H_U:

$$\sigma(H_U) = \{E \in \mathbb{R} : F_U(E) = 0\} . \tag{9}$$

As it turns out, the spectral function of this system takes the form

$$F_U(E) = \det(U) - a(E)\,\mathrm{tr}(U) + b(E)\,\mathrm{tr}(U\sigma_x) + c(E) \tag{10}$$

where σ_x is the first Pauli matrix while $a(E)$, $b(E)$ and $c(E)$ are certain functions of the energy E obtained from the solution of the spectral problem, see [16] for details. In particular, the above spectral function depends on U only trough the quantities $\det(U)$, $\mathrm{tr}(U)$ and $\mathrm{tr}(U\sigma_x)$, which cannot completely characterize the four-dimensional space of parameters of a generic U(2) matrix U. This observation is reflected in a (boundary) symmetry of the spectral function: for any $\lambda \in \mathbb{R}$ we have that

$$F_U(E) = F_{U_\lambda}(E), \tag{11}$$

where

$$U_\lambda = \mathrm{e}^{\mathrm{i}\lambda\sigma_\mathrm{x}}\mathrm{U}\mathrm{e}^{-\mathrm{i}\lambda\sigma_\mathrm{x}}\,. \tag{12}$$

One can easily verify that the U(2) transformation $U \mapsto U_\lambda$ is indeed the most general one which simultaneously preserves the values of $\det(U)$, $\mathrm{tr}(U)$ and $\mathrm{tr}(U\sigma_x)$ appearing in Eq. (10).

Remarkably, the boundary symmetry (11) is ultimately due to a symmetry of the *bulk*, associated with the one-parameter unitary group

$$\mathcal{P}_\lambda = \mathrm{e}^{\mathrm{i}\lambda\mathrm{P}} = \cos(\lambda)\mathrm{I} + \mathrm{i}\sin(\lambda)\mathrm{P}\,, \qquad \lambda \in \mathbb{R}\,, \tag{13}$$

with I the 2×2 identity matrix, generated by the parity operator

$$P\colon \psi(x) \mapsto \psi(-x) \tag{14}$$

on $\mathcal{L}^2(-L/2, L/2)$. For each $\lambda \in \mathbb{R}$ the action of $\mathcal{P}_\lambda$ leaves formally invariant the Hamiltonian (2), that is the Laplacian, but changes its domain according to

$$\mathcal{P}_\lambda H_U \mathcal{P}_\lambda^\dagger = H_{U_\lambda}\,, \tag{15}$$

with U_λ given by Eq. (12). As a consequence of the above unitary equivalence, for each $\lambda \in \mathbb{R}$ we get the following *isospectral relation*:

$$\sigma(H_U) = \sigma\left(H_{\mathrm{e}^{\mathrm{i}\lambda\sigma_\mathrm{x}}\mathrm{U}\mathrm{e}^{-\mathrm{i}\lambda\sigma_\mathrm{x}}}\right). \tag{16}$$

Moreover, by solving Eq. (12) for $U = U_\lambda$, one can find the family of self-adjoint Hamiltonians which are actually symmetric under $\mathcal{P}_\lambda$ (or, equivalently, under P). Such operators belong to the $(\mathrm{U}(1) \times \mathrm{U}(1))/\mathbb{Z}_2$ subgroup of U(2) described by the matrices

$$U(\eta, \theta) = \mathrm{e}^{\mathrm{i}(\eta\mathrm{I}+\theta\sigma_\mathrm{x})} = \mathrm{e}^{\mathrm{i}\eta}\begin{pmatrix} \cos(\theta) & \mathrm{i}\sin(\theta) \\ \mathrm{i}\sin(\theta) & \cos(\theta) \end{pmatrix}, \tag{17}$$

with $\eta \in [0, \pi)$ and $\theta \in [0, 2\pi)$.

This result completes the characterization of the isospectrality problem for this system. The answer is only partially positive: we can indeed hear *some* BCs. As it turns out, there is a two-parameters family of BCs, and thus of Hamiltonians H_U, such that each element is uniquely associated with the corresponding spectrum $\sigma(H_U)$ and vice versa. These BCs, described by the matrices $U(\eta, \theta)$ defined in Eq. (17), are associated with parity-symmetric self-adjoint realizations. Note that, for example, the Robin conditions (6) belong to this family. Conversely, each BC U not belonging to the family (17) generates a U(1)-family of isospectral BCs $\{U_\lambda\}_{\lambda\in\mathbb{R}}$, with U_λ given by Eq. (12). The corresponding Hamiltonians, which are not symmetric under parity, are related by Eqs. (15) and (16) and thus they all "sound" the same. An example of this case is given by the U(1)-family of quasi-periodic conditions

$$\psi\left(\tfrac{L}{2}\right) = \mathrm{i}\cot\left(\tfrac{\alpha}{2} + \tfrac{\pi}{4}\right)\psi\left(-\tfrac{L}{2}\right), \qquad \psi'\left(\tfrac{L}{2}\right) = \mathrm{i}\tan\left(\tfrac{\alpha}{2} + \tfrac{\pi}{4}\right)\psi'\left(-\tfrac{L}{2}\right), \tag{18}$$

corresponding to the matrices

$$U_{\mathrm{qp}}(\alpha) = \begin{pmatrix} -\sin(\alpha) & \mathrm{i}\cos(\alpha) \\ -\mathrm{i}\cos(\alpha) & \sin(\alpha) \end{pmatrix}, \qquad \alpha \in [0, 2\pi), \tag{19}$$

and admitting as particular cases the pseudo-periodic BC with phase $\mathrm{e}^{\mathrm{i}\pi/2}$

$$\psi\left(\tfrac{L}{2}\right) = \mathrm{i}\psi\left(-\tfrac{L}{2}\right), \qquad \psi'\left(\tfrac{L}{2}\right) = \mathrm{i}\psi'\left(-\tfrac{L}{2}\right), \tag{20}$$

for $\alpha = 0$, and the mixed Neumann-Dirichlet BC

$$\psi'\left(-\tfrac{L}{2}\right) = 0, \qquad \psi\left(\tfrac{L}{2}\right) = 0, \tag{21}$$

for $\alpha = \pi/2$. Note in particular that

$$\mathrm{e}^{\mathrm{i}\lambda\sigma_x}\mathrm{U}_{\mathrm{qp}}(\alpha)\mathrm{e}^{-\mathrm{i}\lambda\sigma_x} = \mathrm{U}_{\mathrm{qp}}(\alpha - 2\lambda). \tag{22}$$

As expected, a standard direct computation shows that the above particular cases share indeed the same spectrum, which is given by

$$\sigma\left(H_{U_{\mathrm{qp}}(\alpha)}\right) = \left\{E_n = \frac{\hbar^2\pi^2}{2mL^2}\left(n + \frac{1}{2}\right)^2 : n \in \mathbb{N}\right\}. \tag{23}$$

3 A Relativistic Particle in a Ring with a Junction

We now consider a relativistic (spin-$\frac{1}{2}$) quantum particle, of mass m, free to move in the one-dimensional ring with a junction considered in the previous section. In this case, the relevant kinetic-energy operator is given by the one-dimensional free Dirac operator [26]

$$H_{\mathrm{D}} = -\mathrm{i}\hbar c\alpha \frac{\mathrm{d}}{\mathrm{dx}} + \mathrm{mc}^2\beta \tag{24}$$

acting on $\mathcal{L}^2(-L/2, L/2) \otimes \mathbb{C}^2$, where c is the speed of light while α and β are two 2×2 Hermitian matrices such that[1]

$$\alpha\beta + \beta\alpha = 0\,, \qquad\qquad \alpha^2 = \beta^2 = I\,. \tag{25}$$

Each particular choice of these two matrices is usually called a *representation* of the Dirac operator. From the physical perspective the latter is somewhat immaterial, since any two Dirac Hamiltonians, say H_{D} and $\tilde{H}_{\mathrm{D}}$, corresponding to two different sets of matrices $\{\alpha, \beta\}$ and $\{\tilde{\alpha}, \tilde{\beta}\}$, are related by the unitary transformation $\tilde{H}_{\mathrm{D}} = V H_{\mathrm{D}} V^\dagger$ where V is the unitary matrix such that [28]

$$V\alpha V^\dagger = \tilde{\alpha}\,, \qquad\qquad V\beta V^\dagger = \tilde{\beta}\,. \tag{26}$$

In any case, for definiteness we henceforth consider the Dirac (or standard) representation $\alpha = \sigma_x$ and $\beta = \sigma_z$, that is we fix the Dirac operator to be

$$H_{\mathrm{D}} = -\mathrm{i}\hbar c\sigma_{\mathrm{x}} \frac{\mathrm{d}}{\mathrm{dx}} + \mathrm{mc}^2\sigma_{\mathrm{z}}\,. \tag{27}$$

As explained in the previous section, we need to supply suitable BCs in order to consider a proper self-adjoint realization of the Dirac operator, see also [23, 29, 30]. In Appendix 5 we construct all the self-adjoint extensions of H_{D} by using a representation-independent approach based on the boundary triple machinery. As it turns out, also in the relativistic case the self-adjoint extensions of the Hamiltonian are in one-to-one correspondence with the unitary matrices $U \in \mathrm{U}(2)$, and we denote them by $H_{\mathrm{D},U}$. By further denoting the *spinorial components* of a generic wavefunction as

$$\Psi(x) = \begin{pmatrix}\phi(x) & \chi(x)\end{pmatrix}^{\mathsf{T}}, \tag{28}$$

the domain of $H_{\mathrm{D},U}$ can be written as

$$\mathfrak{D}(H_{\mathrm{D},U}) = \left\{\Psi = \begin{pmatrix}\phi & \chi\end{pmatrix}^{\mathsf{T}} \in \mathcal{H}^1\left(-\tfrac{L}{2}, \tfrac{L}{2}\right) \otimes \mathbb{C}^2 : \Psi_{\mathrm{D},-} = U\Psi_{\mathrm{D},+}\right\}, \tag{29}$$

where $\mathcal{H}^1(\Omega)$ is the first Sobolev space while the vectors $\Psi_{\mathrm{D},\pm}$ of boundary data are now given by

$$\Psi_{\mathrm{D},\pm} = \begin{pmatrix}\phi\left(-\frac{L}{2}\right) \mp \chi\left(-\frac{L}{2}\right) \\ \phi\left(+\frac{L}{2}\right) \pm \chi\left(+\frac{L}{2}\right)\end{pmatrix}, \tag{30}$$

[1] This is the relevant operator that is obtained from the ususal three-dimensional Dirac Hamiltonian by means of dimensional reduction, see e.g. [27].

so that the BCs encoded in $\mathfrak{D}(H_{D,U})$ explicitly read:

$$\begin{pmatrix} \phi(-\frac{L}{2}) + \chi(-\frac{L}{2}) \\ \phi(+\frac{L}{2}) - \chi(+\frac{L}{2}) \end{pmatrix} = U \begin{pmatrix} \phi(-\frac{L}{2}) - \chi(-\frac{L}{2}) \\ \phi(+\frac{L}{2}) + \chi(+\frac{L}{2}) \end{pmatrix}, \qquad U \in \mathrm{U}(2). \tag{31}$$

To make some examples, the diagonal matrix

$$U_{\mathrm{ch}}(\alpha) = \begin{pmatrix} \mathrm{e}^{\mathrm{i}\alpha} & 0 \\ 0 & \mathrm{e}^{\mathrm{i}\alpha} \end{pmatrix}, \qquad \alpha \in [0, 2\pi), \tag{32}$$

leads now to the local *chiral* BCs [31–33]

$$\chi(-\tfrac{L}{2}) = \mathrm{i}\tan(\tfrac{\alpha}{2})\phi(-\tfrac{L}{2}), \qquad \chi(\tfrac{L}{2}) = -\mathrm{i}\tan(\tfrac{\alpha}{2})\phi(\tfrac{L}{2}). \tag{33}$$

These conditions relate, on the boundary, states with negative chirality with those with positive chirality, the latter being associated with wavefunctions that are respectively eigenvectors of σ_x with eigenvalue -1 and $+1$.[2] The anti-diagonal matrix

$$\tilde{U}_{\mathrm{pp}} = \begin{pmatrix} 0 & \mathrm{e}^{-\mathrm{i}\alpha} \\ \mathrm{e}^{\mathrm{i}\alpha} & 0 \end{pmatrix}, \qquad \alpha \in [0, 2\pi), \tag{34}$$

give instead the non-local BCs

$$\phi(\tfrac{L}{2}) = \mathrm{e}^{\mathrm{i}\alpha}\phi(-\tfrac{L}{2}), \qquad \chi(\tfrac{L}{2}) = \mathrm{e}^{\mathrm{i}\alpha}\chi(-\tfrac{L}{2}). \tag{35}$$

In the following, after deriving in Sect. 3.1 the spectral function of $H_{D,U}$ by explicitly solving its eigenvalue equation, in Sect. 3.2 we discuss the isospectrality problem by determining the symmetries of the spectral function.

3.1 Spectral Problem

The eigenvalue equation $(H_{D,U} - E)\Psi(x) = 0$ consists of two coupled differential equations, given by

$$-\mathrm{i}\phi'(\mathrm{x};\epsilon) = (\epsilon + \epsilon_0)\chi(\mathrm{x};\epsilon), \qquad -\mathrm{i}\chi'(\mathrm{x};\epsilon) = (\epsilon - \epsilon_0)\phi(\mathrm{x};\epsilon), \tag{36}$$

where here $\epsilon = E/(\hbar c)$ and $\epsilon_0 = mc/\hbar$ denote the rescaled energy and rest (or mass) energy, respectively. The space of solutions for a given $\epsilon \in \mathbb{R}$ is two-dimensional, and the general solution can be written as

[2] The Pauli matrix σ_x plays here the role of the chiral gamma matrix γ^5 [27, 31].

$$\Psi(x;\epsilon) = c_1 \begin{pmatrix} \phi_1(x;\epsilon) \\ \chi_1(x;\epsilon) \end{pmatrix} + c_2 \begin{pmatrix} \phi_2(x;\epsilon) \\ \chi_2(x;\epsilon) \end{pmatrix}, \tag{37}$$

where $c_1, c_2 \in \mathbb{C}$. In order to solve Eq. (36) with the proper BCs, we note that by inserting the above solution in the expression (30) of the boundary values $\Psi_{\mathrm{D},\pm}$ and by setting

$$\Psi_{\mathrm{D},\pm} = A_\pm(\epsilon) \begin{pmatrix} c_1 \\ c_2 \end{pmatrix}, \tag{38}$$

where

$$A_\pm(\epsilon) = \begin{pmatrix} \phi_1(-\frac{L}{2};\epsilon) \mp \chi_1(-\frac{L}{2};\epsilon) & \phi_2(-\frac{L}{2};\epsilon) \mp \chi_2(-\frac{L}{2};\epsilon) \\ \phi_1(+\frac{L}{2};\epsilon) \pm \chi_1(+\frac{L}{2};\epsilon) & \phi_2(+\frac{L}{2};\epsilon) \pm \chi_2(+\frac{L}{2};\epsilon) \end{pmatrix}, \tag{39}$$

imposing the BCs (31) corresponds to requiring the vanishing of the following spectral function:

$$F_{\mathrm{D},U}(\epsilon) = \det\big(B(\epsilon) - U\big) = 0, \qquad B(\epsilon) = A_-(\epsilon) A_+^{-1}(\epsilon). \tag{40}$$

Thus, also the spectrum of the Dirac Hamiltonian can be obtained by finding the real zeros of the corresponding spectral function:

$$\sigma(H_{\mathrm{D},U}) = \left\{ E \in \mathbb{R} : F_{\mathrm{D},U}\left(\frac{E}{\hbar c}\right) = 0 \right\}. \tag{41}$$

Before finding the explicit solution of the spectral problem, let us note that we can manipulate the expression of $F_{U,\mathrm{D}}(\epsilon)$ by making use of the relation

$$\det(M - N) = \det(M) + \det(N) + \mathrm{tr}(MN) - \mathrm{tr}(M)\,\mathrm{tr}(N), \tag{42}$$

which is generally true for any pair of 2×2 matrices [34], obtaining that

$$F_{\mathrm{D},U}(\epsilon) = \det(B(\epsilon)) + \det(U) + \mathrm{tr}(B(\epsilon)U) - \mathrm{tr}(B(\epsilon))\,\mathrm{tr}(U). \tag{43}$$

The structure of the eigenvalue equation (36), and hence that of its solution (37), depends on the value of the wavenumber

$$k = k(\epsilon) = \mathrm{e}^{\mathrm{i}\arg(\epsilon^2 - \epsilon_0^2)/2} \sqrt{|\epsilon^2 - \epsilon_0^2|} \tag{44}$$

which is a real number outside of the mass gap, that is for $|\epsilon| \geq \epsilon_0$, and purely imaginary inside of the mass gap, i.e. for $\epsilon \in (-\epsilon_0, \epsilon_0)$. In particular, one should distinguish the analysis of the *mass modes*, that is the solutions with $\epsilon = \pm\epsilon_0$ (so that $k = 0$), from the other modes having a non-vanishing wavenumber. When $\epsilon = \epsilon_0$ or $\epsilon = -\epsilon_0$ we get, respectively, the coupled equations

$$\begin{cases} -\mathrm{i}\phi'(\mathrm{x};\epsilon_0) = 2\epsilon_0\chi(\mathrm{x};\epsilon_0) \\ -\mathrm{i}\chi'(\mathrm{x};\epsilon_0) = 0 \end{cases}, \quad \begin{cases} -\mathrm{i}\phi'(\mathrm{x};-\epsilon_0) = 0 \\ -\mathrm{i}\chi'(\mathrm{x};-\epsilon_0) = -2\epsilon_0\phi(\mathrm{x};-\epsilon_0) \end{cases}. \tag{45}$$

The above systems have, respectively, the following solutions

$$\Psi(x;\epsilon_0) = \begin{pmatrix} c_1 + c_2 x/L \\ -\mathrm{i}c_2/(2\epsilon_0\mathrm{L}) \end{pmatrix}, \qquad \Psi(x;-\epsilon_0) = \begin{pmatrix} \mathrm{i}c_2/(2\epsilon_0\mathrm{L}) \\ c_1 + c_2 x/L \end{pmatrix}, \tag{46}$$

from which we obtain the matrices

$$A_\pm(\epsilon_0) = \begin{pmatrix} 1 & -\frac{1}{2}\left(1 \mp \frac{\mathrm{i}}{\epsilon_0 L}\right) \\ 1 & +\frac{1}{2}\left(1 \mp \frac{\mathrm{i}}{\epsilon_0 L}\right) \end{pmatrix}, \qquad A_\pm(-\epsilon_0) = \begin{pmatrix} \mp 1 & \frac{1}{2}\left(\frac{\mathrm{i}}{\epsilon_0 L} \pm 1\right) \\ \pm 1 & \frac{1}{2}\left(\frac{\mathrm{i}}{\epsilon_0 L} \pm 1\right) \end{pmatrix}. \tag{47}$$

Note that each of the above matrices is invertible, thus we compute

$$B(\pm\epsilon_0) = \pm\frac{1}{\epsilon_0 L \mp \mathrm{i}} \begin{pmatrix} \epsilon_0 L & -\mathrm{i} \\ -\mathrm{i} & \epsilon_0 L \end{pmatrix}. \tag{48}$$

When $\epsilon \neq \pm\epsilon_0$, instead, we can rearrange the equations (36) as

$$\phi''(x;\epsilon) = -k^2\phi(x;\epsilon), \qquad \chi(x;\epsilon) = \frac{-\mathrm{i}}{\epsilon+\epsilon_0}\phi'(x;\epsilon), \tag{49}$$

from which we readily obtain the general solution

$$\Psi(x;\epsilon) = \begin{pmatrix} c_1 \mathrm{e}^{\mathrm{ikx}} + \mathrm{c}_2 \mathrm{e}^{-\mathrm{ikx}} \\ \sqrt{\frac{\epsilon-\epsilon_0}{\epsilon+\epsilon_0}}\left(c_1 \mathrm{e}^{\mathrm{ikx}} - \mathrm{c}_2 \mathrm{e}^{-\mathrm{ikx}}\right) \end{pmatrix}, \tag{50}$$

and the corresponding matrices

$$A_\pm(\epsilon) = \begin{pmatrix} \mathrm{e}^{-\mathrm{ikL}/2}\left(1 \mp \sqrt{\frac{\epsilon-\epsilon_0}{\epsilon+\epsilon_0}}\right) & \mathrm{e}^{+\mathrm{ikL}/2}\left(1 \pm \sqrt{\frac{\epsilon-\epsilon_0}{\epsilon+\epsilon_0}}\right) \\ \mathrm{e}^{+\mathrm{ikL}/2}\left(1 \pm \sqrt{\frac{\epsilon-\epsilon_0}{\epsilon+\epsilon_0}}\right) & \mathrm{e}^{-\mathrm{ikL}/2}\left(1 \mp \sqrt{\frac{\epsilon-\epsilon_0}{\epsilon+\epsilon_0}}\right) \end{pmatrix}. \tag{51}$$

As the reader can check by inspecting

$$\det\left(A_\pm(\epsilon)\right) = \frac{-4\mathrm{i}}{\epsilon+\epsilon_0}\left[\epsilon \sin\left(kL\right) \mp \mathrm{ik}\cos\left(\mathrm{kL}\right)\right], \tag{52}$$

both $A_\pm(\epsilon)$ are invertible if and only if $\epsilon \neq \pm\epsilon_0$. Under such assumption we can thus compute

$$B(\epsilon) = a_{\mathrm{D}}(\epsilon)I + b_{\mathrm{D}}(\epsilon)\sigma_x\,, \tag{53}$$

where we defined

$$a_{\mathrm{D}}(\epsilon) = \frac{\epsilon_0 \sin(kL)}{\epsilon \sin(kL) - \mathrm{ik}\cos(\mathrm{kL})}, \tag{54}$$

$$b_{\mathrm{D}}(\epsilon) = \frac{-\mathrm{ik}}{\epsilon \sin(kL) - \mathrm{ik}\cos(\mathrm{kL})}. \tag{55}$$

Similarly to the non-relativistic case [16], although the limit matrices

$$\lim_{\epsilon \to +\epsilon_0} A_{\pm}(\epsilon) = \begin{pmatrix} 1 & 1 \\ 1 & 1 \end{pmatrix}, \tag{56}$$

$$\lim_{\epsilon \to -\epsilon_0^-} A_{\pm}(\epsilon) = \begin{pmatrix} \mp\infty & \pm\infty \\ \pm\infty & \mp\infty \end{pmatrix} = \lim_{\epsilon \to -\epsilon_0^+} \mathrm{i}A_{\pm}(\epsilon), \tag{57}$$

are clearly not invertible and a fortiori different from the matrices $A_{\pm}(\epsilon_0)$ and $A_{\pm}(-\epsilon_0)$, the other two limits

$$\lim_{\epsilon \to \pm\epsilon_0} B(\epsilon) = B(\pm\epsilon_0) \tag{58}$$

coincide with the corresponding quantities obtained for the mass modes.

3.2 Symmetries of the Spectral Function

At this point, by inserting the explicit expression (53) of $B(\epsilon)$ in the formula (43) of $F_{\mathrm{D},U}(\epsilon)$ we obtain the final result

$$F_{\mathrm{D},U}(\epsilon) = \det(U) - a_{\mathrm{D}}(\epsilon)\,\mathrm{tr}(U) + b_{\mathrm{D}}(\epsilon)\,\mathrm{tr}(U\sigma_x) + c_{\mathrm{D}}(\epsilon) \tag{59}$$

where

$$c_{\mathrm{D}}(\epsilon) = \det(B(\epsilon)) = a_{\mathrm{D}}^2(\epsilon) - b_{\mathrm{D}}^2(\epsilon) = \frac{\epsilon \sin(kL) + \mathrm{ik}\cos(\mathrm{kL})}{\epsilon \sin(kL) - \mathrm{ik}\cos(\mathrm{kL})}. \tag{60}$$

As the reader can see, the isospectrality problem of the Dirac Hamiltonian mirrors that of the Schrödinger Hamiltonian: also in this case the spectral function, and hence the spectrum, depends on U only through $\det(U)$, $\mathrm{tr}(U)$ and $\mathrm{tr}(U\sigma_x)$. The emerging isospectrality can be explained in terms of a symmetry of the Dirac Hamiltonian, analogously to the non-relativistic case. Here, the role of parity is played by the operator

$$\sigma_z P : \begin{pmatrix} \phi(x) \\ \chi(x) \end{pmatrix} \mapsto \begin{pmatrix} \phi(-x) \\ -\chi(-x) \end{pmatrix}, \tag{61}$$

which generates the one-parameter unitary group

$$\mathcal{P}_{\mathrm{D},\lambda} = \mathrm{e}^{\mathrm{i}\lambda\sigma_z \mathrm{P}} = \cos(\lambda)\mathrm{I} + \mathrm{i}\sin(\lambda)\sigma_z \mathrm{P}\,, \qquad \lambda \in \mathbb{R}\,. \tag{62}$$

Although the latter leaves formally invariant the Dirac Hamiltonian, it changes its domain according to

$$\mathcal{P}_\lambda H_{\mathrm{D},U} \mathcal{P}_\lambda^\dagger = H_{\mathrm{D},U_\lambda}\,, \tag{63}$$

with U_λ given by Eq. (12), namely

$$U_\lambda = \mathrm{e}^{\mathrm{i}\lambda\sigma_\mathrm{x}} \mathrm{U} \mathrm{e}^{-\mathrm{i}\lambda\sigma_\mathrm{x}}\,. \tag{64}$$

As a consequence, for each $\lambda \in \mathbb{R}$ we get the following isospectral relation,

$$\sigma(H_{\mathrm{D},U}) = \sigma(H_{\mathrm{D},\mathrm{e}^{\mathrm{i}\lambda\sigma_\mathrm{x}}\mathrm{U}\mathrm{e}^{-\mathrm{i}\lambda\sigma_\mathrm{x}}})\,, \tag{65}$$

which fully characterize the isospectrality problem for the Dirac Hamiltonian. In particular, the BCs that can be heard are the same, in terms of the unitary matrices U to which they correspond, that one finds in the non-relativistic case:

$$U(\eta,\theta) = \mathrm{e}^{\mathrm{i}(\eta\mathrm{I}+\theta\sigma_\mathrm{x})} = \mathrm{e}^{\mathrm{i}\eta}\begin{pmatrix}\cos(\theta) & \mathrm{i}\sin(\theta)\\ \mathrm{i}\sin(\theta) & \cos(\theta)\end{pmatrix}\,, \tag{66}$$

with $\eta \in [0,\pi)$ and $\theta \in [0, 2\pi)$, compare with Eq. (17).

To complete this section, we make two observations. The first one concerns the mass modes. By parametrizing $U \in \mathrm{U}(2)$ as

$$U = \mathrm{e}^{\mathrm{i}\eta}(\mathrm{m}_0\mathrm{I} + \mathrm{i}\mathbf{m}\cdot\boldsymbol{\sigma}) = \mathrm{e}^{\mathrm{i}\eta}\begin{pmatrix} m_0 + \mathrm{i}m_3 & m_2 + \mathrm{i}m_1\\ -m_2 + \mathrm{i}m_1 & m_0 - \mathrm{i}m_3\end{pmatrix}\,, \tag{67}$$

where $\boldsymbol{\sigma} = (\sigma_x, \sigma_y, \sigma_z)$, $\eta \in [0,\pi)$ while the four real parameters m_0 and $\mathbf{m} = (m_1, m_2, m_3)$ are subjected to the $\mathbb{S}^3$ constraint $m_0^2 + |\mathbf{m}|^2 = 1$, we have that

$$\det(U) = \mathrm{e}^{\mathrm{i}2\eta}\,, \qquad \mathrm{tr}(U) = 2\,\mathrm{e}^{\mathrm{i}\eta}\mathrm{m}_0\,, \qquad \text{and} \qquad \mathrm{tr}(U\sigma_x) = 2\mathrm{i}\mathrm{e}^{\mathrm{i}\eta}\mathrm{m}_1\,. \tag{68}$$

Accordingly, we find that the mass modes $\epsilon = \pm\epsilon_0$ belong to the spectrum $\sigma(H_{\mathrm{D},U})$ if and only if, respectively,

$$m_1 + \sin(\eta) = \epsilon_0 L\big(m_0 \mp \cos(\eta)\big)\,. \tag{69}$$

The second observation regards instead the relativistic spectral space, that is the set of *distinct* spectra

$$\Sigma_\mathrm{D} = \{\sigma(H_{\mathrm{D},U}) \,:\, U \in \mathrm{U}(2)\}\,. \tag{70}$$

We mention that, exactly as it happens in the non-relativistic case, Σ_D is given by a twisted solid torus, inheriting thus the non-trivial topology of the group U(2), see [16] for a detailed discussion. Remarkably, the non-trivial topology of Σ_D signals the presence of spectral anholonomies and geometric phases [35, 36].

4 Conclusions

In the previous section we completely characterized the isospectrality problem for the Dirac Hamiltonian on the ring with arbitrary self-adjoint BCs. In particular, we determined all the BCs which can be heard, being uniquely associated with the spectrum of the corresponding Hamiltonian. These turned out to define all the self-adjoint extensions which are symmetric under the parity operator introduced in Eq. (61), and they are given by the two-parameter family (66). All the extensions which are not symmetric under parity, instead, are associated with a U(1) family of isospectral (actually, unitarily equivalent) Dirac Hamiltonians, according to Eq. (63). Remarkably, this result exactly mirrors the one that has been found for the non-relativistic Schrödinger Hamiltonian in [16]. This may be a general feature related to the fact that the Dirac Hamiltonian "reduces" to the Schrödinger Hamiltonian in the non-relativistic limit, see e.g.[26, 29]. Note however that analogous systems with a different domain Ω present different isospectrality structures [16, 36].

An interesting question, which is still open, is if it is possible to discern the various energy spectra belonging to the same isospectrality family by using other physical observables (that is, other self-adjoint operators) beside the Hamiltonian. We may investigate this problem in the future, both in the non-relativistic and the relativistic setting. Another interesting extension of our analysis would be to consider higher dimensional drums $\Omega \subset \mathbb{R}^n$ with $n > 1$. In this case, indeed, the defect indices of the Hamiltonian are typically infinite, and the unitary matrix characterizing its self-adjoint extensions is actually a unitary operator acting on a infinite-dimensional Hilbert space.

Acknowledgements We thank Professors Paolo Facchi and Giuseppe Marmo for useful discussions. This work was partially supported by Istituto Nazionale di Fisica Nucleare (INFN) through the project "QUANTUM", by Regione Puglia and QuantERA ERA-NET Cofund in Quantum Technologies (Grant No. 731473), project PACE-IN, and by the Italian National Group of Mathematical Physics (GNFM-INdAM).

5 Boundary Conditions for the Dirac Hamiltonian

In this Appendix we derive all the possible BCs which render the Dirac Hamiltonian

$$H_{\mathrm{D}} = -\mathrm{i}\hbar c\alpha \frac{\mathrm{d}}{\mathrm{dx}} + \mathrm{mc}^2\beta \tag{71}$$

a self-adjoint operator, for an arbitrary pair of 2×2 Hermitian matrices α and β satisfying the properties (25). We will adopt the standard procedure related to the boundary triple formalism, see e.g. [37–39], also partially following the construction of [23, 40]. When defined on $C_0^\infty(-L/2, L/2) \otimes \mathbb{C}^2$, H_{D} is symmetric but not self-adjoint, $H_{\mathrm{D}}^\dagger$ being defined on $\mathcal{H}^1(-L/2, L/2) \otimes \mathbb{C}^2$. An easy calculation shows that its deficiency indices $n_\pm = \dim\ker(H_{\mathrm{D}}^\dagger \pm \mathrm{i})$ are both $n_\pm = 2$. To see this, one can work in a specific representation (such as the Dirac one $\alpha = \sigma_x$ and $\beta = \sigma_z$) to compute the indices, and then invoke the unitary equivalence (26) to come back to the general case. As a consequence, the self-adjoint extensions of H_{D} can be parametrized by a 2×2 unitary matrix $U \in \mathrm{U}(2)$, and will be henceforth denoted by $H_{\mathrm{D},U}$. Moreover, by setting

$$\Psi_1(x) = \begin{pmatrix} \phi_1(x) \\ \chi_1(x) \end{pmatrix}, \qquad \Psi_2(x) = \begin{pmatrix} \phi_2(x) \\ \chi_2(x) \end{pmatrix}, \tag{72}$$

we can express the boundary form of H_{D} in the following way:

$$\Lambda(\Psi_1, \Psi_2) = \langle H_{\mathrm{D}}^\dagger \Psi_1 | \Psi_2 \rangle_{\mathcal{H}} - \langle \Psi_1 | H_{\mathrm{D}}^\dagger \Psi_2 \rangle_{\mathcal{H}} \tag{73}$$

$$= \mathrm{i}\hbar \mathrm{c} \int_{-\mathrm{L}/2}^{+\mathrm{L}/2} \langle'_1(\mathrm{x})|\alpha_2(\mathrm{x})\rangle_{\mathbb{C}^2} + \langle_1(\mathrm{x})|\alpha'_2(\mathrm{x})\rangle_{\mathbb{C}^2}\,\mathrm{dx} \tag{74}$$

$$= \mathrm{i}\hbar \mathrm{c} \sum_{\mathrm{s}=\pm\frac{\mathrm{L}}{2}} \langle_1(\mathrm{s})|\alpha_{\mathrm{n}}(\mathrm{s})_2(\mathrm{s})\rangle_{\mathbb{C}^2}\,, \tag{75}$$

where $\mathcal{H} = \mathcal{L}^2(-L/2, L/2) \otimes \mathbb{C}^2$ and $\alpha_{\mathrm{n}}(\pm L/2) = \pm\alpha$. Since $\alpha^2 = I$, we have the spectral decomposition

$$\alpha_{\mathrm{n}}(s) = |e_+(s)\rangle\langle e_+(s)| - |e_-(s)\rangle\langle e_-(s)| \tag{76}$$

where the (normalized) eigenvectors $e_\pm(s) \in \mathbb{C}^2$ satisfy

$$\alpha_{\mathrm{n}}(s) e_\pm(s) = \pm e_\pm(s)\,. \tag{77}$$

At this point we define the maps

$$\Gamma_\pm \colon \mathcal{H}^1\big(-\tfrac{L}{2},\tfrac{L}{2}\big)\otimes\mathbb{C}^2 \to \mathbb{C}^2\,, \qquad \Psi(x)\mapsto \begin{pmatrix} \big\langle e_\pm\big(-\tfrac{L}{2}\big)\big|\Psi\big(-\tfrac{L}{2}\big)\big\rangle_{\mathbb{C}^2} \\ \big\langle e_\pm\big(+\tfrac{L}{2}\big)\big|\Psi\big(+\tfrac{L}{2}\big)\big\rangle_{\mathbb{C}^2} \end{pmatrix}. \tag{78}$$

which, since $e_\pm(-L/2) = e_\mp(+L/2) = e_\mp$, simplify to

$$\Gamma_\pm\Psi(x) = \begin{pmatrix} \big\langle e_\mp\big|\Psi\big(-\tfrac{L}{2}\big)\big\rangle_{\mathbb{C}^2} \\ \big\langle e_\pm\big|\Psi\big(+\tfrac{L}{2}\big)\big\rangle_{\mathbb{C}^2} \end{pmatrix}. \tag{79}$$

A direct calculation shows then that

$$\Lambda(\Psi_1,\Psi_2)/a = \langle\Gamma_-\Psi_1|\Gamma_-\Psi_2\rangle_{\mathbb{C}^2} - \langle\Gamma_+\Psi_1|\Gamma_+\Psi_2\rangle_{\mathbb{C}^2}\,, \qquad a = -\mathrm{i}\hbar c\,, \tag{80}$$

i.e. that $(\mathbb{C}^2,\Gamma_-,\Gamma_+)$ is a boundary triple for the Dirac Hamiltonian H_{D}, in the sense of Definition 7.1.11 of [38]. We can then conclude, by e.g. Theorem 7.1.13 of [38], that for each $U\in\mathrm{U}(2)$ the self-adjoint extension $H_{\mathrm{D},U}$ is defined in the domain

$$\mathfrak{D}(H_{\mathrm{D},U}) = \{\Psi\in\mathcal{H}^1\big(-\tfrac{L}{2},\tfrac{L}{2}\big)\otimes\mathbb{C}^2 : \Gamma_-\Psi = U\Gamma_+\Psi\}\,. \tag{81}$$

The relation $\Gamma_-\Psi = U\Gamma_+\Psi$ represents thus an α-depending BC, expressed explicitly in terms of the eigenvectors of α by Eq. (79). In particular, for $\alpha=\sigma_x$ we have the eigenvectors $e_\pm = \frac{1}{\sqrt{2}}\,(1\ \pm 1)^{\mathsf{T}}$, so that the quantities

$$\Gamma_\pm\begin{pmatrix}\phi(x)\\ \chi(x)\end{pmatrix} = \frac{1}{\sqrt{2}}\begin{pmatrix}\phi\big(-\tfrac{L}{2}\big)\mp\chi\big(-\tfrac{L}{2}\big)\\ \phi\big(+\tfrac{L}{2}\big)\pm\chi\big(+\tfrac{L}{2}\big)\end{pmatrix} \tag{82}$$

correspond, beside a constant multiplicative factor, to the boundary data vectors $\Psi_{\mathrm{D},\pm}$ introduced in Eq. (30).

References

1. Kac, M.: Can one hear the shape of a drum? Am. Math. Monthly **73**, 1 (1966)
2. Giraud, O., Thas, K.: Hearing shapes of drums: Mathematical and physical aspects of isospectrality. Rev. Mod. Phys. **82**, 2213–2255 (2010)
3. Zelditch, S.: The inverse spectral problem. Surv. Diff. Geom. **9**, 401–467 (2004)
4. Gordon, C., Webb, D., Wolpert, S.: Isospectral plane domains and surfaces via Riemannian orbifolds. Invent. Math. **110**, 1–22 (1992)
5. Zelditch, S.: Spectral determination of analytic bi-axisymmetric plan domains. Geom. Funct. Anal. **10**, 50 (2000)
6. Brossard, J., Carmona, R.: Can one hear the dimension of a fractal? Commun. Math. Phys. **104**, 103–122 (1986)
7. Yu, P., Dietz, B., Xu, H.-Y., Ying, L., Huang, L., Lai, Y.-C.: Kac's isospectrality question revisited in neutrino billiards. Phys. Rev. E **101**, 032215 (2020)
8. Gutkin, B., Smilansky, U.: Can one hear the shape of a graph? J. Phys. A: Math. Gen. **34**, 6061–6068 (2001)

9. Rueckriemen, R.: Quasi-isospectrality on quantum graphs. J. Geom. Anal. **25**, 306–316 (2015)
10. Ławniczak, M., Kurasov, P., Bauch, S., Białous, M., Yunko, V., Sirko, L.: Hearing Euler characteristic of graphs. Phys. Rev. E **101**, 052320 (2020)
11. Ławniczak, M., Kurasov, P., Bauch, S., Białous, M., Akhshani, A., Sirko, L.: A new spectral invariant for quantum graphs. Sci. Rep. **11**, 15342 (2021)
12. Fabila-Carrasco, J.S., Lledó, F., Post, O.: A geometric construction of isospectral magnetic graphs. Anal. Math. Phys. **13**, 64 (2023)
13. Park, S., Lee, I., Kim, J., Park, N., Yu, S.: Hearing the shape of a drum for light: isospectrality in photonics. Nanophotonics **11**, 11 (2022)
14. Sridhar, S., Kudrolli, A.: Experiments on not "hearing the shape" of drums. Phys. Rev. Lett. **72**, 2175–2178 (1994)
15. Even, C., Pieranski, P.: On "hearing the shape of drumsâŁž: An experimental study using vibrating smectic films. Europhys. Lett. **47**, 531–537 (1999)
16. Angelone, G., Facchi, P., Marmo, G.: Hearing the shape of a quantum boundary condition. Mod. Phys. Lett. A **37**, 2250114 (2022)
17. Vion, D., Aassime, A., Cottet, A., Joyez, P., Pothier, H., Urbina, C., Esteve, D., Devoret, M.H.: Manipulating the quantum state of an electrical circuit. Science **296**, 886–889 (2002)
18. Poletto, S., Chiarello, F., Castellano, M.G., Lisenfeld, J., Lukashenko, A., Cosmelli, C., Torrioli, G., Carelli, P., Ustinov, A.V.: Coherent oscillations in a superconducting tunable flux qubit manipulated without microwaves. New J. Phys. **11**, 013009 (2009)
19. Paauw, F.G., Fedorov, A., Harmans, C., Mooij, J.E.: Tuning the gap of a superconducting flux qubit. Phys. Rev. Lett. **102**, 090501 (2009)
20. Friedman, J.R., Patel, V., Chen, W., Tolpygo, S.K., Lukens, J.E.: Quantum superposition of distinct macroscopic states. Nature **406**, 43–46 (2000)
21. Haroche, S., Raimond, J.-M.: Exploring the Quantum: Atoms, Cavities, and Photons. Oxford University Press (2006)
22. Paul, W.: Electromagnetic traps for charged and neutral particles. Rev. Mod. Phys. **62**, 531–540 (1990)
23. Asorey, M., Ibort, A., Marmo, G.: The topology and geometry of selfadjoint and elliptic boundary conditions for Dirac and Laplace operators. Int. J. Geom. Methods Mod. Phys. **12**, 1561007 (2015)
24. Asorey, M., Ibort, A., Marmo, G.: Global theory of quantum boundary conditions and topology change. Int. J. Mod. Phys. A **20**, 1001–1025 (2005)
25. Kurasov, P.: Distribution theory for discontinuous test functions and differential operators with generalized coefficients. J. Math. Anal. Appl. **201**, 297 (1996)
26. Thaller, B.: The Dirac Equation. Springer, Berlin, Heidelberg (1992)
27. Angelone, G., Ercolessi, E., Facchi, P., Lonigro, D., Maggi, R., Marmo, G., Pascazio, S., Pepe, F.V.: Dimensional reduction of the Dirac theory. J. Phys. A: Math. Theor. **56**, 065201 (2023)
28. Good, R.H.: Properties of the Dirac matrices. Rev. Mod. Phys. **27**, 187 (1955)
29. Alonso, V., De Vincenzo, S.: General boundary conditions for a Dirac particle in a box and their non-relativistic limits. J. Phys. A: Math. Gen. **30**, 8573–8585 (1997)
30. Beneventano, C.G., Santangelo, E.M., Wipf, A.: Spectral asymmetry for bag boundary conditions. J. Phys. A: Math. Gen. **35**, 9343–9354 (2002)
31. Jaffe, R.L., Manohar, A.: Bound states of the Dirac equation outside a hard sphere. Ann. Phys. **192**(2), 321–330 (1989)
32. Hosaka, A., Toki, H.: Chiral bag model for the nucleon. Phys. Rep. **277**(2–3), 65–188 (1996)
33. Rohim, A., Yamamoto, K.: Effects of chiral MIT boundary conditions for a Dirac particle in a box. Prog. Theor. Exp. Phys. **2021**(11), 113B01 (2021)
34. Bonneau, G., Faraut, J., Valent, G.: Self-adjoint extensions of operators and the teaching of quantum mechanics. Am. J. Phys. **69**, 322–331 (2001)
35. Shapere, A., Wilczek, F., eds.: Geometric Phases in Physics. World Scientific (1989)
36. Tsutsui, I., Fülöp, T., Cheon, T.: Möbius structure of the spectral space of Schrödinger operators with point interaction. J. Math. Phys. **42**, 5687–5697 (2001)

37. Brüning, J., Geyler, V., Pankrashkin, K.: Spectra of self-adjoint extensions and applications to solvable Schrödinger operators. Rev. Math. Phys. **20**, 1–70 (2008)
38. de Oliveira, C.R.: Intermediate spectral theory and quantum dynamics. Progress in Mathematical Physics v. 54, Birkhäuser (2009)
39. Schmüdgen, K.: Unbounded Self-Adjoint Operators on Hilbert Space. Springer, Netherlands (2012)
40. Asorey, M., Balachandran, A.P., Pérez-Pardo, J.M.: Edge states: topological insulators, superconductors and QCD chiral bags. J. High Energ. Phys. **2013**, 73 (2013)

Bosonized Momentum Distribution of a Fermi Gas via Friedrichs Diagrams

Sascha Lill

Abstract Recently [1], Benedikter and the author proved an approximate formula for the momentum distribution of a 3d fermionic gas interacting by a short-range pair potential in the mean-field regime, within a trial state close to the ground state. Here, we derive an exact formula for the momentum distribution in this trial state, using a diagrammatic formalism due to Friedrichs. We further demonstrate how the formula of Benedikter and the author arises from a restriction of the contributing diagrams to those corresponding to a bosonization approximation.

Keywords Fermi gas · Many-body quantum mechanics · Momentum distribution · Bosonization · Friedrichs diagrams · Diagrammatic expansions

2020 Mathematics Subject Classification: 81V70 · 81V74 · 81Q12 · 81S05

1 Introduction

In recent years, there has been a surge of interest in the mathematical research on three-dimensional (3d) fermionic quantum gases, employing bosonization techniques. Generally, such a system comprises N fermions on a 3d torus, $\mathbb{T}^3 := [0, L]^3$, whose state is described by a vector

$$\begin{aligned} &\psi \in L^2_{\mathrm{a}}(\mathbb{T}^{3N}) := \\ &\left\{\psi \in L^2(\mathbb{T}^{3N}) \;\middle|\; \psi(\ldots, x_i, \ldots, x_j, \ldots) = -\psi(\ldots, x_j, \ldots, x_i, \ldots)\ \forall 1 \leq i < j \leq n\right\}. \end{aligned} \tag{1}$$

S. Lill (✉)
Dipartimento di Matematica Federigo Enriques, Università degli Studi di Milano, Milan, Italy
e-mail: sascha.lill@unimi.it

B. Cassano et al. (eds.), *Singularities, Asymptotics, and Limiting Models*, Springer INdAM Series 64, https://doi.org/10.1007/978-981-96-3584-9_6

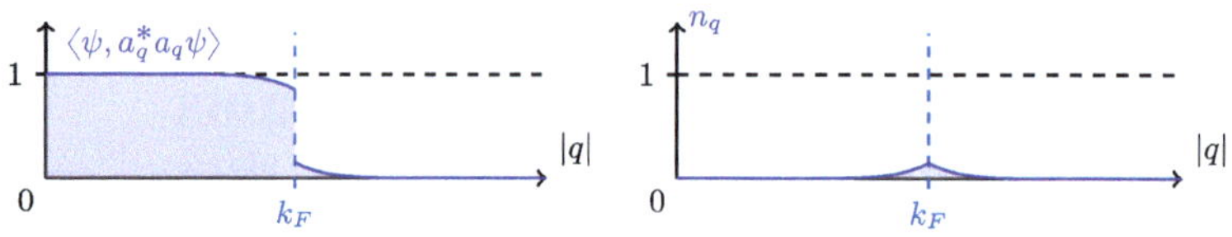

Fig. 1 Left: Schematic depiction of the momentum distribution $\langle\psi, a_q^* a_q \psi\rangle$, where the Fermi momentum k_{F} is the radius of the Fermi ball $B_{\mathrm{F}} = \overline{B_{k_{\mathrm{F}}}(0)}$. Right: Excitation density n_q for the same trial state ψ

The system is modeled by a Hamiltonian operator of the form

$$H_N := \sum_{j=1}^{N} -\hbar^2 \Delta_{x_j} + \lambda \sum_{i<j}^{N} V(x_i - x_j) , \tag{2}$$

with a sufficiently regular pair potential $V : \mathbb{R}^3 \to \mathbb{R}$, and where $\hbar > 0$ is a semi-classical constant[1] and $\lambda > 0$ the coupling constant. For a given density $\rho := \frac{N}{L^3}$, we require the energy to be extensive, which fixes $\hbar \sim \rho^{-\frac{1}{3}}$ and $\lambda \sim \rho^{-1}$.

Since physical systems often feature large particle numbers, it is of particular interest to derive mathematical statements in the limit $N = \rho L^3 \to \infty$. Evidently, there are several choices of sequences in (ρ, L) for which $N \to \infty$, where we will focus on the **mean-field limit** characterized by

$$L := 2\pi , \qquad \hbar := N^{-\frac{1}{3}} , \qquad \lambda := N^{-1} , \qquad \rho \to \infty . \tag{3}$$

We consider a sequence of trial states $\psi_N = \psi$, first introduced in [2] and close to the ground state. The quantity we are interested in is the **momentum distribution** $q \mapsto \langle\psi, a_q^* a_q \psi\rangle \in [0, 1]$, with momentum $q \in \mathbb{Z}^3$ and where a_q^*, a_q are the fermionic creation and annihilation operators defined below. Physically, $\langle\psi, a_q^* a_q \psi\rangle$ describes the probability that a momentum mode q is occupied. It is well-known that for $\lambda = 0$, this probability is 1 if q is inside the Fermi ball $B_{\mathrm{F}} := \overline{B_{k_{\mathrm{F}}}(0)}$ for some Fermi momentum k_{F}, and 0 outside B_{F}. The deviation from this profile for $\lambda > 0$ is called **excitation density**

$$n_q := \begin{cases} \langle\psi, a_q^* a_q \psi\rangle & \text{for } q \in B_{\mathrm{F}}^c \\ 1 - \langle\psi, a_q^* a_q \psi\rangle & \text{for } q \in B_{\mathrm{F}} \end{cases} , \tag{4}$$

see Fig. 1. n_q is expected to be small for small λ, and it fully characterizes the momentum distribution. Our **main result**, Theorem 1, is now an exact formula for n_q in terms of so-called Friedrichs diagrams. Morally, it reads

[1] We remark that $\hbar$ is not introduced to track any semiclassical limit $\hbar \to 0$, but purely rescales H such that the ground state energy becomes extensive.

$$n_q = \sum_{\text{diagrams}} \text{value of the diagram} , \tag{5}$$

with the sum being infinite but convergent. Friedrichs diagrams are a convenient tool to facilitate computationally intensive (anti-)commutator evaluations. They first appeared in [18] and soon found a widespread application in constructive quantum field theory (CQFT) [17, 22–24]. For further information on Friedrichs diagrams, we refer the interested reader to [8, 15].

Theorem 1 complements a recent result in [1], where an approximate excitation density $n_q^{(b)}$ was derived using a bosonization technique, and proven to be the correct leading order expression for n_q. We will demonstrate in Proposition 1 how $n_q^{(b)}$ arises from an intuitive restriction to certain "bosonized diagrams" as

$$n_q^{(b)} = \sum_{\text{bosonized diagrams}} \text{value of the diagram} . \tag{6}$$

Let us remark that a similar diagrammatic formalism was applied in the early physics literature for deriving the ground state energy of a Fermi gas with Coulomb interaction (Jellium) at high densities [14, 19, 29]. In particular, [14] employs a Feynman–Hellmann argument to obtain formulas for the momentum distribution, which agree with the one implied by $n_q^{(b)}$ for short-ranged interactions, see [1, Appendix B]. However, the formalism in [14, 19, 29] corresponds to Feynman diagrams rather than Friedrichs diagrams: Lines represent propagators rather than Kronecker deltas and vertices represent interactions in the interaction picture at distinct times.

Other results in the physics literature on the ground state energy of a 3d Fermi gas include [6, 7, 12, 13, 25, 30]. From [25], further formulas for the momentum distribution involving spin were derived by a Feynman–Hellmann argument in [26].

The trial state we use is of the form $\psi = RT\Omega$ with R being a particle-hole transformation and T an "almost bosonic" Bogoliubov transformation, based on a patch construction, which we describe below in Sect. 2. It was first used in [2] to derive an upper bound on the ground state energy of a 3d mean-field Fermi gas, with[2] $\operatorname{supp}\hat{V}$ being compact. The same construction was shortly afterwards used to provide a matching lower bound for weak interactions (compact $\operatorname{supp}\hat{V}$ and $\|\hat{V}\|_\infty$ small) [3] and stronger interactions ($\sum_k |k|\hat{V}(k) < \infty$) [5], as well as for a description of the dynamics generated by H_N with $\operatorname{supp}\hat{V}$ compact [4]. Another almost bosonic Bogoliubov transformation similar to T, but not depending on a patch construction, was introduced in [9], allowing for an upper and lower bound on the ground state energy in case $\sum_k |k|\hat{V}(k) < \infty$. This transformation was employed to determine the lower excitation spectrum [10] and proving an upper bound on the ground state energy [11], both in case $\sum_k \hat{V}(k)^2 < \infty$. This also covers the Coulomb case $\hat{V}(k) = |k|^{-2}$ for $k \neq 0$ and $\hat{V}(0) = 0$, also called Jellium model. Note that all references above in this paragraph assume $\hat{V} \geq 0$.

[2] Here, $\hat{V}$ denotes the Fourier transform of V.

Another way to send $N \to \infty$ for a 3d Fermi gas (2) is given by the thermodynamic limit, where one fixes ρ and takes $L \to \infty$. Here, in the dilute regime ($\rho \ll 1$), further "almost bosonic" Bogoliubov transformations have been introduced, which are based on solutions of the scattering equation. They allow for proving upper and lower bounds on the ground state energy for smooth $V \geq 0$ [16, 20, 21].

The proof of ground state energy formulas in the thermodynamic limit for large ρ remains a challenging open problem. Also, the rigorous establishment of momentum distribution formulas for the true ground state, rather than just some trial state ψ, is an interesting open task.

Let us further remark that in the 1d Luttinger model [28], the exact momentum distribution for the ground state has been obtained long ago [27], as the model is exactly solvable.

The rest of this article is structured as follows. In Sect. 2 we introduce the mathematical notation that is needed to define the trial state ψ in (12). Section 3 contains our main result, Theorem 1, as well as Proposition 2, preceded by the minimal definitions needed to write down the diagrammatic contributions. The actual Friedrichs diagram formalism is introduced in Sect. 4. It allows us to prove Theorem 1 in Sect. 5, where it also becomes clear how the diagrammatic contributions arise from actual diagrams. In Sect. 6, we heuristically motivate why the largest contributions to n_q come from the "bosonized" diagrams in $n_q^{(b)}$, and we finally prove Proposition 1.

2 Bosonized Operators and Trial State

We mostly adopt the setting of [1]. It is well-known that in the non-interacting case, $V = 0$, one ground state of H_N is given by a Slater determinant, called Fermi ball state

$$\psi_{\mathrm{FB}}(x_1, x_2, \ldots, x_N) := \frac{1}{\sqrt{N!}} \det \left(\frac{1}{(2\pi)^{3/2}} e^{\mathrm{i} k_j \cdot x_i} \right)_{j,i=1}^{N}, \tag{7}$$

with disjoint momenta $(k_j)_{j=1}^N \subset \mathbb{Z}^3$ minimizing the kinetic energy $\sum_{j=1}^N \hbar^2 |k_j|^2$. More precisely, the k_j occupy the Fermi ball

$$B_{\mathrm{F}} := \{k \in \mathbb{Z}^3 \mid |k| \leq k_{\mathrm{F}}\} \quad \text{for some Fermi momentum} \quad k_{\mathrm{F}} > 0\,, \tag{8}$$

where in particular, we assume[3] $N = |B_{\mathrm{F}}|$.

For $V \neq 0$, the true ground state of H_N has a highly non-trivial structure. We will therefore approximate it by the (N-dependent) trial state $\psi = \psi_N$ introduced in [2]. To define ψ, it is convenient to work in the language of second quantization. To this end, we introduce the fermionic Fock space

[3] That is, when taking $N \to \infty$, we restrict to a sequence $(N_n)_{n\in\mathbb{N}}$ such that $N_n = |\overline{B_{k_{\mathrm{F}}^{(n)}}(0)} \cap \mathbb{Z}^3|$ for some $(k_{\mathrm{F}}^{(n)})_{n\in\mathbb{N}} \subset \mathbb{R}$.

$$\mathcal{F} := \bigoplus_{n=0}^{\infty} L^2_{\mathrm{a}}(\mathbb{T}^{3n}) ,$$

and, for $f \in L^2(\mathbb{T}^3)$, the creation/annihilation operators $a^*(f), a(f) : \mathcal{F} \to \mathcal{F}$, which act on $\psi = (\psi^{(0)}, \psi^{(1)}, \psi^{(2)}, \ldots) \in \mathcal{F}$ as

$$\begin{aligned}(a^*(f)\psi)^{(n)}(x_1, \ldots, x_n) &:= \sum_{j=1}^{n} \frac{(-1)^{j-1}}{\sqrt{n}} f(x_j)\psi^{(n-1)}(x_1, \ldots, x_{j-1}, x_{j+1}, \ldots, x_n) ,\\ (a(f)\psi)^{(n)}(x_1, \ldots, x_n) &:= \sqrt{n+1} \int \mathrm{d}x \overline{f(x)} \psi^{(n+1)}(x, x_1, \ldots, x_n) .\end{aligned} \tag{9}$$

Further, we define the orthonormal plane wave basis $(f_q)_{q\in\mathbb{Z}^3} \subset L^2(\mathbb{T}^3)$ with $f_q(x) := (2\pi)^{-\frac{3}{2}} e^{\mathrm{i}q\cdot x}$, $q \in \mathbb{Z}^3$, and the corresponding creation and annihilation operators

$$a^*_q, a_q : \mathcal{F} \to \mathcal{F} , \qquad a^*_q := a^*(f_q) , \qquad a_q := a(f_q) ,$$

which satisfy the canonical anticommutation relations (CAR)

$$\{a_q, a^*_{q'}\} = \delta_{q,q'} , \qquad \{a_q, a_{q'}\} = \{a^*_q, a^*_{q'}\} = 0 \qquad \text{for all } q, q' \in \mathbb{Z}^3 . \tag{10}$$

The vacuum vector $\Omega \in \mathcal{F}$ is given by $\Omega := (1, 0, 0, \ldots)$ and satisfies $a_q\Omega = 0$ for all $q \in \mathbb{Z}^3$. Further, the Fermi ball state ψ_{FB} (7) can now conveniently be written as $\psi_{\mathrm{FB}} = R\Omega$ with $R = R^* : \mathcal{F} \to \mathcal{F}$ being a unitary **particle–hole transformation** defined via

$$R^* a^*_q R := \begin{cases} a^*_q & \text{if } q \in B^c_{\mathrm{F}} \\ a_q & \text{if } q \in B_{\mathrm{F}} \end{cases} . \tag{11}$$

By contrast, our **trial state** is of the form

$$\psi := RT\Omega , \qquad \psi \in L^2_{\mathrm{a}}(\mathbb{T}^{3N}) \subset \mathcal{F} , \tag{12}$$

where $T : \mathcal{F} \to \mathcal{F}$ is a unitary "almost bosonic Bogoliubov transformation", which we define in the following.

The first step in the definition of T consists of constructing M **patches** $B_1, \ldots, B_M \subset \mathbb{R}^3$ around the Fermi surface ∂B_{F}, see Fig. 2. Details of the construction are given in [2, 3] and involve some fixed constant $R > 0$. Each patch covers approximately the same surface area of $\frac{4\pi k_{\mathrm{F}}^2}{M}$, has a thickness of $2R$, and is separated from the neighboring patches by a corridor of width $> R$. Its edges are assumed to be regular in the sense that

$$\operatorname{diam}(B_\alpha) \leq C N^{\frac{1}{3}} M^{-\frac{1}{2}} ,$$

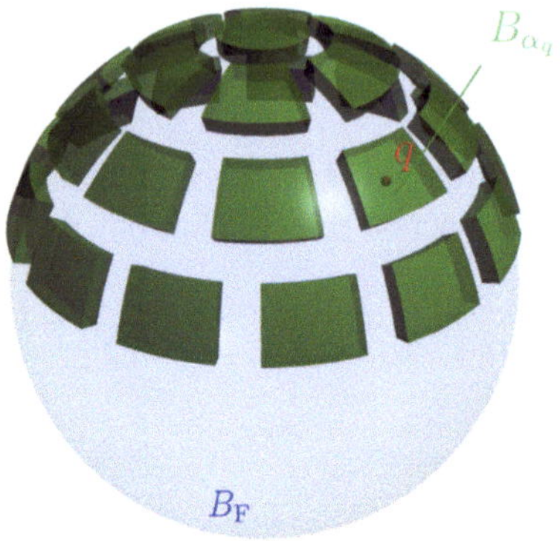

Fig. 2 Patches on the Fermi ball in momentum space

and the patch number M is chosen such that $N^{2\delta} \ll M \ll N^{\frac{2}{3}-2\delta}$ for some fixed $0 < \delta < \frac{1}{6}$. More precisely, in [1], we fix $M = N^{\frac{1}{3}}$ and $\delta = \frac{1}{12}$.

Denoting the center of a patch B_α by $\omega_\alpha \in \mathbb{R}^3$ with corresponding unit vector $\hat{\omega}_\alpha := \frac{\omega_\alpha}{|\omega_\alpha|}$, we define for every fixed $k \in \mathbb{Z}^3$ the index sets

$$\begin{aligned} \mathcal{I}_k^+ &:= \left\{\alpha \in \{1, \ldots, M\} \,\middle|\, k \cdot \hat{\omega}_\alpha \geq N^{-\delta}\right\} , \\ \mathcal{I}_k^- &:= \left\{\alpha \in \{1, \ldots, M\} \,\middle|\, k \cdot \hat{\omega}_\alpha \leq -N^{-\delta}\right\} , \\ \mathcal{I}_k &:= \mathcal{I}_k^+ \cup \mathcal{I}_k^- . \end{aligned} \tag{13}$$

The patch construction is assumed to be symmetric under replacement $k \mapsto -k$, so we have $|\mathcal{I}_k^+| = |\mathcal{I}_k^-| \leq \frac{M}{2}$. The symmetry allows us to restrict our focus to momenta k within the northern half-sphere

$$H^{\mathrm{nor}} := \left\{k \in \mathbb{R}^3 \,\middle|\, k_3 > 0 \text{ or } (k_3 = 0 \text{ and } k_2 > 0) \text{ or } (k_3 = k_2 = 0 \text{ and } k_1 > 0)\right\} . \tag{14}$$

Let us further adopt the following conventions: All momenta k, q, p, h are assumed to be elements of $\mathbb{Z}^3$. For $p \in \mathbb{Z}^3$ ("particle"), the condition $p \in B_{\mathrm{F}}^c \cap B_\alpha$ is abbreviated as $p : \alpha$ (read as "p is compatible with B_α") and for $h \in \mathbb{Z}^3$ ("hole"), the condition $h \in B_{\mathrm{F}} \cap B_\alpha$ is abbreviated as $h : \alpha$. Further, we introduce the notation

$$\pm k := \begin{cases} +k & \text{if } \alpha \in \mathcal{I}_k^+ \\ -k & \text{if } \alpha \in \mathcal{I}_k^- \end{cases} \quad \text{and} \quad \mp k := \begin{cases} -k & \text{if } \alpha \in \mathcal{I}_k^+ \\ +k & \text{if } \alpha \in \mathcal{I}_k^- \end{cases} .$$

Given a momentum transfer

$$k \in \Gamma^{\mathrm{nor}} := H^{\mathrm{nor}} \cap \mathbb{Z}^3 \cap B_R(0) , \tag{15}$$

the number of particle–hole pairs with $p = h \pm k$ in patch $B_\alpha \in \mathcal{I}_k$ is then

$$n_{\alpha,k}^2 := \sum_{p,h:\alpha} \delta_{p,h\pm k} = \sum_{\substack{p:p\in B_F^c\cap B_\alpha \\ p\mp k\in B_F\cap B_\alpha}} 1 \,. \tag{16}$$

We now define for $k \in \Gamma^{\text{nor}}$ and $B_\alpha \in \mathcal{I}_k$ the **almost bosonic creation operator**

$$c_\alpha^*(k) := \frac{1}{n_{\alpha,k}} \sum_{p,h:\alpha} \delta_{p,h\pm k} a_p^* a_h^* = \frac{1}{n_{\alpha,k}} \sum_{\substack{p:p\in B_F^c\cap B_\alpha \\ p\mp k\in B_F\cap B_\alpha}} a_p^* a_{p\mp k}^* \,, \tag{17}$$

with its adjoint being $c_\alpha(k)$. The normalization factor $n_{\alpha,k}$ is precisely chosen such that c^*, c satisfy the **approximate canonical commutation relations** (CCR) [1–3]

$$[c_\alpha(k), c_\beta^*(\ell)] = \begin{cases} 0 & \text{if } \alpha \neq \beta \\ \delta_{k,\ell} + \mathcal{E}_\alpha(k,\ell) & \text{if } \alpha = \beta \end{cases} \,, \tag{18}$$

with commutation error

$$\mathcal{E}_\alpha(k,\ell) := -\sum_{p,h_1,h_2:\alpha} \frac{\delta_{h_1,p\mp k}\delta_{h_2,p\mp\ell}}{n_{\alpha,k}n_{\alpha,\ell}} a_{h_2}^* a_{h_1} - \sum_{p_1,p_2,h:\alpha} \frac{\delta_{h,p_1\mp k}\delta_{h,p_2\mp\ell}}{n_{\alpha,k}n_{\alpha,\ell}} a_{p_2}^* a_{p_1} \,. \tag{19}$$

The **almost bosonic Bogoliubov transformation** $T : \mathcal{F} \to \mathcal{F}$ generating the trial state $\psi = RT\Omega$ is then of the form

$$T := e^{-S} \,, \qquad S := -\frac{1}{2} \sum_{k\in\Gamma^{\text{nor}}} \sum_{\alpha,\beta\in\mathcal{I}_k} K(k)_{\alpha,\beta} \big(c_\alpha^*(k) c_\beta^*(k) - \text{h.c.} \big) \,, \tag{20}$$

with $K(k) \in \mathbb{R}^{|\mathcal{I}_k|\times|\mathcal{I}_k|}$ being a symmetric matrix defined below. So $S^* = -S$ and

$$\begin{aligned} S =& S_+ + S_- \,, \quad \text{where} \quad S_- = -(S_+)^* \quad \text{and} \\ S_+ :=& -\frac{1}{2} \sum_{k\in\Gamma^{\text{nor}}} \sum_{\alpha,\alpha'\in\mathcal{I}_k} K(k)_{\alpha,\alpha'} \sum_{\substack{p,h:\alpha \\ p',h':\alpha'}} \frac{\delta_{p,h\pm k}\delta_{p',h'\pm k}}{n_{\alpha,k}n_{\alpha',k}} a_p^* a_h^* a_{p'}^* a_{h'}^* \,. \end{aligned} \tag{21}$$

The motivation behind the definition of T is to "almost-diagonalize" some effective quadratic bosonic Hamiltonian, see [2, 3]. This requires the following choice of k-dependent matrices:

$$K := \log |S_1^T| \,, \tag{22}$$
$$S_1 := (D + W - \widetilde{W})^{\frac{1}{2}} E^{-\frac{1}{2}} \,,$$
$$E := \left((D + W - \widetilde{W})^{\frac{1}{2}} (D + W + \widetilde{W})(D + W - \widetilde{W})^{\frac{1}{2}} \right)^{\frac{1}{2}} \,,$$

with the $\mathbb{R}^{|\mathcal{I}_k| \times |\mathcal{I}_k|}$ symmetric block matrices

$$D = \begin{pmatrix} d & 0 \\ 0 & d \end{pmatrix} \,, \quad W = \begin{pmatrix} b & 0 \\ 0 & b \end{pmatrix} \,, \quad \widetilde{W} = \begin{pmatrix} 0 & b \\ b & 0 \end{pmatrix} \,,$$

where $d, b \in \mathbb{R}^{|\mathcal{I}_k^+| \times |\mathcal{I}_k^+|}$ are given by

$$d := \sum_{\alpha \in \mathcal{I}_k^+} |\hat{k} \cdot \hat{\omega}_\alpha| \, |\alpha\rangle\langle\alpha| \,, \qquad b := \sum_{\alpha, \beta \in \mathcal{I}_k^+} \frac{\hat{V}(k)}{2\hbar\kappa N |k|} n_{\alpha,k} n_{\beta,k} \, |\alpha\rangle\langle\beta| \,. \tag{23}$$

Here, $|\alpha\rangle \in \mathbb{R}^{|\mathcal{I}_k^+|}$ is the α-th canonical basis vector, we have $\hat{k} := k/|k|$ and $\kappa := k_F N^{-\frac{1}{3}} \approx \left(\frac{3}{4\pi}\right)^{\frac{1}{3}}$. This concludes the construction of the trial state ψ.

3 Momentum Distribution

In [1, Thm. 3.1], it was proved that for $\hat{V} \geq 0$ compactly supported, the excitation density n_q (4) for $q \in \alpha_q$ with some $1 \leq \alpha_q \leq M$ is approximately given by some **bosonized excitation density** [1, (5.7)] of the form

$$n_q^{(b)} := \frac{1}{2} \sum_{k \in \tilde{C}^q \cap \mathbb{Z}^3} \frac{1}{n_{\alpha_q,k}^2} \big(\cosh(2\,K(k)) - 1 \big)_{\alpha_q,\alpha_q} \,, \tag{24}$$

in the sense that for fixed V and most q,

$$|n_q - n_q^{(b)}| \leq C_\varepsilon N^{-\frac{2}{3} - \frac{1}{12}}, \tag{25}$$

with $C_\varepsilon > 0$ depending only on $\varepsilon > 0$ and not on N. Here $\tilde{C}^q$ [1, (3.1)] is defined such that

$$\tilde{C}^q \cap \mathbb{Z}^3 = \begin{cases} \left\{ k \in \Gamma^{\text{nor}} \;\middle|\; \alpha_q \in \mathcal{I}_k \text{ and } q \mp k : \alpha_q \right\} & \text{if } q \in B_F^c \\ \left\{ k \in \Gamma^{\text{nor}} \;\middle|\; \alpha_q \in \mathcal{I}_k \text{ and } q \pm k : \alpha_q \right\} & \text{if } q \in B_F \end{cases} . \tag{26}$$

The approximation $n_q^{(b)}$ arises when evaluating the multicommutator series

$$n_q = \langle \Omega, T^* a_q^* a_q T \Omega \rangle = \langle \Omega, e^S a_q^* a_q e^{-S} \Omega \rangle = \sum_{n=0}^{\infty} \frac{1}{n!} \langle \Omega, \mathrm{ad}_S^n (a_q^* a_q) \Omega \rangle \,, \qquad (27)$$

with $\mathrm{ad}_A^n(B) = [A, \ldots, [A, [A, B]] \ldots]$ being the n-fold multicommutator.

We will now provide an exact formula (39) for n_q, which requires introducing some notation. It turns out that only even n render contributions to n_q in (27). For such n, we denote the momenta of the n involved S-operators, see (21), by $\{p_j, p'_j, h_j, h'_j\}_{j=1}^n$. For later convenience we write

$$a_q^* a_q = \begin{cases} \sum_{p_0, p'_0} \delta_{q,p_0} \delta_{q,p'_0} \; a^*_{p_0} a_{p'_0} & \text{for } q \in B_F^c \\ \sum_{h_0, h'_0} \delta_{q,h_0} \delta_{q,h'_0} \; a^*_{h_0} a_{h'_0} & \text{for } q \in B_F \end{cases} \,, \qquad (28)$$

which allows to write the involved momentum indices in the multicommutator as

$$\begin{aligned} \boldsymbol{P} &:= (p_0, p_1, \ldots, p_n) \,, & \boldsymbol{P}' &:= (p'_0, p'_1, \ldots, p'_n) \,, \\ \boldsymbol{H} &:= (h_1, \ldots, h_n) \,, & \boldsymbol{H}' &:= (h'_1, \ldots, h'_n) \,, \end{aligned} \qquad (29)$$

for $q \in B_F^c$. In case $q \in B_F$, we analogously include h_0 and h'_0 into $\boldsymbol{H}$ and $\boldsymbol{H}'$. We will further split each commutator with $S = S_+ + S_-$ (21) into one commutator with S_+ and one with S_-, where the choices between S_+ or S_- are tracked by a map

$$\xi \in \Xi_n := \Big\{ \xi : \{1, \ldots, n\} \mapsto \{1, -1\} \Big| \sum_{j=1}^{n} \xi(j) = 0 \Big\} \,. \qquad (30)$$

Here, $\xi(j) = 1$ means that S_+ and $\xi(j) = -1$ that S_- is chosen for the j-th commutator. According to $\xi \in \Xi_n$, we may split the momentum indices into those belonging to creation and annihilation parts. In case $q \notin B_F$,

$$\begin{aligned} \boldsymbol{P} &= \boldsymbol{P}_+ \cup \boldsymbol{P}_- \,, & \boldsymbol{P}_+ &:= p_0 \cup (p_j : \xi(j) = 1) \,, & \boldsymbol{P}_- &:= (p_j : \xi(j) = 0) \,, \\ \boldsymbol{H} &= \boldsymbol{H}_+ \cup \boldsymbol{H}_- \,, & \boldsymbol{H}_+ &:= (h_j : \xi(j) = 1) \,, & \boldsymbol{H}_- &:= (h_j : \xi(j) = 0) \,, \\ \boldsymbol{P}' &= \boldsymbol{P}'_+ \cup \boldsymbol{P}'_- \,, & \boldsymbol{P}'_+ &:= (p'_j : \xi(j) = 1) \,, & \boldsymbol{P}'_- &:= p'_0 \cup (p'_j : \xi(j) = 0) \,, \\ \boldsymbol{H}' &= \boldsymbol{H}'_+ \cup \boldsymbol{H}'_- \,, & \boldsymbol{H}'_+ &:= (h'_j : \xi(j) = 1) \,, & \boldsymbol{H}'_- &:= (h'_j : \xi(j) = 0) \,. \end{aligned} \qquad (31)$$

In case $q \in B_F$, we analogously include h_0 in $\boldsymbol{H}_+$ and h'_0 in $\boldsymbol{H}'_-$.

The contractions will now each be between one creation- $(+)$ and one annihilation momentum index $(-)$. We track them by two bijective maps

$$\pi_p : \boldsymbol{P}_- \cup \boldsymbol{P}'_- \mapsto \boldsymbol{P}_+ \cup \boldsymbol{P}'_+ \,, \qquad \pi_h : \boldsymbol{H}_- \cup \boldsymbol{H}'_- \mapsto \boldsymbol{H}_+ \cup \boldsymbol{H}'_+ \,. \qquad (32)$$

All contractions are subject to the constraint that for each $S_\pm$-operator, at least one momentum index is contracted to an "earlier" index:

$$\forall j \in \{1, \dots n\}\ \exists \ell \in \{0, \dots, j-1\} : \\ \begin{cases} \{\pi_p(p_\ell), \pi_p(p'_\ell), \pi_h(h_\ell), \pi_h(h'_\ell)\} \cap \{p_j, p'_j, h_j, h'_j\} \neq \emptyset & \text{if } \xi(j) = 1 \\ \{\pi_p(p_j), \pi_p(p'_j), \pi_h(h_j), \pi_h(h'_j)\} \cap \{p_\ell, p'_\ell, h_\ell, h'_\ell\} \neq \emptyset & \text{if } \xi(j) = -1 \end{cases} . \tag{33}$$

The emergence of this constraint will become apparent later in Sect. 5. We then denote the set of **admissible contraction choices** by

$$\Pi_n^{(\xi)} := \{(\pi_p, \pi_h) \mid (33) \text{ holds}\} . \tag{34}$$

To each contraction choice, we will now associate a sign factor $\text{sgn}(\xi, \pi_p, \pi_h) \in \{1, -1\}$. In order to define this sign factor, let us introduce the sets of creation- and annihilation associated momenta

$$\boldsymbol{Q}_+ := \boldsymbol{P}_+ \cup \boldsymbol{P}'_+ \cup \boldsymbol{H}_+ \cup \boldsymbol{H}'_+ , \qquad \boldsymbol{Q}_- := \boldsymbol{P}_- \cup \boldsymbol{P}'_- \cup \boldsymbol{H}_- \cup \boldsymbol{H}'_- , \tag{35}$$

as well as two ordering relations $<$ on $\boldsymbol{Q}_-$ and on $\boldsymbol{Q}_+$, respectively, defined by

$$\ell < j \;\Rightarrow\; p_\ell, h_\ell, p'_\ell, h'_\ell < p_j, h_j, p'_j, h'_j \quad \text{and} \quad p_j < h_j < p'_j < h'_j . \tag{36}$$

To each contraction choice, we associate a sign factor

$$\text{sgn}(\xi, \pi_p, \pi_h) := \prod_{q_- \in \boldsymbol{Q}_-} \prod_{\substack{q'_- \in \boldsymbol{Q}_- \\ q'_- > q_-,\ \pi_\sharp(q'_-) < \pi_\sharp(q_-)}} (-1) , \tag{37}$$

with $\sharp \in \{p, h\}$, according to whether q_-, q'_- are particle or hole momenta. Finally, let us abbreviate the momentum transfer and patch indices involved in the n operators S by

$$\boldsymbol{K} := (k_1, \dots, k_n) , \qquad \boldsymbol{\alpha} := (\alpha_1, \dots, \alpha_n) , \qquad \boldsymbol{\alpha}' := (\alpha'_1, \dots, \alpha'_n) . \tag{38}$$

We are now in the position to write down the exact formula for n_q.

Theorem 1 (Main result, exact excitation density) *Let* $q \in \mathbb{Z}^3$. *If* $q \in B_{\mathrm{F}}^c$, *then the excitation density in the trial state* $\psi = RT\Omega$ *(12) from [2] is given by*

$$\begin{aligned} n_q = \sum_{\substack{n=2 \\ n:\text{even}}}^{\infty} \frac{1}{2^n n!} \sum_{\boldsymbol{K}} \sum_{\boldsymbol{\alpha},\boldsymbol{\alpha}'} \sum_{\substack{\boldsymbol{P},\boldsymbol{P}' \\ \boldsymbol{H},\boldsymbol{H}'}} \left(\prod_{j=1}^{n} \frac{\delta_{p_j, h_j \pm k_j} \delta_{p'_j, h'_j \pm k_j}}{n_{\alpha_j, k_j} n_{\alpha'_j, k_j}} K(k_j)_{\alpha_j, \alpha'_j} \right) \times \\ \times \sum_{\xi \in \Xi_n} \sum_{(\pi_p, \pi_h) \in \Pi_n^{(\xi)}} \left(\prod_{p \in \boldsymbol{P}_- \cup \boldsymbol{P}'_-} \delta_{p, \pi_p(p)} \right) \left(\prod_{h \in \boldsymbol{H}_- \cup \boldsymbol{H}'_-} \delta_{h, \pi_h(h)} \right) \times \\ \times \delta_{q, p_0} \delta_{q, p'_0} \text{sgn}(\xi, \pi_p, \pi_h) , \end{aligned} \tag{39}$$

where the sum in $\boldsymbol{K}$ *runs over* $(\Gamma^{\text{nor}})^n \subset \mathbb{Z}^{3n}$ *(see (15)), the sums in* $\boldsymbol{\alpha}, \boldsymbol{\alpha}'$ *are such that* $\alpha_j, \alpha'_j \in \mathcal{I}_{k_j}$ *(see (13)), and the sums over* $\boldsymbol{P}, \boldsymbol{P}', \boldsymbol{H}, \boldsymbol{H}'$ *are such that* $p_j, h_j : \alpha_j$ *and* $p'_j, h'_j : \alpha'_j$*, that is,*

$$\begin{aligned} p_j &\in B_{\mathrm{F}}^c \cap B_{\alpha_j} \quad \text{and} \quad & h_j &\in B_{\mathrm{F}} \cap B_{\alpha_j} \quad \forall j \in \{1, \ldots, n\} , \\ p'_j &\in B_{\mathrm{F}}^c \cap B_{\alpha'_j} \quad \text{and} \quad & h'_j &\in B_{\mathrm{F}} \cap B_{\alpha'_j} \quad \forall j \in \{1, \ldots, n\} . \end{aligned} \tag{40}$$

In case $q \in B_{\mathrm{F}}$*, (39) remains valid after a replacement of* $\delta_{q,p_0}\delta_{q,p'_0}$ *by* $\delta_{q,h_0}\delta_{q,h'_0}$*.*

The proof is given in Sect. 5 using Friedrichs diagrams. There, it will also become clear how the rather involved term in (39) arises from a step-by-step back-translation of Friedrichs diagrams into a mathematical expression.

Remark 1 (*Convergence of the diagrammatic expansion*) It is not difficult to show that the sum over n in (39) converges absolutely. However, absolute convergence of the joint sum over (n, ξ, π_p, π_h) is not obvious and might even be false.

To see the absolute convergence of the sum over n, first observe that the number of diagrams per order n is finite. We may thus execute the sums in (ξ, π_p, π_h) for each n separately, resulting in a unique finite value. And one can easily show that the sum in n over these finite values is absolutely convergent: (39) is equal to the multicommutator expansion (27), so the sum over n in (39) converges if and only if (27) converges. We have $\|a_q^* a_q\| = 1$, so if we can show that S is bounded, then we obtain absolute convergence of the commutator series (27) as

$$\sum_{n=0}^{\infty} \frac{|\langle \Omega, \mathrm{ad}_S^n(a_q^* a_q)\Omega\rangle|}{n!} \leq \sum_{n=0}^{\infty} \frac{\|\mathrm{ad}_S^n(a_q^* a_q)\|}{n!} \leq \sum_{n=0}^{\infty} \frac{\|2\,S\|^n \|a_q^* a_q\|}{n!} = e^{\|2\,S\|} . \tag{41}$$

In fact, recalling (20) we have

$$\|S\| \leq \sum_{k \in \Gamma^{\text{nor}}} \sum_{\alpha,\beta \in \mathcal{I}_k} |K(k)_{\alpha,\beta}| \|c_\alpha^\sharp(k)\| \|c_\beta^\sharp(k)\| . \tag{42}$$

With [5, Lemma 7.1], we estimate $|K(k)_{\alpha,\beta}| \leq C M^{-1} \hat{V}(k)$. Further, by (17),

$$\|c_\alpha^*(k)\| \leq \frac{1}{n_{\alpha,k}} \sum_{\substack{p: p \in B_{\mathrm{F}}^c \cap B_\alpha \\ p \mp k \in B_{\mathrm{F}} \cap B_\alpha}} \|a_p^* a_{p \mp k}^*\| \leq \frac{1}{n_{\alpha,k}} \sum_{\substack{p: p \in B_{\mathrm{F}}^c \cap B_\alpha \\ p \mp k \in B_{\mathrm{F}} \cap B_\alpha}} 1 = n_{\alpha,k} , \tag{43}$$

and the same bound holds for $\|c_\alpha(k)\|$. With $n_{\alpha,k}^2 \leq C N^{\frac{2}{3}} M^{-1}$, which follows from the patch construction, and using that the sums $\sum_\alpha$, $\sum_\beta$ run over $\leq M$ elements, we conclude

$$\|2\,S\| \leq C M^{-1} \sum_{k \in \Gamma^{\text{nor}}} \hat{V}(k) \sum_{\alpha,\beta \in \mathcal{I}_k} n_{\alpha,k} n_{\beta,k} \leq C N^{\frac{2}{3}} \sum_{k \in \Gamma^{\text{nor}}} \hat{V}(k) < \infty , \tag{44}$$

since Γ^{nor} comprises finitely many lattice points. So the sum in n converges absolutely.

However, if we take the absolute value of each diagram indexed by (n, ξ, π_p, π_h) separately, and then sum these values up, we will lose cancellations between positive and negative diagrams, making a proof of convergence much more challenging. In that case, the number of diagrams contributing at each order n becomes difficult to control. If any of the $(n!)^2$ contractions between particle- and hole connectors was allowed for any of the $n!$ orders to introduce vertices into the diagram, then the number of diagrams would scale like $(n!)^3$. While constraints from the building process of the diagrams significantly reduce this number, it may still be much larger than the factor of $n!$ in the denominator, potentially making the sum divergent.

Remark 2 (Allowed potentials and scaling limits) In Theorem 1, we did not specify any conditions on V. Indeed, our result holds for *any potential* V, provided that $\hat{V} : \mathbb{Z}^3 \to \mathbb{R}$ exists. This may first seem somewhat surprising and is owed to the particular choice of the trial state ψ, involving finite sums $\sum_{k \in \Gamma^{\text{nor}}}$ in S (20), and $\sum_p$ in $c^{\sharp}_{\alpha}(k)$ (17). However, if the potential $\hat{V}(k)$ does not satisfy the requirements of [5] that $\sum_{k \in \mathbb{Z}^3} \hat{V}(k)|k| < \infty$ and $\hat{V} \geq 0$, then there is no guarantee that ψ is a good approximation of the ground state (in terms of energy). So while Theorem 1 is still correct, its physical significance then still has to be demonstrated.

Likewise, we could pick any side length L of the torus and analogously construct a trial state ψ using the refined momentum lattice $\frac{2\pi}{L}\mathbb{Z}^3$. Then, sums will still be finite and Theorem 1 is still valid. In particular, Theorem 1 holds for any element of a sequence of trial states $(\psi_N)_{N \in \mathbb{N}}$, constructed as above, in *any scaling limit*, including the thermodynamic limit at high density ρ. However, there is no guarantee that these ψ_N will be good approximations for the ground state.

Remark 3 (*Orders of the contributing diagrams*) As we explain in Sect. 6.1, the contractions $\delta_{p,\pi_p(p)}$ and $\delta_{h,\pi_h(h)}$ eliminate certain sums over α_j, α'_j by setting patch indices equal. As each sum runs over $\sim M$ elements, we expect the diagrams, indexed by (ξ, π_p, π_h), to be of different orders, depending on how many sums survive. More precisely, we expect (39) to result in an expansion of the form

$$n_q = f_0 + f_1 M^{-1} + f_2 M^{-2} + f_3 M^{-3} + \dots , \tag{45}$$

where the coefficients $f_j \in \mathbb{R},\ j \in \mathbb{N}$ have identical scaling in N. The leading order is expected to be $f_0 = n_q^{(\text{b})}$, which scales like $\sim N^{-\frac{2}{3}}$ for "most" q, as explained in [1, Remark 2]. For the choice $M = \frac{1}{3}$ and $\delta = \frac{1}{12}$ as above, the optimal relative error of the approximation $n_q \approx n_q^{(\text{b})}$ is thus expected to be $M^{-1} = N^{-\frac{1}{3}}$, which is much smaller than the existing relative error bound $N^{-\frac{1}{12}}$.

We believe that an expansion like (45) can be made rigorous, once suitable combinatorial bounds on the number of occurring diagrams have been derived.

We are able to interpret the bosonized excitation density $n_q^{(\mathrm{b})} \approx n_q$ obtained in [1, Theorem 3.1] as coming from a subset of bosonized Friedrichs diagrams, which we explain in Sect. 6. The subset is characterized by a restriction on (π_p, π_h). First, we require that the two $c_\alpha^\sharp(k)$-operators, whose momenta are contracted to $a_q^* a_q$, have their second momenta directly contracted to each other. Second, for all other $c_\alpha(k)$-operators, both momenta must be contracted at the same time to the same $c_\alpha^*(k)$-operator. So in case $q \in B_\mathrm{F}^c$, with $\sharp_1, \sharp_2 \in \{\cdot, '\}$ (so $p^{\sharp_1}$ is either p or p'), the **bosonization constraint** reads

$$\begin{aligned} \pi_p^{-1}(p_0) = p_j^{\sharp_1} \text{ and } \pi_p(p_0') = p_\ell^{\sharp_2} \quad &\Rightarrow \quad \pi_h(h_j^{\sharp_1}) = h_\ell^{\sharp_2} \\ \forall j, \ell \geq 1 : \quad \pi_p(p_j^{\sharp_1}) = p_\ell^{\sharp_2} \quad &\Rightarrow \quad \pi_h(h_j^{\sharp_1}) = h_\ell^{\sharp_2} \, . \end{aligned} \tag{46}$$

The diagrammatic interpretation of this constraint is explained in Sect. 6.1 and depicted in Fig. 10. The restricted set of **bosonized contraction choices** is then

$$\Pi_{n,(\mathrm{b})}^{(\xi)} := \left\{ (\pi_p, \pi_h) \in \Pi_n^{(\xi)} \;\middle|\; (46) \text{ holds} \right\} . \tag{47}$$

Proposition 1 (Bosonized excitation density) *The bosonized excitation density $n_q^{(\mathrm{b})}$ (24) from [1] amounts to a restriction of (39) to bosonized diagrams, given by replacing $\Pi_n^{(\xi)}$ with $\Pi_{n,(\mathrm{b})}^{(\xi)}$. That is, for $q \in B_\mathrm{F}^c$,*

$$\begin{aligned} n_q^{(\mathrm{b})} = &\sum_{\substack{n=2 \\ n:\text{even}}}^{\infty} \frac{1}{2^n n!} \sum_{\boldsymbol{K}} \sum_{\alpha, \alpha'} \sum_{\substack{\boldsymbol{P}, \boldsymbol{P}' \\ \boldsymbol{H}, \boldsymbol{H}'}} \left(\prod_{j=1}^{n} \frac{\delta_{p_j, h_j \pm k_j} \delta_{p_j', h_j' \pm k_j}}{n_{\alpha_j, k_j} n_{\alpha_j', k_j}} K(k_j)_{\alpha_j, \alpha_j'} \right) \times \\ &\times \sum_{\xi \in \Xi_n} \sum_{(\pi_p, \pi_h) \in \Pi_{n,(\mathrm{b})}^{(\xi)}} \left(\prod_{p \in \boldsymbol{P}_- \cup \boldsymbol{P}_-'} \delta_{p, \pi_p(p)} \right) \left(\prod_{h \in \boldsymbol{H}_- \cup \boldsymbol{H}_-'} \delta_{h, \pi_h(h)} \right) \times \\ &\times \delta_{q, p_0} \delta_{q, p_0'} \operatorname{sgn}(\xi, \pi_p, \pi_h) \, , \end{aligned} \tag{48}$$

and the same holds for $q \in B_\mathrm{F}$ after a replacement of $\delta_{q,p_0}\delta_{q,p_0'}$ by $\delta_{q,h_0}\delta_{q,h_0'}$.

We illustrate and explain the meaning of (48) in Sect. 6.1 and give the diagrammatic proof of Proposition 1 in Sect. 6.2.

4 Friedrichs Diagram Formalism

Let us quickly recap the diagrammatic formalism by Friedrichs [18], for which we use the same notation as in [8]. Since we will need to evaluate both bosonic and fermionic commutators, we will assume in this section that a_q^*, a_q can describe both

Fig. 3 Left: A vertex with connectors, representing f and its momenta q_j, q'_j. Right: A Friedrichs diagram with one vertex, representing A in (50)

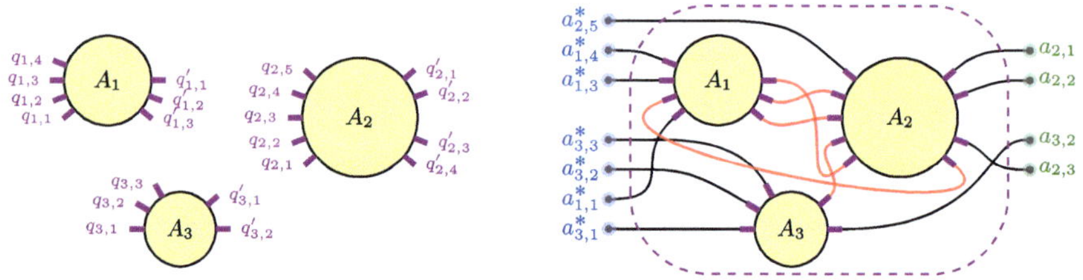

Fig. 4 Left: Vertices and connectors of a Friedrichs diagram with 3 vertices. Right: A Friedrichs diagram with 3 vertices

a species of fermionic or of bosonic creation/annihilation operators. That is, they satisfy either the CAR (10) or the CCR

$$[a_q, a^*_{q'}] = \delta_{q,q'} , \qquad [a_q, a_{q'}] = [a^*_q, a^*_{q'}] = 0 \qquad \text{for all } q, q' \in \mathbb{Z}^3 . \tag{49}$$

In the formalism of Friedrichs diagrams, an operator of the form

$$A = \sum_{\substack{q_1,\ldots,q_n \\ q'_1,\ldots,q'_m}} f(q_1, \ldots, q_n, q'_1, \ldots, q'_m) a^*_{q_n} \ldots a^*_{q_1} a_{q'_1} \ldots a_{q'_m} \tag{50}$$

is represented by a vertex, see Fig. 3, that encodes the function ("kernel") $f \in \ell^2(\mathbb{Z}^{3(n+m)})$ with n legs pointing to the left and m legs pointing to the right.

When taking multicommutators as in (27), the CCR/CAR will produce Kronecker deltas of the kind $\delta_{q,q'}$, which we represent by contracted legs. To be precise, a general Friedrichs diagram as in Fig. 4 consists of:

- V vertices, indexed by $v \in \{1, \ldots, V\}$, representing $f_v : \mathbb{Z}^{3(n_v+m_v)} \to \mathbb{C}$.
- n_v left-connectors and m_v right-connectors on each vertex, representing the momenta $q_{v,1}, \ldots, q_{v,n_v}$ and $q'_{v,1}, \ldots, q'_{v,m_v}$, respectively. So the total index sets are

$$\mathcal{J} = \bigcup_{v=1}^{V} \mathcal{J}_v, \qquad \mathcal{J}_v := \{(v,1), \ldots, (v,n_v)\} \qquad \text{and}$$
$$\mathcal{J}' = \bigcup_{v=1}^{V} \mathcal{J}'_v, \qquad \mathcal{J}'_v := \{(v,1), \ldots, (v,m_v)\} \, . \tag{51}$$

- $C \leq \min\{|\mathcal{J}|, |\mathcal{J}'|\}$ contractions between a left- and a right-connector. We formally keep track of them by two maps[4] $\pi : \{1, \ldots, C\} \to \mathcal{J}$ and $\pi' : \{1, \ldots, C\} \to \mathcal{J}'$, where a contraction goes from connector $\pi'(c)$ to $\pi(c)$.
- $|\mathcal{J}| + |\mathcal{J}'| - 2C$ external legs, one for each uncontracted connector, representing creation operators a^*_q (for left-connectors) or annihilation operators $a_{q'}$ (for right-connectors). We formally keep track of the operator orderings by maps $\sigma : \{1, \ldots, |\mathcal{J}| - C\} \to \mathcal{J}$ and $\sigma' : \{1, \ldots, |\mathcal{J}'| - C\} \to \mathcal{J}'$.

For brevity, we introduce the momentum vectors $\boldsymbol{Q}_v := (q_{v,1}, \ldots, q_{v,n_v})$, $\boldsymbol{Q}'_v := (q'_{v,1}, \ldots, q'_{v,m_v})$ and $\boldsymbol{Q} = (\boldsymbol{Q}_1, \ldots, \boldsymbol{Q}_V)$, $\boldsymbol{Q}' = (\boldsymbol{Q}'_1, \ldots, \boldsymbol{Q}'_V)$ and abbreviate $a^*_{q_{v,\ell}} =: a^*_{v,\ell}$ and $a_{q'_{v,\ell}} =: a_{v,\ell}$. A Friedrichs diagram can then be translated into an operator

$$G = \sum_{\boldsymbol{Q},\boldsymbol{Q}'} \left(\prod_{v=1}^{V} f_v(\boldsymbol{Q}_v, \boldsymbol{Q}'_v) \right) \left(\prod_{c=1}^{C} \delta_{q_{\pi(c)}, q'_{\pi'(c)}} \right) \left(\prod_{\ell=1}^{|\mathcal{J}|-C} a_{\sigma(\ell)} \right)^* \left(\prod_{\ell'=1}^{|\mathcal{J}'|-C} a_{\sigma'(\ell')} \right) . \tag{52}$$

Note that (52) is again an operator of the form (50), so it can alternatively be written as a Friedrichs diagram with a single vertex.

Commutators between two bosonic or two fermionic operators A_1 and A_2 of type (50) can now be expressed in terms of so-called attached products A_1—○—A_2 (bosonic) and A_1—○═A_2 (fermionic): Loosely speaking, those are "sums over all ways to contract A_1 with A_2 from left to right", possibly including signs. Mathematically, we may track the "ways to contract" by the set of contraction configurations

$$\begin{aligned} \mathcal{C} := \big\{ (\pi, \pi') \,\big|\, & \pi : \{1, \ldots, C\} \to \mathcal{J}_2, \ \pi' : \{1, \ldots, C\} \to \mathcal{J}'_1, \\ & 1 \leq C \leq \min(m_1, n_2), \ |\mathrm{imag}(\pi')| = C, \ \pi(1) > \ldots > \pi(C) \big\} , \end{aligned} \tag{53}$$

where each (π, π') renders two sets of contractible but uncontracted connectors

$$\begin{aligned} \mathcal{U} :=& \big\{ (2, j) \in \mathcal{J}_2 \ \big|\, \nexists\, c \in \{1, \ldots, C\} : \pi(c) = (2, j) \big\} , \\ \mathcal{U}' :=& \big\{ (1, k) \in \mathcal{J}'_1 \ \big|\, \nexists\, c \in \{1, \ldots, C\} : \pi'(c) = (1, k) \big\} . \end{aligned} \tag{54}$$

---

[4] We remark that π, π' do not play the same role as π_p, π_h above. Here, π indexes contracted connectors on the left and π' those on the right, while π_p, π_h directly associate connectors on the left to those on the right. In fact, the notation with π, π' is more general. Above, we chose the more compact notation with π_p, π_h, since there we are in a special case where we know that all connectors have to be contracted.

The **bosonic attached product** is then defined as

$$A_1 \multimap\!\!\circ\!\!- A_2 := \sum_{(\pi,\pi')\in\mathcal{C}} \sum_{\boldsymbol{Q},\boldsymbol{Q}'} f_1(\boldsymbol{Q}_1, \boldsymbol{Q}_1') f_2(\boldsymbol{Q}_2, \boldsymbol{Q}_2') \left(\prod_{c=1}^{C} \delta_{q_{\pi(c)}, q'_{\pi'(c)}} \right) \times \\ \times \left(\prod_{\ell=1}^{n_1} a_{1,\ell} \right)^* \left(\prod_{u\in\mathcal{U}} a_u \right)^* \prod_{u'\in\mathcal{U}'} a_{u'} \prod_{\ell'=1}^{m_2} a_{2,\ell'} \,. \tag{55}$$

For fermions, the attached product is defined analogously, up to a change of sign in front of certain contributions: Let σ, σ' be the permutations of $\mathcal{J}_2$ and $\mathcal{J}_1'$ that take the diagram into a **maximally crossed** form, while preserving the order of the uncontracted connectors. Here, by maximally crossed, we mean that the first right-connector of A_1 from the bottom, $(1, m_1)$, is connected to the first left-connector of A_2 from the top, $(2, n_2)$, the second to the second, and so on. That means,

$$\begin{aligned} \sigma(\pi(c)) &= (2, n_2 - c + 1) \quad \text{and} \quad u_1 < u_2 \Rightarrow \sigma(u_1) < \sigma(u_2) \quad \forall\, u_1, u_2 \in \mathcal{U}\,, \\ \sigma(\pi'(c)) &= (1, m_1 - c + 1) \quad \text{and} \quad u_1' < u_2' \Rightarrow \sigma'(u_1') < \sigma'(u_2') \quad \forall\, u_1', u_2' \in \mathcal{U}'\,. \end{aligned} \tag{56}$$

See also Fig. 5. The sign of a contraction configuration $(\pi, \pi') \in \mathcal{C}$ is then given by

$$\operatorname{sgn}(\pi, \pi') := (-1)^{(m_1 - C)(n_2 - C)} \operatorname{sgn}(\sigma) \operatorname{sgn}(\sigma') \tag{57}$$

and the **fermionic attached product** is defined as

$$A_1 \multimap\!\!\circ\!\!= A_2 := \sum_{(\pi,\pi')\in\mathcal{C}} \operatorname{sgn}(\pi, \pi') \sum_{\boldsymbol{Q},\boldsymbol{Q}'} f_1(\boldsymbol{Q}_1, \boldsymbol{Q}_1') f_2(\boldsymbol{Q}_2, \boldsymbol{Q}_2') \prod_{c=1}^{C} \delta_{q_{\pi(c)}, q'_{\pi'(c)}} \times \\ \times \left(\prod_{\ell=1}^{n_1} a_{1,\ell} \right)^* \left(\prod_{u\in\mathcal{U}} a_u \right)^* \prod_{u'\in\mathcal{U}'} a_{u'} \prod_{\ell'=1}^{m_2} a_{2,\ell'} \,. \tag{58}$$

For a motivation of the sign factor $\operatorname{sgn}(\pi, \pi')$, see [8, Appendix A].

(Anti-)commutators can now conveniently be expressed in terms of attached products.

Fig. 5 Left: A maximally crossed Friedrichs diagram. Right: In this diagram, a permutation σ' with 3 swaps is necessary to achieve a maximally crossed form. So $\operatorname{sgn}(\sigma') = -1$

Proposition 2 ([8, Theorems 3.1 and 3.2]) *For bosonic operators of the form (50), we have*

$$[A_1, A_2] = A_1\!-\!\!\circ\!\!-\!A_2 - A_2\!-\!\!\circ\!\!-\!A_1 \,. \tag{59}$$

For fermionic operators of the form (50), we have

$$\begin{aligned} [A_1, A_2] &= A_1\!-\!\!\circ\!\!=\!A_2 - A_2\!-\!\!\circ\!\!=\!A_1 \qquad && \text{if } (m_1+n_1)(m_2+n_2) \text{ is even} \,, \\ \{A_1, A_2\} &= A_1\!-\!\!\circ\!\!=\!A_2 + A_2\!-\!\!\circ\!\!=\!A_1 \qquad && \text{if } (m_1+n_1)(m_2+n_2) \text{ is odd} \,. \end{aligned} \tag{60}$$

With these commutator formulas at hand, we are ready to evaluate the multicommutator in (27) diagrammatically.

5 Multicommutator Evaluation via Friedrichs Diagrams

Using Friedrichs diagrams, we will now evaluate the multicommutator in (27) to derive the formula (39) for n_q as claimed in Theorem 1.

Proof (Proof of Theorem 1). Recall (27):

$$n_q = \sum_{n=0}^{\infty} \frac{1}{n!} \langle \Omega, \mathrm{ad}_S^n(a_q^* a_q) \Omega \rangle \,.$$

First, let us specify how to represent the operators $a_q^* a_q, c_\alpha^*(k), c_\alpha(k)$, and S diagrammatically, see Fig. 6.

- The operator $a_q^* a_q$ is characterized by the kernel $f(p_0, p_0') = \delta_{q,p_0}\delta_{q,p_0'}$ in case $q \in B_\mathrm{F}^c$ and $f(h_0, h_0') = \delta_{q,h_0}\delta_{q,h_0'}$ if $q \in B_\mathrm{F}$. We represent it by a small vertex.
- Each operator $c_\alpha^*(k)$ is characterized by a kernel

$$f(p, h) = d_{\alpha,k}(p, h) := \delta_{p,h\pm k} \frac{1}{n_{\alpha,k}} \chi_{B_\mathrm{F}^c \cap B_\alpha}(p) \chi_{B_\mathrm{F} \cap B_\alpha}(h) \,, \tag{61}$$

compare (17), where we adopted the notation of [1, (4.2) and (4.1)]. We also represent $c_\alpha^*(k)$ by a small vertex, whose two legs are pointing left.

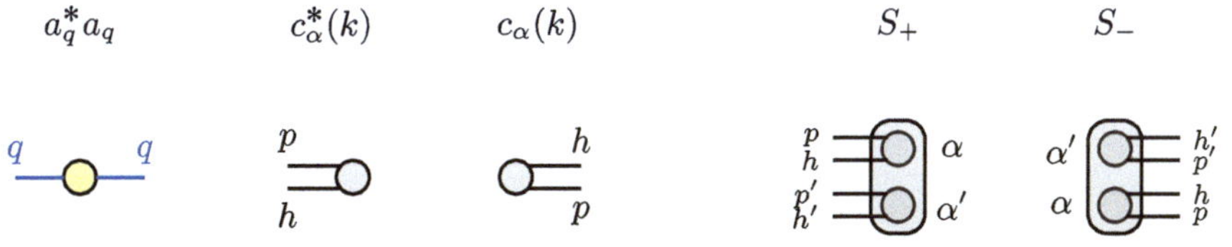

Fig. 6 From left to right: The operators $a_q^* a_q, c_\alpha^*(k), c_\alpha(k)$ and $S = S_+ + S_-$ are translated into Friedrichs diagrams

- Likewise, $c_\alpha(k)$ is represented by a small vertex with two legs pointing to the right. The directions in which the legs are pointing makes it clear, which operator is meant by which small vertex.
- We represent the creation part S_+ of S by a large rectangular vertex including two smaller c^*-vertices. The large vertex can then be translated into a factor of $\frac{1}{2}K(k)_{\alpha,\alpha'}$, where sums over k, α, and α' are implicitly assumed. Further, the large vertex fixes both momentum transfers inside the small vertices $c^*_\alpha(k)$ and $c^*_{\alpha'}(k)$ to be the same vector $k \in \Gamma^{\mathrm{nor}} \subset \mathbb{Z}^3$.
  Likewise, S_- is represented by a large vertex with 4 legs pointing to the right.

We would like to apply Proposition 2 for evaluating the multicommutators $\mathrm{ad}^n_S(a^*_q a_q)$ in (27). In each commutator $\mathrm{ad}^n_S(a^*_q a_q) = [S_+, \mathrm{ad}^{n-1}_S(a^*_q a_q)] + [S_-, \mathrm{ad}^{n-1}_S(a^*_q a_q)]$, the leg numbers of the first vertex $A_1 := S_\pm$ are $(n_1, m_1) = (4, 0)$ or $(0, 4)$, respectively. So irrespective of the leg numbers of the diagrams in $A_2 := \mathrm{ad}^{n-1}_S(a^*_q a_q)$, we have $(m_1 + n_1)(m_2 + n_2) = 4(m_2 + n_2)$, which is even. Thus, (60) indeed renders a formula for a commutator.

Now, following Proposition 2, the multicommutators $\mathrm{ad}^n_S(a^*_q a_q)$ in (27) correspond to diagrams, which are built by starting with an $a^*_q a_q$-vertex and successively contracting n vertices of type $S_\pm$ into the diagram.

After taking the vacuum expectation value $\langle \Omega, \mathrm{ad}^n_S(a^*_q a_q)\Omega\rangle$, any diagram with external legs will vanish as it yields linear combinations of terms of the form $\langle \Omega, a^*_{q_n} \dots a^*_{q_1} a_{q'_1} \dots a_{q'_m} \Omega\rangle$, and we have $a_q \Omega = 0$. So we only need to consider diagrams where all $4n + 2$ legs have been contracted. As contractions always connect a left- and a right-connector, we need to have $2n + 1$ connectors of either kind. So only diagrams with $\frac{n}{2}$ vertices of type S_+ and $\frac{n}{2}$ vertices of type S_- contribute. In particular,

$$\langle \Omega, \mathrm{ad}^n_S(a^*_q a_q)\Omega\rangle = 0 \qquad \text{if } n \text{ is odd} , \tag{62}$$

and the sum in (27) reduces to even n.

In order to derive (39), let us back-translate the corresponding diagrams. Irrespective of the contractions, the S-vertices contribute the sums $\sum_{\boldsymbol{K} \in (\Gamma^{\mathrm{nor}})^n}$ and $\sum_{\boldsymbol{\alpha},\boldsymbol{\alpha}'}$, see (38), such that $\alpha_j, \alpha'_j \in \mathcal{I}_{k_j}$, as claimed in Theorem 1.

The $4n + 2$ momenta of the connectors are tracked in $\boldsymbol{P}, \boldsymbol{P}', \boldsymbol{H}, \boldsymbol{H}'$, see (29), where the condition (40) encodes the factors $\chi_{B^c_{\mathrm{F}} \cap B_\alpha}(p)$ and $\chi_{B_{\mathrm{F}} \cap B_\alpha}(h)$ from the $c^\sharp_\alpha(k)$-vertices.

Now, for $1 \le j \le n$, every S_--vertex also contributes a factor of $\frac{1}{2}K(k_j)_{\alpha_j,\alpha'_j}$ and every S_+-vertex a factor of $-\frac{1}{2}K(k_j)_{\alpha_j,\alpha'_j}$. As S_+ contains creation operators, it gets contracted "from the right", yielding a contribution $\mathrm{ad}^{j-1}_S(a^*_q a_q)$—○=$S_+$ which comes with an additional minus sign, see (60). So including the sign from (60), both S_+ and S_- effectively contribute $\frac{1}{2}K(k_j)_{\alpha_j,\alpha'_j}$.

Further, the two $c^\sharp_\alpha(k)$-vertices in $S_\pm$ contribute a factor of $\delta_{p_j, h_j \pm k_j} n^{-1}_{\alpha_j,k_j}$ and $\delta_{p'_j, h'_j \pm k_j} n^{-1}_{\alpha'_j,k_j}$. Recalling that each order $n \in \mathbb{N}$ comes with a factor of $(n!)^{-1}$ in the series (27), this reproduces the first line of (39).

The second line of (39) now accounts for contractions in the diagrams. As explained around (30), the map $\xi \in \Xi_n$ tracks whether S_+ or S_- has been chosen for contraction, which is unique for each diagram.

Accordingly, we split the connectors into those on the right ($\boldsymbol{P}_+$, $\boldsymbol{P}'_+$, $\boldsymbol{H}_+$, $\boldsymbol{H}'_+$), and those on the left ($\boldsymbol{P}_-$, $\boldsymbol{P}'_-$, $\boldsymbol{H}_-$, $\boldsymbol{H}'_-$) as in (31).

The $2n+1$ contractions are tracked by the bijective maps π_p, π_h as in (32). Here, π_p associates to every particle-connector p on the left a particle-connector $\pi_p(p)$ on the right to which p is contracted. Every such contraction results in a contribution of $\delta_{p,\pi_p(p)}$. Likewise, every hole-connector h on the left is contracted to $\pi_h(h)$ on the right, resulting in a contribution of $\delta_{h,\pi_h(h)}$.

As the attached product —○= must include at least one contraction, we require the j-th S-vertex to be contracted to an existing ℓ-th vertex ($\ell < j$). This is exactly constraint (33), resulting in the contraction sum running over $(\pi_p, \pi_h) \in \Pi_n^{(\xi)}$.

Finally, the factor of $\delta_{q,p_0}\delta_{q,p'_0}$ or $\delta_{q,h_0}\delta_{q,h'_0}$ is just the kernel of the $a^*_q a_q$-vertex, as explained above.

The only remaining step is to evaluate the sign factors $\mathrm{sgn}(\pi, \pi')$ as in (57) and to show that their product amounts to the factor $\mathrm{sgn}(\xi, \pi_p, \pi_h)$ in (37). We start with considering the first contraction, which appears in $\mathrm{ad}^1_S(a^*_q a_q) = [S_+, a^*_q a_q] + [S_-, a^*_q a_q]$. Here, one q-connector gets contracted to any of the 4 connectors of $S_\pm$, say, the m-th one, counted from the top for S_+ and from the bottom for S_-. The diagram then arises from a maximally crossed one by employing $m-1$ swaps, see Fig. 7. If we now swap the remaining uncontracted $a^\sharp_q$-operator into the position, where the contracted operator of $S_\pm$ used to be (second step in Fig. 7), we have to apply further $m-1$ swaps. After this move, we end up with an operator that has the same structure as an $S_\pm$-operator, while the sign factor we pick up is $(-1)^{2m-2} = 1$. So for the sign evaluation, we can proceed as if the $a^*_q a_q$-vertex didn't exist.

In the following contraction steps, we then successively add S_--diagrams from the bottom to the top on the left-hand side and S_+-diagrams from the top to the bottom on the right-hand side, see Fig. 8. Note that the order (bottom to top or vice versa) is enforced by the ordering prescription introduced in (50) and below. Also, observe that the ordering relation $q'_- > q_-$ in (36) and (37) just means that the connector q'_- is above q_- in the diagram, while $\pi_\sharp(q'_-) < \pi_\sharp(q_-)$ means that the connector $\pi_\sharp(q'_-)$ is above $\pi_\sharp(q_-)$, see also Fig. 8. So the sign factor $\mathrm{sgn}(\xi, \pi_p, \pi_h)$ in (37) essentially counts how many swaps would be necessary to take the diagram into maximally

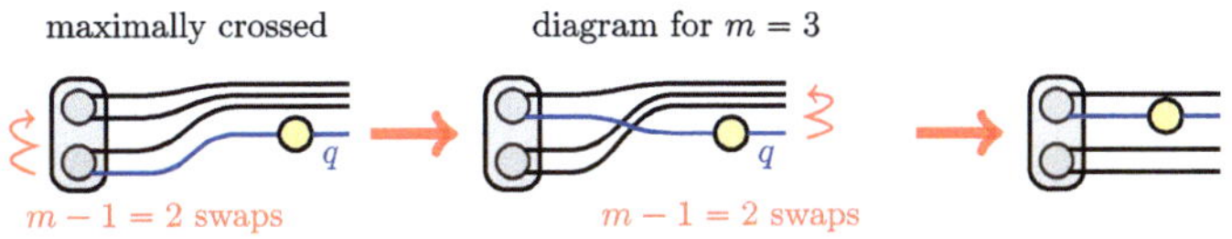

Fig. 7 After the first contraction, the diagram can be brought into the same structure as $S_\pm$, while picking up a sign factor of $(-1)^{2m-2} = 1$

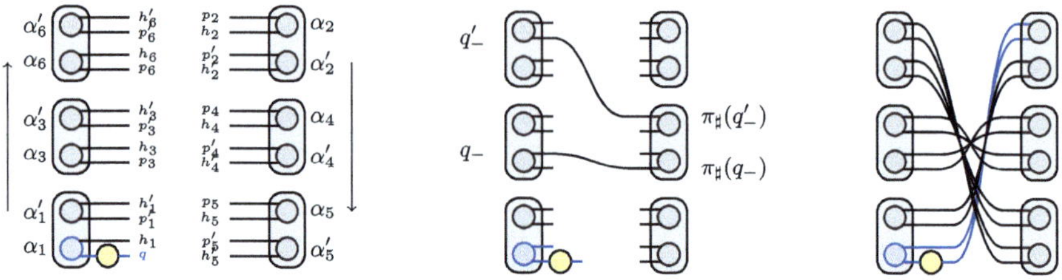

Fig. 8 Left: Example of a diagram with $n = 6$ vertices with only the first contraction drawn. Middle: A situation with $q'_- > q_-$ and $\pi_\sharp(q'_-) < \pi_\sharp(q_-)$, making a swap necessary to finally achieve maximal crossing. Right: A maximally crossed diagram with $n = 6$

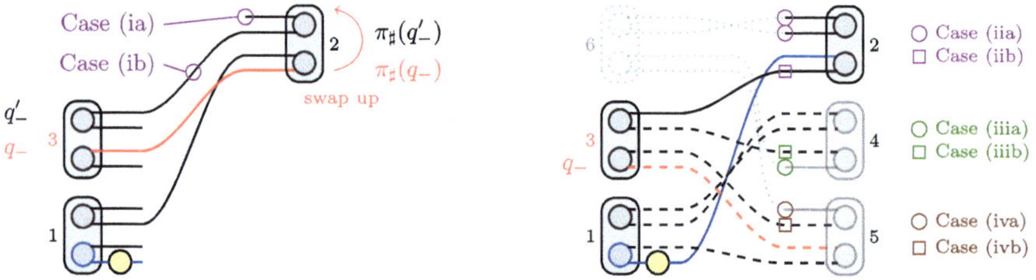

Fig. 9 Left: Cases (ia) and (ib), in which q_- is accounted for a factor of (-1). Right: Cases (ii), (iii) and (iv). The numbers 1–6 indicate the order in which $S_\pm$-vertices enter the diagram. Here, $j = 3$

crossed form[5] as depicted in Fig. 8, while ignoring the $a_q^* a_q$-vertex. Thus, we may finish the proof by establishing the following claim.

Claim: The product of all sign factors $\operatorname{sgn}(\pi, \pi')$ appearing in the multicommutator evaluation for every diagram indexed by (ξ, π_p, π_h) is identical to the sign factor $\operatorname{sgn}(\xi, \pi_p, \pi_h)$ we would need to make the diagram, including n vertices $S_\pm$ and ignoring $a_q^* a_q$, maximally crossed.

Proof of the Claim: Following (57), a factor of (-1) in $\operatorname{sgn}(\pi, \pi')$ enters if and only if within the contraction of a new $S_\pm$-vertex:

(*A*) A connector gets contracted, and it has to be swapped with another connector in order to achieve maximal crossing. This rule accounts for the factor of $\operatorname{sgn}(\sigma)\operatorname{sgn}(\sigma')$ in (57).

(*B*) A connector of some operator a is not contracted and has to "jump" over another uncontracted operator a^* to achieve normal ordering. This accounts for $(-1)^{(m_1 - C)(n_2 - C)}$ in (57).

We now re-distribute the sign factors (-1) coming from (A) and (B) among the $2n$ connectors on the left-hand side, $q_- \in \{p_j, p'_j, h_j, h'_j\} \subset \boldsymbol{Q}_-$, by making every q_- accountable for the following factors of (-1), see Fig. 9:

[5] Note that the maximally crossed diagram in Fig. 8 does not contribute to n_q due to the constraint (33).

(i) If q_- gets contracted to $\pi_\sharp(q_-)$ immediately as the j-th vertex $S_\pm = S_-$ enters, we make it accountable for all (-1) of type (A) caused by $\pi_\sharp(q_-)$ being swapped upwards past all connectors above $\pi_\sharp(q_-)$ that are

(ia) uncontracted or
(ib) contracted to some connector q'_- above q_-.

(ii) If q_- does not immediately get contracted as the j-th vertex $S_\pm = S_-$ enters, we make it accountable for

(iia) all (-1) of type (B) caused by q_- jumping over uncontracted connectors on the right and
(iib) all (-1) of type (A) caused by connectors $q'_- > q_-$ being swapped down below q_-.

(iii) As an ℓ-th vertex $S_\pm = S_+$, $\ell > j$ enters, and q_- is still not contracted, we make it accountable for

(iiia) all (-1) of type (B) caused by q_- jumping over uncontracted connectors in S_- and
(iiib) all (-1) of type (A) caused by connectors $q'_- > q_-$ that got contracted to vertex ℓ and have to be swapped down past q_-.

(iv) As an ℓ-th vertex $S_\pm = S_+$, $\ell > j$ enters, and q_- is contracted to $\pi_\sharp(q_-)$ within this step, we make q_- accountable for $\pi_\sharp(q_-)$ being swapped past

(iva) all uncontracted connectors in the ℓ-th $S_\pm$ above $\pi_\sharp(q_-)$
(ivb) all connectors in the ℓ-th $S_\pm$ above $\pi_\sharp(q_-)$ contracted to some $q'_- > q_-$.

Indeed, multiplying all sign factors (-1) from (i)–(iv) for all q_- recovers exactly contributions (A) and (B): It is easy to see that, as the j-th S-vertex enters, and it is of type S_-, then the contributions (iia) from all uncontracted q_- in this S_- make up all sign factors of type (B) occurring in this step. Further, (ia) and (ib) account for all swaps on the right and (iib) for all swaps on the left that are needed to achieve maximal crossing in this step, rendering all sign factors of (A).

By contrast, if an S_+-vertex enters, then (iiia) will yield all sign factors of type (B) in this step. Then, (iiib) accounts for all swaps on the left, and (iva) and (ivb) for all swaps on the right that yield maximal crossing after the step, rendering the factors (A). So all factors (i)–(iv) from all q_- indeed render the total product of all sign factors $\mathrm{sgn}(\pi, \pi')$ appearing in the n contractions.

Now, let us determine the sign factor $\mathrm{sgn}(\xi, \pi_p, \pi_h)$ appearing when swapping the entire diagram into maximally crossed form, as in Fig. 8. The transition to maximal crossing can be achieved by successively considering connectors $q_- \in \boldsymbol{Q}_-$ and swapping $\pi_\sharp(q_-)$ on the right past all $\pi_\sharp(q'_-) < \pi_\sharp(q_-)$ with $q'_- > q_-$.

First, assume that q_- is contracted immediately as the j-th vertex $S_\pm$ enters. Any $q'_- > q_-$ with $\pi_\sharp(q'_-) < \pi_\sharp(q_-)$ is either in the j-th vertex, rendering (ib), or above the j-th vertex, so $\pi_\sharp(q'_-)$ is not yet contracted in step j, rendering (ia).

Conversely, assume that q_- is not immediately contracted as the j-th vertex $S_\pm$ enters, but only in step $\ell > j$. Then, for the connectors $\pi_\sharp(q'_-)$ swapped with $\pi_\sharp(q_-)$ there are the following options: $\pi_\sharp(q'_-)$ could be present in the diagram after step j and at this point either already contracted to some $q'_- > q_-$, rendering (iib), or uncontracted, so it gets later contracted to q'_- in some S_--vertex above q_-, rendering contribution (iia). Or, $\pi_\sharp(q'_-)$ could join the diagram in steps $j+1$ through ℓ, rendering contributions (iiib) and (ivb) if it is immediately contracted to some $q'_- > q_-$, or contributions (iiia) and (iva) if it is yet uncontracted upon entering and gets later contracted to a q'_- in some vertex above the j-th. After step ℓ, only connectors $\pi_\sharp(q'_-)$ below $\pi_\sharp(q_-)$ enter, which do not contribute to the swaps encoded in (37).

Concluding both cases, we observe that multiplying all sign factors (i)–(iv) for all q_- also renders exactly the factor needed to swap the entire diagram into maximally crossed form, viz. $\mathrm{sgn}(\xi, \pi_p, \pi_h)$. This establishes the claim and concludes the proof. □

6 Bosonization Approximation in Friedrichs Diagrams

Formula (39) is rather bulky, as it contains many contributions of various forms. Therefore, it makes sense to restrict to those Friedrichs diagrams, which we expect to make the largest contribution. These are exactly the diagrams corresponding to the bosonization approximation in [1], as we will see in this section.

6.1 *Heuristic Motivation*

To get a heuristic intuition of which diagrams make the largest contribution, consider the first two Friedrichs diagrams in Fig. 10, which both contribute to $\langle \Omega, \mathrm{ad}^{6s}(a_q^* a_q)\Omega\rangle$. As a contraction $\delta_{q_-,\pi_\sharp(q_-)}$ sets the momenta of the two adjacent connectors q_- and $\pi_\sharp(q_-)$ equal, it also sets the patch index of the corresponding $c^\sharp$-vertices equal to some α_j. The set of $c^\sharp$-vertices thus decays into subsets of size ≥ 2 with identical patch indices. In each subset of m $c^\sharp$-vertices, the m adjacent contractions form a single loop that successively runs through all vertices. For instance, the first diagram in Fig. 10 contains 3 loops, which run through $m = 6$, 4 and 2 $c^\sharp$-vertices, respectively. The loop with 6 $c^\sharp$-vertices also runs through $a_q^* a_q$, so its patch index is fixed to α_q. The two shorter loops carry patch indices α_1 and α_2, over which we have to take a double sum $\sum_{\alpha_1,\alpha_2}$.

As each sum $\sum_{\alpha_j}$ contains $\sim M$ terms, we expect the largest contributions to come from diagrams with the largest loop number. This is achieved if all loops[6] have length 2, resulting in n loops, as depicted in the second diagram in Fig. 10.

[6] The length of a loop refers to the number of $c^\sharp$-vertices it runs through, not taking into consideration $a_q^* a_q$.

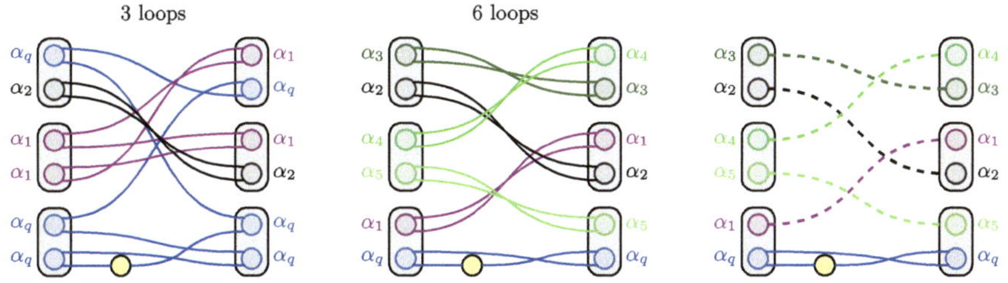

Fig. 10 Left: A generic Friedrichs diagram with 3 loops of lengths 6, 4, and 2. Middle: A Friedrichs diagram with maximal number of loops 6, all of length 2. Right: Bosonization—fermionic contraction pairs have been replaced by bosonic contractions

The diagrammatic contribution then contains an $(n-1)$-fold sum $\sum_{\alpha_1,\dots,\alpha_{n-1}}$. In that case, the two contractions in each loop (except the one running through the $a_q^* a_q$-vertex) form a pair and effectively act like a single bosonic contraction, as depicted in the third diagram in Fig. 10.

We can thus think of the restriction to diagrams where only loops of length 2 are permitted as some kind of bosonization. These diagrams turn out to be particularly easy to evaluate and are expected to give the largest contribution to n_q.

6.2 Evaluating the Bosonized Multicommutator Diagrammatically

Proof (Proof of Proposition 1) We directly evaluate the right-hand side of (48) diagrammatically and show that it amounts to the cosh-term in (24). First, notice that the constraint (46), in the language of the previous subsection, exactly means that we restrict to bosonized diagrams with only loops including 2 $c^\sharp$-vertices. Thus, the right-hand side of (48) is just a sum over all bosonized diagrams.

Let us evaluate these diagrams (compare also Fig. 10). First, note that there are n loops, where in each loop, the contractions set the adjacent patch indices equal. The patch index of the loop involving the $a_q^* a_q$-vertex is fixed to α_q. So the patch index sum $\sum_{\boldsymbol{\alpha},\boldsymbol{\alpha}'}$ reduces to a sum over $n-1$ loops, namely those in $\boldsymbol{\alpha} \setminus \alpha_j$. The contractions in a loop also set all momentum transfers k_j of the adjacent $S_\pm$-vertices equal. Now observe that all bosonized diagrams are fully connected, since otherwise constraint (33) is violated. Thus, all k_j are set equal to one single momentum transfer $k \in \Gamma^{\text{nor}}$ and the n -fold sum $\sum_{\boldsymbol{K}}$ reduces to a single sum $\sum_k$. The contributing K-matrix elements then become $K(k)_{\alpha_j,\beta(\alpha_j)}$ with $\beta : \boldsymbol{\alpha} \to \boldsymbol{\alpha}$ being an appropriate cyclic permutation.

Further, by the bosonization assumption (46), $p_j^\sharp = \pi_p(p_j^\sharp)$ already implies $h_j^\sharp = \pi_h(h_j^\sharp)$ so the product $\prod_{h\in \boldsymbol{H}_- \cup \boldsymbol{H}'_-} \delta_{h,\pi_h(h)}$ in (48) becomes redundant and can be eliminated. Also, the factor $\delta_{p_j,h_j\pm k_j}\delta_{p'_j,h'_j\pm k_j} = \delta_{p_j,h_j\pm k}\delta_{p'_j,h'_j\pm k}$ in (48) eliminates the sums over $\boldsymbol{H}$, $\boldsymbol{H}'$, while leaving the condition that $h_j = p_j \mp k$ and $h'_j = p'_j \mp k$ be holes in patch α_j. Let us denote these conditions as

$$\chi(\boldsymbol{H}, \boldsymbol{H}' : \boldsymbol{\alpha}) := \prod_{j=1}^{n} \chi(h_j, h'_j : \alpha_j) = \prod_{j=1}^{n} \chi(p_j \mp k \in B_{\alpha_j} \cap B_{\mathrm{F}}) \chi(p'_j \mp k \in B_{\alpha_j} \cap B_{\mathrm{F}}) . \tag{63}$$

So the r.h.s. of (48) becomes

$$\text{r.h.s.} = \sum_{\substack{n=2 \\ n:\text{even}}}^{\infty} \frac{1}{2^n n!} \sum_{\xi \in \Xi_n} \sum_{(\pi_p, \pi_h) \in \Pi^{(\xi)}_{n,(\mathrm{b})}} \sum_{k} \sum_{\boldsymbol{\alpha} \setminus \alpha_q} \sum_{\boldsymbol{P}, \boldsymbol{P}'} \chi(\boldsymbol{H}, \boldsymbol{H}' : \boldsymbol{\alpha}) \times$$
$$\times \left(\prod_{p \in \boldsymbol{P}_- \cup \boldsymbol{P}'_-} \delta_{p, \pi_p(p)} \right) \left(\prod_{j=1}^{n} \frac{1}{n^2_{\alpha_j, k}} K(k)_{\alpha_j, \beta(\alpha_j)} \right) \delta_{q, p_0} \delta_{q, p'_0} \operatorname{sgn}(\xi, \pi_p, \pi_h) . \tag{64}$$

Now, the contractions $\delta_{p,\pi_p(p)}$, not involving p_0, eliminate the sums in $\boldsymbol{P}$, $\boldsymbol{P}'$ over all connectors on the right, that is, over all momentum indices $p_j \in \boldsymbol{P}_+ \setminus p_0$ and $p'_j \in \boldsymbol{P}'_+$, or equivalently, all p_j, p'_j with $\xi(j) = 1$. So only those p_j, p'_j on the left ($\xi(j) = -1$) and p_0, p'_0 survive. The condition $h_j, h'_j : \alpha_j$ is then automatically fulfilled for all j with $\xi(j) = 1$, so we only need to impose it on those j with $\xi(j) = -1$:

$$\begin{aligned} &\sum_{\boldsymbol{P}, \boldsymbol{P}'} \chi(\boldsymbol{H}, \boldsymbol{H}' : \boldsymbol{\alpha}) \left(\prod_{p \in \boldsymbol{P}_- \cup \boldsymbol{P}'_-} \delta_{p, \pi_p(p)} \right) \delta_{q, p_0} \delta_{q, p'_0} \\ &= \sum_{p_0, p'_0} \left(\prod_{j:\xi(j)=-1} \sum_{p_j, p'_j} \chi(h_j, h'_j : \alpha_j) \right) \delta_{\pi_p^{-1}(p_0), p_0} \delta_{q, p_0} \delta_{q, p'_0} . \end{aligned} \tag{65}$$

The sums $\sum_{p_j, p'_j}$ run over all particle–hole pairs and thus amount to $n^2_{\alpha_j, k}$, except for the sum over $\pi_p^{-1}(p_0)$, which is eliminated by $\delta_{\pi_p^{-1}(p_0), p_0} \delta_{q, p_0}$. Now observe that every loop with unique patch index α_j contains exactly one index p_j or p'_j on the left ($\xi(j) = -1$). So we get exactly one factor $n^2_{\alpha_j, k}$ for every α_j, except for $n^2_{\alpha_q, k}$. Thus,

$$\text{r.h.s.} = \sum_{\substack{n=2 \\ n:\text{even}}}^{\infty} \frac{1}{2^n n!} \sum_{\xi} \sum_{(\pi_p, \pi_h)} \sum_{k} \frac{1}{n^2_{\alpha_q, k}} \sum_{\boldsymbol{\alpha} \setminus \alpha_q} \left(\prod_{j=1}^{n} K(k)_{\alpha_j, \beta(\alpha_j)} \right) \operatorname{sgn}(\xi, \pi_p, \pi_h) . \tag{66}$$

Note that the sum over $k \in \Gamma^{\mathrm{nor}}$ here gets reduced to those k with $\alpha_q \in \mathcal{I}_k$ and $q \mp k : \alpha_q$, as otherwise, the contribution vanishes. So comparing with (26), the sum becomes $\sum_k = \sum_{k\in\tilde{C}^q\cap\mathbb{Z}^3}$. As β is cyclic, the sum in $\boldsymbol{\alpha} \setminus \alpha_j$ over the K-matrix elements amounts to an $(n-1)$-fold matrix multiplication, so

$$\text{r.h.s.} = \sum_{\substack{n=2\\ n:\text{even}}}^{\infty} \frac{1}{2^n n!} \sum_{\xi} \sum_{(\pi_p,\pi_h)} \sum_{k\in\tilde{C}^q\cap\mathbb{Z}^3} \frac{1}{n^2_{\alpha_q,k}} (K(k)^n)_{\alpha_q,\alpha_q} \operatorname{sgn}(\xi,\pi_p,\pi_h) \,. \tag{67}$$

Next, we evaluate the sign factor $\operatorname{sgn}(\xi,\pi_p,\pi_h)$, which is the same one needed to bring the entire diagram into maximally crossed form while ignoring $a^*_q a_q$, see Fig. 8. Observe that the maximally crossed form of the diagram obeys the bosonization structure, that is, it satisfies (46). Every other bosonized diagram can be derived from it by a finite number of swaps of two $c^\sharp$-vertices, each amounting to 4 swaps of fermionic connectors. So the total number of fermionic swaps is divisible by 4 and thus even, which immediately yields $\operatorname{sgn}(\xi,\pi_p,\pi_h) = 1$. Finally, it remains to count the admissible diagrams, indexed by (ξ,π_p,π_h), contributing to the r.h.s.. For this, observe that the topological structure of all contributing diagrams is the same in the following sense: We can transform any diagram into any other by changing the order in which $S_\pm$-vertices enter the diagram and swapping the two $c^\sharp$-vertices inside certain $S_\pm$-vertices. In total, there are 2^n ways to select in which of the n vertices $S_\pm$ the two $c^\sharp$-vertices shall be swapped. Next, consider the order in which the $S_\pm$-vertices enter the diagram. At each contraction step $j \in \{1,\dots,n-1\}$, there are two $S_\pm$-vertices which may join the existing diagram at that point, out of which one can be chosen. This renders 2^{n-1} distinct orders for the $S_\pm$-vertices. Thus, $\sum_\xi \sum_{(\pi_p,\pi_h)}$ yields 2^{2n-1} diagrams of identical value contributing to the r.h.s. of (48). So comparing with (24),

$$\begin{aligned} \text{r.h.s.} &= \sum_{\substack{n=2\\ n:\text{even}}}^{\infty} \frac{2^{n-1}}{n!} \sum_{k\in\tilde{C}^q\cap\mathbb{Z}^3} \frac{1}{n^2_{\alpha_q,k}} (K(k)^n)_{\alpha_q,\alpha_q} \\ &= \frac{1}{2} \sum_{k\in\tilde{C}^q\cap\mathbb{Z}^3} \frac{1}{n^2_{\alpha_q,k}} \big(\cosh(2\,K(k)) - 1\big)_{\alpha_q,\alpha_q} = n_q^{(\mathrm{b})} \,, \end{aligned} \tag{68}$$

which establishes (48). □

Acknowledgements This research was supported by the European Research Council (ERC) through the Starting Grant FermiMath, Grant Agreement No. 101040991 of Niels Benedikter, as well as by Gruppo Nazionale per la Fisica Matematica (GNFM) in Italy. The author thanks Niels Benedikter for helpful discussions.

References

1. Benedikter, N., Lill, S.: Momentum Distribution of a Fermi Gas in the Random Phase Approximation (2023). https://arxiv.org/abs/2310.02706
2. Benedikter, N., Nam, P.T., Porta, M., Schlein, B., Seiringer, R.: Optimal upper bound for the correlation energy of a fermi gas in the mean-field regime. Commun. Math. Phys. **374**, 2097–2150 (2020)
3. Benedikter, N., Nam, P.T., Porta, M., Schlein, B., Seiringer, R.: Correlation Energy of a weakly interacting Fermi Gas. Invent. Math. **225**, 885–979 (2021)
4. Benedikter, N., Nam, P.T., Porta, M., Schlein, B., Seiringer, R.: Bosonization of fermionic many-body dynamics. Ann. Henri Poincaré **23**, 1725–1764 (2022)
5. Benedikter, N., Porta, M., Schlein, B., Seiringer, R.: Correlation energy of a weakly interacting fermi gas with large interaction potential. Arch. Ration. Mech. Anal. **247**, 65 (2023)
6. Bohm, D., Huang, K., Pines, D.: Role of subsidiary conditions in the collective description of electron interactions. Phys. Rev. **107**, 71–80 (1957)
7. Bohm, D., Pines, D.: A collective description of electron interactions: III. Coulomb interactions in a degenerate electron gas. Phys. Rev. **92**(3), 609–625 (1953)
8. Brooks, M., Lill, S.: Friedrichs diagrams: bosonic and fermionic. Lett. Math. Phys. **113**, 101 (2023)
9. Christiansen, M.R., Hainzl, C., Nam, P.T.: The random phase approximation for interacting fermi gases in the mean-field regime. Forum Math. Pi **11**, e32 (2023)
10. Christiansen, M.R., Hainzl, C., Nam, P.T.: On the effective quasi-bosonic Hamiltonian of the electron gas: collective excitations and plasmon modes. Lett. Math. Phys. **112**, 114 (2022)
11. Christiansen, M.R., Hainzl, C., Nam, P.T.: The Gell-Mann-Brueckner formula for the correlation energy of the electron gas: a rigorous upper bound in the mean-field regime. Commun. Math. Phys. **401**, 1469–1529 (2023)
12. Coldwell-Horsfall, R.A., Maradudin, A.A.: Zero-Point energy of an electron lattice. J. Math. Phys. **1**(5), 395–404 (1960)
13. Coldwell-Horsfall, R.A., Maradudin, A.A.: Erratum: zero-point energy of an electron lattice. J. Math. Phys. **4**(4), 582 (1963)
14. Daniel, E., Vosko, S.: Momentum distribution of an interacting electron gas. Phys. Rev. **120**(6), 2041–2044 (1960)
15. Dereziński, J., Gérard, C.: Mathematics of Quantization and Quantum Fields. Cambridge University Press (2013)
16. Falconi, M., Giacomelli, E.L., Hainzl, C., Porta, M.: The Dilute Fermi Gas via Bogoliubov Theory. Ann. Henri Poincaré **22**, 2283–2353 (2021)
17. Feldman, J.S., Osterwalder, K.: The Wightman Axioms and the mass gap for weakly coupled $(\phi^4)_3$ quantum field theories. Ann. Phys. **97**, 80–135 (1976)
18. Friedrichs, K.O.: Perturbation of Spectra in Hilbert Space. American Mathematical Soc. (1965)
19. Gell-Mann, M., Brueckner, K.A.: Correlation energy of an electron gas at high density. Phys. Rev. **106**(2), 364–368 (1957)
20. Giacomelli, E.L.: Bogoliubov theory for the dilute Fermi gas in three dimensions. In: Correggi, M., Falconi, M. (eds.), Quantum Mathematics II, Springer INdAM Series 58. Springer, Singapore (2023)
21. Giacomelli, E.L.: An optimal upper bound for the dilute Fermi gas in three dimensions. J. Funct. Anal. **285**(8), 110073 (2023)
22. Glimm, J.: Boson fields with the $:\Phi^4:$ interaction in three dimensions. Commun. Math. Phys. **10**, 1–47 (1968)
23. Glimm, J., Jaffe, A.: Positivity of the φ_3^4 Hamiltonian. Fortschr. Phys. **21**, 327–376 (1973)
24. Hepp, K.: Théorie de la renormalisation, Lecture Notes in Physics. Springer, Berlin (1969)
25. Lam, J.: Correlation energy of the electron gas at Metallic Densities. Phys. Rev. B **3**(6), 1910–1918 (1971)
26. Lam, J.: Momentum distribution and pair correlation of the electron gas at metallic densities. Phys. Rev. B **3**(10), 3243–3248 (1971)

27. Lieb, E.H., Mattis, D.C.: Exact solution of a Many-Fermion system and its associated boson field. J. Math. Phys. **6**(2), 304–312 (1965)
28. Luttinger, J.M.: An exactly soluble model of a many-fermion system. J. Math. Phys. **4**(9), 1154–1162 (1963). https://doi.org/10.1063/1.1704046
29. Sawada, Katuro: Correlation energy of an electron gas at high density. Phys. Rev. **106**(2), 372–383 (1957). https://doi.org/10.1103/PhysRev.106.372
30. Sawada, K., Brueckner, K.A., Fukuda, N., Brout, R.: Correlation energy of an electron gas at high density: plasma oscillations. Phys. Rev. **108**(3), 507–514 (1957). https://doi.org/10.1103/PhysRev.108.507

Self-Adjointness and Domain of Generalized Spin–Boson Models with Mild Ultraviolet Divergences

Sascha Lill and Davide Lonigro

Abstract We provide a rigorous construction of a large class of generalized spin–boson models with ultraviolet-divergent form factors. This class comprises various models of many possibly non-identical atoms with arbitrary but finite numbers of levels, interacting with a boson field. Ultraviolet divergences are assumed to be mild, such that no self-energy renormalization is necessary. Our construction is based on recent results by A. Posilicano, which also allow us to state an explicit formula for the domain of self-adjointness for our Hamiltonians.

Keywords Spin–boson model · Renormalization · Ultraviolet divergence · Self-adjointness domain · Interior–boundary conditions · Scales of Hilbert spaces

2020 Mathematics Subject Classification 46N50 · 47A10 · 47B25 · 81Q10 · 81T16

1 Introduction and Main Result

We consider a quantum mechanical system, described by a common finite-dimensional Hilbert space $\mathfrak{h} := \mathbb{C}^D$, coupled to a single boson field, described by an abstract single-particle Hilbert space $\mathcal{H}$. The Hilbert space of the entire system is given by

S. Lill
Dipartimento di Matematica Federigo Enriques, Università degli Studi di Milano, Milan, Italy
e-mail: sascha.lill@unimi.it

D. Lonigro (✉)
Department Physik, Friedrich-Alexander-Universität Erlangen-Nürnberg, Staudtstraße 7, 91058 Erlangen, Germany
e-mail: davide.lonigro@fau.de

Dipartimento di Matematica, Università degli Studi di Bari Aldo Moro, Bari, Italy

Istituto Nazionale di Fisica Nucleare, Sezione di Bari, Bari, Italy

B. Cassano et al. (eds.), *Singularities, Asymptotics, and Limiting Models*, Springer INdAM Series 64, https://doi.org/10.1007/978-981-96-3584-9_7

$$\mathfrak{H} := \mathfrak{h} \otimes \mathcal{F} , \qquad \mathcal{F} := \mathcal{F}^{+}(\mathcal{H}) = \bigoplus_{n=0}^{\infty} S_n \mathcal{H}^{\otimes n} , \tag{1}$$

with S_n being the symmetrization operator, and $\mathcal{F}$ the *bosonic Fock space*. The Hamiltonian of the non-interacting system is of the form

$$H_{\text{free}} := K \otimes I + I \otimes \mathrm{d}\Gamma(\omega) , \tag{2}$$

where $K \in \mathbb{C}^{D \times D}$ is a self-adjoint matrix describing the energy of the atoms, and $\mathrm{d}\Gamma(\omega)$ is the second quantization of some positive self-adjoint operator $\omega \geq m_0 > 0$ defined on a dense subspace of $\mathcal{H}$. m_0 is also called the *mass of the boson*. Without loss of generality, and for physical convenience, we will fix $\mathcal{H}$ as the space of square-integrable complex-valued functions on some measure space (X, Σ, μ), and ω as the multiplication by some measurable real-valued function $k \in X \mapsto \omega(k) \in \mathbb{R}$.

The interaction between the quantum system and the field will be constructed as follows. Given $N \in \mathbb{N}$, we take a family of matrices $B_j \in \mathbb{C}^{D \times D}$, $j \in \{1, \dots, N\}$, and a family of form factors f_j, $j \in \{1, \dots, N\}$, each being a measurable function $f_j : X \to \mathbb{C}$. We consider an interaction term in the form $A^* + A$, where, formally,

$$A := \sum_{j=1}^{N} B_j^* \otimes a(f_j) , \qquad A^* := \sum_{j=1}^{N} B_j \otimes a^*(f_j) , \tag{3}$$

with $a^*(f_j), a(f_j)$ (formally) being the bosonic creation/annihilation operators on $\mathcal{F}$ defined by requiring

$$\begin{aligned} a^*(f_j) S_n(\phi_1 \otimes \ldots \otimes \phi_n) &:= \sqrt{n+1}\, S_n(f_j \otimes \phi_1 \otimes \ldots \otimes \phi_n) , \\ a(f_j) S_n(\phi_1 \otimes \ldots \otimes \phi_n) &:= \frac{1}{\sqrt{n}} S_n \sum_{\ell=1}^{n} \langle f_j, \phi_\ell \rangle (\phi_1 \otimes \ldots \otimes \phi_{\ell-1} \otimes \phi_{\ell+1} \otimes \ldots \otimes \phi_n) \end{aligned} \tag{4}$$

for any $\phi_1, \dots, \phi_n \in \mathcal{H}$ and $n \geq 0$ (for $a^*(f_j)$) or $n \geq 1$ (for $a(f_j)$), respectively. One easily checks that these operators satisfy the canonical commutation relations (CCR)

$$[a(f), a(g)] = [a^*(f), a^*(g)] = 0 , \qquad [a(f), a^*(g)] = \langle f, g \rangle . \tag{5}$$

Thus, the total Hamiltonian formally reads

$$\begin{aligned} H = H_{f_1, \dots, f_N} :=& H_{\text{free}} + A^* + A \\ =& H_{\text{free}} + \sum_{j=1}^{N} \left[B_j \otimes a^*(f_j) + B_j^* \otimes a(f_j) \right] . \end{aligned} \tag{6}$$

Such models, denoted as *generalized spin–boson* (GSB) *models*, were first introduced, to our best knowledge, in [6] as a generalization of the well-known spin–boson model; their mathematical properties have been extensively addressed in the literature, cf. [4, 5, 14, 23, 38–41, 43, 60] for the single spin case and [8, 9, 22, 67–69] for the general one.

The aim of this work is to give, for a general choice of (N, D) and of the operators $B_1, \ldots, B_N$, a rigorous interpretation of Eq. (6) as a self-adjoint operator on a dense subspace of $\mathfrak{H}$ in the case in which the form factors $f_1, \ldots, f_N$ are not square-integrable—a problem that is also called *non-perturbative renormalization problem.* At this point, we did not specify yet in which function space the form factors f_j have to be accommodated. And indeed, the difficulty of the renormalization problem heavily depends on the growth properties of f_j.

To elucidate how the difficulty of the renormalization problem depends on the growth of f_j, let us state some well-known results in the case $N = D = 1$, also known as the *van Hove model* [17]. In that case, we only have one form factor $f : X \to \mathbb{C}$. Depending on the behavior of f with respect to the dispersion relation ω, one can now essentially distinguish 4 difficulty classes for the renormalization problem.

Case 0: $\int |f|^2 < \infty \Leftrightarrow f \in \mathcal{H}$. This is the simplest case. Both A^* and A are Kato perturbations of H_{free}, so Eq. (6) readily defines a self-adjoint operator H on $\operatorname{Dom}(H) = \operatorname{Dom}(H_{\text{free}})$.

Case 1: $\int |f|^2 = \infty$, but still, $\int \frac{|f|^2}{\omega} < \infty$. In this case, $A^* + A$ is still a form perturbation of H_{free}, so the KLMN theorem allows defining H as a self-adjoint operator. However, the perturbation changes the domain.

Case 2: $\int \frac{|f|^2}{\omega} = \infty$, but still, $\int \frac{|f|^2}{\omega^2} < \infty$. Here, we can no longer make sense of Eq. (6) without further modifications. In fact, one needs a *self-energy renormalization procedure*, to obtain a well-defined Hamiltonian: for some sequence of UV-cutoffs $\Lambda \in \mathbb{R}$, one introduces the regularized form factors $f_\Lambda \in \mathcal{H}$ such that $\lim_{\Lambda\to\infty} f_\Lambda = f$ in a suitable sense. Substituting f by f_Λ in H then yields the regularized Hamiltonians H_Λ, which are well-defined on a dense subspace of $\mathcal{F}$. By subtracting from them the *self-energies* $E_\Lambda := -\int \frac{|f_\Lambda|^2}{\omega} \in \mathbb{R}$, one may then obtain a renormalized Hamiltonian:

$$\widetilde{H} := \lim_{\Lambda\to\infty} (H_\Lambda - E_\Lambda) \,, \tag{7}$$

which is self-adjoint on a dense domain in $\mathcal{F} \simeq \mathfrak{H}$. In fact, using a (unitary) *dressing operator* $W_\Lambda := \exp\big(a(\omega^{-1} f_\Lambda) - a^*(\omega^{-1} f_\Lambda)\big)$, it turns out that

$$W_\Lambda^* H_\Lambda W_\Lambda = H_{\text{free}} + E_\Lambda \,, \tag{8}$$

so W_Λ extracts the self-energy. In particular,

$$\widetilde{H} = W H_{\text{free}} W^* \,, \tag{9}$$

with the unitary dressing operator $W := \exp\big(a(\omega^{-1}f) - a^*(\omega^{-1}f)\big)$, so the dynamics generated by $\widetilde{H}$ are unitarily equivalent to those of H_{free}.

Case 3: $\int \frac{|f|^2}{\omega^2} = \infty$. This is the most singular case. Formally, Eq. (9) would yield a renormalized operator $\widetilde{H}$ that is equivalent to H_{free}, but since W is ill-defined, Eq. (9) makes mathematically no sense. One way out of this problem is to interpret H as in Eq. (6) as an element of a suitably defined *-algebra $\mathcal{A}$ and to replace the conjugation with W by an (algebraic) *Weyl transformation* $\mathcal{V}_W : \mathcal{A} \to \mathcal{A}$ that maps $H \mapsto H_{\text{free}}$. As $\mathcal{V}_W$ preserves the commutation relations, one may see H_{free} as a legitimate renormalized Hamiltonian, which is self-adjoint on a dense domain in $\mathcal{F}$ [24, Sect. 3].[1]

Another way out is to interpret W as a map from a subspace of $\mathcal{F}$ to some Fock space extension $\overline{\mathcal{F}}$, and to define $E := -\int \frac{|f|^2}{\omega}$ as an element of a suitable vector space [50, 51]. This way, one can give a precise mathematical meaning to the equation

$$H_{\text{free}} = W(H - E)W^* , \tag{10}$$

and interpret H_{free} as the renormalized Hamiltonian.

Coming back to the general case, let us mention that in **Case 0**, i.e., when $f_1, \ldots, f_N \in \mathcal{H}$ for $N, D \in \mathbb{N}$, the renormalization problem is easily solved: $A^* + A$ is again a Kato perturbation of H_{free}, so H as in Eq. (6) is self-adjoint with coupling-independent domain $\text{Dom}(H) = \text{Dom}(H_{\text{free}}) = \mathfrak{h} \otimes \text{Dom}(\mathrm{d}\Gamma(\omega))$.

In this work, we address the renormalization problem for general values of $N, D \in \mathbb{N}$, and form factors $f_1, \ldots, f_N$ corresponding to **Case 1**. Our proofs essentially rely on a recent work by Posilicano [59], which introduces an abstract machinery for renormalizing Hamiltonians of the form $H_{\text{free}} + A^* + A$ and finding their domains. This paper thus demonstrates how these results can be successfully applied to generalized spin–boson models. To our best knowledge, the present work is the first rigorous proof of a renormalization in Case 1 for this kind of models with a general choice of $B_1, \ldots, B_N$: while the existence of a self-adjoint operator associated with Eq. (6) is in fact guaranteed by the KLMN theorem (cf. [53, Proposition 4.2]), explicit expressions for the domain were only obtained in particular cases [53, 54, 57].

Within **Case 2**, the application of the abstract results from [59], while still possible, offers new challenges—as it will be elucidated in Sect. 3 for the abstract case, an unambiguous identification of a self-adjoint operator associated with Eq. (6) is no longer viable. One needs to specify a suitably "regularizing operator" T which must be both mathematically admissible and physically meaningful. This problem will be

[1] More precisely, one considers a subalgebra $\mathcal{A}_{\text{obs}} \subset \mathcal{A}$, containing all relevant observables, and constructs a representation π of $\mathcal{A}_{\text{obs}}$ on some domain $\mathcal{D} \subset \mathcal{F}$, in which $\pi(H) = H_{\text{free}}$. So, H acts as if it were a free Hamiltonian. However, in this representation, $\pi(a^*(\varphi))$, $\pi(a(\varphi))$ for $\varphi \in \mathcal{H}$ (if they are defined) do not act like creation/annihilation operators in Eq. (4), and there is also no unitary operator $W : \mathcal{F} \to \mathcal{F}$ that makes $W\pi(a^*(\varphi))W^*$, $W\pi(a(\varphi))W^*$ act like creation/annihilation operators. So π is inequivalent to the standard Fock representation. Equivalently, one also says that π is a non-Fock representation, or that H leaves the Fock space.

investigated elsewhere. It would be highly desirable to also achieve self-adjointness results in **Case 3**; however, the construction of a suitable dressing transformation W for general values of N and D is far more challenging than within the van Hove model. We hope to achieve this goal in future research.

For other models of matter–field interaction, there exists an extensive mathematical literature on the non-perturbative renormalization problem. References about successful renormalization in Cases 1 and 2 include [20, 25, 32, 33, 58, 65]. A successful renormalization in Case 3 was achieved for a related model by Gross [34]. We refer to [51, Sect. 1.3] for a more detailed discussion of the related literature. We also point out that there exists a recently introduced renormalization technique based on the so-called "interior–boundary conditions" (IBC) [45–48, 62, 70, 71], which is closely related to the work of Posilicano and ours. IBC has successfully rendered renormalized Hamiltonians together with their domains in Cases 1 and 2. For a more detailed overview about the recent IBC literature, we refer the reader to [51, Sect. 1.4].

Furthermore, the renormalization problem has been addressed and solved in the context of Constructive Quantum Field Theory (CQFT), e.g., [26–30]. Here, Haag's Theorem [36, 37] requires to "leave the Fock space", as in Case 3 above, while relativistic causality admits the employment of further techniques (e.g., Segal's theorem [63], which circumvents the search for a dressing transformation). We refer to [15, 66] for an extensive overview of the literature on CQFT.

For generalized spin–boson models, there is a plenitude of applications in the physics literature, often requiring UV-divergent form factors [12, 49, 74], common examples being the description of subradiance and superradiance phenomena in quantum optics [18, 35, 73], e.g., bound states in the continuum in photonic waveguides [19, 21, 31, 52, 61, 72], or toy models of quantum Markovian systems [55, 56]. On the mathematical side, there is a rich literature about the spectrum and dynamics of the spin–boson model, see e.g., [7, 10, 16, 23, 38, 44, 69] and the references therein; however, the form factors often contain a UV-cutoff and the renormalization problem is not addressed. Concerning the renormalization of the spin–boson model, we point out a recent result by Dam and Møller [13], proving that certain spin-boson models with $\mathfrak{h} = \mathbb{C}^2$ in Case 3 renormalize to a direct sum of two van Hove Hamiltonians for an appropriately chosen dressing transformation. Finally, an explicit renormalization of a particular class of spin–boson models was recently addressed by one of the authors, first for $N = 1$ in Cases 1 and 2 [53, 54] and then for $N > 1$ in Case 1 [57]. However, the present article is, up to our best knowledge, the first work establishing self-adjointness *and* a formula for the domain in the general case.

To achieve this result, we will resort to the formalism of *scales of Hilbert spaces* [2, 3, 64]. Essentially, one introduces two scales of spaces $\{\mathcal{H}_s\}_{s\in\mathbb{R}}$, $\{\mathcal{F}_s\}_{s\in\mathbb{R}}$, obtained as the completion of a space of sufficiently well-behaved vectors under the norms

$$\|f\|_{\mathcal{H}_s} := \left\|\omega^{s/2} f\right\|_{\mathcal{H}}, \qquad \|\Psi\|_{\mathcal{F}_s} := \left\|(\mathrm{d}\Gamma(\omega) + 1)^{s/2}\,\Psi\right\|_{\mathcal{F}}. \tag{11}$$

Of course, for $s = 0$ one obtains the original spaces, and $\mathcal{H}_{+s} \subset \mathcal{H} \subset \mathcal{H}_{-s}$, as well as $\mathcal{F}_{+s} \subset \mathcal{F} \subset \mathcal{F}_{-s}$ for any $s \geq 0$, with all inclusions being dense. Mathematically, $\mathcal{H}_{-s}$ is a space of non-normalizable functions which, however, satisfy a growth constraint weaker than the one of L^2: a more negative $-s$ corresponds to stronger allowed growths where $\omega(k)$ is large, that is, to stronger UV-singularities. In fact, we can conveniently express the divergence case in terms of the index s:

- **Case 0**: $f_1, \ldots, f_N \in \mathcal{H} = \mathcal{H}_0$;
- **Case 1**: $f_1, \ldots, f_N \in \mathcal{H}_{-1} \setminus \mathcal{H}_0$;
- **Case 2**: $f_1, \ldots, f_N \in \mathcal{H}_{-2} \setminus \mathcal{H}_{-1}$;
- **Case 3**: $f_1, \ldots, f_N \notin \mathcal{H}_{-2}$.

Within this formalism, given $s \geq 1$ and $f \in \mathcal{H}_{-s}$, one can define two continuous linear operators $a(f) \in \mathcal{B}(\mathcal{F}_{+s}, \mathcal{F})$ and $a^*(f) \in \mathcal{B}(\mathcal{F}, \mathcal{F}_{-s})$, whose action is compatible with the "regular" creation and annihilation operators (4) for $f \in \mathcal{H}$, and which can be approximated, in the respective operator norms, by sequences of regular creation and annihilation operators (see e.g., [53, Proposition 3.7]).

1.1 Main Result

Our results are based on a recent work by Posilicano [59], which provides abstract conditions for the self-adjointness of Hamiltonians corresponding to the formal expression $H_{\text{free}} + A^* + A$, see Sect. 3. The conditions on the matrices B_j under which the results of [59] apply to our Hamiltonian H in Eq. (6) can be concluded as follows:

Assumption 1 The interaction matrices B_j, $j \in \{1, \ldots, N\}$, are chosen such that:

(i) every B_j is normal, i.e., $B_j^* B_j = B_j B_j^*$,
(ii) $[B_j, B_\ell] = 0$, $j, \ell = 1, \ldots, N$, and
(iii) $\bigcap_{j=1}^N \operatorname{Ker} B_j = \{0\}$.

As discussed, we will examine the case $f_1, \ldots, f_N \notin \mathcal{H}$. In fact, we will even require the stronger assumption that no linear combination of the f_j lies in $\mathcal{H}$ (except for the trivial one, of course).

Definition 1 (*$\mathcal{H}$-independence*; [2, Definition 3.1.1]) A family $f_1, \ldots, f_N \in \mathcal{H}_{-s} \setminus \mathcal{H}$, $s > 0$ is called *$\mathcal{H}$-independent*, if and only if for all $c_1, \ldots, c_N \in \mathbb{C}$,

$$\sum_{j=1}^{N} c_j f_j \notin \mathcal{H}\,, \quad \text{unless } c_j = 0 \; \forall j\,. \tag{12}$$

A fortiori, an $\mathcal{H}$-independent family of form factors is also a family of linearly independent elements of $\mathcal{H}_s$.

Also notice that there is no loss of generality in requiring the operators B_j's to be linearly independent: if this is not the case, i.e., if there exist $\alpha_1, \dots, \alpha_{N-1} \in \mathbb{C}$ such that (up to permutations) $B_N = \sum_{j=1}^{N-1} \alpha_j B_j$, then

$$A = \sum_{j=1}^{N-1} B_j^* \otimes a(f_j) + \sum_{j=1}^{N-1} \overline{\alpha_j} B_j^* \otimes a(f_N) = \sum_{j=1}^{N-1} B_j^* \otimes a\big(f_j + \alpha_j f_N\big) \; .$$

This request is compatible with the $\mathcal{H}$-independence of the form factors: given an $\mathcal{H}$-independent set $\{f_j\}_{j=1,\dots,N}$, the set of form factors after the rearrangement, $\{f_j + \alpha_j f_N\}_{j=1,\dots,N-1}$, is still $\mathcal{H}$-independent.

Theorem 2 (Main Result) *Let $f_1, \dots, f_N \in \mathcal{H}_{-s} \setminus \mathcal{H}$, $s \leq 1$, be a family of $\mathcal{H}$-independent form factors and let $B_1, \dots, B_N \in \mathbb{C}^{D \times D}$ satisfy Assumption 1. Then the following statements hold true:*

(a) The operator $H = H_{\text{free}} + A^* + A$ as in Eq. (6), *with H_{free} as in Eq.* (2) *and A, A^* as in Eq.* (3), *is well-defined and self-adjoint on the domain*

$$\text{Dom}(H) = \left\{\Psi \in \mathfrak{H} \; : \; \left[1 + (H_{\text{free}} - z_0)^{-1} A^*\right] \Psi \in \text{Dom}(H_{\text{free}})\right\} \; , \qquad (13)$$

where $z_0 \in \varrho(H_{\text{free}}) \cap \mathbb{R}$.

(b) For every $j \in \{1, \dots, N\}$, there exists an approximating sequence $\{f_{j,n}\}_{n\in\mathbb{N}} \subset \mathcal{H}$ with $\|f_{j,n} - f_j\|_{\mathcal{H}_{-s}} \to 0$ as $n \to \infty$, such that, defining

$$A_n := \sum_{j=1}^{N} B_j^* \otimes a(f_{j,n}) \; , \qquad A_n^* := \sum_{j=1}^{N} B_j \otimes a^*(f_{j,n}) \; , \qquad (14)$$

we have

$$(H_{\text{free}} + A_n^* + A_n) \to (H_{\text{free}} + A^* + A) \qquad (15)$$

as $n \to \infty$ in the norm resolvent sense.

In the above statement, it is understood that the annihilation and creation operators in the definition of A and A^*, whenever involving form factors $f_j \notin \mathcal{H}$, are to be interpreted as elements, respectively, of $\mathcal{B}(\mathcal{F}_{+1}, \mathcal{F})$ and $\mathcal{B}(\mathcal{F}, \mathcal{F}_{-1})$, cf. Proposition 1.

Remark 1 (Identical atom spaces) In the case in which the quantum system is composed by N d-dimensional smaller systems (atoms) each separately interacting with the boson field, it is common to associate to each atom a Hilbert space $\mathbb{C}^d$, so $\mathfrak{h} = \mathbb{C}^{d^N}$, i.e., $D = d^N$. Typically, B_j then only acts on the tensor factor of the j-th spin, which means it takes the form

$$B_j = I \otimes \ldots \otimes \underbrace{b_j}_{j\text{-th}} \otimes \ldots \otimes I \qquad (16)$$

for some $b_j \in \mathbb{C}^{d\times d}$. This is a special case of our model (6). All examples introduced in Sect. 2 will fall in this class.

With this choice, (i) in Assumption 1 becomes equivalent to $b_j^* b_j = b_j b_j^*$ and (ii) is trivially satisfied. Point (iii) is fulfilled, if and only if at least one of the matrices $b_1, \ldots, b_N$ has full rank. To see the latter statement, first observe that if some b_j has full rank, then $\operatorname{Ker} b_j = \{0\} \Rightarrow \operatorname{Ker} B_j = \{0\} \Rightarrow \bigcap_j \operatorname{Ker} B_j = \{0\}$. Conversely, if no b_j has full rank, then we can find one $0 \neq u^{(j)} \in \mathbb{C}^d$ for each b_j, such that $b_j u^{(j)} = 0$. But then, $0 \neq u^{(1)} \otimes \ldots \otimes u^{(N)} \in \bigcap_j \operatorname{Ker} B_j = \{0\}$.

Remark 2 (Infinite-dimensional $\mathfrak{h}$) It is not too difficult to see that our result also holds true if $\mathfrak{h}$ is infinite-dimensional and B_j are compact operators satisfying Assumption 1. In fact, the properties of B_j only enter in Propositions 2 and 3, as well as in Eq. (71). In the latter equation, the boundedness of all B_j is exploited to prove that certain operators are closable. Proposition 2 also holds true for B_j being generic bounded operators. In Proposition 3, we construct a common eigenbasis of $\{B_j\}_{j=1}^N$, which would also be possible if B_j were compact operators. It is easy to see that in this case, the proof of Proposition 3 (with an infinite eigenbasis) would still go through.

In the case in which all B_j are bounded and satisfy Assumption 1, the proof of Proposition 3 has to be modified using spectral calculus in order to obtain an analogue of Theorem 2. This generalization is straightforward but technical. To keep the current article short, we postpone it to a future publication.

2 Some Examples

Before starting with the proof of Theorem 2, let us discuss some particular instances of Hamiltonians in the form (6) to which Theorem 2 applies.

The σ_x spin–boson model. Consider the following expression on $\mathbb{C}^2 \otimes \mathcal{F}$:

$$H = \frac{\eta}{2}\sigma_z \otimes I + I \otimes \mathrm{d}\Gamma(\omega) + \sigma_x \otimes \left(a^*(f) + a(f)\right) , \tag{17}$$

where $f \in \mathcal{H}_{-s}$, $s \leq 1$, and $\eta \geq 0$; furthermore, σ_x and σ_z denote the first and third Pauli matrices, respectively:

$$\sigma_x = \begin{bmatrix} 0 & 1 \\ 1 & 0 \end{bmatrix}, \qquad \sigma_z = \begin{bmatrix} 1 & 0 \\ 0 & -1 \end{bmatrix} . \tag{18}$$

This is the model usually referred to as "the" spin–boson model in the mathematical literature, cf. [4, 5, 14, 23, 39–41, 43]. Physically, it describes a two-level system (qubit), whose energy levels differ by a quantity η, undergoing *decay* dynamics as a result of its interaction with the boson field; for example, it is employed in quantum optics to describe the interaction of a two-level atom with an electromagnetic field.

In the monochromatic case ($\mathcal{H} = \mathbb{C}$), it reduces to the well-known Rabi model [11, 42, 75, 76].

This model clearly belongs to the class of operators analyzed in the present work, corresponding to the case $D = 2$, $N = 1$, and with $K = \eta/2\,\sigma_z$, $B = \sigma_x$ (thus $A = \sigma_x \otimes a(f)$). As such, Theorem 2 applies. Using the obvious isomorphism $\mathfrak{H} = \mathbb{C}^2 \otimes \mathcal{F} \simeq \mathcal{F} \oplus \mathcal{F}$—that is, representing the most general element of $\mathfrak{H}$ as

$$\Psi = \begin{bmatrix} \Psi_{\mathrm{e}} \\ \Psi_{\mathrm{g}} \end{bmatrix} : \quad \Psi_{\mathrm{e}}, \Psi_{\mathrm{g}} \in \mathcal{F}\,, \tag{19}$$

with Ψ_{e} and Ψ_{g} respectively playing the role of the "excited" and "ground" wavefunction of the boson field, we can conveniently rewrite Eq. (17) as

$$\begin{aligned} H &= \begin{bmatrix} \mathrm{d}\Gamma(\omega) + \frac{\eta}{2} & a^*(f) + a(f) \\ a^*(f) + a(f) & \mathrm{d}\Gamma(\omega) - \frac{\eta}{2} \end{bmatrix} \\ &= \underbrace{\begin{bmatrix} \mathrm{d}\Gamma(\omega) + \frac{\eta}{2} & \\ & \mathrm{d}\Gamma(\omega) - \frac{\eta}{2} \end{bmatrix}}_{H_{\mathrm{free}}} + \underbrace{\begin{bmatrix} & a^*(f) \\ a^*(f) & \end{bmatrix}}_{A^*} + \underbrace{\begin{bmatrix} & a(f) \\ a(f) & \end{bmatrix}}_{A}\,. \end{aligned} \tag{20}$$

Clearly, Assumption 1 is satisfied since σ_x is Hermitian (thus normal) and invertible. By Theorem 2, the expression above defines a self-adjoint operator on the domain

$$\mathrm{Dom}(H) = \left\{ \begin{bmatrix} \Psi_{\mathrm{e}} \\ \Psi_{\mathrm{g}} \end{bmatrix} : \begin{bmatrix} \Psi_{\mathrm{e}} + \frac{1}{\mathrm{d}\Gamma(\omega) + \frac{\eta}{2} - z_0} a^*(f)\,\Psi_{\mathrm{g}} \\ \Psi_{\mathrm{g}} + \frac{1}{\mathrm{d}\Gamma(\omega) - \frac{\eta}{2} - z_0} a^*(f)\,\Psi_{\mathrm{e}} \end{bmatrix} \in \mathrm{Dom}(H_{\mathrm{free}}) \right\}\,, \tag{21}$$

with $\mathrm{Dom}(H_{\mathrm{free}}) = \mathrm{Dom}(\mathrm{d}\Gamma(\omega)) \oplus \mathrm{Dom}(\mathrm{d}\Gamma(\omega))$ in this representation. Here, z_0 is any fixed real number in the resolvent set of H_{free}.

This expression should be compared with the one in [53, Sects. 5–6], where the *rotating-wave approximation* (RWA) spin–boson model was investigated. The RWA corresponds (cf. [1]) to replacing $\sigma_x \otimes (a(f) + a^*(f))$ by $\sigma_+ \otimes a(f) + \sigma_- \otimes a(f)$, where

$$\sigma_+ = \begin{bmatrix} 0 & 1 \\ 0 & 0 \end{bmatrix} = \sigma_-^*, \qquad \sigma_x = \sigma_+ + \sigma_-\,; \tag{22}$$

physically, counter-rotating terms are removed. The expression of the domain without suppressing such terms is indeed more involved—the "excited" and "ground" components of the total state acquire a *mutual* interdependence, differently from what happens in the rotating-wave case in which only the ground component depends on the excited one. Note that, since σ_+ has a nontrivial kernel and is not normal, Theorem 2 does not apply to the RWA spin–boson model; nevertheless, the self-adjointness domain provided in [53] is in fact compatible with Eq. (13), cf. [54, Sect. 6]. We briefly comment on this point in Sect. 5.

Multi-atom generalization. The discussion above can be immediately generalized to the case of N atoms, cf. Remark 1. Given $N \in \mathbb{N}$, we consider the following expression on $\mathbb{C}^{2^N} \otimes \mathcal{F}$:

$$H = \sum_{j=1}^{N} \frac{\eta_j}{2} \sigma_{z,j} \otimes I + I \otimes \mathrm{d}\Gamma(\omega) + \sum_{j=1}^{N} \sigma_{x,j} \otimes \big(a^*(f_j) + a(f_j)\big) , \tag{23}$$

where $f_1, \dots, f_N$ constitute an $\mathcal{H}$-independent family of form factors in $\mathcal{H}_{-s}, s \leq 1$, and $\eta_1, \dots, \eta_N \geq 0$. The operators $\sigma_{x,j}, \sigma_{z,j}$ are defined as in Eq. (16), i.e.,

$$\sigma_{x,j} = I \otimes \dots \otimes \underbrace{\sigma_x}_{j\text{-th}} \otimes \dots I , \qquad \sigma_{z,j} = I \otimes \dots \otimes \underbrace{\sigma_z}_{j\text{-th}} \otimes \dots I . \tag{24}$$

Again, with this choice of parameters, Assumption 1 is satisfied (cf. Remark 1) and Theorem 2 applies.

Similarly, as in the case $N = 1$, one may find explicit representations of the action of H in matrix form, as well as its self-adjointness domain Dom H in vector form, by exploiting the obvious isomorphism

$$\mathfrak{H} = \mathbb{C}^{2^N} \otimes \mathcal{F} \simeq \overbrace{\mathcal{F} \oplus \dots \oplus \mathcal{F}}^{2^N \text{ times}} . \tag{25}$$

This is a natural choice (though possibly not the most natural one) obtained by representing the tensor product in $\mathbb{C}^{2^N}$ as a Kronecker product. As an example, we shall present the case $N = 2$ and, for the ease of exposition, we shall set $\eta_1 = \eta_2 \equiv \eta$. The most general element of $\mathfrak{H}$ is represented as

$$\Psi = \begin{bmatrix} \Psi_{\mathrm{ee}} \\ \Psi_{\mathrm{eg}} \\ \Psi_{\mathrm{ge}} \\ \Psi_{\mathrm{gg}} \end{bmatrix}, \qquad \Psi_{\mathrm{ee}}, \Psi_{\mathrm{eg}}, \Psi_{\mathrm{ge}}, \Psi_{\mathrm{gg}} \in \mathcal{F} , \tag{26}$$

and, in this representation, Eq. (23) reads

$$\begin{aligned} H &= \begin{bmatrix} \mathrm{d}\Gamma(\omega) + \eta & a^*(f_2) + a(f_2) & a^*(f_1) + a(f_1) & \\ a^*(f_2) + a(f_2) & \mathrm{d}\Gamma(\omega) & & a^*(f_1) + a(f_1) \\ a^*(f_1) + a(f_1) & & \mathrm{d}\Gamma(\omega) & a^*(f_2) + a(f_2) \\ & a^*(f_1) + a(f_1) & a^*(f_2) + a(f_2) & \mathrm{d}\Gamma(\omega) - \eta \end{bmatrix} \\ &= \underbrace{\begin{bmatrix} \mathrm{d}\Gamma(\omega) + \eta & & & \\ & \mathrm{d}\Gamma(\omega) & & \\ & & \mathrm{d}\Gamma(\omega) & \\ & & & \mathrm{d}\Gamma(\omega) - \eta \end{bmatrix}}_{H_{\mathrm{free}}} \end{aligned}$$

$$+\underbrace{\begin{bmatrix} & a^*(f_2)\,a^*(f_1) & \\ a^*(f_2) & & a^*(f_1) \\ a^*(f_1) & & a^*(f_2) \\ & a^*(f_1)\,a^*(f_2) & \end{bmatrix}}_{A^*}+\underbrace{\begin{bmatrix} & a(f_2)\,a(f_1) & \\ a(f_2) & & a(f_1) \\ a(f_1) & & a(f_2) \\ & a(f_1)\,a(f_2) & \end{bmatrix}}_{A}, \tag{27}$$

from which one can reconstruct the analogue of Eq. (21) for $N = 2$ atoms.

We remark that, while we have only considered here the case in which H_{free} does not involve off-diagonal terms (spin–spin interactions), Theorem 2 also covers this possibility—of course, in such a case all expressions, e.g., Eq. (27) in the case $N = 2$, will acquire additional terms and become more involved. Finally, like in the single-atom case, the RWA of this model with mild divergences (which is not covered by Theorem 2) was covered in [57].

The σ_z (dephasing) spin–boson model. Let us consider the following expression:

$$H = \frac{\eta}{2}\sigma_z \otimes I + I \otimes \mathrm{d}\Gamma(\omega) + \sigma_z \otimes \big(a^*(f) + a(f)\big)\,, \tag{28}$$

again with $f \in \mathcal{H}_{-s}$, $s \le 1$, and $\eta \ge 0$. So the (off-diagonal) σ_x-interaction from above has been replaced with a (diagonal) σ_z-interaction. Physically, Eq. (28) describes a two-level system undergoing *decoherence* as a result of its interaction with the boson field. Using the same representation as before, we can write it as

$$\begin{aligned} H &= \begin{bmatrix} \mathrm{d}\Gamma(\omega) + \frac{\eta}{2} + a^*(f) + a(f) & \\ & \mathrm{d}\Gamma(\omega) - \frac{\eta}{2} + a^*(f) + a(f) \end{bmatrix} \\ &= \underbrace{\begin{bmatrix} \mathrm{d}\Gamma(\omega) + \frac{\eta}{2} & \\ & \mathrm{d}\Gamma(\omega) - \frac{\eta}{2} \end{bmatrix}}_{H_{\text{free}}} + \underbrace{\begin{bmatrix} a^*(f) & \\ & a^*(f) \end{bmatrix}}_{A^*} + \underbrace{\begin{bmatrix} a(f) & \\ & a(f) \end{bmatrix}}_{A}. \end{aligned} \tag{29}$$

By Theorem 2, this expression defines a self-adjoint operator on the domain

$$\mathrm{Dom}(H) = \left\{ \begin{bmatrix} \Psi_{\mathrm{e}} \\ \Psi_{\mathrm{g}} \end{bmatrix} : \begin{bmatrix} \Big(1 + \frac{1}{\mathrm{d}\Gamma(\omega) + \frac{\eta}{2} - z_0} a^*(f)\Big)\Psi_{\mathrm{e}} \\ \Big(1 + \frac{1}{\mathrm{d}\Gamma(\omega) - \frac{\eta}{2} - z_0} a^*(f)\Big)\Psi_{\mathrm{g}} \end{bmatrix} \in \mathrm{Dom}(H_{\text{free}}) \right\}, \tag{30}$$

which, differently from the σ_x-case, involves no mutual dependence between the two components Ψ_{e}, Ψ_{g}, due to the inherently diagonal nature of the interaction. In particular, Eq. (29) can be regarded as a direct sum of two van Hove models with different energy offsets $\pm\frac{\eta}{2}$.

Similar expressions can be found for the multi-atom version of the model, following the same arguments presented for the σ_x model—however, while in the previous case the structure of the multi-atom model for $N > 1$ becomes quickly involved, this model will preserve a relatively simple, diagonal structure. For instance, in the case $N = 2$ (and again setting $\eta_1 = \eta_2 \equiv \eta$), one has

$$H = \underbrace{\begin{bmatrix} \mathrm{d}\Gamma(\omega)+\eta & & & \\ & \mathrm{d}\Gamma(\omega) & & \\ & & \mathrm{d}\Gamma(\omega) & \\ & & & \mathrm{d}\Gamma(\omega)-\eta \end{bmatrix}}_{H_{\mathrm{free}}}$$
$$+ \underbrace{\begin{bmatrix} a^*(f_+) & & & \\ & a^*(f_-) & & \\ & & -a^*(f_-) & \\ & & & -a^*(f_+) \end{bmatrix}}_{A^*} + \underbrace{\begin{bmatrix} a(f_+) & & & \\ & a(f_-) & & \\ & & -a(f_-) & \\ & & & -a(f+) \end{bmatrix}}_{A}, \quad (31)$$

where $f_\pm := f_1 \pm f_2$. Of course, the $\mathcal{H}$-independence of the form factors f_1 and f_2 ensures that $f_\pm \in \mathcal{H}_{-1} \setminus \mathcal{H}$, so both combinations are UV-divergent.

3 Self-Adjointness of $H_{\mathrm{free}} + A^* + A$

3.1 Generalized Setting for $H_{\mathrm{free}} + A^* + A$

We shall start by revising the main results obtained in [59] (with a slightly modified notation). These results hold for $\mathfrak{H}$ being any Hilbert space, and H_{free} any densely defined, self-adjoint operator on $\mathfrak{H}$ with domain $\mathrm{Dom}(H_{\mathrm{free}}) \subset \mathfrak{H}$. Besides, we will assume H_{free} to be bounded from below; to simplify the notation, we will then assume $H_{\mathrm{free}} \geq 0$, the general case being recovered by a simple shift. We can then define the Hilbert scale $\{\mathfrak{H}_s\}_{s\in\mathbb{R}}$ associated with H_{free} in the usual way: for any $s \in \mathbb{R}$,

$$\mathfrak{H}_s := \overline{\bigcap_{n\in\mathbb{N}} \mathrm{Dom}(H_{\mathrm{free}}^n)}^{\|\cdot\|_{\mathfrak{H}_s}}, \qquad \|\Psi\|_{\mathfrak{H}^s} := \left\|(H_{\mathrm{free}}+1)^{s/2}\Psi\right\|_{\mathfrak{H}}, \quad (32)$$

with $\mathfrak{H}_{+s} \subset \mathfrak{H}_0 = \mathfrak{H} \subset \mathfrak{H}_{-s}$ for any $s \geq 0$, and the spaces $\mathfrak{H}_{\pm s}$ being naturally dual. Notice that $\mathfrak{H}_{+s} = \mathrm{Dom}(H_{\mathrm{free}}^{s/2})$ for any $s \geq 0$. With a slight abuse of notation, the duality pairing between the two spaces $\mathfrak{H}_{\pm s}$ will sometimes be denoted like the scalar product on $\mathfrak{H}$ by $\langle\cdot,\cdot\rangle$ or $\langle\cdot,\cdot\rangle_{\mathcal{F}_{-s},\mathcal{F}_s}$. Notice that the inclusions

$$\mathcal{B}(\mathfrak{H}_{+s}, \mathfrak{H}_{-s}) \subset \mathcal{B}(\mathfrak{H}_{+s'}, \mathfrak{H}_{-s'}) \quad (33)$$

hold for all $s' > s > 0$. With this formalism, H_{free} can be interpreted as a continuous operator from $\mathfrak{H}_s$ to $\mathfrak{H}_{s-2}$ for any $s \in \mathbb{R}$, i.e., it decreases the regularity by 2. Now, suppose that for some $s' \geq 1$, we have $A \in \mathcal{B}(\mathfrak{H}_{+s'}, \mathfrak{H})$ and $A^* \in \mathcal{B}(\mathfrak{H}, \mathfrak{H}_{-s'})$. These two statements are expected to hold if A^*, A involve linear couplings to a boson field with form factors $f_j \in \mathcal{H}_{-s'}$ (in our case, $f_j \in \mathcal{H}_{-s}, s \leq 1 \Rightarrow f_j \in \mathcal{H}_{-1}$). If they do hold, then the following operator is already well-defined and continuous:

$$(H_{\mathrm{free}} + A^* + A) : \mathfrak{H}_{+s'} \to \mathfrak{H}_{-s'} \,. \quad (34)$$

However, in order to obtain quantum dynamics, we need an operator defined on a dense subspace of $\mathfrak{H}$ with values in $\mathfrak{H}$ (that is, it "does not leave" the Hilbert space $\mathfrak{H}$) and which is self-adjoint. It turns out that in our less singular Case 1 (so $s \in (0, 1]$), this can be achieved by restricting $H_{\text{free}} + A^* + A$ to a suitable domain $\text{Dom}(H)$. The more singular Case 2 (so $s \in (1, 2]$), is expected to require a further modification of Eq. (34), which is formally equivalent to subtracting an infinite self-energy. We comment on this in Remark 4.

The domain $\text{Dom}(H)$ given by [59] is conveniently described in terms of the resolvents

$$R_z := (H_{\text{free}} - z)^{-1}, \qquad z \in \varrho(H_{\text{free}}) \supset \mathbb{C}\,. \tag{35}$$

To simplify notation, as in [59], we fix some $z_0 \in \varrho(H_{\text{free}}) \cap \mathbb{R}$ and set $R := R_{z_0}$. Since $H_{\text{free}} \geq 0$, it is no loss to choose $z_0 = -1$. All resolvents improve regularity by 2, in the sense that they can be interpreted as continuous operators $R_z, R : \mathfrak{H}_s \to \mathfrak{H}_{s+2}$ for any $s \in \mathbb{R}$. Furthermore, the resolvent vanishes as z moves away from the spectrum in the following sense.

Lemma 1 *Let $s \in [0, 2)$, $\{z_n\}_{n\in\mathbb{N}} \subset \mathbb{C}$, $\operatorname{Im} z_n \leq 0$ such that*

$$\lim_{n\to\infty} \text{dist}(z_n, \sigma(H_{\text{free}})) = \infty\,.$$

Then, for any $r \in \mathbb{R}$, we have

$$\lim_{n\to\infty} \|R_{z_n}\|_{\mathfrak{H}_{r-s}\to\mathfrak{H}_r} = 0\,. \tag{36}$$

Proof We first consider the case $r = 0$. By spectral calculus, for $\Psi \in \mathfrak{H}_{-s}$ we have

$$\begin{aligned}
\|R_{z_n}\Psi\|_{\mathfrak{H}} =& \|(H_{\text{free}} - z_n)^{-1}\Psi\|_{\mathfrak{H}} \leq \left\| \frac{(H_{\text{free}}+1)^{\frac{s}{2}}}{H_{\text{free}} - z_n} \right\|_{\mathfrak{H}\to\mathfrak{H}} \|(H_{\text{free}}+1)^{-\frac{s}{2}}\Psi\|_{\mathfrak{H}} \\
\leq& \sup_{\lambda\in\sigma(H_{\text{free}})} \left| \frac{(\lambda+1)^{\frac{s}{2}}}{\lambda - z_n} \right| \|\Psi\|_{\mathfrak{H}_{-s}} \leq \sup_{\lambda} \left| \frac{\lambda+1}{\lambda - z_n} \right|^{\frac{s}{2}} |\lambda - z_n|^{-1+\frac{s}{2}} \|\Psi\|_{\mathfrak{H}_{-s}} \\
\leq& \Bigg(\underbrace{\sup_{\lambda} \frac{\lambda}{|\lambda - z_n|}}_{\leq 1} + \underbrace{\sup_{\lambda} \frac{1}{|\lambda - z_n|}}_{\to 0} \Bigg)^{\frac{s}{2}} \underbrace{\text{dist}(z_n, \sigma(H_{\text{free}}))^{-1+\frac{s}{2}}}_{\to 0} \|\Psi\|_{\mathfrak{H}_{-s}}\,.
\end{aligned} \tag{37}$$

The statement for general $r \in \mathbb{R}$ immediately follows, as for any $\Phi \in \mathfrak{H}_{r-s}$ and $\Psi := (H_{\text{free}} + 1)^{\frac{r}{2}}\Phi \in \mathfrak{H}_{-s}$, we have

$$\|R_{z_n}\Phi\|_{\mathfrak{H}_r} = \|R_{z_n}(H_{\text{free}}+1)^{-\frac{r}{2}}\Psi\|_{\mathfrak{H}_r} = \|(H_{\text{free}}+1)^{-\frac{r}{2}}R_{z_n}\Psi\|_{\mathfrak{H}_r} = \|R_{z_n}\Psi\|_{\mathfrak{H}}\,, \tag{38}$$

and $\|\Psi\|_{\mathfrak{H}_{-s}} = \|\Phi\|_{\mathfrak{H}_{r-s}}$. □

Now, $\text{Dom}(H)$ as in [59] (also cf. [45, 46]) involves vectors of the form

$$\Psi \in \mathfrak{H} : \quad (I + RA^*)\Psi = \Psi_0 \in \mathrm{Dom}(H_{\mathrm{free}}) . \tag{39}$$

The results of [59] are valid in Cases 1 and 2, viz. $s \in (0, 2]$. Now, a direct check shows that a vector Ψ as in Eq. (39) does *not* belong to $\mathfrak{H}_{+s}$ whenever $s > 1$, so that the action of A on Ψ is ill-defined for $s \in (1, 2]$. For that reason, in [59] a modification A_T of A is introduced, which is indexed by a symmetric operator $T : \mathrm{Dom}(T) \subset \mathfrak{H} \to \mathfrak{H}$ and defined as

$$\mathrm{Dom}(A_T) = \left\{\Psi \in \mathfrak{H} : \exists \Phi \in \mathrm{Dom}(T) \text{ s.t. } \Psi + RA^*\Phi \in \mathrm{Dom}(H_{\mathrm{free}})\right\} , \tag{40}$$

$$A_T\Psi = A\left(\Psi + RA^*\Phi\right) + T\Phi . \tag{41}$$

Since $\Psi + RA^*\Phi \in \mathrm{Dom}(H_{\mathrm{free}}) = \mathfrak{H}_{+2} \subset \mathfrak{H}_{+s}$, the operator A_T is well-defined also for $1 < s \leq 2$ (in contrast to A). We conclude that the modified Hamiltonian

$$\mathrm{Dom}(H_T) = \left\{\Psi \in \mathrm{Dom}(T) : (I + RA^*)\Psi = \Psi_0 \in \mathrm{Dom}(H_{\mathrm{free}})\right\} \tag{42}$$

$$\subset \mathrm{Dom}(A_T) \tag{43}$$

$$H_T = H + A^* + A_T \tag{44}$$

is also well-defined. In particular, on $\mathrm{Dom}(H_T)$, the operator A_T defined in [59, Sect. 3] acts as[2]

$$A_T\Psi = A\left(\Psi + RA^*\Psi\right) + T\Psi = A\Psi_0 + T\Psi . \tag{45}$$

As [59] makes statements about H_T, we need to find out which T makes H_T correspond to $H = H_{\mathrm{free}} + A^* + A$ in case $s \in (0, 1]$. A short calculation reveals

$$H_0\Psi = (H_{\mathrm{free}} + A^*)\Psi + A\Psi_0 = (H_{\mathrm{free}} + A^*)\Psi + A\Psi + ARA^*\Psi . \tag{46}$$

Thus, to obtain a renormalized Hamiltonian H_T that physically corresponds to our model Hamiltonian $H_{\mathrm{free}} + A^* + A$, we need to "manually add" the interaction

$$T = -ARA^* \tag{47}$$

to H_0. In fact, for $s \leq 1$ we have $ARA^* \in \mathcal{B}(\mathfrak{H})$, whereas ARA^* is ill-defined for $s > 1$, since we have the mappings and inclusions

$$\mathfrak{H} \xrightarrow{A^*} \mathfrak{H}_{-s} \xrightarrow{R} \mathfrak{H}_{-s+2} \begin{cases} \subset \mathfrak{H}_{+s} \xrightarrow{A} \mathfrak{H}, & 0 \leq s \leq 1 ; \\ \supset \mathfrak{H}_{+s} & 1 < s \leq 2 . \end{cases} \tag{48}$$

---

[2] We remark that the definition of A_T in [59, Sect. 3] requires a unique splitting $\Psi = \Psi_0 - RA^*\Phi$ with $\Psi_0 \in \mathfrak{H}_2$. In case of our multi-spin–boson model (6), this splitting is only unique if $s > 0$. Otherwise, RA^* maps $\mathfrak{H} \to \mathfrak{H}_2$, and we can arbitrarily shift vectors between Ψ_0 and $RA^*\Phi$. So the machinery of [59] strictly applies to Cases 1 and 2 ($s \in (0, 2]$) and not to the less singular Case 0 ($s \leq 0$).

In case $s > 1$, a different choice of T, involving an infinite self-energy, would become necessary, as we explain in Remark 4.

3.2 Self-Adjointness and Convergence Results

In [59], it is now shown that, under suitable assumptions on A, the operator H_T is self-adjoint on $\mathrm{Dom}(H_T)$, as defined in Eqs. (42)–(44), for a large class of choices of T (including all bounded operators). Furthermore, [59] provides conditions for when H_T is the norm resolvent limit of some regularized operators $H_n := H_{\text{free}} + A_n^* + A_n$, as they are used in cutoff renormalization [20, 25, 32, 33, 58, 65]. In the following, we summarize some of these results. Let us introduce the following notation: for every $z \in \varrho(H_{\text{free}})$, we define $M(z)$ as the bounded operator on $\mathfrak{H}$ given by

$$M(z) := A(R_z - R)A^* . \tag{49}$$

This is indeed a bounded operator on $\mathfrak{H}$ whenever $A \in \mathcal{B}(\mathfrak{H}_2, \mathfrak{H})$ (or, a fortiori, $A \in \mathcal{B}(\mathfrak{H}_s, \mathfrak{H})$ for some $s \in (0, 2]$) since, by the second resolvent identity, the difference between the resolvents of H_{free}, evaluated in two distinct points of its resolvent, maps $\mathfrak{H}_{-2}$ into $\mathfrak{H}_{+2}$. With this notation, combining [59, Lemma 3.6] and [59, Corollary 3.7], we now get the following result.

Theorem 3 *Let H_{free} be a self-adjoint operator, $A \in \mathcal{B}(\mathfrak{H}_s, \mathfrak{H})$ for some $s \in (0, 2)$, with $A^* \in \mathcal{B}(\mathfrak{H}, \mathfrak{H}_{-s})$ its dual, and $T : \mathrm{Dom}(T) \subset \mathfrak{H} \to \mathfrak{H}$ self-adjoint. Suppose that*

(i) $\mathrm{Ker}\, A|_{\mathfrak{H}_2}$ and $\mathrm{Ran}\, A|_{\mathfrak{H}_2}$ are dense in $\mathfrak{H}$;
(ii) there exists $z \in \varrho(H_{\text{free}})$ such that $M(z) - T$ has a bounded inverse;
(iii) there exists $t \in [0, 1 - \frac{s}{2})$ such that $\mathrm{Dom}(T) \supset \mathfrak{H}_{2t}$ and $T|_{\mathfrak{H}_{2t}} \in \mathcal{B}(\mathfrak{H}_t, \mathfrak{H})$.

Then, the following statement holds true:

(a') The operator H_T defined by

$$\mathrm{Dom}(H_T) = \left\{\Psi \in \mathfrak{H}_{2-s} : \left(I + RA^*\right)\Psi \in \mathrm{Dom}(H_{\text{free}})\right\} ; \tag{50}$$

$$H_T = H_{\text{free}} + A^* + A_T \tag{51}$$

is self-adjoint, and its resolvent is given by [59, (3.11)]:

$$(H_T - z)^{-1} = (H_{\text{free}} - z)^{-1} - \begin{bmatrix} R_z A^* & R_z \end{bmatrix} \begin{bmatrix} A_T R_z A^* & A R_z + 1 \\ R_z A^* + 1 & R_z \end{bmatrix}^{-1} \begin{bmatrix} A R_z \\ R_z \end{bmatrix} . \tag{52}$$

Note that requirement (ii) is automatically satisfied whenever T itself has a bounded inverse.

Proof [59, Corollary 3.7] implies that (a') holds if we can verify (i) and the assumptions of [59, Lemma 3.6]. The latter lemma, in turn, assumes that (ii) and (iii) hold, as well as $\exists \gamma \in \mathbb{R}$ with $|\gamma|$ large enough, such that $(1 - G_{\pm i\gamma})^{-1}$ and $(1 - G^*_{\pm i\gamma})^{-1}$ are bounded operators on $\mathfrak{H}$, where $G_z := -R_z A^*$. But now,

$$\|G_{\pm i\gamma}\|_{\mathfrak{H}\to\mathfrak{H}} \leq \|A^*\|_{\mathfrak{H}\to\mathfrak{H}_{-s}} \|R_{\pm i\gamma}\|_{\mathfrak{H}_{-s}\to\mathfrak{H}} . \tag{53}$$

By Lemma 1, this norm becomes arbitrarily small as $|\gamma| \to \infty$, eventually dropping below 1, so $(1 - G_{\pm i\gamma})$ has a bounded inverse for large enough $|\gamma|$. The same reasoning applies to $(1 - G^*_{\pm i\gamma})$. So the additional invertibility condition is automatically satisfied. □

Although the above conditions for self-adjointness of H_T are already very useful, [59] provides another incarnation of its main result, which applies even more conveniently to the case of bounded T.

Theorem 4 ([59, Theorem 3.13]) *Let H_{free} be a self-adjoint nonnegative operator, $A \in \mathcal{B}(\mathfrak{H}_{+s}, \mathfrak{H})$ for some $s \in (0, 2)$, with $A^* \in \mathcal{B}(\mathfrak{H}, \mathfrak{H}_{-s})$ its dual, and T : $\mathrm{Dom}(T) \subset \mathfrak{H} \to \mathfrak{H}$ self-adjoint. Suppose that*

(i) $\mathrm{Ker}\, A|_{\mathfrak{H}_2}$ *and* $\mathrm{Ran}\, A|_{\mathfrak{H}_2}$ *are dense in* $\mathfrak{H}$.

Then, (a') in Theorem 3 for $T = 0$ holds true, that is, the operator H_0, defined by Eqs. (50)–(51) *with $T = 0$, is self-adjoint. Furthermore, if some (possibly nonzero) T : $\mathrm{Dom}(T) \subset \mathfrak{H} \to \mathfrak{H}$, symmetric is given, such that*

(ii) T is H_0-bounded,[3]

then, (a') in Theorem 3 holds true also for that T, that is, the operator H_T defined by Eqs. (50)–(51) *is self-adjoint, and its resolvent is given by Eq.* (52).

Finally, the question remains, if the renormalized Hamiltonian $H_T = H_{\text{free}} + A^* + A$ is equivalent to a Hamiltonian obtained by cutoff renormalization. That is: does there exist a sequence of regularized Hamiltonians $H_n := H_{\text{free}} + A_n^* + A_n + E_n$ such that $H_n \to H_T$ in a suitable sense? Here, [59, Theorem 3.10] provides us with conditions for an affirmative answer. The following result is an immediate consequence of [59, Theorem 3.10], using[4] Theorem 4.

Theorem 5 *Let H_{free} a self-adjoint nonnegative operator, $A \in \mathcal{B}(\mathfrak{H}_{+s}, \mathfrak{H})$ for some $s \in (0, 2)$, with $A^* \in \mathcal{B}(\mathfrak{H}, \mathfrak{H}_{-s})$ its dual, and T : $\mathrm{Dom}(T) \subset \mathfrak{H} \to \mathfrak{H}$ self-adjoint. Suppose that*

(i) $\mathrm{Ker}\, A|_{\mathfrak{H}_2}$ *and* $\mathrm{Ran}\, A|_{\mathfrak{H}_2}$ *are dense in* $\mathfrak{H}$;

[3] By *X is Y-bounded* or *X is relatively bounded with respect to Y* we mean $\exists a, b \geq 0$ s.t. $\|X\Psi\| \leq a\|Y\Psi\| + b\|\Psi\|$ $\forall \Psi \in \mathrm{Dom}(Y)$. The constant a is called *relative bound*. For $a < 1$, we call X *Kato-bounded w.r. to Y*. If for every $a > 0$, there exists some b such that the relative bound is true, we call *X infinitesimally Y-bounded*.

[4] Note that the consequences of Theorem 4 allow us to restrict to the case $\mu = 0$ in [59, Theorem 3.10].

(ii) *T is H_0-bounded with relative bound smaller than one;*

besides, suppose that there exists a sequence of closable operators $A_n : \mathrm{Dom}(A_n) \subset \mathfrak{H} \to \mathfrak{H}$ ("regularized annihilation operators"), such that

(iii) *there exists some $s' \in [0, 1]$ such that $\mathrm{Dom}(A_n) \supset \mathfrak{H}_{s'}$ and $A_n|_{\mathfrak{H}_{s'}} \in \mathcal{B}(\mathfrak{H}_{s'}, \mathfrak{H})$;*
(iv) *$A_n^* + A_n$ is H_{free}-bounded with n-independent relative bound smaller than one;*
(v) *$\lim_{n\to\infty} \|A_n - A\|_{\mathfrak{H}_2 \to \mathfrak{H}} = 0$;*
(vi) *there exists a sequence $\{E_n\}_{n\in\mathbb{N}} \subset \mathcal{B}(\mathfrak{H})$, E_n: symmetric, such that, for all $n \in \mathbb{N}$, $A_n R A_n^* + E_n$ is relatively bounded w.r.t. $H_{\mathrm{free}} + A_n^* + A_n + A_n R A_n^*$, with n-independent relative bound smaller than one, and*

$$\lim_{n\to\infty} \left\| A_n R A_n^* + E_n + T \right\|_{\mathrm{Dom}(T)\to\mathfrak{H}} = 0 . \tag{54}$$

Then, we have

(b') *$H_{\mathrm{free}} + A_n^* + A_n - E_n$ converges to $H_{\mathrm{free}} + A^* + A_T$ in the norm resolvent sense.*

Remark 3 (On two different notions of adjoint) Note that we are requiring each A_n to be *closable*, thus admitting an adjoint $A_n^\dagger : \mathrm{Dom}(A_n^\dagger) \subset \mathfrak{H} \to \mathfrak{H}$ with respect to the topology of $\mathfrak{H}$. This is different from the adjoint with respect to the pairing between $\mathfrak{H}_{s'}$ and $\mathfrak{H}_{-s'}$ (for some s' s.t. $\mathfrak{H}_{s'} \subset \mathrm{Dom}(A_n)$), i.e., there are "two different adjoints":

- an unbounded operator $A_n^\dagger : \mathrm{Dom}(A_n^\dagger) \subset \mathfrak{H} \to \mathfrak{H}$ defined by

$$\langle \Psi, A_n \Phi \rangle_{\mathfrak{H}} = \langle A_n^\dagger \Psi, \Phi \rangle_{\mathfrak{H}} \qquad \text{for all } \Phi \in \mathrm{Dom}(A_n) \; ; \tag{55}$$

- a continuous operator $A_n^* : \mathfrak{H} \to \mathfrak{H}_{-s'}$ defined by

$$\langle \Psi, A_n \Phi \rangle_{\mathfrak{H}} = \left\langle A_n^* \Psi, \Phi \right\rangle_{\mathfrak{H}_{-s'}, \mathfrak{H}_{s'}} \qquad \text{for all } \Phi \in \mathfrak{H}_{s'} \, , \Psi \in \mathfrak{H} \, . \tag{56}$$

Clearly A_n^* is an extension of $A_n^\dagger$ to all of $\mathfrak{H}$. With a slight abuse of notation, we will also write A_n^* for the restriction of this operator to $\mathrm{Dom}(A_n^\dagger) \to \mathfrak{H}$ (i.e., for $A_n^\dagger$). It then becomes clear from the context which domain is meant.

Remark 4 (Self-energy renormalization) Theorem 5 is valid for $s \in (0, 2)$, i.e., both in Case 1 and most situations of Case 2. The latter case is the one in which the terms E_n become necessary, which can physically be interpreted as "self-energy counterterms". As described above, to make the renormalized Hamiltonian H_T formally correspond to the physically desired Hamiltonian $H_{\mathrm{free}} + A^* + A$, we would need to "manually add" the interaction $T = -ARA^* = \lim_{n\to\infty} -A_n R A_n^*$. Now, the sequence of operators $A_n R A_n^*$:

- will converge to ARA^*, if $s \le 1$ (Case 1);

- will generally not have a limit, if $s > 1$ (Case 2).

However, in the latter case, several QFT-type models admit a properly chosen family of (self-energy) counterterms $\{E_n\}_{n\in\mathbb{N}}$, such that the operator

$$T = \lim_{n\to\infty} -A_n R A_n^* - E_n \tag{57}$$

is well-defined, within a suitable sense of convergence. For example, for the Nelson model [58] in Case 2, $\{E_n\}_{n\in\mathbb{N}}$ just amounts to a divergent sequence of real numbers. Since adding a constant to the Hamiltonian does not change the Heisenberg equations of motion, the operator T can heuristically be regarded as equivalent to the ill-defined expression $-ARA^*$. Likewise, H_T is heuristically equivalent to the desired model Hamiltonian $H_{\text{free}} + A^* + A$.

Of course, the choice of E_n is *not unique*: different choices of $\{E_n\}_{n\in\mathbb{N}}$ may result in different operators T, and vice versa, various choices of T are reachable by selecting an appropriate sequence of counterterms, resulting in inequivalent quantum dynamics. It is thus important to make a heuristic reasoning, based on physical arguments, to justify which renormalized Hamiltonian shall be the correct one.

Instead, in our Case 1 ($s \in (0, 1]$), we can simply set $E_n = 0$ to obtain the limit $T = -ARA^*$ as required in Eq. (47). The situation in Case 2 will be analyzed elsewhere.

4 Proof of Theorem 2

Our plan is to use Theorems 4 and 5 to establish (a) and (b). We start by collecting some mathematical tools that will turn out useful for the proof.

4.1 Preliminaries

Let $\{\mathcal{H}_s\}_{s\in\mathbb{R}}, \{\mathcal{F}_s\}_{s\in\mathbb{R}}$ be the scales of Hilbert spaces associated respectively with ω and $\mathrm{d}\Gamma(\omega)$, cf. Equation (11) and related discussion. Let us recall the following proposition.

Proposition 1 ([53, Propositions 3.4, 3.5 and 3.7]) *Let $f \in \mathcal{H}_{-s}$ for some $s \geq 1$. Then the following statements are true:*

(i) the restriction of the annihilation operator to $\mathcal{F}_{+s}$ defines a continuous operator $a(f) \in \mathcal{B}(\mathcal{F}_{+s}, \mathcal{F})$;

(ii) its adjoint $a^*(f) := a(f)^* \in \mathcal{B}(\mathcal{F}, \mathcal{F}_{-s})$ with respect to the duality pairing between $\mathcal{F}_{+s}$ and $\mathcal{F}_{-s}$ is a continuous operator whose action on $\Psi \in \mathcal{F}_{+1}$ agrees with the creation operator defined in Eq. (4);

(iii) there exists a sequence $\{f^i\}_{i\in\mathbb{N}} \subset \mathcal{H}$ *such that*

$$\lim_{i\to\infty} \left\|a(f) - a(f^i)\right\|_{\mathcal{F}_s\to\mathcal{F}} = 0\,, \qquad \lim_{i\to\infty} \left\|a^*(f) - a^*(f^i)\right\|_{\mathcal{F}\to\mathcal{F}_{-s}} = 0\,, \tag{58}$$

and this happens if and only if $\|f - f^i\|_{-s} \to 0$.

In words: If $f \in \mathcal{H}_{-s} \setminus \mathcal{H}$, then the annihilation operator $a(f)$, while still a legitimate densely defined operator on $\mathcal{F}$, fails to be closable and thus to have an adjoint with respect to the topology of $\mathfrak{H}$. However, its restriction to $\mathcal{F}_{+s}$, interpreted as an operator in $\mathcal{B}(\mathcal{F}_{+s}, \mathcal{F})$, always admits an adjoint $a^*(f) \in \mathcal{B}(\mathcal{F}, \mathcal{F}_{-s})$ with respect to the pairing. If $f \in \mathcal{H}$, following the same line of reasoning as in Remark 3, the "singular" creation operator defined as the adjoint with respect to the pairing corresponds to an extension of the "regular" one for Eq. (4).

We shall need the following definitions (see, e.g., [45, Sect. 6.1]). Every single-particle vector $g \in \mathcal{H}$ is associated with a *(non-normalized) coherent* vector $\varepsilon(g) \in \mathcal{F}$ defined by

$$\varepsilon(g) := \left\{\varepsilon(g)^{(n)}\right\}_{n\in\mathbb{N}}\,, \qquad \varepsilon(g)^{(n)} = \frac{1}{\sqrt{n!}} g^{\otimes n}\,. \tag{59}$$

In particular, if $g \in \mathcal{H}_{+s}$, then $\varepsilon(g) \in \mathcal{F}_{+s}$ and, given $f \in \mathcal{H}_{-s}$, $a(f)\,\varepsilon(g) = \langle f, g\rangle_{\mathcal{H}_{-s},\mathcal{H}_{+s}}\,\varepsilon(g)$. With this definition, given any subset $\mathcal{D} \subset \mathcal{H}$, the *coherent domain over* $\mathcal{D}$ is the subspace $\mathcal{E}(\mathcal{D}) \subset \mathcal{F}$ defined by

$$\mathcal{E}(\mathcal{D}) = \operatorname{Span}\{\varepsilon(g) :\ g \in \mathcal{D}\}\,, \tag{60}$$

and it can be proven that $\mathcal{E}(\mathcal{D})$ is dense in $\mathcal{F}$ whenever $\mathcal{D}$ is dense in $\mathcal{H}$, and that $\varepsilon(g_n) \to \varepsilon(g)$ whenever $g_n \to g$.

4.2 Denseness of Kernel and Range

Statement (a) will follow from (a') in Theorem 4. As discussed above, in case $s \le 1$, the operator $T = -ARA^*$ is bounded, and thus also H_0-bounded with relative bound zero. It remains to establish the denseness of $\operatorname{Ker} A|_{\mathfrak{H}_2}$ and $\operatorname{Ran} A|_{\mathfrak{H}_2}$, which we do in this subsection.

Lemma 2 *Let* $f \in \mathcal{H}_{-s} \setminus \mathcal{H}$ *for* $s \in (0, 2]$, *and* $a(f) \in \mathcal{B}(\mathcal{F}_{+s}, \mathcal{F})$. *Then*

(i) $\operatorname{Ker} a(f)\,|_{\mathcal{F}_{+2}}$ *is dense in* $\mathcal{F}$;
(ii) $\operatorname{Ran} a(f)\,|_{\mathcal{F}_{+2}}$ *is dense in* $\mathcal{F}$.

Proof It suffices to prove the claims in the case $s = 2$; the claims for $s < 2$ are then true *a fortiori*. To simplify the notation, in the following we will omit the subscript in the the duality pairing $\langle\cdot,\cdot\rangle_{\mathcal{H}_{-2},\mathcal{H}_{+2}}$ between $\mathcal{H}_{-2}$ and $\mathcal{H}_{+2}$.

(i) Since the functional $\varphi \in \mathcal{H}_{+2} \mapsto \langle f, \varphi\rangle$ is unbounded, its kernel, i.e., the space $\mathcal{D}_0 = \{g \in \mathcal{H}_{+2} : \langle f, g\rangle = 0\}$, is dense in $\mathcal{H}$ [2, Lemma 1.2.3], whence its associated coherent domain $\mathcal{E}(\mathcal{D}_0)$ is dense in $\mathcal{F}$. But, given any $g \in \mathcal{D}_0$, its corresponding coherent vector $\varepsilon(g)$ satisfies $a(f)\,\varepsilon(g) = \langle f, g\rangle\, \varepsilon(g) = 0$, whence the same holds for all elements of $\mathcal{E}(\mathcal{D}_0)$. This proves $\mathcal{E}(\mathcal{D}_0) \subset \operatorname{Ker} a(f)\,|_{\mathcal{F}_{+2}}$. Since the former is dense in $\mathcal{F}$, so is the latter.

(ii) Since $f \neq 0$, the set $\mathcal{D} = \{g \in \mathcal{H}_{+2} : \langle f, g\rangle \neq 0\}$ is also dense in $\mathcal{H}$, thus $\mathcal{E}(\mathcal{D})$ is again dense in $\mathcal{F}$. Given $g \in \mathcal{D}$, we then have $a(f)\left(\langle f, g\rangle^{-1}\, \varepsilon(g)\right) = \varepsilon(g)$. This proves $\mathcal{E}(\mathcal{D}) \subset \operatorname{Ran} a(f)\,|_{\mathcal{F}_{+2}}$ and thus, again, the latter is dense in $\mathcal{F}$. □

We remark that, while the denseness of the range of $a(f)$ holds under the only condition $f \neq 0$, the denseness of its kernel additionally requires $f \notin \mathcal{H}$.

Proposition 2 *Let $s \leq 2$, $f_1, \ldots, f_N \in \mathcal{H}_{-s} \setminus \mathcal{H}$ a family of $\mathcal{H}$-independent form factors (see Definition 1), and $B_1, \ldots, B_N \in \mathbb{C}^{D\times D}$. Then, for A defined in Eq.* (3), $\operatorname{Ker} A|_{\mathfrak{H}_{+2}}$ *is dense in $\mathfrak{H}$.*

Proof Again, it suffices to prove the claim for $s = 2$. Since $\{f_j\}_{j=1}^N$ are $\mathcal{H}$-independent, the space of single-particle states that are in the kernel of *all* form factors,

$$\mathcal{D}_0 = \left\{g \in \mathcal{H}_{+2} : \langle f_j, g\rangle_{\mathcal{H}_{-2},\mathcal{H}_{+2}} = 0 \ \forall j = 1, \ldots, N\right\} \tag{61}$$

is dense in $\mathcal{H}$, see [2, Lemma 3.1.1]. Then the coherent domain, $\mathcal{E}(\mathcal{D}_0)$, is also dense in $\mathcal{F}$, and likewise $\mathfrak{h} \otimes \mathcal{E}(\mathcal{D}_0)$ is dense in $\mathfrak{H}$. Now, given any coherent vector $\varepsilon(g) \in \mathcal{E}(\mathcal{D}_0)$, one has $a\big(f_j\big)\,\varepsilon(g) = 0$, so for any $u \in \mathfrak{h}$,

$$A\,(u \otimes \varepsilon(g)) = \sum_{j=1}^{N} (B_j^* u) \otimes \left(a\big(f_j\big)\,\varepsilon(g)\right) = 0\,, \tag{62}$$

whence, by linearity, $\mathfrak{h} \otimes \mathcal{E}(\mathcal{D}_0) \subset \operatorname{Ker} A|_{\mathfrak{H}_{+2}}$. Since the former is dense in $\mathfrak{H}$, so is the latter. □

We remark that the denseness of $\operatorname{Ker} A|_{\mathfrak{H}_{+2}}$ is irrespective of the choice of $B_1, \ldots, B_N \in \mathbb{C}^{D\times D}$. No point of Assumption 1 enters here. However, the $\mathcal{H}$-independence of the form factors is crucial.

Proposition 3 *Let $s \leq 2$, $f_1, \ldots, f_N \in \mathcal{H}_{-s} \setminus \mathcal{H}$ be a family of $\mathcal{H}$-independent form factors (see Definition 1), and let $B_1, \ldots, B_N \in \mathbb{C}^{D\times D}$ satisfy Assumption 1. Then, for A defined in Eq.* (3), $\operatorname{Ran} A|_{\mathfrak{H}_{+2}}$ *is dense in $\mathfrak{H}$.*

Proof Because of points (i) and (ii) in Assumption 1, the matrices $B_1, \ldots, B_N$ share a common orthonormal eigenbasis: there exists a set of orthonormal vectors $\{u^{(a)}\}_{a=1}^D \subset \mathfrak{h}$ such that

$$B_j u^{(a)} = b_j^{(a)} u^{(a)}\,, \qquad \forall j = 1, \ldots, N\,,\ \ a = 1, \ldots, D\,, \tag{63}$$

with $b_j^{(a)} \in \mathbb{C}$. By normality, $\{u^{(a)}\}_{a=1}^D$ is also a common eigenbasis for their adjoints, with

$$B_j^* u^{(a)} = \overline{b_j^{(a)}} u^{(a)} , \qquad \forall j = 1, \ldots, N , \quad a = 1, \ldots, D . \tag{64}$$

We can thus conveniently decompose the Hilbert space as

$$\mathfrak{H} \simeq \bigoplus_{a=1}^{D} \mathfrak{H}^{(a)} , \qquad \mathfrak{H}^{(a)} = \mathrm{Span}\{u^{(a)}\} \otimes \mathcal{F} \simeq \mathcal{F} . \tag{65}$$

This decomposition is preserved by A, i.e., $A\mathfrak{H}^{(a)} \subset \mathfrak{H}^{(a)}$, since for any $\Psi \in \mathcal{F}$,

$$\begin{aligned} A\left(u^{(a)} \otimes \Psi\right) &= \sum_{j=1}^{N} B_j^* u^{(a)} \otimes a\big(f_j\big) \Psi \\ &= \sum_{j=1}^{N} \overline{b_j^{(a)}} u^{(a)} \otimes a\big(f_j\big) \Psi \\ &= u^{(a)} \otimes a\left(\sum_{j=1}^{N} b_j^{(a)} f_j\right) \Psi . \end{aligned} \tag{66}$$

So we can also decompose

$$A = \bigoplus_{a=1}^{D} A^{(a)} , \qquad A^{(a)} = a\left(\sum_{j=1}^{D} b_j^{(a)} f_j\right) . \tag{67}$$

Obviously, $\mathrm{Ran}\, A = \bigoplus_a \mathrm{Ran}\, A^{(a)}$. So it suffices to prove that $\mathrm{Ran}\, A^{(a)}$ is dense in $\mathfrak{H}^{(a)}$ for each a, in order to establish the proposition. Now, $A^{(a)}$ is just an annihilation operator on $\mathfrak{H}^{(a)} \simeq \mathcal{F}$, so Lemma 2 applies, provided that

$$\sum_{j=1}^{N} b_j^{(a)} f_j \in \mathcal{H}_{-2} \setminus \mathcal{H} . \tag{68}$$

This requirement is fulfilled because of the $\mathcal{H}$-independence of $f_1, \ldots, f_N$, provided that there exists at least one $j = 1, \ldots, N$ such that $b_j^{(a)} \neq 0$. The latter statement means that none of the vectors $u^{(a)}$ is such that $B_j u^{(a)} = 0$ for all $j = 1, \ldots, N$ simultaneously, which is guaranteed by point (iii) of Assumption 1, viz. $\bigcap_{j=1}^{N} \mathrm{Ker}\, B_j = \{0\}$. Therefore Proposition 2 indeed applies, rendering denseness of $\mathrm{Ran}\, A^{(a)}|_{\mathcal{F}_{+2}} \subset \mathrm{Ran}\, A^{(a)}$ in $\mathfrak{H}^{(a)}$, which establishes the proposition. □

4.3 Finishing the Proof

Proof of Theorem 2 (a) We would like to apply Theorem 4. Since $f_j \in \mathcal{H}_{-s} \setminus \mathcal{H}$ for $s \leq 1$, we have $f_j \in \mathcal{H}_{-1} \setminus \mathcal{H}$. By Proposition 1, this implies $a(f_j) \in \mathcal{B}(\mathcal{F}_1, \mathcal{F})$. In case of the spin–boson model (6) we are considering, it is then immediate to see that $A \in \mathcal{B}(\mathfrak{H}_{+1}, \mathfrak{H})$, and by duality, $A^* \in \mathcal{B}(\mathfrak{H}, \mathfrak{H}_{-1})$. The requirement that $\operatorname{Ker} A|_{\mathfrak{H}_2}$ and $\operatorname{Ran} A|_{\mathfrak{H}_2}$ be dense in $\mathfrak{H}$ have been proven in Propositions 2 and 3, respectively.

Now, recall that, for $s \in (0, 1]$, the choice $T = -ARA^*$ is needed to make $H = H_{\text{free}} + A^* + A$ agree with H_T. This T is bounded, see Eq. (48), so it is in particular Kato-bounded with respect to H_0. Thus, Theorem 4 applies, rendering statement (a).

(b) We would like to apply Theorem 5. Conditions (i) and (ii) have already been verified. Proposition 1 now provides us, for each form factor f_j, with a sequence of regularized form factors $\{f_{j,n}\}_{n\in\mathbb{N}} \subset \mathcal{H}$ such that $\|f_{j,n} - f_j\|_{\mathcal{H}_{-s}} \to 0$ as $n \to \infty$, and

$$\|a(f_{j,n}) - a(f_j)\|_{\mathcal{F}_s \to \mathcal{F}} \to 0\,, \qquad \|a^*(f_{j,n}) - a^*(f_j)\|_{\mathcal{F} \to \mathcal{F}_{-s}} \to 0\,. \tag{69}$$

Correspondingly, we define the regularized operators as in Eq. (14)

$$A_n := \sum_{j=1}^{N} B_j^* \otimes a(f_{j,n})\,, \qquad A_n^* := \sum_{j=1}^{N} B_j \otimes a^*(f_{j,n})\,, \tag{70}$$

which are obviously closable as $f_{j,n} \in \mathcal{H}$. Furthermore, for $\Psi \in \mathfrak{H}_1$, introducing the number operator $\mathcal{N} := \mathrm{d}\Gamma(1)$, we have

$$\begin{aligned} \|A_n^\sharp \Psi\| &\leq \sum_{j=1}^{N} \|B_j\| \|(I \otimes a^\sharp(f_{j,n}))\Psi\| \leq \sum_{j=1}^{N} \|B_j\| \|f_{j,n}\|_{\mathcal{H}} \|(\mathcal{N}+1)^{\frac{1}{2}}\Psi\| \\ &\leq \sum_{j=1}^{N} \|B_j\| \|f_{j,n}\|_{\mathcal{H}} (\min\{m_0, 1\})^{-\frac{1}{2}} \|(H_{\text{free}}+1)^{\frac{1}{2}}\Psi\| \leq c\|\Psi\|_{\mathfrak{H}_1} \end{aligned} \tag{71}$$

for some $c > 0$, so $A_n|_{\mathfrak{H}_1} \in \mathcal{B}(\mathfrak{H}_1, \mathfrak{H})$ and $\operatorname{Dom}(A_n) \supset \mathfrak{H}_1$, which establishes requirement (iii). Furthermore, Eq. (71) implies that there is some $C > 0$, such that for any $\varepsilon > 0$:

$$\|A_n \Psi\| \leq C\|(H_{\text{free}}+1)^{\frac{1}{2}}\Psi\| \leq C\big(\varepsilon\|(H_{\text{free}}+1)\Psi\| + \varepsilon^{-1}\|\Psi\|\big)\,. \tag{72}$$

So $A_n^* + A_n$ is infinitesimally Kato-bounded against $(H_{\text{free}} + 1)$ and thus also against H_{free}, which establishes requirement (iv). Concerning requirement (v), the convergence (69) implies

$$\|A_n - A\|_{\mathfrak{H}_s \to \mathfrak{H}} \leq \sum_{j=1}^{N} \|B_j\| \|a(f_{j,n}) - a(f_j)\|_{\mathcal{F}_s \to \mathcal{F}} \to 0 \,, \tag{73}$$

as $n \to \infty$. So in particular, $\|A_n - A\|_{\mathfrak{H}_2 \to \mathfrak{H}} \to 0$, which is requirement (v).

Finally, for (vi), we just choose $E_n = 0$. By the same mapping property argument as in Eq. (48), $A_n R A_n^*$ is bounded, so it is relatively bounded with respect to any symmetric operator with relative bound 0. Further, boundedness of T implies $\mathrm{Dom}(T) = \mathfrak{H}$ and we have

$$\begin{aligned} &\|A_n R A_n^* + E_n + T\|_{\mathfrak{H} \to \mathfrak{H}} = \|A_n R A_n^* - A R A^*\|_{\mathfrak{H} \to \mathfrak{H}} \\ &\leq \underbrace{\|A_n - A\|_{\mathfrak{H}_s \to \mathfrak{H}}}_{\to 0} \underbrace{\|R A_n^*\|_{\mathfrak{H} \to \mathfrak{H}_s}}_{<\infty} + \underbrace{\|A R\|_{\mathfrak{H}_{-s} \to \mathfrak{H}}}_{<\infty} \underbrace{\|A_n^* - A^*\|_{\mathfrak{H} \to \mathfrak{H}_{-s}}}_{\to 0} \to 0 \,, \end{aligned} \tag{74}$$

where $\|A_n^* - A^*\|_{\mathfrak{H} \to \mathfrak{H}_{-s}} \to 0$ follows from Eq. (69) by the same arguments as in (73). This also establishes (vi).

Consequently, Theorem 5 applies, and we obtain the desired convergence (15).□

5 Concluding Remarks

In this work, we have discussed the ultraviolet renormalization problem for generalized spin–boson models, exploiting the abstract framework provided by Posilicano in [59]. By this technique, under suitable requirements, we were able to provide an explicit expression for the self-adjointness domain of the Hamiltonian. The requirements include the choice of form factors $f_1, \ldots, f_N \in \mathcal{H}_{-1}$ (Case 1, see Eq. (11)), so they exhibit "mild" ultraviolet divergences, as well as Assumption 1 on the operators $B_1, \ldots, B_N$ modulating the interaction between the quantum system and the boson field. Furthermore, it was shown that such models can be obtained as the norm resolvent limit of their regularized versions (e.g., those in which an ultraviolet cutoff is imposed). This generalizes the results presented in [53, Sect. 4], where the existence of a self-adjoint realization of generalized spin–boson models was proven through perturbative methods.

The research initiated in this paper offers, among others, two clear routes of continuation. First, the conditions on the matrices $B_1, \ldots, B_N$, while reasonably general and including many cases of physical interest, are likely prone to be relaxed. This is strongly suggested by the results in [53, Sect. 5 and 6] and [57], where an explicit expression for the self-adjointness domain fully compatible with the one provided here (cf. [57, Remark 3.8]) was obtained for a class of models corresponding to a choice of $B_1, \ldots, B_N$ which *violates* Assumption 1. An extension of our results to the case of an infinite-dimensional quantum system interacting with the field is also foreseeable along the lines depicted in Remark 2.

Second, the abstract framework developed by Posilicano is naturally suited to treat the case of "strong" ultraviolet divergences belonging to Case 2. As already discussed (cf. Remark 4), additional care is required in this case in order to properly choose the operator T entering the statements of Theorems 3–5: while for mild divergences the “minimal” choice $T = -ARA^*$ is available (and bounded), novel mathematical intricacies appear when renormalization enters the game. Future research will be devoted to this topic.

Acknowledgements This research was supported by the European Research Council (ERC) through the Starting Grant FERMIMATH, Grant Agreement No. 101040991 of Niels Benedikter, by “Istituto Nazionale di Fisica Nucleare” (INFN) through the project “QUANTUM”, by the Italian National Group of Mathematical Physics (GNFM–INdAM), and by European Union - NextGenerationEU (CN00000013 - “National Center for HPC, Big Data and Quantum Computing”). D.L. thanks the Department of Mathematics at “Università degli Studi di Milano” for its hospitality.

References

1. Agarwal, G.: Rotating-wave approximation and spontaneous emission. Phys. Rev. A **4**, 1778 (1971)
2. Albeverio, S., Kurasov, P.: Singular Perturbations of Differential Operators: Solvable Schrödinger-Type Operators. Cambridge University Press (2000)
3. Albeverio, S., Kuzhel, S., Nizhnik, L.: Singularly perturbed self-adjoint operators in scales of Hilbert spaces. Ukrainian Math. J. **59**, 787–810 (2007)
4. Amann, A.: Ground states of a spin-boson model. Ann. Phys. **208**, 414–448 (1991)
5. Arai, A.: An asymptotic analysis and its application to the nonrelativistic limit of the Pauli-Fierz and a spin-boson model. J. Math. Phys. **31**, 2653–2663 (1990)
6. Arai, A., Hirokawa, M.: On the existence and uniqueness of ground states of a generalized spin-boson model. J. Funct. Anal. **151**, 455–503 (1997)
7. Arai, A., Hirokawa, M., Hiroshima, F.: On the absence of eigenvectors of Hamiltonians in a class of massless quantum field models without infrared cutoff. J. Funct. Anal. **168**, 470–497 (1999)
8. Arai, A.: Essential spectrum of a self-adjoint operator on an abstract Hilbert space of Fock type and applications to quantum field Hamiltonians. J. Math. Anal. Appl. **246**, 189–216 (2000)
9. Arai, A., Hirokawa, M.: Ground states of a general class of quantum field Hamiltonians. Rev. Math. Phys. **12**, 1085–1135 (2000)
10. Bach, V., Ballesteros, M., Könenberg, M., Menrath, L.: Existence of ground state eigenvalues for the spin-boson model with critical infrared divergence and multiscale analysis. J. Math. Anal. Appl. **453**, 773–797 (2017)
11. Braak, D.: Integrability of the Rabi model. Phys. Rev. Lett. **107**, 100401 (2011)
12. Breuer, H., Petruccione, F.: The Theory of Open Quantum Systems. Oxford University Press on Demand (2002)
13. Dam, T., Møller, J.: Asymptotics in Spin-Boson type models. Commun. Math. Phys. **374**, 1389–1415 (2020)
14. Davies, E.: Symmetry breaking for molecular open systems. Ann. De L'I.H.P. Physique Théorique **35**, 149–171 (1981)
15. Dedushenko, M.: Snowmass white paper: the quest to define QFT. Int. J. Mod. Phys. A **38**, 2330002 (2023)
16. Dereziński, J., Jakšić, V.: Spectral theory of Pauli-Fierz operators. J. Funct. Anal. **180**, 243–327 (2001)

17. Dereziński, J.: Van Hove Hamiltonians—exactly solvable models of the infrared and ultraviolet problem. Ann. Henri Poincaré **4**, 713–738 (2003)
18. Dicke, R.: Coherence in spontaneous radiation processes. Phys. Rev. **93**, 99 (1954)
19. Dorner, U., Zoller, P.: Laser-driven atoms in half-cavities. Phys. Rev. A **66**, 023816 (2002)
20. Eckmann, J.: A model with persistent vacuum. Commun. Math. Phys. **18**, 247–264 (1970)
21. Facchi, P., Lonigro, D., Pascazio, S., Pepe, F., Pomarico, D.: Bound states in the continuum for an array of quantum emitters. Phys. Rev. A **100**, 023834 (2019)
22. Falconi, M.: Self-adjointness criterion for operators in Fock spaces. Math. Phys. Anal. Geom. **18**, 2 (2015)
23. Fannes, M., Nachtergaele, B., Verbeure, A.: The equilibrium states of the spin-boson model. Commun. Math. Phys. **114**, 537–548 (1988)
24. Fewster, C., Rejzner, K.: Algebraic Quantum Field Theory—an introduction. In: Finster, F., Giulini, D., Kleiner, J., Tolksdorf, J. (eds.), Progress and Visions in Quantum Theory in View of Gravity: Bridging Foundations of Physics and Mathematics (2020)
25. Fröhlich, J.: On the infrared problem in a model of scalar electrons and massless, scalar bosons. Ann. De L'I.H.P. Physique Théorique **19**, 1–103 (1973)
26. Glimm, J., Jaffe, A.: A $\lambda\phi^4$ quantum field theory without cutoffs. I. Phys. Rev. **176**, 1945 (1968)
27. Glimm, J., Jaffe, A.: The $\lambda\varphi_2^4$ quantum field theory without cutoffs: II. The field operators and the approximate vacuum. Ann. Math. 362–401 (1970)
28. Glimm, J., Jaffe, A.: The $\lambda\varphi_2^4$ quantum field theory without cutoffs: III. Phys. Vacuum. Acta Math. **125**, 203–267 (1970)
29. Glimm, J., Jaffe, A.: The Yukawa_2 quantum field theory without cutoffs. J. Funct. Anal. **7**, 323–357 (1971)
30. Glimm, J., Jaffe, A.: Positivity of the φ_3^4 Hamiltonian. Fortschritte Der Physik **21**, 327–376 (1973)
31. González-Tudela, A., Paulisch, V., Kimble, H., Cirac, J.: Efficient multiphoton generation in waveguide quantum electrodynamics. Phys. Rev. Lett. **118**, 213601 (2017)
32. Griesemer, M., Wünsch, A.: Self-adjointness and domain of the Fröhlich Hamiltonian. J. Math. Phys. **57**, 021902 (2016)
33. Griesemer, M., Wünsch, A.: On the domain of the Nelson Hamiltonian. J. Math. Phys. **59**, 042111 (2018)
34. Gross, L.: The relativistic polaron without cutoffs. Commun. Math. Phys. **31**, 25–73 (1973)
35. Gross, M., Haroche, S.: Superradiance: an essay on the theory of collective spontaneous emission. Phys. Rep. **93**, 301–396 (1982)
36. Haag, R.: On quantum field theories. Danske Videnskabernes Selskab Matematisk-Fysiske Meddelelser **29** (1955)
37. Haag, R.: Local Quantum Physics: Fields, Particles, Algebras. Springer Science and Business Media (1996)
38. Hasler, D., Hinrichs, B., Siebert, O.: On existence of ground states in the spin boson model. Commun. Math. Phys. **388**, 419–433 (2021)
39. Hirokawa, M.: An expression of the ground state energy of the Spin-Boson model. J. Funct. Anal. **162**, 178–218 (1999)
40. Hirokawa, M.: Remarks on the ground state energy of the spin-boson model: an application of the Wigner-Weisskopf model. Rev. Math. Phys. **13**, 221–251 (2001)
41. Hübner, M., Spohn, H.: Radiative decay: nonperturbative approaches. Rev. Math. Phys. **7**, 363–387 (1995)
42. Hwang, M., Puebla, R., Plenio, M.: Quantum phase transition and universal dynamics in the Rabi model. Phys. Rev. Lett. **115**, 180404 (2015)
43. Hübner, M., Spohn, H.: Spectral properties of the spin-boson Hamiltonian. Ann. De L'I.H.P. Physique Théorique **62**, 289–323 (1995)
44. Jakšić, V., Kritchevski, E., Pillet, C.: Mathematical theory of the Wigner-Weisskopf atom. Large Coulomb Systems: Lecture Notes On Mathematical Aspects Of QED, pp. 145–215 (2006)

45. Lampart, J., Schmidt, J., Teufel, S., Tumulka, R.: Particle creation at a point source by means of interior-boundary conditions. Math. Phys. Anal. Geom. **21**, 1–37 (2018)
46. Lampart, J., Schmidt, J.: On Nelson-type Hamiltonians and abstract boundary conditions. Commun. Math. Phys. **367**, 629–663 (2019)
47. Lampart, J.: A nonrelativistic quantum field theory with point interactions in three dimensions. Ann. Henri Poincaré **20**, 3509–3541 (2019)
48. Lampart, J.: The renormalized Bogoliubov-Fröhlich Hamiltonian. J. Math. Phys. **61**, 101902 (2020)
49. Leggett, A., Chakravarty, S., Dorsey, A., Fisher, M., Garg, A., Zwerger, W.: Dynamics of the dissipative two-state system. Rev. Mod. Phys. **59**, 1 (1987)
50. Lill, S.: Extended State Space for describing renormalized Fock spaces in QFT. Rev. Math. Phys. **36**(8), 2450029 (2024)
51. Lill, S.: Time Dynamics in Quantum Field Theory Systems. U. Tübingen (2022)
52. Lonigro, D., Facchi, P., Pascazio, S., Pepe, F., Pomarico, D.: Stationary excitation waves and multimerization in arrays of quantum emitters. New J. Phys. **23**, 103033 (2021). https://doi.org/10.1088/1367-2630/ac2ce0
53. Lonigro, D.: Generalized spin-boson models with non-normalizable form factors. J. Math. Phys. **63**, 072105 (2022)
54. Lonigro, D.: Renormalization of spin-boson interactions mediated by singular form factors. In: Correggi, M., Falconi, M. (eds.), Quantum Mathematics II. INdAM 2022. Springer INdAM Series, vol 58. Springer, Singapore (2023)
55. Lonigro, D., Chruściński, D.: Quantum regression beyond the Born-Markov approximation for generalized spin-boson models. Phys. Rev. A **105**, 052435 (2022)
56. Lonigro, D., Chruściński, D.: Quantum regression in dephasing phenomena. J. Phys. A Math. Theor. **55**, 225308 (2022)
57. Lonigro, D.: Self-Adjointness of a class of multi-spin-boson models with ultraviolet divergences. Math. Phys. Anal. Geom. **26**, 15 (2023)
58. Nelson, E.: Interaction of nonrelativistic particles with a quantized scalar field. J. Math. Phys. **5**, 1190–1197 (1964)
59. Posilicano, A.: On the self-adjointness of H+A*+A. Math. Phys. Anal. Geom. **23**, 1–31 (2020)
60. Reker, J.: Existence of resonances for the spin-boson model with critical coupling function. J. Math. Anal. Appl. **483**, 123628 (2020)
61. Sánchez-Burillo, E., Zueco, D., Martin-Moreno, L., Garcia-Ripoll, J.: Dynamical signatures of bound states in waveguide QED. Phys. Rev. A **96**, 023831 (2017)
62. Schmidt, J.: The massless Nelson Hamiltonian and its domain. Mathematical Challenges Of Zero-Range Physics: Models, Methods, Rigorous Results, Open Problems, pp. 57–80 (2021)
63. Segal, I.: Notes toward the construction of nonlinear relativistic quantum fields, I. The Hamiltonian in two space-time dimensions as the generator of a C*-automorphism group. Proc. Natl. Acad. Sci. **57**, 1178–1183 (1967)
64. Simon, B.: Spectral analysis of rank one perturbations and applications. CRM Lect. Notes **8**, 109–149 (1995)
65. Sloan, A.: The polaron without cutoffs in two space dimensions. J. Math. Phys. **15**, 190–201 (1974)
66. Summers, S.: A perspective on constructive quantum field theory (2012). https://arxiv.org/abs/1203.3991
67. Takaesu, T.: On generalized spin-boson models with singular perturbations. Hokkaido Math. J. **39**, 317–349 (2010)
68. Teranishi, N.: Self-adjointness of the generalized spin-boson Hamiltonian with a quadratic boson interaction. Hokkaido Math. J. **44**, 409–423 (2015)
69. Teranishi, N.: Absence of ground states of generalized spin-boson models. Kyushu J. Math. **72**, 1–14 (2018)
70. Teufel, S., Tumulka, R.: Avoiding ultraviolet divergence by means of interior-boundary conditions. In: Quantum Mathematical Physics: a Bridge Between Mathematics and Physics, pp. 293–311 (2016). https://doi.org/10.1007/978-3-319-26902-3_14

71. Teufel, S., Tumulka, R.: Hamiltonians without ultraviolet divergence for quantum field theories. Quantum Stud.: Math. Found. **8**, 17–35 (2021)
72. Tufarelli, T., Ciccarello, F., Kim, M.: Dynamics of spontaneous emission in a single-end photonic waveguide. Phys. Rev. A **87**, 013820 (2013)
73. Van Loo, A., Fedorov, A., Lalumiere, K., Sanders, B., Blais, A., Wallraff, A.: Photon-mediated interactions between distant artificial atoms. Science **342**, 1494–1496 (2013)
74. Weiss, U.: Quantum dissipative systems. World Sci. (2012)
75. Xie, Q., Zhong, H., Batchelor, M., Lee, C.: The quantum Rabi model: solution and dynamics. J. Phys. A Math. Theor. **50**, 113001 (2017)
76. Zhong, H., Xie, Q., Batchelor, M., Lee, C.: Analytical eigenstates for the quantum Rabi model. J. Phys. A Math. Theor. **46**, 415302 (2013)

Random Linear Systems with Quadratic Constraints: from Random Matrix Theory to Replicas and Back

Pierpaolo Vivo

Abstract I present here a pedagogical introduction to the works by Rashel Tublin and Yan V. Fyodorov on random linear systems with quadratic constraints, using tools from Random Matrix Theory and replicas [1, 2]. These notes illustrate and complement the material presented at the Summer School organised within the Puglia Summer Trimester 2023 in Bari (Italy). Consider a system of M linear equations in N unknowns, $\sum_{j=1}^{N} A_{kj}x_j = b_k$ for $k = 1, \ldots, M$, subject to the constraint that the solutions live on the N-sphere, $x_1^2 + \cdots + x_N^2 = N$. Assume that both the coefficients A_{ij} and the parameters b_i be independent Gaussian random variables with zero mean. Using two different approaches—based on Random Matrix Theory and on a replica calculation—it is possible to compute whether a large linear system subject to a quadratic constraint is *typically* solvable or not, as a function of the ratio $\alpha = M/N$ and the variance σ^2 of the b_i's. This is done by defining a quadratic loss function $H(\boldsymbol{x}) = \frac{1}{2}\sum_{k=1}^{M}\left[\sum_{j=1}^{N} A_{kj}x_j - b_k\right]^2$ and computing the statistics of its minimal value on the sphere, $\mathcal{E}_{min} = \min_{||\boldsymbol{x}||^2=N} H(\boldsymbol{x})$, which is zero if the system is compatible, and larger than zero if it is incompatible. One finds that there exists a *compatibility threshold* $0 < \alpha_c < 1$, such that systems with $\alpha > \alpha_c$ are typically incompatible. This means that even weakly under-complete linear systems could become typically incompatible if forced to additionally obey a quadratic constraint.

Keywords Quadratic optimisation · Random matrix theory · Replica method · Random linear systems

Prologue

These lecture notes are based on a short course I gave at the Summer School organised within the Puglia Summer Trimester 2023 in Bari (Italy). There is no new research material in here. All the credit goes to Rashel Tublin and Yan V. Fyodorov, who

P. Vivo (✉)
Department of Mathematics, King's College London, London, UK
e-mail: pierpaolo.vivo@kcl.ac.uk

B. Cassano et al. (eds.), *Singularities, Asymptotics, and Limiting Models*, Springer INdAM Series 64, https://doi.org/10.1007/978-981-96-3584-9_8

wrote two insightful papers on the subject [1, 2], which–along with Rashel's Ph.D. work at King's College London [3]—I have here drawn liberally and *verbatim* from. I wish to thank both of them for writing such clear papers and for allowing me access to their deep knowledge and insight about this problem. I take this opportunity to warmly thank all the organisers of the Summer School—Biagio Cassano, Fabio Deelan Cunden, Matteo Gallone, Marilena Ligabò, Alessandro Michelangeli, and Domenico Pomarico—for the flawless organisation, for putting together such a rich and diverse programme, and for many inspiring conversations. I cherish their friendship dearly. I would also like to thank my fellow speakers at the School, Nilanjana Datta and Riccardo Adami, from whom I learnt so much in such a short period of time. I enjoyed their company immensely. Finally, I extend my gratitude to all the students and early-career researchers who participated in the School for their engagement and the interest they showed in what I had to say.

I chose to prepare my short course on the topic of optimisation on the sphere as it allowed me the rare luxury of being able to present the same calculation—the average minimal loss, signalling typical compatibility/incompatibility of a random linear system of equations with a quadratic constraint—in two alternative ways: one based on the synergy between Lagrange optimisation and Random Matrix Theory, the other based on the heuristic "replica method" from the physics of disordered systems. I tried to explain all the steps in full detail and make these notes as self-contained—and hopefully instructive—as possible.

1 Introduction

Scenario 1. Consider the following linear system of two equations ($M = 2$) in three unknowns ($N = 3$)

$$x_1 + x_2 - x_3 = b_1 \tag{1}$$

$$x_1 + 2x_2 + x_3 = b_2 \ , \tag{2}$$

for b_1, b_2 real parameters. Clearly, any triple of the form $(2b_1 - b_2 + 3z, -b_1 + b_2 - 2z, z)$ with $z \in \mathbb{R}$ is a solution of the system.

What happens if we add the constraint that the solution vector $\boldsymbol{x} = (x_1, x_2, x_3)$ must live on the sphere of square radius $r^2 = N = 3$, namely

$$x_1^2 + x_2^2 + x_3^2 = 3 \ ? \tag{3}$$

The system now only admits real solutions if

$$-3 \leqslant b_1 \leqslant 3 \quad \text{and} \quad \frac{2b_1}{3} - \frac{1}{3}\sqrt{14}\sqrt{9 - b_1^2} \leqslant b_2 \leqslant \frac{1}{3}\sqrt{14}\sqrt{9 - b_1^2} + \frac{2b_1}{3} \ , \tag{4}$$

i.e. for "small enough" known coefficients. This simple example shows that adding a quadratic constraint to a linear system that would otherwise have infinitely many solutions may turn it into an incompatible system if the known values of b_i in the right-hand sides of linear equations are "large enough".

Scenario 2. Consider now the following *oblique Procrustes problem*:

Given a $M \times N$, $M > N$ matrix A and a "target structure" matrix B of the same dimension, find a $N \times N$ matrix X such that

$$B = AX \tag{5}$$

holds with maximal precision, and columns of X are of unit norm.

We ask the following questions:

1. Is there a relation between the two scenarios?
2. What is the significance of these problems?

"Matrix fitting" problems as the one described in Scenario 2 fall under the umbrella label of **Procrustes problems**. According to Greek mythology, Procrustes, son of Poseidon, was a rogue smith and a bandit, who had a bed in his stronghold, in which he invited every passer-by to spend the night. Using his smith's hammer, he would then get to work on his guests to stretch or mutilate them until they fit his bed. Nobody ever fit the bed exactly. The "target structure" matrix B in Eq. (5) plays the role of Procrustes' bed, the matrix A of the passer-by, and X of the smith's hammer.

A prominent application of the oblique Procrustes problem is in the context of multiple *factor analysis*, a field of psychological research pioneered by Spearman [4] and Thurstone [5, 6]. In quite general terms, the correlation between a number of observed variables in an experiment or test—say, the exam scores by N students in M subjects—may be explained by a much smaller (but unobserved) number of variables (the *factors*), for example different types of "intelligence" (scientific, linguistic, social etc.). Once the number of factors are posited or guessed based on external evidence, one seeks for the *loading matrix* that best describes the data, i.e. the makeup and weighted combination of intelligence traits that best defines each student given their exam results. It turns out that—under quite general assumptions—the problem of determining the best loading matrix can be cast in the oblique Procrustes form given above.

Another Procrustes problem—the so-called *orthogonal* version—with applications in computer vision [7], analytical chemistry [8], speech translation [9, 10] and many others, seeks for the "best" orthogonal matrix $\tilde{\Omega}$ that connects two clouds of points (red and blue) on the plane, where the blue points have been first rigidly rotated with respect to the red ones, and then smeared on the plane by adding some small Gaussian noise (see Fig. 1 for an illustration, and check [11] for a very pedagogical explanation of the setting, the analytical solution, and a Julia computer code to implement the analytical solution).

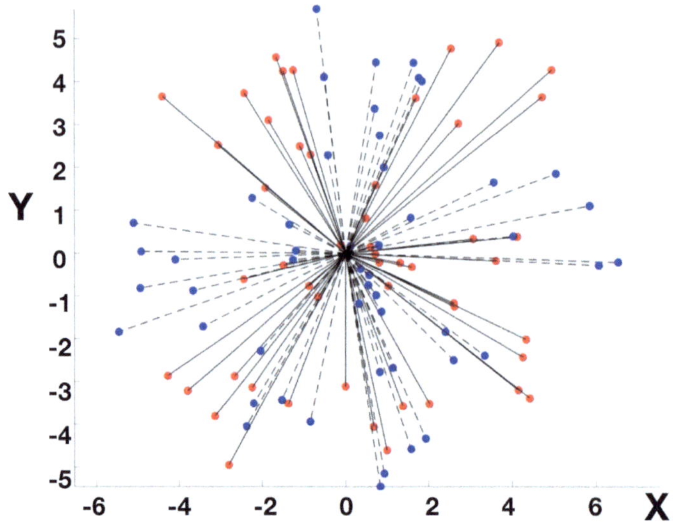

Fig. 1 Setting of the orthogonal Procrustes problem. A cloud of red points is first rigidly rotated clockwise via an orthogonal matrix Ω, and some small Gaussian noise is then added to the resulting rotated points, yielding the final blue points. Given the two clouds (red and blue points), one seeks for the "best" orthogonal matrix $\tilde{\Omega}$ that would bring the blue points "as close as possible" to the initial red points

Coming back to the *oblique* Procrustes problem described in Scenario 2, we may write the system of Eq. (5) for the columns $\{\boldsymbol{x}\}$ of X and $\{\boldsymbol{b}\}$ of B—and set the norm of $\boldsymbol{x}$ to N instead of 1 for later convenience. This way, we can reformulate the oblique Procrustes problem by asking for a solution $\boldsymbol{x}$—a N-dimensional vector—of the over-complete linear system

$$A\boldsymbol{x} = \boldsymbol{b}\ , \tag{6}$$

with A a $M \times N$ matrix, supplemented with the condition $||\boldsymbol{x}||^2 = \boldsymbol{x}^T\boldsymbol{x} = N$, where $||\boldsymbol{x}||^2 = x_1^2 + \cdots + x_N^2$ is the squared norm of the vector, and $(\cdot)^T$ denotes vector/matrix transposition. This way, we cast the problem described in Scenario 2 precisely in the form of Scenario 1.

As is easy to see, the system of equations in (6) is over-complete for $M > N$—i.e., it has more equations than unknowns—so in general, one cannot find a vector $\boldsymbol{x}$ that satisfies it exactly. To find an approximate solution, the best one can do is to minimise some cost function that penalises deviations from the relation $A\boldsymbol{x} = \boldsymbol{b}$, where $\boldsymbol{x}$ and $\boldsymbol{b}$ denote the corresponding columns of X and B respectively. In this setting, the *least square fitting* is one of the most natural and frequently used minimisations, corresponding to the *cost function*

$$H(\boldsymbol{x}) = \frac{1}{2}||A\boldsymbol{x} - \boldsymbol{b}||^2\ , \tag{7}$$

to be minimised over the sphere $||\boldsymbol{x}||^2 = N$. If a vector $\boldsymbol{x}$ with squared norm equal to N can be found for which $H(\boldsymbol{x}) = 0$, then the system (6) is precisely compatible, with solution $\boldsymbol{x}$. If this is not possible, then $\boldsymbol{x} = \text{argmin } H(\boldsymbol{x})$—i.e. the vector that minimises $H(\boldsymbol{x})$ on the sphere—is the best approximation (in the sense of squared norm) that we could find for the incompatible linear system (6). This problem attracted considerable attention starting from the work [12], see e.g. [13, 14].

So far, we have not specified the matrix A and the vector of coefficients $\boldsymbol{b}$ forming the linear system (6) that we look for an exact solution of-or the best approximation to. And we are not going to do that!

The papers [1, 2] indeed analyse a *random* version of the Procrustes problem, where both the matrix A and the vector $\boldsymbol{b}$ are drawn at random from a Gaussian probability density function. The reason for doing this is to analyse *average/typical* properties of the linear system (6) supplemented with a quadratic constraints, which do not depend on the fine details of individual instances of the problem.

More precisely, consider the $M \times N$ entries A_{ij} to be independently drawn at random from a Gaussian probability density function with mean zero, and variance $1/N$,

$$\langle A_{ij} \rangle_A = 0 \qquad \langle A_{kj} A_{\ell m} \rangle_A = (1/N)\delta_{k\ell}\delta_{jm} , \tag{8}$$

while the M entries b_k of the vector $\boldsymbol{b}$ are independently drawn from a Gaussian probability density function with mean zero, and variance σ^2,

$$\langle b_k \rangle_{\boldsymbol{b}} = 0 \qquad \langle b_k b_\ell \rangle_{\boldsymbol{b}} = \sigma^2 \delta_{k\ell} , \tag{9}$$

where

$$\langle \cdot \rangle_A = \int \mathrm{d}A \; P_A(A)(\cdot) \tag{10}$$

denotes averaging over the joint probability density (jpd) $P_A(A) := P_A(A_{11}, \ldots, A_{MN})$ of the matrix A (and similarly for the vector $\boldsymbol{b}$).

The cost function

$$H(\boldsymbol{x}) = \frac{1}{2}||A\boldsymbol{x} - \boldsymbol{b}||^2 := \frac{1}{2}\sum_{k=1}^{M}\left[\sum_{j=1}^{N} A_{kj}x_j - b_k\right]^2 \tag{11}$$

is therefore a random variable, as it depends on the realisation of the randomness (*disorder*) in A and $\boldsymbol{b}$. We are interested in the statistics of the minimal cost

$$\mathcal{E}_{min} = H(\boldsymbol{x}_{min}) = \min_{||\boldsymbol{x}||^2=N} H(\boldsymbol{x}) , \tag{12}$$

corresponding to the "best" approximation to the solution of the linear system (6) on the sphere, for large N, M, as a function of the parameters $\alpha = M/N$ (ratio between number of equations and number of unknowns) and σ^2 (variance of the "noise" term).

Following [1, 2], to achieve this goal we will use two different (and largely complementary) approaches. The first is based on the method of Lagrange multipliers combined with tools from Random Matrix Theory (RMT) and is developed in Sect. 2 for the case $M, N \to \infty$ with $\alpha = M/N > 1$. The outcome of this calculation is a precise formula (see Eq. (43)) for

$$\lim_{N,M\to\infty} \frac{\langle \mathcal{E}_{min} \rangle_{A,\boldsymbol{b}}}{N}, \tag{13}$$

where the average $\langle \cdot \rangle_{A,\boldsymbol{b}}$ is taken with respect to the Gaussian disorder in A and $\boldsymbol{b}$. While derived only in the regime $\alpha > 1$, the formula (43) confirms the natural expectation that over-complete systems with an additional quadratic constraint are typically incompatible, but hints towards the existence of an intermediate regime for "weakly" under-complete systems ($0 < \alpha_c < \alpha < 1$) that are also typically incompatible. This is at odds with what happens without spherical constraint, where under-complete systems $\alpha < 1$ are always typically compatible (see Appendix 5).

A second method—used in Sect. 3—exploits ideas from Statistical Mechanics for directly searching and characterising $\mathcal{E}_{min}$. When doing this one can dispose of the restriction $\alpha = M/N > 1$. The method is based on setting up an auxiliary thermodynamical system of particles subject to a quadratic potential and in equilibrium at inverse temperature β: the zero-temperature free energy of the system can be easily related to $\mathcal{E}_{min}$. Taking the average over the disorder in A and $\boldsymbol{b}$ is the main technical challenge, as it involves a so-called *quenched* average of the logarithm of the associated partition function of the system. The technical challenge can be overcome using a heuristic method—developed in the field of disordered systems and spin glasses—called the *replica method*.

2 Lagrange Multiplier Method

We start with presenting how to set up the Lagrange multiplier method assuming $M > N$ (over-complete system, with more equations than unknowns). Following the standard idea of a constrained minimisation, one uses the cost function (11) to build the associated Lagrangian

$$\mathcal{L}_\lambda(\boldsymbol{x}) = H(\boldsymbol{x}) - \frac{\lambda}{2}\boldsymbol{x}^T\boldsymbol{x} = \frac{1}{2}\sum_{k=1}^{M}\left[\sum_{j=1}^{N} A_{kj}x_j - b_k\right]^2 - \frac{\lambda}{2}\sum_k x_k^2, \tag{14}$$

with $\lambda \in \mathbb{R}$ being the Lagrange multiplier taking care of the spherical constraint.

The stationarity conditions $\nabla\mathcal{L}_\lambda(\boldsymbol{x}) = 0$ entails computing the partial derivative $\partial_{x_r}\mathcal{L}_\lambda(\boldsymbol{x})$ w.r.t. the component x_r, which reads explicitly

$$\frac{1}{2} \times 2\sum_k \left[\sum_{j=1}^N A_{kj}x_j - b_k\right] A_{kr} - \frac{\lambda}{2} \times 2x_r = 0\,. \tag{15}$$

This equation can be cast in the form

$$A^T[A\boldsymbol{x} - \boldsymbol{b}] = \lambda\boldsymbol{x} \Rightarrow A^T A\boldsymbol{x} - A^T\boldsymbol{b} = \lambda\boldsymbol{x} \Rightarrow (A^T A - \lambda\mathbb{I}_N)\boldsymbol{x} = A^T\boldsymbol{b}\,, \tag{16}$$

where $\mathbb{I}_N$ is the $N \times N$ identity matrix.

Solving for $\boldsymbol{x}$, we get

$$\boxed{\boldsymbol{x} = (W - \lambda\mathbb{I}_N)^{-1}A^T\boldsymbol{b}}\,, \tag{17}$$

where $W = A^T A$ is the so-called *Wishart* matrix of size N—see Appendix 6 for generalities on the Wishart ensemble.

Imposing now the spherical constraint $\boldsymbol{x}^T\boldsymbol{x} = N$ yields from Eq. (17)

$$\boxed{\boldsymbol{b}^T A(W - \lambda\mathbb{I}_N)^{-2}A^T\boldsymbol{b} = N}\,, \tag{18}$$

where we used the transposition rule for matrices $(AB)^T = B^T A^T$.

The way to interpret this equation is as follows: fix an instance A from the random matrix ensemble, and an instance $\boldsymbol{b}$ from the random vector ensemble, then find all the real solutions $\{\lambda\}$ of Eq. (18). Every such solution is an acceptable Lagrange multiplier corresponding to a stationary point[1] of the cost function. Then, insert each of these Lagrange multipliers into Eq. (17) and compute the corresponding vector $\boldsymbol{x}_\lambda$, which provides the "location" of the corresponding stationary point of $H(\boldsymbol{x})$. An important property proved in [12] (see Appendix 7 for a sketch of the proof) is that the order of Lagrange multipliers exactly corresponds to the order of values taken by the cost function at the corresponding point $\boldsymbol{x}$. Namely, denoting $\mathcal{N}$ the total number of Lagrange multipliers, assumed to be distinct and ordered as $\lambda_1 < \lambda_2 < \cdots < \lambda_{\mathcal{N}}$, such order implies $H(\boldsymbol{x}_1) < H(\boldsymbol{x}_2) < \cdots < H(\boldsymbol{x}_{\mathcal{N}})$. Thus the minimal loss[2] is always given by

$$\mathcal{E}_{min} = H(\boldsymbol{x}_{min})\,, \tag{19}$$

where $\boldsymbol{x}_{min}$ corresponds to $\lambda_1 := \lambda_{min}$.

[1] This stationary point may be a minimum, a maximum, or a saddle-point, though.

[2] Since the N-sphere is compact and $H(\boldsymbol{x})$ is continuous, it follows from the Extreme Value Theorem that the absolute minimum exists and is reached. Given Browne's theorem, λ_1 must therefore correspond to the absolute minimum (not to a saddle, or to a maximum).

To develop some intuition about the number of solutions of Eq. (18) as a function of the noise variance parameter σ, it is useful to write another representation of the matrix

$$A(W - \lambda \mathbb{I}_N)^{-2} A^T \tag{20}$$

appearing on the left-hand side of Eq. (18) which turns out to be more insightful. To this end, alongside with the Wishart $N \times N$ matrix $W = A^T A$, we define the associated *anti-Wishart* $M \times M$ matrix $W^{(a)} = AA^T$ [15, 16]. Recalling $M > N$, the two matrices share the same set of nonzero eigenvalues $\{s_i\}_{i=1}^N$ and $W^{(a)}$ has another set of $M - N$ eigenvalues equal exactly to 0 (see Appendix 6 for a proof). Note that $\{s_i\}_{i=1}^N$ are all non-negative, as W is a correlation matrix (symmetric and positive semi-definite), and the typical scale of the Wishart eigenvalues – if the variance of the entries of A is $\sim \mathcal{O}(1/N)$ as per Eq. (8)—is $\sim \mathcal{O}(1)$.

The main ingredients are:

1. The formal "geometric series" expansion

$$(W - \lambda \mathbb{I}_N)^{-2} = \sum_{k=0}^{\infty} \frac{(k+1)W^k}{\lambda^{k+2}} , \tag{21}$$

which can be proven by differentiating w.r.t. μ the formal identity

$$(\mathbb{I}_N - \mu W)^{-1} = \sum_{k=0}^{\infty} \mu^k W^k , \tag{22}$$

and setting $\mu = 1/\lambda$ at the end.

2. the rewriting

$$AW^k A^T = \underbrace{A(A^T}\underbrace{A)(A^T}A) \cdots (A^T \underbrace{A)A^T} = [W^{(a)}]^{k+1} , \tag{23}$$

obtained by grouping the matrices differently.

3. the spectral decomposition $W^{(a)} = \sum_{i=1}^N s_i \boldsymbol{v}_i \boldsymbol{v}_i^T$ where $\{\boldsymbol{v}_i\}_{i=1}^M$ are the normalised eigenvectors of $W^{(a)}$ corresponding to its non-zero eigenvalues s_i, $i = 1, \ldots, N$, as well as its corollary $\varphi(W^{(a)}) = \sum_{i=1}^N \varphi(s_i) \boldsymbol{v}_i \boldsymbol{v}_i^T$ for any analytic function $\varphi(z)$.

Starting from Eq. (20), we have

$$\begin{aligned} A(W - \lambda \mathbb{I}_N)^{-2} A^T &\underbrace{=}_{\uparrow \atop (1.)} \sum_{k=0}^{\infty} \frac{(k+1)AW^k A^T}{\lambda^{k+2}} \underbrace{=}_{\uparrow \atop (2.)} \sum_{k=0}^{\infty} \frac{(k+1)[W^{(a)}]^{k+1}}{\lambda^{k+2}} \\ &= W^{(a)}(\lambda \mathbb{I}_M - W^{(a)})^{-2} \underbrace{=}_{\uparrow \atop (3.)} \sum_{i=1}^{N} \frac{s_i}{(\lambda - s_i)^2} \boldsymbol{v}_i \boldsymbol{v}_i^T . \end{aligned} \tag{24}$$

Sandwiching both sides between $\boldsymbol{b}^T$ and $\boldsymbol{b}$ (see Eq. (18)), we obtain the spherical constraint equality in the form

$$\sum_{i=1}^{N} \frac{s_i}{(\lambda - s_i)^2} (\boldsymbol{b}^T \boldsymbol{v}_i)(\boldsymbol{v}_i^T \boldsymbol{b}) = N \Rightarrow \boxed{\sum_{i=1}^{N} \frac{s_i}{(\lambda - s_i)^2} (\boldsymbol{\xi}^T \boldsymbol{v}_i)^2 = \frac{N}{\sigma^2}}, \tag{25}$$

where we set $\boldsymbol{b} = \sigma^2 \boldsymbol{\xi}$, with $\boldsymbol{\xi}$ having standard Gaussian entries (zero mean and variance one).

It is easy to see that the left-hand side is a positive function of λ having a single minimum between every consecutive pair of eigenvalues of W, see Fig. 2 for $N = 5$ below.

This implies there are typically 0 or 2 solutions of Eq. (25) (and 1 solution with probability zero at exceptional points) for λ between every consecutive pair of eigenvalues, plus two more solutions: the minimal one $\lambda_{min} \in (-\infty, s_1)$ and the maximal one $\lambda_{max} \in (s_N, \infty)$. Note that the latter two solutions exist for any value of $\sigma \in [0, \infty]$, whereas by changing σ one changes the number of solutions available between consecutive eigenvalues. In particular, in the limit of vanishing noise (i.e. $\sigma \to 0$, hence $\boldsymbol{b} \to 0$), every stationary point solution for the Lagrange multiplier corresponds to an eigenvalue s_k of the Wishart matrix, with $\boldsymbol{x} = \pm \boldsymbol{e}_k$ being the associated eigenvectors (hence there are $2N$ stationary points). On the other hand when $\sigma \to \infty$ the ratio N/σ^2 in the right-hand side becomes smaller than the global minimum of the left-hand side in $[s_1, s_N]$. Then only two stationary points survive:

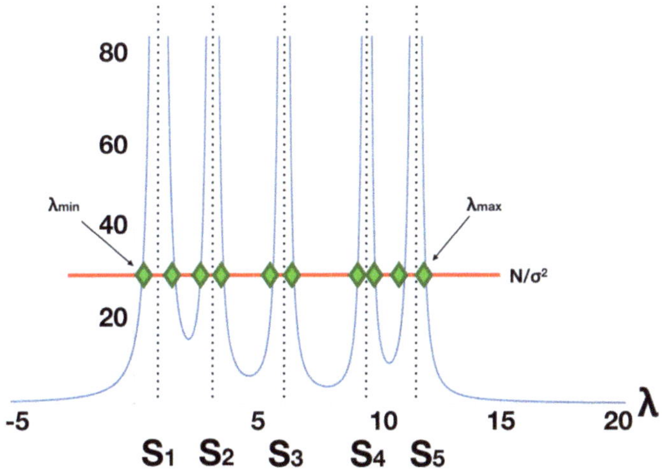

Fig. 2 Sketch of the plot of the left hand side of Eq. (25) as a function of λ for $N = 5$. The horizontal red line signals the right hand side constant value N/σ^2, and green points highlight the intersections (solutions of the equation (25)). The leftmost intersection (to the left of the smallest eigenvalue s_1) corresponds to the minimal cost

$\lambda_N := \lambda_{max}$ and $\lambda_1 := \lambda_{min}$—which correspond to the global maximum and the global minimum of $H(\boldsymbol{x})$.

Combining Eq. (19) with Eqs. (11) and (17), we have

$$\mathcal{E}_{min} = H(\boldsymbol{x}_{min}) = H\left((W - \lambda_1 \mathbb{I}_N)^{-1} A^T \boldsymbol{b}\right) = \frac{1}{2}\left|\left|A(W - \lambda_1 \mathbb{I}_N)^{-1} A^T \boldsymbol{b} - \boldsymbol{b}\right|\right|^2 , \tag{26}$$

where λ_1 is the minimal solution of Eq. (25), and $W = A^T A$. We stress that $\mathcal{E}_{min}$ is a random variable, which depends on the realisation of the disorder in the matrix A and the parameters $\boldsymbol{b}$ both directly, and through the minimal value of the Lagrange multiplier λ_1.

To compute the average minimal loss, $\langle \mathcal{E}_{min} \rangle_{A,\boldsymbol{b}}$, we proceed in two steps: (i) we compute the average/typical value $\lambda^\star$ of λ_1, and we replace the random variable λ_1 with $\lambda^\star$ in (26). Then (ii) we compute the average $\langle \mathcal{E}_{min} \rangle_{A,\boldsymbol{b}}$, where any indirect dependence on $A, \boldsymbol{b}$ has been removed. This two-step procedure of course hinges on a *self-averaging* assumption, whereby the random variable λ_1 quickly converges to (and is therefore well approximated by) its mean value $\lambda^\star$.

2.1 Calculation of $\lambda^\star$

We show in this section that the average value $\lambda^\star$ of the smallest solution λ_1 of Eq. (25) is given by

$$\boxed{\lambda^\star = \frac{(\sqrt{\alpha} - \sqrt{1+\sigma^2})(\sqrt{\alpha(1+\sigma^2)} - 1)}{\sqrt{1+\sigma^2}}} , \tag{27}$$

in the limit $N, M \to \infty$ with $\alpha = M/N$ fixed—where M is the number of equations and N the number of unknowns—and σ^2 is the variance of the noise terms b_k.

We arrive at Eq. (27) starting from Eq. (25), which is rewritten as

$$g(\lambda) = \frac{1}{N} \sum_{i=1}^{N} \frac{s_i}{(\lambda - s_i)^2} (\boldsymbol{\xi}^T \boldsymbol{v}_i)^2 = \frac{1}{\sigma^2} . \tag{28}$$

We wish to take the average $\langle g(\lambda) \rangle_{A,\boldsymbol{\xi}}$ and find the minimal solution of $\langle g(\lambda) \rangle_{A,\boldsymbol{\xi}} = 1/\sigma^2$ for large N, M.

First, taking the average over the variables $\boldsymbol{\xi}$ reads

$$\langle g(\lambda) \rangle_{\boldsymbol{\xi}} = \frac{1}{N} \sum_{i=1}^{N} \frac{s_i}{(\lambda - s_i)^2} \sum_{\ell,m} \underbrace{\langle \xi_\ell \xi_m \rangle_{\boldsymbol{\xi}}}_{\delta_{\ell m}} v_{i\ell} v_{im} = \frac{1}{N} \sum_{i=1}^{N} \frac{s_i}{(\lambda - s_i)^2} \underbrace{\sum_{\ell} v_{i\ell}^2}_{=1} , \tag{29}$$

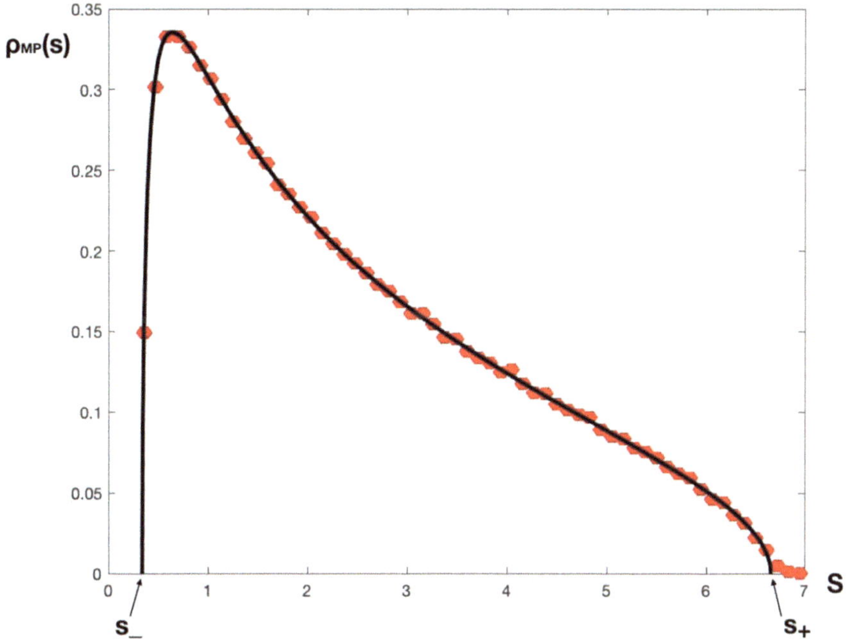

Fig. 3 Red points: histogram of eigenvalues s_i of a Wishart ensemble generated from 500 "data" matrices A each of size $M \times N$, with $M = 500$, $N = 200$ (corresponding to $\alpha = 2.5$) and filled with i.i.d. Gaussian entries with mean zero and variance $1/N$. Solid black line: the Marčenko-Pastur law (31) over the compact support $[s_-, s_+]$, with $s_- = (\sqrt{\alpha} - 1)^2 \approx 0.338$ and $s_+ = (\sqrt{\alpha} + 1)^2 \approx 6.662$, showing excellent agreement

where we first used that $\boldsymbol{\xi}$ are i.i.d. standard Gaussian variables, and then the normalisation of eigenvectors of $W^{(a)}$.

Next, we use the property

$$\lim_{\substack{N,M\to\infty \\ M/N=\alpha}} \left\langle \frac{1}{N} \mathrm{Tr}\, \varphi(W) \right\rangle_A = \lim_{\substack{N,M\to\infty \\ M/N=\alpha}} \left\langle \frac{1}{N} \sum_{i=1}^{N} \varphi(s_i) \right\rangle_A = \int_{s_-}^{s_+} \mathrm{d}s\, \rho_{MP}(s) \varphi(s) \,, \tag{30}$$

where $\{s_i\}$ are the eigenvalues of the $N \times N$ Wishart matrix $W = A^T A$, and

$$\rho_{MP}(s) = \frac{2}{\pi s} \frac{1}{(\sqrt{s_+} - \sqrt{s_-})^2} \sqrt{(s_+ - s)(s - s_-)} \tag{31}$$

is the *Marčenko-Pastur* probability density function[3] [17], having compact support between the edge points $s_\pm = (\sqrt{\alpha} \pm 1)^2$ on the positive real axis. In Fig. 3, we test numerically the Marčenko-Pastur law on randomly generated Wishart matrices.

[3] It is easy to check that $\int_{s_-}^{s_+} \mathrm{d}s\, \rho_{MP}(s) = 1$.

The lower edge point s_- corresponds to the average location of the smallest eigenvalue of the Wishart matrix in the large N, M limit. Since we are interested in the smallest intersection $\lambda^\star$ of the function $g(\lambda)$ with the horizontal line $1/\sigma^2$, we should be looking for this solution to the left of s_- (see Fig. 2).

Combining (29) and (30), we have that

$$\lim_{\substack{N,M\to\infty \\ M/N=\alpha}} \langle g(\lambda)\rangle_{A,\xi} = \int_{s_-}^{s_+} \mathrm{d}s\ \rho_{MP}(s)\frac{s}{(\lambda-s)^2}\ . \tag{32}$$

This integral is computed in Appendix 8 to give

$$\lim_{\substack{N,M\to\infty \\ M/N=\alpha}} \langle g(\lambda)\rangle_{A,\xi} = \frac{1}{(\sqrt{s_+}-\sqrt{s_-})^2}\frac{(\sqrt{s_+-\lambda}-\sqrt{s_--\lambda})^2}{\sqrt{(s_+-\lambda)(s_--\lambda)}}\ , \tag{33}$$

for $\lambda < s_-$. The value $\lambda^\star$ will therefore be obtained from Eq. (28) by equating (33) to $1/\sigma^2$, which leads after simple algebra to Eq. (27).

Note from Eq. (27) that:

1. For $\sigma^2 \to 0$, we have $\lambda^\star \to (\sqrt{\alpha}-1)^2 \equiv s_-$. Indeed, in this limit (see Fig. 2), the smallest Lagrange multiplier should coincide with the minimal eigenvalue of the Wishart matrix.
2. From Fig. 2, one may see that by increasing σ for fixed α, N one makes the lowest Lagrange multiplier change sign from positive to negative. Our calculation confirms this intuition and shows that this sign change typically happens when $\alpha = 1+\sigma^2$.

2.2 Calculation of $\langle \mathcal{E}_{min}\rangle_{A,\boldsymbol{b}}$

Having computed the typical value $\lambda^\star$ of the minimal solution of Eq. (28) (averaged over the disorder) we insert this fixed, non-random value into Eq. (26) and proceed to compute the average of the loss function as follows:

$$\begin{aligned}
\langle \mathcal{E}_{min}\rangle_{A,\boldsymbol{b}} &= \frac{1}{2}\left\langle \left|\left|A(W-\lambda^\star\mathbb{I}_N)^{-1}A^T\boldsymbol{b}-\boldsymbol{b}\right|\right|^2\right\rangle_{A,\boldsymbol{b}} \\
&= \frac{1}{2}\left\langle \left(A(W-\lambda^\star\mathbb{I}_N)^{-1}A^T\boldsymbol{b}-\boldsymbol{b}\right)^T\left(A(W-\lambda^\star\mathbb{I}_N)^{-1}A^T\boldsymbol{b}-\boldsymbol{b}\right)\right\rangle_{A,\boldsymbol{b}} \\
&= \frac{1}{2}\left\langle \left(\boldsymbol{b}^T A(W-\lambda^\star\mathbb{I}_N)^{-1}A^T-\boldsymbol{b}^T\right)\left(A(W-\lambda^\star\mathbb{I}_N)^{-1}A^T\boldsymbol{b}-\boldsymbol{b}\right)\right\rangle_{A,\boldsymbol{b}} \\
&= \frac{1}{2}\Big\langle \boldsymbol{b}^T A(W-\lambda^\star\mathbb{I}_N)^{-1}\underbrace{A^TA}_{W}(W-\lambda^\star\mathbb{I}_N)^{-1}A^T\boldsymbol{b}-2\boldsymbol{b}^T A(W-\lambda^\star\mathbb{I}_N)^{-1}A^T\boldsymbol{b} \\
&\qquad +\boldsymbol{b}^T\boldsymbol{b}\Big\rangle_{A,\boldsymbol{b}}\ .
\end{aligned}$$

Replacing W with $(W - \lambda^\star \mathbb{I}_N) + \lambda^\star \mathbb{I}_N$ in the middle of the first contribution leads to

$$\langle \mathcal{E}_{min} \rangle_{A,\boldsymbol{b}} = \frac{1}{2} \Big\langle \boldsymbol{b}^T A (W - \lambda^\star \mathbb{I}_N)^{-1} A^T \boldsymbol{b} + \lambda^\star \underbrace{\boldsymbol{b}^T A (W - \lambda^\star \mathbb{I}_N)^{-2} A^T \boldsymbol{b}}_{=N}$$
$$-2\boldsymbol{b}^T A (W - \lambda^\star \mathbb{I}_N)^{-1} A^T \boldsymbol{b} + \boldsymbol{b}^T \boldsymbol{b} \Big\rangle_{A,\boldsymbol{b}} , \tag{34}$$

where we used Eq. (18). Combining the first and third contribution together, we eventually get to

$$\langle \mathcal{E}_{min} \rangle_{A,\boldsymbol{b}} = \frac{1}{2} \left\langle \boldsymbol{b}^T A (\lambda^\star \mathbb{I}_N - W)^{-1} A^T \boldsymbol{b} + \lambda^\star N + |\boldsymbol{b}|^2 \right\rangle_{A,\boldsymbol{b}} . \tag{35}$$

We first perform the average over the i.i.d. Gaussian distributed $\boldsymbol{b}$ vector with variance σ^2, using

$$\langle |\boldsymbol{b}|^2 \rangle_{\boldsymbol{b}} = \sum_{i=1}^{M} \langle b_i^2 \rangle_{\boldsymbol{b}} = M\sigma^2 \tag{36}$$

$$\langle \boldsymbol{b}^T R \boldsymbol{b} \rangle_{\boldsymbol{b}} = \sum_{i,j} R_{ij} \langle b_i b_j \rangle_{\boldsymbol{b}} = \sigma^2 \sum_{i,j} R_{ij} \delta_{ij} = \sigma^2 \mathrm{Tr} R , \tag{37}$$

where Tr denotes the matrix trace.

Setting $R = A(\lambda^\star \mathbb{I}_N - W)^{-1} A^T$ and looking at the limit $N, M \to \infty$ with $\alpha = M/N$ fixed we get from Eq. (35)

$$\lim_{N,M\to\infty} \frac{\langle \mathcal{E}_{min} \rangle_{A,\boldsymbol{b}}}{N} = \frac{1}{2} \left[\lambda^\star + \alpha\sigma^2 + \sigma^2 \left\langle \frac{1}{N} \mathrm{Tr} \frac{W}{\lambda^\star \mathbb{I}_N - W} \right\rangle_A \right] . \tag{38}$$

The average over A of the last term can be performed in the large N, M limit appealing again to the formula (30) to yield

$$\left\langle \frac{1}{N} \mathrm{Tr} \frac{W}{\lambda^\star \mathbb{I}_N - W} \right\rangle_A = \int_{s_-}^{s_+} \mathrm{d}s \, \rho_{MP}(s) \frac{s}{\lambda^\star - s} , \tag{39}$$

which can be computed along the same lines as in Appendix 8 to give

$$\left\langle \frac{1}{N} \mathrm{Tr} \frac{W}{\lambda^\star \mathbb{I}_N - W} \right\rangle_A = \frac{2}{\pi} \frac{(s_+ - s_-)^2}{(\sqrt{s_+} - \sqrt{s_-})^2 (\lambda - s_-)} \int_0^1 \mathrm{d}t \, \frac{\sqrt{t(1-t)}}{1 + \gamma t} \tag{40}$$

$$= -\frac{\left(\sqrt{s_-} + \sqrt{s_+}\right)^2}{-2\lambda^\star + 2\sqrt{(s_- - \lambda^\star)(s_+ - \lambda)} + s_- + s_+} , \tag{41}$$

where

$$\gamma = \frac{s_+ - s_-}{s_- - \lambda^\star} > 0 \, . \tag{42}$$

Inserting this result back into Eq. (38), and substituting the value of $\lambda^\star$ as given in Eq. (27) as well as $s_\pm = (\sqrt{\alpha} \pm 1)^2$, we obtain after simplifications

$$\boxed{\lim_{N,M\to\infty} \frac{\langle \mathcal{E}_{min} \rangle_{A,\boldsymbol{b}}}{N} = \frac{1}{2} \left[\sqrt{\alpha(1+\sigma^2)} - 1 \right]^2} \, . \tag{43}$$

This result, obtained under the assumption[4] $\alpha > 1$, confirms the natural expectation that over-complete systems with an additional quadratic constraint are typically incompatible, as the average cost function is strictly positive. However, the expression in (43) vanishes for

$$\alpha = \alpha_c = \frac{1}{1+\sigma^2} < 1 \, , \tag{44}$$

so—assuming that the result in (43) remains valid for $\alpha < 1$ as well—we may conclude that there are *three* regimes for this problem:

- Over-complete systems ($\alpha > 1$), which are typically incompatible. Here, the spherical constraint does not add much to what would happen anyway for "standard" linear systems (see Appendix 5 for the calculation of the average minimal cost for linear systems without the spherical constraint).
- Strongly under-complete systems ($0 < \alpha < \alpha_c < 1$), which are typically compatible even with the additional spherical constraint.
- Weakly under-complete systems ($\alpha_c < \alpha < 1$), which are typically incompatible. This is a new regime, where the additional spherical constraint is responsible for turning a linear system that would be perfectly compatible into an incompatible one due to the additional quadratic requirement to be satisfied by the unknowns.

It is interesting to notice that the critical threshold α_c shrinks with increasing variance σ^2 of the "noise" variables b_k. This means that the incompatibility region widens, and even systems with a handful of equations and many more unknowns become hard to satisfy as the known coefficients become more and more "blurred".

[4] The assumption $\alpha > 1$ (equivalently, $M > N$) was at play around Eq. (23) and (24) (in the case $\alpha < 1$ the matrix W of size $N \times N$ would effectively be an anti-Wishart matrix, with $M < N$ nonzero eigenvalues). Also the evaluation of the integral (30) would require the spectral density of the anti-Wishart ensemble, given in Eq. (147). It is likely that the RMT treatment of the case $\alpha < 1$ could be carried out in full as well, however this is left for future work.

3 Replica Calculation

To verify the formula (43) for the minimal cost independently (and in this way also to justify some assumptions used to arrive to it), we now follow [18] and reformulate the minimisation problem from the viewpoint of Statistical Mechanics of disordered systems. Namely, to look for the minimum of a loss/cost function $H(\boldsymbol{x})$ over the sphere $||\boldsymbol{x}||^2 = N$ one introduces an auxiliary non-negative parameter $\beta > 0$ (called *the inverse temperature*) and uses it to define the so-called *canonical partition function* $Z(\beta)$ via

$$Z(\beta) = \int\limits_{||\boldsymbol{x}||^2=N} \mathrm{d}\boldsymbol{x}\, \mathrm{e}^{-\beta H(\boldsymbol{x})} . \tag{45}$$

Suppose the minimum $\mathcal{E}_{min}$ of the loss/cost function $H(\boldsymbol{x})$ is achieved at some $\boldsymbol{x}_{min}$. For $\beta \to \infty$, the integral in Eq. (45) can be evaluated using Laplace's method (summarised in Appendix 9). Applying this method gives the leading exponential behaviour in the form

$$Z(\beta) \propto \mathrm{e}^{-\beta \mathcal{E}_{min}} ,$$

hence the minimal value can be recovered in the limit $\beta \to \infty$ from the logarithm of the partition function:

$$\mathcal{E}_{min} = - \lim_{\beta\to\infty} \frac{1}{\beta} \log Z(\beta) . \tag{46}$$

The canonical partition function is nothing but the normalisation factor of the Gibbs-Boltzmann probability density function

$$P_\beta(\boldsymbol{x}) = \frac{\mathrm{e}^{-\beta H(\boldsymbol{x})}}{Z(\beta)} \tag{47}$$

governing the probability of finding a thermodynamical system of particles around the point $\boldsymbol{x}$ in space, if their total energy is given by $H(\boldsymbol{x})$ and they are in equilibrium with a heat bath at inverse temperature β. In this language, the *free energy* of the system is given by

$$F(\beta) = -\frac{1}{\beta} \log Z(\beta) , \tag{48}$$

from which it follows that $\mathcal{E}_{min}$ is nothing but the free energy at zero temperature of the associated particle system, and $\boldsymbol{x}_{min}$ the most likely zero-temperature configuration of particles at equilibrium.

The relation (46) is very general (in particular, it does not assume the global minimum is achieved at only a single $\boldsymbol{x}_{min}$, there may be several such points), but for problems involving random cost/loss functions its actual usefulness crucially depends on our ability to characterise the behaviour of the log on the right-hand side.

3.1 Average $\langle \mathcal{E}_{min} \rangle_{A,b}$

The simplest, yet already highly nontrivial task is to find the mean value $\langle \mathcal{E}_{min} \rangle_{A,\boldsymbol{b}}$ for the minimal cost. Obviously, one needs to average the logarithm $\log Z(\beta)$ on the right-hand side of (46)

$$\langle \log Z(\beta) \rangle_{A,\boldsymbol{b}} = ? \tag{49}$$

What is the problem here? Writing the average more explicitly, we are after the multiple integral

$$\int \mathrm{d}A\, \mathrm{d}\boldsymbol{b}\; P_A(A) P_{\boldsymbol{b}}(\boldsymbol{b}) \log \left(\int_{||\boldsymbol{x}||^2=N} \mathrm{d}\boldsymbol{x}\; \mathrm{e}^{-\beta H(\boldsymbol{x})} \right) , \tag{50}$$

where $P_A(A)$ and $P_{\boldsymbol{b}}(\boldsymbol{b})$ are the joint probability density function of the entries of the matrix A, and of the vector $\boldsymbol{b}$, respectively[5]. Equation (50) defines a so called *quenched*[6] average: there are two nested sources of randomness, (i) the disorder in A and $\boldsymbol{b}$, and (ii) the Gibbs-Boltzmann distribution at inverse temperature β of the auxiliary degrees of freedom $\boldsymbol{x}$. A single instance of A and of $\boldsymbol{b}$ is selected and kept fixed, while the $\boldsymbol{x}$ degrees of freedom equilibrate at inverse temperature β, before another instance of the $A, \boldsymbol{b}$ disorder is picked, and the process is repeated. In thermodynamical terms, a quenched average rests on the separation between two temporal scales, the "fast" equilibration of the $\boldsymbol{x}$ degrees of freedom at fixed temperature and at fixed instance of the disorder, and a "slower" scale of change of the background disorder. This should be contrasted with the *annealed*[7] scheme, which is only approximate but way easier to handle analytically, consisting in treating the averages over the disorder and over the Gibbs-Boltzmann distribution on the same footing, and simultaneously. We will not consider annealed calculations here—for more information, see [19, 20].

Clearly, the only obvious way to crack the integral (50) is to solve the $\boldsymbol{x}$-integral *first*, take the logarithm of the result, and take the A and $\boldsymbol{b}$ integrals *last*. Unfortunately, this procedure fails in most cases, because the $\boldsymbol{x}$-integral cannot in general be solved exactly for a fixed instance of A and $\boldsymbol{b}$ (or–for different models—of the so called *disorder*, namely the randomness inherent to the parameters entering the definition of the cost function $H(\boldsymbol{x})$—in our case, the entries of A and $\boldsymbol{b}$).

It would therefore be very helpful to make some headway if we were able somehow to take the average over the disorder (A and $\boldsymbol{b}$) *first*, and the $\boldsymbol{x}$-integral *last*—in the hope that swapping the order of integration would make the integrals more friendly to tackle.

---

[5] In our setting, these are just given by product of independent Gaussian functions, one for each entry.

[6] From Merriam-Webster dictionary. *Quench* [transitive verb]: to cool (something, such as heated metal) suddenly by immersion (as in oil or water).

[7] *Anneal* [transitive verb]: to heat and then cool slowly (a material, such as steel or glass) usually for softening and making less brittle.

A heuristic recipe to do just that originated in the "theory of spin glasses" [19] in the 70 s. Doing this swapping of integrals in a fully rigorously manner is quite challenging even in the simplest possible instances, though considerable progress has been achieved in the last decades in evaluating such averages in a mathematically controllable way in the case when the cost function is normally distributed, see e.g. [21–23]. Unfortunately, in our case the cost function is not normally distributed, but rather represents a sum of squared normally distributed pieces. In such a case the rigorous theory has not been yet developed, but progress is still possible within the powerful but heuristic method of Theoretical Physics, known as the "replica trick", see e.g. [24].

The main idea rests on the exact identity

$$\langle \log z \rangle = \lim_{n\to 0} \frac{1}{n} \log\langle z^n \rangle \,, \tag{51}$$

which can be proven by noting that $\langle z^n \rangle = \langle 1 + n\log z + o(n) \rangle = \langle 1 \rangle + n\langle \log z \rangle + \cdots = 1 + n\langle \log z \rangle + \ldots$, where we used linearity of expectations, and normalisation $\langle 1 \rangle = 1$.

While the identity (51) is mathematically fully rigorous, and requires n to be real and in the vicinity of zero, the way it is implemented in replica calculations is as follows: assume first that n is an integer.

This approach assumes that the mean value we are after can be found not from directly calculating the average, but by considering the expectation of the integer moments of the partition function, frequently called in the physical literature the "replicated" disorder averaged partition function $\langle Z^n(\beta) \rangle$ and subsequently taking the limit[8] $n \rightsquigarrow 0$ to recover the averaged log

$$\langle \mathcal{E}_{min} \rangle_{a,\boldsymbol{b}} = \lim_{\beta\to\infty} \frac{-1}{\beta} \lim_{n\rightsquigarrow 0} \frac{1}{n} \log \langle Z^n(\beta) \rangle_{A,\boldsymbol{b}} \tag{52}$$

and eventually for large N, M with $\alpha = M/N$ constant, we wish to compute

$$\lim_{N\to\infty} \frac{\langle \mathcal{E}_{min} \rangle_{A,\boldsymbol{b}}}{N} = \lim_{N\to\infty} \frac{1}{N} \lim_{\beta\to\infty} \frac{-1}{\beta} \lim_{n\rightsquigarrow 0} \frac{1}{n} \log \langle Z^n(\beta) \rangle_{A,\boldsymbol{b}} \tag{53}$$

to be compared with the corresponding RMT result in Eq. (43).

We now show how to evaluate the replicated moments of the finite-temperature partition function induced by a general least-square random cost landscape of the form

$$H(\boldsymbol{x}) = \frac{1}{2} \sum_{k=1}^{M} [V_k(\boldsymbol{x})]^2 \,, \tag{54}$$

---

[8] We use the wiggly arrow $\rightsquigarrow$ to denote the "replica limit" procedure as follows: first convert $n \in \mathbb{R}$ to an integer. Then evaluate the corresponding observable as an explicit function of the integer n. Then pretend that n could be analytically continued in the vicinity of zero without ambiguities.

restricted to the sphere $||\boldsymbol{x}||^2 = N$, where $V_k(\boldsymbol{x})$ are chosen to be Gaussian-distributed functions of the vector $\boldsymbol{x}$ with expectations $\langle V_k(\boldsymbol{x})\rangle_V = 0$ and covariance

$$\langle V_k(\boldsymbol{x}_a)V_\ell(\boldsymbol{x}_b)\rangle_V = \delta_{k\ell} f\Big(\frac{\boldsymbol{x}_a \cdot \boldsymbol{x}_b}{N}\Big) . \tag{55}$$

Here and in the following, $\boldsymbol{u} \cdot \boldsymbol{v}$ denotes the dot product between vectors.

The particular case $V_k(\boldsymbol{x}) = \sum_{j=1}^N A_{kj}x_j - b_k$ treated here (linear system of equations) corresponds to the choice

$$f(u) = \sigma^2 + u \tag{56}$$

(see Appendix 10 for the full derivation), but the calculation can be performed for any covariance of the form (55) and essentially follows the method of [25]. The non-linear version of the model was considered in [3] and then fully solved in [26] in the context of a model of rigidity transition in confluent systems, where a full Replica Symmetry Breaking structure was uncovered and the properties around the transition were studied in detail (see also [27] for an insightful dynamical mean-field theory of the non-linear model, and [28] for a quantum version of it).

Starting with the partition function $Z(\beta)$

$$Z(\beta) = \int\limits_{||\boldsymbol{x}||^2=N} \mathrm{d}\boldsymbol{x}\ \mathrm{e}^{-\frac{1}{2}\beta\sum\limits_{k=1}^{M} V_k^2(\boldsymbol{x})} , \tag{57}$$

we aim to arrive to a convenient integral representation for the averaged partition function of n copies of the same system $\langle Z^n(\beta)\rangle$. To facilitate the averaging, we introduce new auxiliary variables of integration and employ the standard Gaussian integral identity (sometimes called in physics literature the Hubbard-Stratonovich identity)

$$\mathrm{e}^{-\frac{\beta}{2}\xi^2} = \frac{1}{\sqrt{2\pi}}\int_{\mathbb{R}} \mathrm{d}u\ \exp\left[-\frac{1}{2}u^2 - \mathrm{i}\sqrt{\beta}\xi u\right] , \tag{58}$$

applying it M times, once for every index $k = 1, \ldots, M$. The effect of this identity is to lower the power of the quantity ξ (in our case, each ξ_k is equal to $V_k(\boldsymbol{x})$), from 2 to 1—the price to pay is that one needs to introduce an extra integration over an auxiliary Gaussian variable u.

Proceeding in this way the partition function $Z(\beta)$ now can be expressed in terms of the integration over a vector $\boldsymbol{u} = (u_1, \ldots, u_M)^T$ as

$$Z(\beta) = \int\limits_{\mathbb{R}^M} \frac{\mathrm{d}\boldsymbol{u}}{(2\pi)^{M/2}}\ \mathrm{e}^{-\frac{1}{2}(\boldsymbol{u}\cdot\boldsymbol{u})} \int\limits_{||\boldsymbol{x}||^2=N} \mathrm{d}\boldsymbol{x}\ \mathrm{e}^{-\mathrm{i}\sqrt{\beta}\sum\limits_{k=1}^{M} u_k V_k(\boldsymbol{x})} . \tag{59}$$

Now we take products of n identical copies of $Z(\beta)$ (which we number with the index $a = 1, 2, \ldots, n$) and aim to evaluate

$$\langle Z^n(\beta)\rangle_V = \int_{\mathbb{R}^M} \prod_{a=1}^{n} \frac{\mathrm{d}\boldsymbol{u}_a}{(2\pi)^{M/2}} \mathrm{e}^{-\frac{1}{2}\sum_{a=1}^{n}(\boldsymbol{u}_a\cdot\boldsymbol{u}_a)} \times$$
$$\times \int_{D_N} \prod_{a=1}^{n} \mathrm{d}\boldsymbol{x}_a \prod_{k=1}^{M} \left\langle \mathrm{e}^{-\mathrm{i}\sqrt{\beta}\sum_{a=1}^{n}[\boldsymbol{u}_a]_k V_k(\boldsymbol{x}_a)} \right\rangle_V, \tag{60}$$

where rectangular brackets $[\boldsymbol{u}_a]_k$ stand for the k-th entry of the vector $\boldsymbol{u}_a$, and $\langle\cdot\rangle_V$ stands for averaging over the random variable V. Here we used the fact that for different k the functions $V_k(\boldsymbol{x})$ are independent, hence the corresponding average of the product factorises in the product of averages. The domain of integration of the $\boldsymbol{x}$ variables is the union of n spheres $D_N = \{\, ||\boldsymbol{x}_a||^2 = N,\ \forall a\}$.

Note that replicating the partition function an *integer* number of times n has effectively allowed us to get rid of the annoying logarithm in (49) and offers now a way to perform the average over the randomness in A and $\boldsymbol{b}$ (through V) directly.

To perform the average over V_k we use that the combination $z = -\mathrm{i}\sqrt{\beta}\sum_{a=1}^{n}[\boldsymbol{u}_a]_k V_k(\boldsymbol{x}_a)$ is Gaussian—being a sum of Gaussian-distributed terms V_k—with mean zero and variance

$$\langle z^2\rangle_V = -\beta \sum_{a,b=1}^{n} [\boldsymbol{u}_a]_k[\boldsymbol{u}_b]_k \langle V_k(\boldsymbol{x}_a)V_k(\boldsymbol{x}_b)\rangle_V = -\beta \sum_{a,b=1}^{n} [\boldsymbol{u}_a]_k[\boldsymbol{u}_b]_k f\left(\frac{\boldsymbol{x}_a\cdot\boldsymbol{x}_b}{N}\right), \tag{61}$$

where we have used the covariance of the V's in Eq. (55).

Using now the following rule for computing the expectation of the exponential of a Gaussian variable

$$\langle \mathrm{e}^z\rangle_{z\sim\mathcal{N}(\mu,\sigma^2)} = \mathrm{e}^{\mu+\frac{\sigma^2}{2}} \tag{62}$$

and taking the product over $k = 1, \ldots, M$, we arrive at

$$\prod_{k=1}^{M}\left\langle \mathrm{e}^{-\mathrm{i}\sqrt{\beta}\sum_{a=1}^{n}[\boldsymbol{u}_a]_k V_k(\boldsymbol{x}_a)}\right\rangle_V = \prod_{k=1}^{M}\exp\left[-\frac{1}{2}\beta\sum_{a,b=1}^{n}[\boldsymbol{u}_a]_k[\boldsymbol{u}_b]_k f\left(\frac{\boldsymbol{x}_a\cdot\boldsymbol{x}_b}{N}\right)\right]$$
$$= \exp\left[-\frac{1}{2}\beta\sum_{a,b=1}^{n}\left(\sum_{k=1}^{M}[\boldsymbol{u}_a]_k[\boldsymbol{u}_b]_k\right) f\left(\frac{\boldsymbol{x}_a\cdot\boldsymbol{x}_b}{N}\right)\right]$$
$$= \exp\left[-\frac{1}{2}\beta(\boldsymbol{u}_a\cdot\boldsymbol{u}_b) f\left(\frac{\boldsymbol{x}_a\cdot\boldsymbol{x}_b}{N}\right)\right]. \tag{63}$$

Substituting this back into the integral representation for $\langle Z^n(\beta)\rangle_V$ in Eq. (60), and writing the dot products more explicitly we get

$$\langle Z^n(\beta)\rangle_V = \int_{D_N}\prod_{a=1}^n \mathrm{d}\boldsymbol{x}_a \int_{\mathbb{R}^{nM}}\prod_{a=1}^n \frac{\mathrm{d}\boldsymbol{u}_a}{(2\pi)^{M/2}}\exp\left[-\frac{1}{2}\sum_{k=1}^M\sum_{a=1}^n([\boldsymbol{u}_a]_k)^2 \right.$$
$$\left. -\frac{1}{2}\beta\sum_{k=1}^M\sum_{a,b=1}^n([\boldsymbol{u}_a]_k[\boldsymbol{u}_b]_k)\, f\left(\frac{\boldsymbol{x}_a\cdot\boldsymbol{x}_b}{N}\right)\right]. \tag{64}$$

One notices by inspection that the exponential factorises into the product of M identical exponentials, therefore we can write

$$\langle Z^n(\beta)\rangle_V = \int_{D_N}\prod_{a=1}^n \mathrm{d}\boldsymbol{x}_a \left\{\int_{\mathbb{R}^n}\prod_{a=1}^n \frac{\mathrm{d}u_a}{(2\pi)^{M/2}}\exp\left[-\frac{1}{2}\sum_{a=1}^n u_a^2 \right.\right.$$
$$\left.\left. -\frac{1}{2}\beta\sum_{a,b=1}^n u_a f\left(\frac{\boldsymbol{x}_a\cdot\boldsymbol{x}_b}{N}\right)u_b\right]\right\}^M. \tag{65}$$

We notice now that the integral over $\{u_a\}$ is simply a multivariate n-dimensional Gaussian integration[9], yielding

$$\langle Z^n(\beta)\rangle_V \propto \int_{D_N}\prod_{a=1}^n \mathrm{d}\boldsymbol{x}_a \left(\det\left[\mathbb{I}_n + \beta f\left(\frac{\boldsymbol{x}_a\cdot\boldsymbol{x}_b}{N}\right)\right]\right)^{-\frac{M}{2}}$$
$$\propto \int_{D_N^{(Q)}} \mathrm{d}Q(\det Q)^{\frac{N-n-1}{2}}\left(\det\left[\mathbb{I}_n + \beta\hat{f}(Q)\right]\right)^{-\frac{\alpha N}{2}}. \tag{66}$$

Here we omitted the exact proportionality constants (which are known but redundant for our goals), and we have replaced M with αN in view of a large N asymptotics.

The change of variables in the last line from the set of vectors $\boldsymbol{x}_a$ to the positive (semi)definite matrix Q of size $n\times n$ with entries defined as[10]

$$q_{ab} = \frac{1}{N}(\boldsymbol{x}_a\cdot\boldsymbol{x}_b) \tag{67}$$

follows the idea of the paper [29]. The details of this transformations, including evaluation of the involved Jacobian determinant factor appearing in the above are explained in detail in [30], Eq. (47). The hat over $\hat{f}(Q)$ serves as a reminder that this is an $n\times n$ matrix as well, with entries $f_{ab} := f(q_{ab})$. The domain of integration

[9] The formula reads $\int_{\mathbb{R}^n}\mathrm{d}\boldsymbol{u}\ \mathrm{e}^{-\frac{1}{2}\boldsymbol{u}^T M\boldsymbol{u}} = \sqrt{\frac{(2\pi)^n}{\det M}}$, for M a $n\times n$ symmetric and positive definite matrix.

[10] The values q_{ab} are known in the spin-glass literature as the *overlaps* between replicas, see [19] for details.

is then over such positive semi-definite matrices Q with the constraint $q_{aa} = 1$ on diagonal entries coming from (67) and the spherical constraint, that is given by

$$Q = \begin{pmatrix} 1 & & q_{ab} \\ & \ddots & \\ q_{ab} & & 1 \end{pmatrix}. \tag{68}$$

In short, the integration goes over the domain $D_N^{(Q)} = \{Q \geqslant 0, q_{aa} = 1\,\forall a\}$, and the integration measure is $\mathrm{d}Q \equiv \prod_{a<b} \mathrm{d}q_{ab}$.

Up to this step, no approximation has been used. Unfortunately, the matrix integral in Eq. (66) is not easy to compute in closed form for any N—result that would be needed to compute the various limits in the right hand side of Eq. (53), in the exact order in which they appear (first n, then β, and only at the end N). Therefore, we are forced to walk another non-rigorous path, by changing the order of limits, and performing the $N \to \infty$ limit *first*.

To this end, we rewrite the integral $\langle Z^n(\beta)\rangle_V$ in the form convenient for approximating it in the limit $N \gg 1$ by Laplace's method:

$$\langle Z^n(\beta)\rangle_V \propto \int_{D_N^{(Q)}} \mathrm{d}Q\,(\det Q)^{-\frac{n+1}{2}}\,\mathrm{e}^{-\frac{N}{2}\Phi_{n,\beta}(Q)}, \tag{69}$$

$$\Phi_{n,\beta}(Q) = \alpha \log\det[\mathbb{I}_n + \beta \hat{f}(Q)] - \log\det Q\,, \tag{70}$$

which we shall analyse for the particular choice $f(u) = \sigma^2 + u$, corresponding to our problem (see Eq. (56)). In our setting, the function $\Phi_{n,\beta}(Q)$ therefore specialises to

$$\Phi_{n,\beta}(Q) \equiv \alpha \log\det[\mathbb{I}_n + \beta(Q + \sigma^2 E_n)] - \log\det Q\,, \tag{71}$$

where we recall that $\alpha = M/N$ is the ratio between number of equations and number of unknowns of our system, σ^2 is the variance of the "noise" terms b_k in the linear system, E_n stands for the $n \times n$ matrix with all entries equal to unity, whereas $\mathbb{I}_n$ stands for the $n \times n$ identity matrix.

Due to the presence of large factor N in the exponent under thc integral over Q in (69), we can apply the Laplace's approximation, which then implies that for some extremising matrix argument Q_{extr}

$$\langle Z^n(\beta)\rangle_V \approx \mathrm{e}^{-\frac{N}{2}\Phi_{n,\beta}(Q_{extr})}\,, \tag{72}$$

where we keep only the leading exponential term, as this is the only one needed to compute $\langle \mathcal{E}_{min}\rangle_{A,\boldsymbol{b}}$ to the leading order (see Eq. (53) after exchanging the order of limits there)

$$\lim_{N\to\infty}\frac{\langle\mathcal{E}_{min}\rangle_{A,\boldsymbol{b}}}{N} = \lim_{N\to\infty}\frac{1}{N}\lim_{\beta\to\infty}\frac{-1}{\beta}\lim_{n\rightsquigarrow 0}\frac{1}{n}\log\langle Z^n(\beta)\rangle_{A,\boldsymbol{b}}$$

$$\simeq \lim_{\beta\to\infty}\frac{-1}{\beta}\lim_{n\rightsquigarrow 0}\frac{1}{n}\lim_{N\to\infty}\frac{1}{N}\log\langle Z^n(\beta)\rangle_{A,\boldsymbol{b}} \tag{73}$$

$$\simeq \lim_{\beta\to\infty}\frac{-1}{\beta}\lim_{n\rightsquigarrow 0}\frac{1}{n}\lim_{N\to\infty}\frac{1}{N}\log \mathrm{e}^{-\frac{N}{2}\Phi_{n,\beta}(Q_{extr})} \tag{74}$$

$$\simeq \lim_{\beta\to\infty}\frac{1}{2\beta}\lim_{n\rightsquigarrow 0}\frac{1}{n}\Phi_{n,\beta}(Q_{extr})\ , \tag{75}$$

where we first exchanged the order of integration (non-rigorous step), and then we inserted the asymptotic behaviour (72) coming from the Laplace approximation.

We now have to determine Q_{extr} using the stationarity conditions

$$\frac{\partial\Phi_{n,\beta}(Q)}{\partial q_{ab}} = 0, \qquad \forall a < b\ , \tag{76}$$

where we restrict to the upper triangle of the matrix Q since it is symmetric.

Looking at Eq. (71), we need to use the formula

$$\frac{\partial}{\partial M_{ab}}\log\det M = [M^{-1}]_{ba}\ , \tag{77}$$

valid for an invertible square matrix M. A proof is provided in Appendix 11.

Applying this identity to Eq. (71) and setting the result to zero, we obtain an equation for the ab entry of two inverse matrices, both related to Q_{extr}

$$\left(Q_{extr}^{-1}\right)_{ab} = \alpha\beta\left[\left(\mathbb{I}_n + \beta(Q_{extr} + \sigma^2 E_n)\right)^{-1}\right]_{ab}\ . \tag{78}$$

To solve this equation we use the so-called *Replica Symmetric* (RS) ansatz, which amounts to assuming that all diagonal entries of the matrix Q_{extr} are equal to one and all other entries are the same number q. At first sight, no other ansatz would seem to make any sense, as all replicas are "created equal"—therefore, two different entries $q_{ab} = \boldsymbol{x}_a \cdot \boldsymbol{x}_b/N$ and $q_{cd} = \boldsymbol{x}_c \cdot \boldsymbol{x}_d/N$ of the matrix Q_{extr} do not have any reason to behave differently from each other at the saddle point. Surprisingly, it was shown by Parisi [31] in the context of the solution of the Sherrington-Kirkpatrick model of spin glasses that the RS scheme may actually fail to produce physical results in certain cases, and should be replaced by an elaborate hierarchy of more sophisticated ansätze where the seemingly obvious symmetry between replicas is broken at various levels. Fortunately, the RS ansatz is sufficient for our problem here—why this is so, and what to do when it is not, goes beyond the scope of these lecture notes and will not be addressed here.

Therefore

$$Q_{extr} = \begin{pmatrix} 1 & & & & \\ & 1 & & q & \\ & & \ddots & & \\ & q & & 1 & \\ & & & & 1 \end{pmatrix} . \tag{79}$$

In order to ensure that the $n \times n$ matrix Q_{extr} be positive semi-definite, we apply an extended version of the Sylvester's criterion [32] that states that a symmetric matrix is positive-semidefinite if and only if all its *principal minors*[11] are nonnegative. This results in the conditions

$$1 - q \geqslant 0 \qquad \text{and} \qquad \frac{1}{2-n} \leqslant q \leqslant 1 . \tag{80}$$

Given that—at this stage—the size of the matrix is an *arbitrary* integer, it follows that we should restrict the value q of the off-diagonal entries to the range $0 \leqslant q \leqslant 1$.

We now make the following ansatz for its inverse

$$Q_{extr}^{-1} = \begin{pmatrix} \gamma & & & & \\ & \gamma & & \eta & \\ & & \ddots & & \\ & \eta & & \gamma & \\ & & & & \gamma \end{pmatrix} . \tag{81}$$

It follows that

$$[Q_{extr} Q_{extr}^{-1}]_{aa} = \gamma + (n-1)q\eta \tag{82}$$

$$[Q_{extr} Q_{extr}^{-1}]_{ab} = q\gamma + \eta + (n-2)q\eta , \tag{83}$$

which should be set to 1 and 0, respectively. Solving for γ and η, we find

$$\gamma = \frac{1 + q(n-2)}{(1-q)(1+q(n-1))} \tag{84}$$

$$\eta = \frac{-q}{(1-q)(1+q(n-1))} . \tag{85}$$

Similarly, it is easy to find the inverse of the matrix $R = \mathbb{I}_n + \beta(Q_{extr} + \sigma^2 E_n)$ appearing on the right hand side of Eq. (78) since this matrix has diagonal entries all equal to $R_{aa} = 1 + \beta(1+\sigma^2) =: r_d$, $\forall a = 1, \ldots, n$ and all off-diagonal entries

[11] A principal minor of a matrix is the determinant of the sub-matrix obtained by erasing corresponding sets of rows and columns (e.g. rows 1 and 6, and columns 1 and 6).

equal to $R_{ab} = \beta(q + \sigma^2) =: r,\ \forall a < b$. Using a similar matrix ansatz as in Eq. (81), we get

$$\left(R^{-1}\right)_{aa} = \frac{(r_d + r(n-2))}{(r_d - r)(r_d + r(n-1))}, \quad \left(R^{-1}\right)_{a<b} = -\frac{r}{(r_d - r)(r_d + r(n-1))}. \tag{86}$$

Hence the stationarity condition Eq. (78) for off-diagonal elements takes the form

$$\frac{q}{(1-q)(1+q(n-1))} = \alpha \frac{\beta^2(q+\sigma^2)}{(1+\beta-\beta q)(1+\beta-\beta q+n\beta(q+\sigma^2))}. \tag{87}$$

We also have to compute $\Phi_{n,\beta}(Q_{extr})$ from Eq. (71). To do so, we have to compute the determinant of matrices with the following structure

$$\det_{n\times n}\begin{pmatrix} \gamma & \eta & \cdots & \eta \\ \vdots & \vdots & \ddots & \vdots \\ \eta & \eta & \cdots & \gamma \end{pmatrix} = (\gamma - \eta)^{n-1}\left[\gamma + (n-1)\eta\right], \tag{88}$$

a result that can be easily proven by induction.

Applying this formula to the two terms in Eq. (71) after inserting the replica-symmetric ansatz Eq. (79), we have

$$\begin{aligned} \Phi_{n,\beta}(Q_{extr}) &= \alpha n \log\left(1 + \beta(1-q)\right) + \alpha \log\left(1 + \frac{\beta(q+\sigma^2)n}{1+\beta(1-q)}\right) \\ &- n\log(1-q) - \log\left(1 + q\frac{n}{1-q}\right), \end{aligned} \tag{89}$$

where q is solution of the saddle-point Eq. (87).

So far, none of the steps infringed on the integer nature of the replica index n. However, in Eq. (89), n now appears as a *parameter*, and no longer as the *integer* number of integrals (as in Eq. (60)) or the *integer* size of a matrix (as in Eq. (66))[12]. This is good news, as in the end we are seeking to perform the replica limit $n \rightsquigarrow 0$. Treating now n as a real parameter that can get arbitrarily close to zero, we get

$$\lim_{n\rightsquigarrow 0}\frac{1}{n}\Phi_{n,\beta}(Q_{extr}) = \alpha\log\left(1+\beta(1-q)\right) + \frac{\alpha\beta(q+\sigma^2)}{1+\beta(1-q)} - \log(1-q) - \frac{q}{1-q}, \tag{90}$$

where we used the standard asymptotics $\log(1+\epsilon) \sim \epsilon$ for small ϵ.

We can also seek for the solution of the saddle-point Eq. (87) directly in the replica limit setting $n = 0$ in the equation, giving:

[12] One of the most interesting aspects of replica calculations is indeed the fact that the replica parameter n changes nature and morphs into something else as the calculation proceeds.

$$\frac{q}{(1-q)^2} = \alpha \frac{\beta^2(q+\sigma^2)}{(1+\beta-\beta q)^2}, \tag{91}$$

to be solved for $q \in [0, 1)$, where q is the generic off-diagonal element of the extremising matrix Q_{extr}.

For finite β the equation is equivalent to a cubic one. To simplify our considerations further, we recall that to find the average minimum of the cost function for large N from (75) we only need to know q in the limit $\beta \to \infty$. Two different situations may happen in this limit

1. q tends in the limit $\beta \to \infty$ to a non-negative value smaller than unity (this range is dictated by positivity of the matrix Q).
2. Alternatively, in such a limit q tends to unity from below in such a way that $v = \lim_{\beta\to\infty} \beta(1-q)$ remains finite.

We can analyse the two situations separately.

1. Computing the limit $\beta \to \infty$ on the right hand side and simplifying the denominator $(1-q)^2$, we obtain the equation

$$q = \alpha(q+\sigma^2) \Rightarrow q = \frac{\alpha\sigma^2}{1-\alpha}. \tag{92}$$

Imposing the condition

$$0 \leqslant \frac{\alpha\sigma^2}{1-\alpha} < 1 \Rightarrow \boxed{0 < \alpha < \alpha_c := \frac{1}{1+\sigma^2} < 1}, \tag{93}$$

where the same critical threshold α_c was obtained via a completely different route earlier on (see Eq. (44)).
Invoking Eq. (75) in the form

$$\lim_{N\to\infty} \frac{\langle \mathcal{E}_{min} \rangle_{A,b}}{N} = \lim_{\beta\to\infty} \frac{1}{2\beta} \lim_{n\rightsquigarrow 0} \frac{1}{n} \Phi_{n,\beta}(Q_{extr})$$
$$= \lim_{\beta\to\infty} \frac{1}{2\beta} \left[\alpha \log(1+\beta(1-q)) + \frac{\alpha\beta(q+\sigma^2)}{1+\beta(1-q)} - \log(1-q) - \frac{q}{1-q} \right] = 0, \tag{94}$$

where we used the result in Eq. (90), shows that the system of equations with a quadratic constraint is generally compatible if the number of equations is smaller than the number of unknowns (which is expected, and no different from the case with no quadratic constraint), but only up to a critical ratio $\alpha_c < 1$. What happens beyond α_c is determined by the next case below.

2. We now seek for a solution of the saddle-point equation Eq. (91) for $\beta \to \infty$ and $q \to 1^-$, in such a way that the combination $\beta(1-q) = v > 0$ remains finite. Substituting $\beta = v/(1-q)$ in the right hand side of Eq. (91), and setting $q \to 1^-$ we get

$$1 = \alpha(1+\sigma^2)\frac{v^2}{(1+v)^2} \Rightarrow v = (\sqrt{\alpha(1+\sigma^2)} - 1)^{-1} , \tag{95}$$

where we have kept only the positive root of the equation, provided that $\alpha > \alpha_c = 1/(1+\sigma^2)$. Substituting now $\beta = v/(1-q)$ in the right hand side of Eq. (90), setting $q \to 1$, and inserting the result into Eq. (75), we obtain

$$\lim_{N\to\infty} \frac{\langle \mathcal{E}_{min}\rangle_{A,\boldsymbol{b}}}{N} = \frac{\alpha(1+\sigma^2)}{2(1+v)} - \frac{1}{2v}\Big|_{v=(\sqrt{\alpha(1+\sigma^2)}-1)^{-1}} = \frac{1}{2}\left[\sqrt{\alpha(1+\sigma^2)} - 1\right]^2 , \tag{96}$$

result to be compared with Eq. (43) obtained by a completely different method based on Lagrange multipliers and Random Matrix Theory, but only valid for $\alpha > 1$.

To summarise, the quenched replica calculation described in this section and valid for any value α of the ratio between the number M of equations and the number N of the unknowns of a large random linear system $A\boldsymbol{x} = \boldsymbol{b}$ supplemented by the spherical constraint yields the following scenarios depending on the variance σ^2 of the parameters b_k on the right hand side. For $\alpha < \alpha_c = 1/(1+\sigma^2) < 1$, i.e. for "strongly" under-complete systems, the average cost function is zero, signalling a system that is typically compatible. For $\alpha > 1$, both the Random Matrix and the replica calculations show that over-complete systems are typically incompatible (as the cost function is on average positive, see Eq. (96))—which is not surprising, as this would be the case even in absence of the spherical constraint. The most interesting situation happens for $\alpha_c < \alpha < 1$, for which the replica calculation shows that such "weakly" under-complete systems are also typically incompatible, a new phenomenon entirely due to the extra spherical constraint.

3.2 Full Distribution of $\mathcal{E}_{min}$

In fact, as was observed in [18], the replica trick sometimes can be used not only to calculate the mean value of the global minimum but also to characterise fluctuations around it for large $N \gg 1$, employing the Large Deviations approach (for a crash course on large deviations, see Appendix 12). Below we briefly give an account of that idea following the aforementioned paper.

One starts from assuming that the random variable $\mathcal{E}_{min}$ be characterised by a probability density function $\mathcal{P}_N(\mathcal{E}_{min})$, which has for large $N \gg 1$ a Large Deviations form

$$\mathcal{P}_N(\mathcal{E}_{min}) \approx R(x)\mathrm{e}^{-N\mathcal{L}(x)}, \quad x = \mathcal{E}_{min}/N . \tag{97}$$

Here, we included the leading rate function $\mathcal{L}(x)$ and also—for completeness—a sub-leading pre-factor $R(x)$ that is however not accessible with the method presented here. In (97), the symbol $\approx$ stands for the precise asymptotics $\lim_{N\to\infty} \frac{-1}{N} \log \mathcal{P}_N(Nx) = \mathcal{L}(x)$.

On the other hand, consider again the "replicated" disorder averaged partition function $\langle Z^n(\beta)\rangle_V$, but instead of taking the limits $n \to 0$ and $\beta \to \infty$ separately, perform the double scaling limit $n \to 0$, $\beta \to \infty$ with the product $n\beta =: s$ fixed. In this way, we may write

$$\lim_{n=s/\beta,\beta\to\infty} \langle Z^n(\beta)\rangle_V = \lim_{n=s/\beta,\beta\to\infty} \langle \mathrm{e}^{n\log Z(\beta)}\rangle_V = \lim_{\beta\to\infty} \langle \mathrm{e}^{\frac{s}{\beta}\log Z(\beta)}\rangle_V = \langle \mathrm{e}^{-Nsx}\rangle_V \;, \tag{98}$$

where in the last step we used the relation (46). Now we can use the large deviation form (97) and rewrite the above as:

$$\lim_{n=s/\beta,\beta\to\infty} \langle Z^n(\beta)\rangle_V = \langle \mathrm{e}^{-Nsx}\rangle_V = \int \mathrm{d}x\, \mathcal{P}_N(x)\mathrm{e}^{-Nsx} \approx \int \mathrm{d}x\, R(x)\mathrm{e}^{-N(sx+\mathcal{L}(x))} \;, \tag{99}$$

which obviously suggests evaluating the integral by the Laplace method for $N \gg 1$, giving

$$\lim_{n=s/\beta,\beta\to\infty} \langle Z^n(\beta)\rangle_V \approx g(s)\mathrm{e}^{N\phi(s)} \;, \tag{100}$$

where

$$\phi(s) = -\min_x (sx + \mathcal{L}(x)) \;. \tag{101}$$

We see that the large deviation rate $\mathcal{L}(x)$ is related by the so-called Legendre transform to the function $\phi(s)$ (see [33] for an excellent discussion of the Legendre transform in the context of large deviations). Hence, if one could—by independent means—find $\phi(s)$, one may recover the rate function $\mathcal{L}(x)$, governing the large deviation decay of the full probability density $\mathcal{P}_N(\mathcal{E}_{min})$ for large N by the inverse Legendre transform:

$$\boxed{\mathcal{L}(x) = -\left(x s_* + \phi(s_*)\right)} \;, \tag{102}$$

where $s_* := s_*(x)$ is determined by the following implicit equation

$$x = -\phi'(s_*) \;. \tag{103}$$

The ability to find the large deviation rate for the minimum of the cost/loss function explicitly hinges on being able to evaluate the above limit and to perform the Legendre inversion in closed form. Technically, this is a difficult task, which can be rarely performed successfully. It turns out that our problem is one of these rare and fortunate cases.

Recalling Eq. (72)

$$\langle Z^n(\beta)\rangle_V \propto \mathrm{e}^{-\frac{N}{2}\Phi_{n,\beta}(Q_{extr})} \;, \tag{104}$$

and comparing the right hand side with Eq. (100), it is clear that

$$\phi(s) = -\frac{1}{2} \lim_{\beta\to\infty} \Phi_{\frac{s}{\beta},\beta}(Q_{extr}) \,, \tag{105}$$

hence we need to evaluate the function $\Phi_{n,\beta}(Q_{extr})$ in the limit $n \to 0$ and $\beta \to \infty$ keeping $n\beta = s$ fixed.

The starting point of this procedure is the expression (89) for $\Phi_{n,\beta}(Q_{extr})$

$$\begin{aligned}\Phi_{n,\beta}(Q_{extr}) &= \alpha n \log\left(1+\beta(1-q)\right) + \alpha \log\left(1 + \frac{\beta(q+\sigma^2)n}{1+\beta(1-q)}\right) \\ &- n\log\left(1-q\right) - \log\left(1 + q\frac{n}{1-q}\right) \,, \end{aligned} \tag{106}$$

as well as the corresponding stationarity condition (87)

$$\frac{q}{(1-q)\,(1+q(n-1))} = \alpha \frac{\beta^2(q+\sigma^2)}{(1+\beta-\beta q)(1+\beta-\beta q+n\beta(q+\sigma^2))} \,. \tag{107}$$

From the previous analysis for $\alpha > \alpha_c$ we expect to have $q \to 1$ as $\beta \to \infty$ in such a way that $v = \beta(1-q)$ remains finite[13]. Correspondingly we substitute $n = s/\beta, q = 1 - v/\beta$ into (106) and set $\beta \to \infty$, resulting in the following expression for the functional appearing on the right hand side of (105)

$$\Phi(s,v) := \lim_{\beta\to\infty} \Phi_{\frac{s}{\beta},\beta}(Q_{extr}) = \alpha \log\left(1 + \frac{s(1+\sigma^2)}{1+v}\right) - \log\left(1+\frac{s}{v}\right) \,, \tag{108}$$

whereas the stationarity condition (107) takes the form

$$\frac{1}{v(v+s)} = \frac{\alpha(1+\sigma^2)}{(1+v)\left(1+v+s(1+\sigma^2)\right)} \,, \tag{109}$$

which is in fact equivalent to the condition $\frac{\partial}{\partial v}\Phi(s,v) = 0$. From this point onwards, one needs to find $v(s)$ solving (109), which can be equivalently written as

$$\alpha(1+\sigma^2)v(v+s) = (1+v)\left(1+v+s(1+\sigma^2)\right) \tag{110}$$

and in this way we first get the function $\phi(s) = -\frac{1}{2}\Phi(s, v(s))$ from Eqs. (105) and (108). After that, one performs the Legendre transform over the variable s to obtain the large deviation rate via Eq. (102)

---

[13] We do not consider the case $\alpha < \alpha_c$ here, because in that regime we have that $\langle \mathcal{E}_{min} \rangle_{A,\boldsymbol{b}} \to 0$ for large N, and $\mathcal{E}_{min} \geqslant 0$. It follows from Markov's inequality that $\lim_{N\to\infty} \text{Prob}[\mathcal{E}_{min} = 0] = 1$ for $\alpha < \alpha_c$.

$$\mathcal{L}(x) = -xs_* + \frac{1}{2}\Phi\left(s_*, v(s_*)\right) = \frac{1}{2}\left[\Phi\left(s_*, v(s_*)\right) - 2xs_*\right], \tag{111}$$

where s_* as a function of x is found by solving the equation

$$x = \frac{1}{2}\frac{\mathrm{d}\Phi}{\mathrm{d}s}\bigg|_{s_*} = \frac{1}{2}\frac{\partial\Phi(s,v)}{\partial s}\bigg|_{s_*}, \tag{112}$$

where the total derivative of Φ with respect to s coincides with the partial derivative because of the above-mentioned stationarity $\frac{\partial}{\partial v}\Phi(s,v) = 0$ with respect to any indirect dependence on s through $v(s)$.

Differentiating (108) we find, using (109), that

$$x = \frac{1}{2}\left(\frac{\alpha(1+\sigma^2)}{1+v+s_*(1+\sigma^2)} - \frac{1}{v+s_*}\right) = \frac{1}{2}\left(\frac{1+v}{v(v+s_*)} - \frac{1}{v+s_*}\right) = \frac{1}{2v(v+s_*)}, \tag{113}$$

or equivalently

$$v(v+s_*) = \frac{1}{2x}. \tag{114}$$

Using the above, one can rewrite (110) as

$$\alpha\left(1+\sigma^2\right)\frac{1}{2x} = (1+v)\left(1+v+s_*(1+\sigma^2)\right). \tag{115}$$

Further expressing s_* from (110) as

$$s_*(1+\sigma^2) = \frac{v^2\left(\alpha(1+\sigma^2)-1\right) - 2v - 1}{1 - v(\alpha-1)} \tag{116}$$

and substituting back into (115) we get the closed-form equation for v as a function of x

$$\begin{aligned}\alpha(1+\sigma^2)\frac{1}{2x} &= (1+v)\left\{1+v+\frac{v^2\left(\alpha(1+\sigma^2)-1\right)-2v-1}{1-v(\alpha-1)}\right\}\\ &= \frac{\alpha(1+v)v(v\sigma^2-1)}{1-v(\alpha-1)},\end{aligned} \tag{117}$$

which can be equivalently rewritten as a cubic equation

$$\boxed{\sigma^2 v^3 + v^2(\sigma^2-1) - v\left(1 - a(\alpha-1)\right) - a = 0\,, \quad a := \frac{1+\sigma^2}{2x}}. \tag{118}$$

Solving this equation, one gets the value of v corresponding to a given x. It turns out that the knowledge of $v(x)$ is enough to get the large deviation rate $\mathcal{L}(x)$, which can be expressed solely via such v. The simplest way to proceed is by multiplying Eq. (109) by $(1+v)^2$, taking logarithms of both sides, and further multiplying by α to get

$$\alpha \log\left(\frac{1+v+s(1+\sigma^2)}{1+v}\right) = \alpha \log\left(\frac{v(v+s)}{(1+v)^2}\alpha(1+\sigma^2)\right), \quad (119)$$

which allows to rewrite the expression (108) first as

$$\Phi(s,v) = \alpha \log\left(\frac{v(v+s)}{(1+v)^2}\alpha(1+\sigma^2)\right) - \log(v(v+s)) + 2\log v\,, \quad (120)$$

and then using (114) as

$$\Phi(s_*(x),v) = -(\alpha-1)\log(2x) + \alpha\log\left(\alpha(1+\sigma^2)\right) + 2\log v - 2\alpha\log(1+v)\,. \quad (121)$$

This should be combined with the relation

$$xs_* = \frac{1}{2v} - xv \quad (122)$$

following from (114). In this way, one completely solves the problem of explicitly inverting the Legendre transform and providing the required Large Deviation Rate function (111) for the minimal cost as a function of x.

To summarise the operational steps, one first fixes the requested variance of the noise terms σ^2 and the ratio $\alpha > \alpha_c = 1/(1+\sigma^2)$ between the number of equations and the number of unknowns. Then, one constructs a mesh $\mathcal{M} = \{x_1, \ldots, x_K\}$ of $x > 0$ values, which should include the special value

$$x_* := \frac{1}{2}\left[\sqrt{\alpha(1+\sigma^2)} - 1\right]^2 \quad (123)$$

for which we know that the corresponding value of v must be $v_* = 1/\sqrt{2x_*}$ (see Eqs. (95) and (96)). This piece of information is useful to select the corresponding root of the cubic equation in Eq. (118). Following by continuity the solution of the cubic equation along the x-mesh $\mathcal{M}$, one construct a set of corresponding v-values $\{v_1, \ldots, v_K\}$. Using Eq. (122), one can then construct the corresponding set of s_* values, $\{(s_*)_1, \ldots, (s_*)_K\}$. Finally, invoking Eqs. (111) and (121), we get the values of the rate function corresponding to the x-mesh points as

$$\mathcal{L}(x_j) = -x_j(s_*)_j + \frac{1}{2}\left[-(\alpha-1)\log(2x_j) + \alpha\log\left(\alpha(1+\sigma^2)\right) + 2\log v_j \right.$$
$$\left. -2\alpha\log(1+v_j)\right] \quad (124)$$

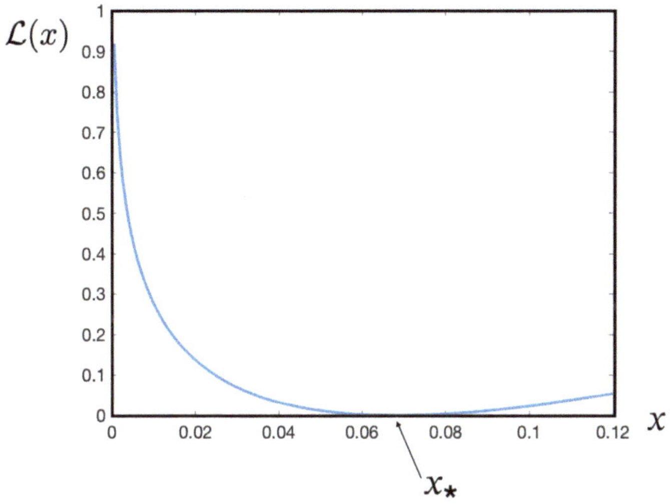

Fig. 4 Rate function $\mathcal{L}(x)$ for $\alpha = 1.5$ and $\sigma = 0.5$, obtained numerically starting from the cubic equation Eq. (118) and using the procedure detailed above

for $j = 1, \ldots, K$. The rate function $\mathcal{L}(x)$ is plotted in Fig. 4. However, determining the precise domain of validity of $\mathcal{L}(x)$ away from its minimum x_* is a difficult problem, which is still largely unsolved, and for which a mathematically rigorous treatment of this problem is very much called for. What can be computed from the expression for $\mathcal{L}(x)$, though, is the quadratic behaviour around its minimum, which provides the variance of $\mathcal{E}_{min}$ in the regime of typical fluctuations.

Combining Eqs. (121), (122) and (111), we can write

$$\mathcal{L}(x, v(x)) = \frac{1}{2}\left(\alpha \log\left(\alpha\left(\sigma^2 + 1\right)\right) - 2\alpha \log(v(x) + 1) + 2\log(v(x))\right.$$
$$\left. + (1 - \alpha)\log(2x)\right) + x v(x) - \frac{1}{2v(x)} . \tag{125}$$

Taylor-expanding the rate function up to second order around $x = x_*$ (see Eq. (123)), we get

$$\mathcal{L}(x, v(x)) = \mathcal{L}(x_*, v(x_*)) + (x - x_*) f_1(x_*) + (x - x_*)^2 f_2(x_*) + \mathcal{O}\left((x - x_*)^3\right) ,$$

where

$$f_1(x_*) = \left(\frac{1}{2}\left(-\frac{2\alpha v'(x_*)}{v(x_*) + 1} + \frac{2v'(x_*)}{v(x_*)} + \frac{1 - \alpha}{x_*}\right) + x_* v'(x_*) + \frac{v'(x_*)}{2v(x_*)^2} + v(x_*)\right) \tag{126}$$

$$f_2(x_*) = \left[\frac{1}{2}\left(\alpha\left(\frac{v'(x_*)^2}{(v(x_*)+1)^2} - \frac{v''(x_*)}{v(x_*)+1}\right) + \frac{v(x_*)v''(x_*) - v'(x_*)^2}{v(x_*)^2} + \frac{\alpha-1}{2{x_*}^2}\right)\right.$$
$$\left. + \frac{v(x_*)v''(x_*) - 2v'(x_*)^2}{4v(x_*)^3} + v'(x_*) + \frac{1}{2}x_* v''(x_*)\right] . \quad (127)$$

The first and second derivative of $v(x)$, evaluated at x_*, can be computed differentiating Eq. (118) once and twice with respect to x, respectively. After some tedious algebra, we find that $\mathcal{L}(x_*, v(x_*)) = f_1(x_*) = 0$ as expected (as the rate function should have a minimum (a zero) at the minimal expected value of $\mathcal{E}_{min}/N = x_*$), while the second-order term reads

$$f_2(x_*) = \frac{1}{\sigma^2\left(\sqrt{\alpha\left(1+\sigma^2\right)} - 1\right)^2} = \frac{1}{2x_*\sigma^2} . \quad (128)$$

Inserting this result into Eq. (97), we find that

$$\mathcal{P}_N(\mathcal{E}_{min}) \approx \exp\left[-N\frac{1}{2x_*\sigma^2}\left(\frac{\mathcal{E}_{min}}{N} - x_*\right)^2\right]$$
$$= \exp\left[-\frac{1}{2(x_*\sigma^2 N)}\left(\mathcal{E}_{min} - Nx_*\right)^2\right] , \quad (129)$$

corresponding to Gaussian typical fluctuations of $\mathcal{E}_{min}$ for large N and $\alpha > \alpha_c$ around its average value Nx_*, and with variance

$$\mathrm{Var}(\mathcal{E}_{min}) = x_*\sigma^2 N . \quad (130)$$

4 Conclusions and Perspectives

These lecture notes constitute an attempt to illustrate and advertise a series of works by Yan V. Fyodorov and Rashel Tublin on a rich and intriguing problem, namely the statistical study of solutions of linear systems in presence of an additional quadratic constraint [1–3]. This problem is interesting especially for its pedagogical and instructive nature: it can be tackled with two largely complementary methods, and lends itself to a number of interesting generalisations and follow-ups, most of which have not been tackled yet.

The first method we discussed in these notes starts from the definition of the cost function $H(\boldsymbol{x})$ in Eq. (7), which is zero when the linear system is compatible, and larger than zero when it is incompatible. Using one Lagrange multiplier to enforce the spherical constraint on the solution vector, $||\boldsymbol{x}||^2 = N$, one may determine the location of the critical points of $H(\boldsymbol{x})$ (see Eq. (17)), as well as the equation for the

Lagrange multiplier (see Eqs. (18) and (25)). In a random/disordered setting—where the matrix A and the vector of coefficients $\boldsymbol{b}$ are Gaussian variables—the average of the minimal Lagrange multiplier can be computed exactly for large N and M, with $M/N = \alpha > 1$ fixed (see Eq. (27)). In turn, this result can be used—under a self-averaging assumption—to compute the average $\langle \mathcal{E}_{min} \rangle_{A,\boldsymbol{b}}$ of the minimal cost, which signals whether the corresponding linear system is expected to be typically compatible or incompatible. While in the absence of the spherical constraint, over-complete systems ($\alpha > 1$) are typically incompatible, and under-complete systems ($\alpha < 1$) are typically compatible (see Appendix 5 for a detailed calculation), the spherical constraint induces a further intermediate regime for $0 < \alpha_c < \alpha < 1$ where "weakly" under-complete systems may still be typically incompatible, and only for "strongly" under-complete systems the presence of the spherical constraint does not typically hinder the possibility to find a solution.

These findings are independently confirmed using an entirely different line of reasoning, inspired by an intriguing connection between the search for the average of the minimum of a random variable, and the free energy of an associated thermodynamical system at zero temperature. Computing the quenched average of the free energy using the celebrated *replica trick* from the physics of disordered systems yields an independent confirmation of a transition between compatible and incompatible linear systems, taking place at a value $\alpha_c < 1$, which can be precisely characterised (see Eq. (93)). As a further (and rare) luxury afforded to us by this problem, the replica calculation in this case allows us access not only to the average of the minimal cost $\langle \mathcal{E}_{min} \rangle_{A,\boldsymbol{b}}$, but also to its full distribution $\mathcal{P}_N(\mathcal{E}_{min})$ (in a large deviation sense)—see discussion in Sect. 3.2.

Given the heuristic nature of the replica calculation, a mathematically rigorous justification of the steps contained in Sect. 3 is still lacking. It is worth mentioning, though, that many questions related to a simpler optimisation problem on the sphere studied originally in [18] by a combination of rigorous Kac-Rice and heuristic replica approaches were subsequently successfully put on firm mathematical grounds in the series of papers [34–38], therefore it is likely that some of those techniques can be also useful in the context of present model as well.

In terms of future perspectives, some interesting generalisations and follow-ups come to mind:

1. The function $f(u) = \sigma^2 + u$ in Eq. (56) is specific to the linear nature of the equations of the system. Other choices are possible, for instance a quadratic $f(u)$ would correspond to a system of quadratic equations with a spherical constraint. This extended class of problems has been addressed only very recently [26–28], with a number of exciting results and conjectures, and intriguing applications for example to the so-called "phase retrieval" problem, see [39, 40].
2. Lifting the restriction that both A and $\boldsymbol{b}$ be Gaussian distributed would be very welcome. For instance, what is the effect on the compatibility threshold of having "fat-tailed" coefficients of the linear system? Relinquishing Gaussianity is

always tricky, though, especially in the context of replica calculations, where the evaluation of the average over the disorder (see (60) and (63)) may then lead to a multivariate integral much less amenable to a Laplace (saddle-point) evaluation. Overcoming this technical hurdle is likely to open up an interesting new field of investigations.

3. The quest for a rigorous—or at least, less non-rigorous—way to perform the replica calculation in Sect. 3 may actually require letting Random Matrix Theory back in the game. The idea would be to attempt an exact evaluation of the integral (66) for any N and n, without any approximation. The key idea—pioneered by Kanzieper [41] and later implemented in the context of spherical spin glasses by Akhanjee and Rudnick [42]—would be to diagonalise the symmetric matrix Q as $Q = OMO^{-1}$, with M the diagonal matrix containing the n eigenvalues $\{\mu_1, \ldots, \mu_n\}$ and O the orthogonal matrix of eigenvectors, and use the fact that the integrand of (66) is "almost" rotationally invariant, as it depends only weakly on the eigenvectors of Q. Reducing the Q integral to a n-fold integral over the eigenvalues of Q could prelude an exact evaluation of the replicated partition function for any N and n in terms of Painlevé transcendents. This research programme could shed light on the interplay between replica symmetry (or breaking thereof) and spectral properties of the replicated partition function without any large N approximation.

5 Appendix: Minimal Cost Function Without Spherical Constraint

In this Appendix, we wish to show how the spherical constraint is indeed the sole responsible for the new phenomenon of "weakly under-complete" systems, by computing the average minimal loss function in the absence of such constraint. The calculation can be performed exactly without replicas.

The partition function Eq. (45) for a single instance of the "disorder" (randomness) encoded in A and $\boldsymbol{b}$ reads in this case

$$Z(\beta) = \int_{\mathbb{R}^N} \mathrm{d}\boldsymbol{x}\ \mathrm{e}^{-\frac{1}{2}\beta \sum_{k=1}^{M}[\sum_{j=1}^{N} A_{kj}x_j - b_k]^2}\ , \tag{131}$$

where we defer considerations about the convergence of this integral to a later stage, and we did not include any spherical constraint here.

Expanding the exponent, we have

$$-\frac{1}{2}\beta\sum_{k=1}^{M}\left[\sum_{j=1}^{N}A_{kj}x_j-b_k\right]^2=-\frac{1}{2}\beta\sum_{k=1}^{M}\left[\sum_{j=1}^{N}A_{kj}x_j-b_k\right]\left[\sum_{\ell=1}^{N}A_{k\ell}x_\ell-b_k\right]$$

$$=-\frac{1}{2}\beta\left[\sum_{j,\ell}x_i\left(\sum_k A_{kj}A_{k\ell}\right)x_\ell-\sum_j\left(2\sum_k A_{kj}b_k\right)x_j+\sum_k b_k^2\right]$$

$$=-\frac{1}{2}\boldsymbol{x}^T W_\beta \boldsymbol{x}+\boldsymbol{h}_\beta^T\boldsymbol{x}-\frac{1}{2}\beta\sum_k b_k^2\,, \tag{132}$$

where

$$W_\beta=\beta A^T A \tag{133}$$

$$\boldsymbol{h}_\beta=\beta A^T\boldsymbol{b}\,. \tag{134}$$

Hence, for the integral (141) to converge, we need the matrix W_β to be positive definite, which requires the $N\times N$ matrix A^TA to be of Wishart (and not of anti-Wishart) type. This happens if $M>N$ ($\alpha>1$). The case $M<N$—for which the matrix A^TA has zero eigenvalues—must be treated separately (see below).

Therefore, for $\alpha>1$ we have

$$Z(\beta)=\exp\left[-\frac{1}{2}\beta\sum_k b_k^2\right]\int_{\mathbb{R}^N}\mathrm{d}\boldsymbol{x}\;\mathrm{e}^{-\frac{1}{2}\boldsymbol{x}^T W_\beta\boldsymbol{x}+\boldsymbol{h}_\beta^T\boldsymbol{x}}$$

$$=\sqrt{\frac{(2\pi)^N}{\det W_\beta}}\exp\left[-\frac{1}{2}\beta\sum_k b_k^2\right]\exp\left[\frac{1}{2}\boldsymbol{h}_\beta^T W_\beta^{-1}\boldsymbol{h}_\beta\right]\,. \tag{135}$$

Recalling formula (46)

$$\mathcal{E}_{min}=-\lim_{\beta\to\infty}\frac{1}{\beta}\log Z(\beta)\,, \tag{136}$$

we have explicitly

$$\mathcal{E}_{min}=\frac{1}{2}\sum_{k=1}^{M}b_k^2-\frac{1}{2}\boldsymbol{b}^T A(A^TA)^{-1}A^T\boldsymbol{b}\,. \tag{137}$$

We can now compute the average of $\mathcal{E}_{min}$ over $\boldsymbol{b}$ first, using the identities (36) and (37)

$$\langle \mathcal{E}_{min} \rangle_{\boldsymbol{b}} = \frac{1}{2} M\sigma^2 - \frac{1}{2}\sigma^2 \mathrm{Tr}\left(A(A^T A)^{-1} A^T\right) \tag{138}$$

$$= \frac{1}{2} M\sigma^2 - \frac{1}{2}\sigma^2 \mathrm{Tr}\mathbb{I}_N = \frac{1}{2}\sigma^2 (M - N)\ , \tag{139}$$

where we used the cyclic property of the trace. Interestingly, the result holds irrespective of what the A matrix is—a clear indication that the nature of the coefficient matrix is irrelevant in determining whether a linear system with more equations than unknowns is compatible (it typically isn't). Therefore, in the large N limit, we can write

$$\lim_{N\to\infty} \frac{\langle \mathcal{E}_{min} \rangle_{A,\boldsymbol{b}}}{N} = \frac{1}{2}\sigma^2(\alpha - 1)\ , \tag{140}$$

which is indeed positive for $\alpha > 1$, signalling that an over-complete system of equations (even in the absence of the spherical constraint) is typically incompatible, as it should.

For $\alpha < 1$ (more unknowns than equations), the integral defining the partition function (141) does not converge, but it can be regularised as follows

$$Z_\lambda(\beta) = \int_{\mathbb{R}^N} \mathrm{d}\boldsymbol{x}\ \mathrm{e}^{-\frac{1}{2}\beta \sum_{k=1}^{M} [\sum_{j=1}^{N} A_{kj} x_j - b_k]^2 - \lambda \sum_{j=1}^{N} x_j^2}\ , \tag{141}$$

which has the effect of penalising solutions $\boldsymbol{x}$ with large norms for $\lambda > 0$—it is essentially a "soft" version of the "hard" spherical constraint used in the main text. This method is called *L_2-regularisation* [43].

Therefore, for $\alpha < 1$ we can write

$$Z_\lambda(\beta) = \exp\left[-\frac{1}{2}\beta \sum_k b_k^2\right] \int_{\mathbb{R}^N} \mathrm{d}\boldsymbol{x}\ \mathrm{e}^{-\frac{1}{2}\boldsymbol{x}^T W_{\beta,\lambda}\boldsymbol{x} + \boldsymbol{h}_\beta^T \boldsymbol{x}}$$

$$= \sqrt{\frac{(2\pi)^N}{\det W_{\beta,\lambda}}} \exp\left[-\frac{1}{2}\beta \sum_k b_k^2\right] \exp\left[\frac{1}{2}\boldsymbol{h}_\beta^T W_{\beta,\lambda}^{-1} \boldsymbol{h}_\beta\right]\ , \tag{142}$$

where

$$W_{\beta,\lambda} = \beta A^T A + 2\lambda \mathbb{I}_N\ , \tag{143}$$

which is positive definite, being the sum of a positive semi-definite matrix (the 'anti-Wishart' matrix $\beta A^T A$ [15, 16]) and a positive definite matrix (a positive multiple of the identity).

Recalling formula (46) again—suitably deformed to remove the regulariser

$$\mathcal{E}_{min} = -\lim_{\lambda\to 0} \lim_{\beta\to\infty} \frac{1}{\beta} \log Z_\lambda(\beta)\ , \tag{144}$$

we have explicitly

$$\mathcal{E}_{min} = \frac{1}{2}\sum_{k=1}^{M} b_k^2 - \frac{1}{2}\lim_{\lambda\to 0}\lim_{\beta\to\infty} \beta \boldsymbol{b}^T A(\beta A^T A + 2\lambda \mathbb{I}_N)^{-1} A^T \boldsymbol{b} \,. \tag{145}$$

We can now compute the average of $\mathcal{E}_{min}$ over $\boldsymbol{b}$ first, using the identities (36) and (37)

$$\langle \mathcal{E}_{min} \rangle_{\boldsymbol{b}} = \frac{1}{2} M\sigma^2 - \frac{1}{2}\sigma^2 \lim_{\lambda\to 0}\lim_{\beta\to\infty} \beta \mathrm{Tr}\left((\beta A^T A + 2\lambda \mathbb{I}_N)^{-1} A^T A\right) \,, \tag{146}$$

where we used the cyclic property of the trace again. The average over the matrix A can be performed recalling that the $N \times N$ anti-Wishart matrix $W^{(a)} = A^T A$ for $M < N$ has the following density of eigenvalues for large N, M with fixed ratio $\alpha = M/N < 1$

$$\rho(s) = (1-\alpha)\delta(s) + \alpha \rho_{MP}(s) \,, \tag{147}$$

which signals the fact that $W^{(a)} = A^T A$ has exactly $N - M$ eigenvalues equal to zero (see Eq. (155)). Therefore

$$\left\langle \frac{1}{N} \mathrm{Tr}\left((\beta W^{(a)} + 2\lambda \mathbb{I}_N)^{-1} W^{(a)}\right) \right\rangle_{W^{(a)}} \to \alpha \int_{s_-}^{s_+} \mathrm{d}s \frac{s}{\beta s + 2\lambda} \rho_{MP}(s) \,, \tag{148}$$

where we have used the property (30) to compute the average trace of a matrix as an integral over its eigenvalue density, and observed that the Dirac delta term in Eq. (147) does not contribute to the integral.

Multiplying the right hand side up and down by β, and adding and subtracting 2λ in the integrand, we get

$$\begin{aligned}\left\langle \frac{1}{N} \mathrm{Tr}\left((\beta W^{(a)} + 2\lambda \mathbb{I}_N)^{-1} W^{(a)}\right) \right\rangle_{W^{(a)}} \to \frac{\alpha}{\beta} \int_{s_-}^{s_+} \rho_{MP}(s) \\ - \frac{2\lambda\alpha}{\beta} \int_{s_-}^{s_+} \mathrm{d}s \frac{1}{\beta s + 2\lambda} \rho_{MP}(s) \,. \end{aligned} \tag{149}$$

Clearly, the first term on the right hand side is equal to α/β since the Marčenko-Pastur density is normalised to one, while the second term is of order $1/\beta^2$ for large β at fixed λ, and therefore will not contribute once multiplied by an extra β in Eq. (146).

Therefore, in the large N limit, we can write from Eq. (146)

$$\lim_{N\to\infty} \frac{\langle \mathcal{E}_{min} \rangle_{A,\boldsymbol{b}}}{N} = \frac{1}{2}\sigma^2\alpha - \frac{1}{2}\sigma^2\alpha = 0 \,, \tag{150}$$

signalling that an under-complete system of equations is typically compatible, as it should.

6 Appendix: Generalities on Wishart Ensemble

In this Appendix, we collect some salient facts about the Wishart ensemble of random matrices.

1. If A_{ij} are i.i.d. real Gaussian variables with mean zero and variance $1/N$, $P_A(A) = \prod_{i=1}^{M}\prod_{j=1}^{N} p(A_{ij}=x)$, where $p(A_{ij}=x) = \exp(-Nx^2/2)/\sqrt{2\pi/N}$, with $M > N$, then the $N \times N$ symmetric matrix $W = A^T A$ belongs to the Wishart ensemble.
2. The entries of W are no longer independent, and their joint probability density (jpd) $P_W(W) := P(W_{11}, \dots, W_{NN})$ is

$$P_W(W) = C_{N,M}(\det W)^{\frac{M-N-1}{2}} \mathrm{e}^{-\frac{N}{2}\mathrm{Tr}W} , \tag{151}$$

   where the normalisation factor $C_{N,M}$ is known in closed form.
3. The N eigenvalues $\{s_1, \dots, s_N\}$ of W are real and non-negative. Their jpd $P(\boldsymbol{s})$ follows from (151) as

$$P(\boldsymbol{s}) = \hat{C}_{N,M}\mathrm{e}^{-\frac{N}{2}\sum_{k=1}^{N} s_k} \prod_{k=1}^{N} s_k^{\frac{M-N-1}{2}} |\Delta(\boldsymbol{s})| , \tag{152}$$

   where $\Delta(\boldsymbol{s}) = \prod_{j<k}(s_j - s_k)$—the Vandermonde determinant—is the Jacobian of the change of variables $W \to OSO^{-1}$, where O is the orthogonal matrix formed by the eigenvectors of W, and S is the diagonal matrix of eigenvalues.
4. The density of eigenvalues $\rho_N(s) = \left\langle \frac{1}{N}\sum_{i=1}^{N} \delta(s - s_i)\right\rangle$ (normalised to 1) converges for large N and M, with $M/N = \alpha > 1$, to the Marčenko-Pastur law

$$\lim_{N\to\infty} \rho_N(s) = \rho_{MP}(s) , \tag{153}$$

   given by Eq. (31)

$$\rho_{MP}(s) = \frac{2}{\pi s}\frac{1}{(\sqrt{s_+} - \sqrt{s_-})^2}\sqrt{(s_+ - s)(s - s_-)} \tag{154}$$

   having compact support between the edge points $s_\pm = (\sqrt{\alpha} \pm 1)^2$ on the positive real axis (see Fig. 3). The density of eigenvalues $\rho_N(s)$ for finite N is known in closed form [44].
5. The *anti-Wishart* $M \times M$ matrix $W^{(a)} = AA^T$ [15, 16] shares the same set of non-negative eigenvalues $\{s_i\}_{i=1,\dots,N}$ with A, plus $M - N$ exactly zero eigenvalues. *Proof:* if $A^T A\boldsymbol{x}_i = s_i\boldsymbol{x}_i$, multiply to the left by A. The spectral density of the

anti-Wishart ensemble is therefore given by the Marčenko-Pastur law, supplemented by a delta function at zero eigenvalue, i.e.

$$\rho(s) = (1-\alpha)\delta(s) + \alpha\rho_{MP}(s)\ . \tag{155}$$

7 Appendix: Sketch of Browne's Theorem

We sketch here the main gist of the proof of Browne's theorem [12], essentially identical to Gander's [13].

Consider two distinct solutions $\boldsymbol{x}_1$ and $\boldsymbol{x}_2$ of the stationarity condition (16)

$$(A^T A - \lambda_1 \mathbb{I}_N)\boldsymbol{x}_1 = A^T \boldsymbol{b}\ , \tag{156}$$

$$(A^T A - \lambda_2 \mathbb{I}_N)\boldsymbol{x}_2 = A^T \boldsymbol{b}\ , \tag{157}$$

corresponding to two distinct Lagrange multipliers λ_1 and λ_2. We aim to compute the following difference

$$\Delta(\boldsymbol{x}_1, \boldsymbol{x}_2; \lambda_1, \lambda_2) = H(\boldsymbol{x}_2) - H(\boldsymbol{x}_1) = ||A\boldsymbol{x}_2 - \boldsymbol{b}||^2 - ||A\boldsymbol{x}_1 - \boldsymbol{b}||^2\ . \tag{158}$$

First, multiply Eq. (156) by $\boldsymbol{x}_1^T$ to the left, and Eq. (157) by $\boldsymbol{x}_2^T$ to get

$$||A\boldsymbol{x}_1||^2 - \boldsymbol{x}_1^T A^T \boldsymbol{b} = \lambda_1 ||\boldsymbol{x}_1||^2 \tag{159}$$

$$||A\boldsymbol{x}_2||^2 - \boldsymbol{x}_2^T A^T \boldsymbol{b} = \lambda_2 ||\boldsymbol{x}_2||^2\ . \tag{160}$$

Transposing the above equations

$$||A\boldsymbol{x}_1||^2 - \boldsymbol{b}^T A\boldsymbol{x}_1 = \lambda_1 ||\boldsymbol{x}_1||^2 \tag{161}$$

$$||A\boldsymbol{x}_2||^2 - \boldsymbol{b}^T A\boldsymbol{x}_2 = \lambda_2 ||\boldsymbol{x}_2||^2\ , \tag{162}$$

and subtracting the first from the second, we get

$$||A\boldsymbol{x}_2||^2 - ||A\boldsymbol{x}_1||^2 - \boldsymbol{b}^T A(\boldsymbol{x}_2 - \boldsymbol{x}_1) = \lambda_2 ||\boldsymbol{x}_2||^2 - \lambda_1 ||\boldsymbol{x}_1||^2\ . \tag{163}$$

Now, multiply Eq. (156) by $\boldsymbol{x}_2^T$ to the left, and Eq. (157) by $\boldsymbol{x}_1^T$ to get

$$\boldsymbol{x}_2^T A^T A\boldsymbol{x}_1 - \boldsymbol{x}_2^T A^T \boldsymbol{b} = \lambda_1 \boldsymbol{x}_2^T \boldsymbol{x}_1 \tag{164}$$

$$\boldsymbol{x}_1^T A^T A\boldsymbol{x}_2 - \boldsymbol{x}_1^T A^T \boldsymbol{b} = \lambda_2 \boldsymbol{x}_1^T \boldsymbol{x}_2\ . \tag{165}$$

Transposing the above equations, noting that $(A\boldsymbol{x}_2)^T(A\boldsymbol{x}_1) = (A\boldsymbol{x}_1)^T(A\boldsymbol{x}_2)$ and $\boldsymbol{x}_2^T\boldsymbol{x}_1 = \boldsymbol{x}_1^T\boldsymbol{x}_2$ since the dot product is symmetric, and subtracting the second equation from the first, we get

$$-\boldsymbol{b}^T A(\boldsymbol{x}_2 - \boldsymbol{x}_1) = (\lambda_1 - \lambda_2)\boldsymbol{x}_1^T\boldsymbol{x}_2 \,. \tag{166}$$

Summing up Eqs. (163) and (166)

$$||A\boldsymbol{x}_2||^2 - ||A\boldsymbol{x}_1||^2 - 2\boldsymbol{b}^T A(\boldsymbol{x}_2 - \boldsymbol{x}_1) = \lambda_2||\boldsymbol{x}_2||^2 - \lambda_1||\boldsymbol{x}_1||^2 + (\lambda_1 - \lambda_2)\boldsymbol{x}_1^T\boldsymbol{x}_2 \,. \tag{167}$$

Now, expanding Eq. (158), we have

$$\Delta(\boldsymbol{x}_1, \boldsymbol{x}_2; \lambda_1, \lambda_2) = ||A\boldsymbol{x}_2||^2 - ||A\boldsymbol{x}_1||^2 - 2\boldsymbol{b}^T A(\boldsymbol{x}_2 - \boldsymbol{x}_1) \,, \tag{168}$$

which is identical to the left hand side of Eq. (167). It follows that

$$\Delta(\boldsymbol{x}_1, \boldsymbol{x}_2; \lambda_1, \lambda_2) = \lambda_2||\boldsymbol{x}_2||^2 - \lambda_1||\boldsymbol{x}_1||^2 + (\lambda_1 - \lambda_2)\boldsymbol{x}_1^T\boldsymbol{x}_2 = \gamma(\lambda_2 - \lambda_1) \,, \tag{169}$$

where

$$\gamma = N - \boldsymbol{x}_1^T\boldsymbol{x}_2 > 0 \,. \tag{170}$$

Here, we have used that both vectors $\boldsymbol{x}_1$ and $\boldsymbol{x}_2$ live on the sphere of squared radius N, and—if they are distinct—their dot product must be smaller than N due to Cauchy-Schwarz inequality. It follows that the cost function $H(\boldsymbol{x}_2)$ is larger than the cost function $H(\boldsymbol{x}_1)$ if the corresponding Lagrange multipliers have the same relation ($\lambda_2 > \lambda_1$), which is in essence the statement of Browne's theorem.

8 Appendix: Calculation of the Integral (32)

We wish to show that

$$I(s_-, s_+, \lambda) = \int_{s_-}^{s_+} \mathrm{d}s\, \rho_{MP}(s)\frac{s}{(\lambda - s)^2} = \frac{1}{(\sqrt{s_+} - \sqrt{s_-})^2}\frac{(\sqrt{s_+ - \lambda} - \sqrt{s_- - \lambda})^2}{\sqrt{(s_+ - \lambda)(s_- - \lambda)}} \tag{171}$$

for $\lambda < s_-$, where

$$\rho_{MP}(s) = \frac{2}{\pi s}\frac{1}{(\sqrt{s_+} - \sqrt{s_-})^2}\sqrt{(s_+ - s)(s - s_-)} \,. \tag{172}$$

Making the change of variables $t = (s - s_-)/(s_+ - s_-)$, we get

$$I(s_-, s_+, \lambda) = \frac{2}{\pi}\frac{(s_+ - s_-)^2}{(\sqrt{s_+} - \sqrt{s_-})^2(\lambda - s_-)^2}\int_0^1 \mathrm{d}t\, \frac{\sqrt{t(1 - t)}}{(1 + \gamma t)^2} \,, \tag{173}$$

where

$$\gamma = \frac{s_+ - s_-}{s_- - \lambda} > 0\,. \tag{174}$$

The integral in (173) admits a long but elementary antiderivative, which leads to

$$\int_0^1 \mathrm{d}t\,\frac{\sqrt{t(1-t)}}{(1+\gamma t)^2} = \frac{\pi\left(\gamma - 2\sqrt{\gamma+1}+2\right)}{2\gamma^2\sqrt{\gamma+1}}\,. \tag{175}$$

Inserting (175) into (173) and after simplifications, in order to establish (171) we need to prove that

$$\frac{(\sqrt{s_+ - \lambda} - \sqrt{s_- - \lambda})^2}{\sqrt{(s_+ - \lambda)(s_- - \lambda)}} = \frac{\left(\gamma - 2\sqrt{\gamma+1}+2\right)}{\sqrt{\gamma+1}}\,. \tag{176}$$

Calling $\sqrt{s_+ - \lambda} = x$ and $\sqrt{s_- - \lambda} = y$, we have $\gamma = x^2/y^2 - 1$. Therefore, the relation (176) is equivalent to

$$\frac{(x-y)^2}{xy} = \frac{x}{y} + \frac{y}{x} - 2\,, \tag{177}$$

which is very easily verified.

9 Appendix: Laplace Method for the Asymptotic Evaluation of Integrals

The Laplace method is a powerful technique used in the field of asymptotic analysis for approximating integrals. This method is particularly useful when dealing with integrals that are difficult to evaluate using standard techniques. The Laplace method is based on the principle of approximating the integral of an exponentially decaying function by the function's value at its extremal point(s).

Consider an integral of the form

$$I(T) = \int_a^b \mathrm{e}^{Tf(x)} g(x)\,\mathrm{d}x, \tag{178}$$

where T is a large parameter, and $f(x)$ and $g(x)$ are smooth functions. The objective is to find an asymptotic approximation of $I(T)$ as $T \to \infty$.

The Laplace method is based on the observation that, for large T, the main contribution to the integral comes from the neighbourhood of the point where $f(x)$ attains its maximum value inside the interval (a, b). Assume this maximum occurs at a point $x_0 \in (a, b)$ such that $f''(x_0) < 0$.

Expanding $f(x)$ around x_0 using Taylor's theorem, we get

$$f(x) \approx f(x_0) + \frac{1}{2} f''(x_0)(x - x_0)^2 + \dots , \tag{179}$$

where we used the fact that $f'(x_0) = 0$ being a maximum. We can also similarly expand the function $g(x)$ to get

$$g(x) \approx g(x_0) + g'(x_0)(x - x_0) + \dots . \tag{180}$$

Substituting these expansions into the integral, and keeping only the first terms, we obtain

$$I(T) \approx \mathrm{e}^{Tf(x_0)} g(x_0) \int_a^b \mathrm{e}^{-\frac{1}{2}T|f''(x_0)|(x-x_0)^2} \left[1 + \frac{g'(x_0)}{g(x_0)}(x - x_0) + \dots\right] \mathrm{d}x \,. \tag{181}$$

Making a change of variables $\sqrt{T}(x - x_0) = y$, we get

$$I(T) \approx \mathrm{e}^{Tf(x_0)} g(x_0) \int_{-\sqrt{T}(x_0-a)}^{\sqrt{T}(b-x_0)} \mathrm{e}^{-\frac{1}{2}|f''(x_0)|y^2} \left[1 + \frac{g'(x_0)}{g(x_0)} \frac{y}{\sqrt{T}} + \dots\right] \mathrm{d}y \,. \tag{182}$$

For large T, the exponential function rapidly decays away from x_0, allowing us to extend the limits of integration to infinity. Ignoring also the sub-leading terms in square brackets, we get to leading order for large T

$$I(T) \approx \mathrm{e}^{Tf(x_0)} g(x_0) \int_{-\infty}^{\infty} \mathrm{e}^{-\frac{1}{2}|f''(x_0)|y^2} \, \mathrm{d}y = \mathrm{e}^{Tf(x_0)} g(x_0) \sqrt{\frac{2\pi}{T|f''(x_0)|}} \,, \tag{183}$$

where we have evaluated the Gaussian integral exactly. In Sect. 3.1 we basically used a multivariate version of the Laplace approximation and we retained only the leading exponential behaviour.

10 Appendix: Covariance Calculation Leading to Eq. (56)

In this Appendix, we evaluate explicitly the covariance $\langle V_k(\boldsymbol{x}^{(a)}) V_\ell(\boldsymbol{x}^{(b)}) \rangle_{A,\boldsymbol{b}}$ between the potential $V(\boldsymbol{x})$ evaluated at two different points.

If $V_k(\boldsymbol{x}) = \sum_j A_{kj}x_j - b_k$ for $k = 1, \ldots, M$ then

$$\begin{aligned}
&\left\langle \left(\sum_j A_{kj}x_j^{(a)} - b_k\right)\left(\sum_m A_{\ell m}x_m^{(b)} - b_\ell\right)\right\rangle_{A,\boldsymbol{b}} \\
&= \sum_{j,m} x_j^{(a)}x_m^{(b)}\langle A_{kj}A_{\ell m}\rangle_A - \sum_j x_j^{(a)}\cancel{\langle A_{kj}\rangle_A\langle b_\ell\rangle_{\boldsymbol{b}}} \\
&- \sum_m x_m^{(b)}\cancel{\langle A_{\ell m}\rangle_A\langle b_k\rangle_{\boldsymbol{b}}} + \langle b_k b_\ell\rangle_{\boldsymbol{b}} = \delta_{k\ell} f\left(\frac{\sum_m x_m^{(a)}x_m^{(b)}}{N}\right),
\end{aligned} \tag{184}$$

where $f(u) = \sigma^2 + u$ and we used $\langle A_{ij}\rangle_A = \langle b_k\rangle_{\boldsymbol{b}} = 0$, $\langle A_{kj}A_{\ell m}\rangle_A = (1/N)\delta_{k\ell}\delta_{jm}$ and $\langle b_k b_\ell\rangle_{\boldsymbol{b}} = \sigma^2\delta_{k\ell}$.

11 Appendix: Log-Det Identity (Eq. (77))

We give here a proof of the identity

$$\boxed{\frac{\partial}{\partial M_{ab}} \log\det M = [M^{-1}]_{ba}}\ . \tag{185}$$

Proof: By the chain rule

$$\frac{\partial}{\partial M_{ab}} \log\det M = \frac{1}{\det M}\frac{\partial}{\partial M_{ab}}\det M\ . \tag{186}$$

Using the cofactor expansion of the determinant along the a-th row

$$\det M = \sum_{k=1}^n M_{ak}C_{ak}\ , \tag{187}$$

where the cofactor matrix C is

$$C_{ij} = (-1)^{i+j}T_{ij} \tag{188}$$

and T_{ij} is a *minor* of M, i.e. the determinant of the $(n-1)\times(n-1)$ matrix obtained removing the i-th row and j-th column of M.

Hence

$$\frac{\partial}{\partial M_{ab}}\det M = \sum_{k=1}^n \left[\underbrace{\frac{\partial M_{ak}}{\partial M_{ab}}}_{\delta_{kb}} C_{ak} + M_{ak}\underbrace{\frac{\partial C_{ak}}{\partial M_{ab}}}_{=0}\right], \tag{189}$$

where the last term vanishes as the elements in row a do not affect the corresponding cofactor.

It follows from (186) and (189) that

$$\frac{\partial}{\partial M_{ab}} \log \det M = \frac{C_{ab}}{\det M} = \frac{[\mathrm{adj}(M)]_{ba}}{\det M}\ , \tag{190}$$

where we use the fact that the adjugate matrix $\mathrm{adj}(M)$ is the transpose of the cofactor matrix. The right hand side of (190) is readily recognised as the element ba of the inverse matrix of M.

12 Appendix: Large Deviations for Coin Tossing

I will provide here a quick crash-course on large deviations for those who have not met the concept and ideas before. I strongly recommend the review [33] by Hugo Touchette for a very thorough introduction to large deviations for physicists. Consider the following question: what is the probability of getting M Heads out of N (fair) coin tosses?

The exact formula for any M, N is given by the following binomial distribution

$$P(M, N) = \binom{N}{M}\frac{1}{2^N}\ , \tag{191}$$

where the binomial factor counts the number of possible arrangements of the M Heads within the sequence of N tosses.

Clearly, we can now compute the average and variance of the number M of Heads as

$$\mu = \sum_{M=0}^{N} M\binom{N}{M}\frac{1}{2^N} = \frac{N}{2} \tag{192}$$

$$\sigma^2 = \sum_{M=0}^{N} M^2\binom{N}{M}\frac{1}{2^N} - \mu^2 = \frac{N}{4}\ , \tag{193}$$

and appealing to Central Limit considerations, we expect that for large N the probability will converge to a Gaussian centred around μ and with fluctuations given by σ^2, i.e.

$$P(M, N) \to \frac{1}{\sqrt{\pi(N/2)}} \exp\left[-\frac{2}{N}\left(M - \frac{N}{2}\right)^2\right]. \tag{194}$$

On the other hand, it is easy to compute exactly the probability of an *anomalous event* characterised by *all* N tosses coming up Heads:

$$P(M = N, N) = \frac{1}{2^N} = \exp(-N \log 2) . \tag{195}$$

Comparing the two results for large N in (195) and (194) clearly shows that the two formulae are inconsistent for $M \simeq N$. In other word, the very natural Gaussian fluctuation law in Eq. (194) is only valid for *typical* fluctuations of the order of $\sim \sqrt{N}$ around the mean, but is inadequate to describe *large* (anomalous) fluctuations where M deviates from the average $N/2$ by an amount of order N.

Reconciling the two results requires introducing the *large deviation* (or rate) function, which governs the precise way the probability distribution decays when N is large, both in the vicinity and away from the most likely value.

Take $P(M = Nx, N)$ for $0 \leqslant x \leqslant 1$ and expand the right hand side for large N using Stirling's formula for the factorials. One obtains (neglecting pre-factors)

$$P(M = Nx, N) \approx \exp\left(-NI(x)\right) , \tag{196}$$

where

$$I(x) = x \log x + (1 - x) \log(1 - x) + \log 2 , \tag{197}$$

where the symbol $\approx$ stands for the precise asymptotics

$$\lim_{N\to\infty} \frac{-1}{N} \log P(M = Nx, N) = I(x) . \tag{198}$$

The rate function $I(x)$ is plotted in Fig. 5.

For $x \to 0$ (or $x \to 1$, symmetrically) the rate function converges to the value $\log 2$, which perfectly reproduces the exact result in Eq. (195). The rate function has a minimum (a zero) at $x = 1/2$, the "most likely" value (corresponding to $N/2$ Heads in N tosses), and then increases on either side of $x = 1/2$ leading $P(M = Nx, N)$ to correspondingly decay exponentially fast away from the most likely occurrence.

Interestingly enough, the rate function also provides important information about the *typical* fluctuations of the random variable M (= number of Heads) around its most likely value $M = N/2$. To see this, we can Taylor-expand $I(x)$ around $x = 1/2$ to get

$$I(x) = 2\left(x - \frac{1}{2}\right)^2 + \mathcal{O}\left(\left(x - \frac{1}{2}\right)^3\right) . \tag{199}$$

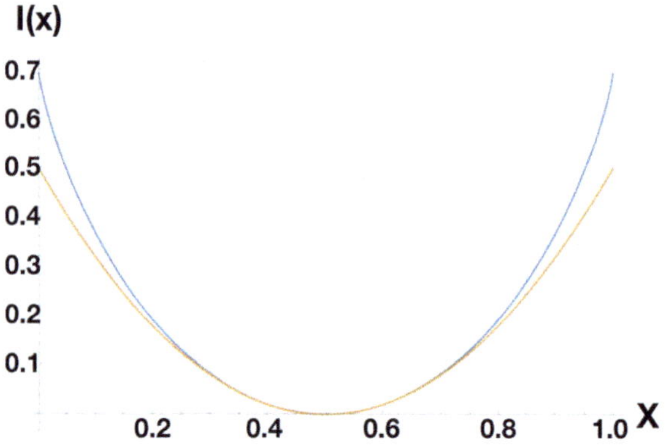

Fig. 5 Solid blue line: rate function $I(x)$ in Eq. (197). Solid yellow line: quadratic behaviour $2(x - 1/2)^2$ around the minimum (see Eq. (199)). This plot—and the visible deviation between the two curves at the edges—clearly shows that the Gaussian behaviour around the minimum is inadequate to characterise anomalous events characterised by a very large or small number of Heads in a series of coin tosses

Inserting this quadratic behaviour back into Eq. (196) gives

$$P(M = Nx, N) \approx \exp\left(-2N\left(\frac{M}{N} - \frac{1}{2}\right)^2\right) = \exp\left(-\frac{2}{N}\left(M - \frac{N}{2}\right)^2\right), \tag{200}$$

which precisely reproduces the Gaussian fluctuations of M around its mean value $N/2$ in Eq. (194).

The rate function $I(x)$ is therefore arguably a more fundamental and richer object than the Gaussian law (194), as it includes the latter but provides more accurate information about larger (anomalous) fluctuations much farther away from the mean value.

References

1. Fyodorov, Y.V., Tublin, R.: Optimization landscape in the simplest constrained random least-square problem. J. Phys. A: Math. Theor. **55**, 244008 (2022)
2. Fyodorov, Y.V., Tublin, R.: Counting stationary points of the loss function in the simplest constrained least-square optimization. Acta Phys. Pol. B **51**, 1663–1672 (2020)
3. Tublin, R.: A Few Results in Random Matrix Theory and Random Optimization. Ph.D. thesis at King's College London (2022). https://kclpure.kcl.ac.uk/ws/portalfiles/portal/182845766/2022_Tublin_Rashel_1210445_ethesis.pdf
4. Spearman, C.: General intelligence objectively determined and measured. Am. J. Psychol. **15**(2), 201–293 (1904)
5. Thurstone, L.: Multiple factor analysis. Psychol. Rev. **38**(5), 406–427 (1931)
6. Thurstone, L.: The Vectors of Mind. Psychol. Rev. **41**, 1–32 (1934)

7. Crosilla, F., et al.: Advanced Procrustes Analysis Models in Photogrammetric Computer Vision. CISM International Centre for Mechanical Sciences, vol. 590. Springer (2019)
8. Andrade, J.M., et al.: Procrustes rotation in analytical chemistry, a tutorial. Chemom. Intell. Lab. Syst. **72**(2), 123–132 (2004)
9. Conneau, A., et al.: Word Translation Without Parallel Data. ICLR 2018 (2018). https://arxiv.org/abs/1710.04087
10. Kementchedjhieva, Y., Ruder, S., Cotterell, R., Søgaard, A.: Generalizing procrustes analysis for better bilingual dictionary induction. In: Proceedings of the 22nd Conference on Computational Natural Language Learning, pp. 211–220. Brussels, Belgium. Association for Computational Linguistics (2018). https://doi.org/10.18653/v1/K18-1021
11. Simon, C.: (2018). https://simonensemble.github.io/posts/2018-10-27-orthogonal-procrustes/
12. Browne, M.W.: On oblique procrustes rotation. Psychometrica **32**(2), 125–132 (1967)
13. Gander, W.: Least squares with a quadratic constraint. Numer. Math. **36**, 291–307 (1981)
14. Golub, G.H., von Matt, U.: Quadratically constrained least squares and quadratic problems. Numerische Mathematik **59**(1), 561–580 (1991)
15. Janik, R.A., Nowak, M.A.: Wishart and anti-Wishart random matrices. J. Phys. A: Math. Gen. **36**(12), 3629 (2003)
16. Yu, Y.-K., Zhang, Y.-C.: On the Anti-Wishart distribution. Physica A: Stat. Mech. Appl. **312**(1–2), 1–22 (2002)
17. Marčenko, V.A., Pastur, L.A.: Distribution of eigenvalues in certain ensembles of random matrices. Mat. Sb. **72**, 507–536 (1967)
18. Fyodorov, Y.V., Le Doussal, P.: Topology Trivialization and large deviations for the minimum in the simplest random optimization. J. Stat. Phys. **154**(1–2), 466–490 (2014)
19. Castellani, T., Cavagna, A.: Spin-glass theory for pedestrians. J. Stat. Mech. **2005**(05), P05012 (2005)
20. Zamponi, F.: Mean field theory of spin glasses (2010). arXiv: 1008.4844
21. Talagrand, M.: Mean Field Models for Spin Glasses. I: Basic Examples. Springer, Berlin, Heidelberg (2011)
22. Bovier, A.: Statistical Mechanics of Disordered Systems: A Mathematical Perspective. Cambridge University Press, Cambridge (2006)
23. Panchenko, D.: The Sherrington-Kirkpatrick Model. Springer Monographs in Mathematics (2013)
24. Parisi, G., Urbani, P., Zamponi, F.: Theory of Simple Glasses: Exact Solution in Infinite Dimensions. Cambridge University Press (2020)
25. Fyodorov, Y.V.: A spin glass model for reconstructing nonlinearly encrypted signals corrupted by noise. J. Stat. Phys. **175**(5), 789–818 (2019)
26. Urbani, P.: A continuous constraint satisfaction problem for the rigidity transition in confluent tissues. J. Phys. A: Math. Theor. **56**, 115003 (2023)
27. Kamali, P.J., Urbani, P.: Dynamical mean field theory for models of confluent tissues and beyond. SciPost Phys. **15**, 219 (2023)
28. Urbani, P.: Quantum exploration of high-dimensional canyon landscapes. Pierfrancesco Urbani J. Stat. Mech. 083301 (2024). https://doi.org/10.1088/1742-5468/ad0635
29. Fyodorov, Y.V.: Negative moments of characteristic polynomials of random matrices: ingham-siegel integral as an alternative to Hubbard-Stratonovich transformation. Nucl. Phys. B [PM] **621**, 643–674 (2002)
30. Fyodorov, Y.V.: Multifractality and freezing phenomena in random energy landscapes: an introduction. Physica A: Stat. Theor. Phys. **389**(20), 4229–4254 (2010)
31. Parisi, G.: The order parameter for spin glasses: a function on the interval $0 - 1$. J. Phys. A: Math. Gen. **13**, 1101–1112 (1980)
32. Prussing, J.E.: The Principal Minor test for semidefinite matrices. J. Guid. Control Dyn. **9**(1), 121–122 (1986)
33. Touchette, H.: The large deviation approach to statistical mechanics. Phys. Rep. **478**(1–3), 1–69 (2009)

34. Dembo, A., Zeitouni, O.: Matrix optimization under random external fields. J. Stat. Phys. **159**(6), 1306–1326 (2015)
35. Kivimae, P.: Critical fluctuations for the spherical Sherrington-Kirkpatrick model in an external field (2020). arXiv: 1908.07512v2
36. Landon, B., Sosoe, P.: Fluctuations of the 2-spin SSK model with magnetic field (2022). arXiv: 2009.12514v2
37. Baik, J., Collins-Woodfin, E., Le Doussal, P., Wu, H.: Spherical spin glass model with external field. J. Stat. Phys. **183**, 31 (2021)
38. Belius, D., Černý, J., Nakajima, S., Schmidt, M.: Triviality of the geometry of mixed p-spin spherical Hamiltonians with external field. J. Stat. Phys. **186**, 12 (2022)
39. Jaganathan, K., Eldar, Y.C., Hassibi, B.: Phase Retrieval: an overview of recent developments. Chapter 13 in Optical Compressive Imaging. CRS Press (2016)
40. Kamali, P.J., Urbani, P.: Stochastic Gradient Descent outperforms Gradient Descent in recovering a high-dimensional signal in a glassy energy landscape (2023). arXiv:2309.04788v2
41. Kanzieper, E.: Replica field theories, Painlevé transcendents, and exact correlation functions. Phys. Rev. Lett. **89**, 250201 (2002)
42. Akhanjee, S., Rudnick, J.: Spherical Spin-Glass-Coulomb-Gas duality: solution beyond mean-field theory. Phys. Rev. Lett. **105**, 047206 (2010)
43. Boyd, S., Vandenberghe, L.: Convex Optimization. Cambridge University Press, Cambridge (2004)
44. Livan, G., Vivo, P.: Moments of Wishart-Laguerre and Jacobi ensembles of random matrices: application to the quantum transport problem in chaotic cavities. Acta Physica Polonica B **42**, 1081 (2011)

New Analytical and Geometrical Aspects on Trudinger-Moser Type Inequality in 2D

Natalino Borgia, Silvia Cingolani, and Gabriele Mancini

Abstract The present survey is devoted to results on Trudinger-Moser inequalities in two dimension. We give a brief overview of the history of these celebrated inequalities and, starting from the geometric problem that motivated Moser's original work, we discuss the connection between Onofri's inequality for the unit sphere and sharp inequalities on Euclidean domains. Finally, we present recent results and new insights into nonlocal interaction energy functionals in two dimension, involving logarithmic kernels.

Keywords Trudinger-Moser inequality · Two dimension · Nirenberg problem · Nonlocal interaction energy functional · Logarithmic kernel

1 Introduction

If Ω is a bounded open subset of $\mathbb{R}^2$, the Sobolev space $H_0^1(\Omega)$ is a borderline case for the Sobolev embedding since it is continuously embedded into $L^q(\Omega)$ for any $1 \le q < \infty$, but $H_0^1(\Omega) \not\hookrightarrow L^\infty(\Omega)$. If Ω is the unit disk, a counterexample is given by the function $u(x) = \ln(1 - \ln |x|)$.

In 1967, Trudinger [86] proved that $H_0^1(\Omega)$ is embedded into the Orlicz space $L_\phi(\Omega)$ corresponding to the Young measure $\phi(t) = e^{t^2} - 1$ (see also Yudovich [90] and Pohozaev [73]). In 1971, Moser [68] established a celebrated result which we recall for the two-dimensional case.

N. Borgia · S. Cingolani (✉) · G. Mancini
Dipartimento di Matematica, Università degli Studi di Bari, Via E. Orabona 4, Bari, Italy
e-mail: silvia.cingolani@uniba.it

N. Borgia
e-mail: natalino.borgia@uniba.it

G. Mancini
e-mail: gabriele.mancini@uniba.it

B. Cassano et al. (eds.), *Singularities, Asymptotics, and Limiting Models*, Springer INdAM Series 64, https://doi.org/10.1007/978-981-96-3584-9_9

Theorem 1 (Trudinger-Moser inequality in 2D) Let $\Omega \subset \mathbb{R}^2$ be a bounded domain. Let $\mathcal{H} := \{u \in H_0^1(\Omega) \mid \|\nabla u\|_2 \le 1\}$, where $\|\cdot\|_2$ denotes the usual L^2 norm. Then

(1) $\forall\, u \in H_0^1(\Omega): \quad e^{u^2} \in L^1(\Omega)$.
(2) There exists a constant $C \ge 1$ depending on the dimension only, such that

$$\sup_{u \in \mathcal{H}} \frac{1}{|\Omega|} \int_\Omega e^{4\pi u^2}\, dx \le C. \tag{1}$$

(3) If $\alpha > 4\pi$, then

$$\sup_{u \in \mathcal{H}} \frac{1}{|\Omega|} \int_\Omega e^{\alpha u^2}\, dx = +\infty.$$

From Theorem 1, it follows that the maximal allowed growth is of exponential type and that for any $p > 1$ there exists a sequence $(u_\epsilon)_{\epsilon>0}$ of functions in $H_0^1(\Omega)$ with $\|\nabla u_\epsilon\|_2 = 1$, verifying $\int_\Omega e^{4\pi p u_\epsilon^2}\, dx \to +\infty$ as $\epsilon \to 0$. On the other hand, for all $u \in H_0^1(\Omega)$, $\int_\Omega e^{4\pi p u^2}\, dx < \infty$ for all $p > 0$.

In 1985, Lions [57] proved that if (u_ϵ) is a sequence of functions in $H_0^1(\Omega)$ with $\|\nabla u_\epsilon\|_2 = 1$ such that $u_\epsilon \to u_0$ weakly in $H_0^1(\Omega)$, then for any $p < 1/(1 - \|\nabla u_0\|_2^2)$ it results

$$\limsup_{\epsilon \to 0} \int_\Omega e^{4\pi p u_\epsilon^2}\, dx < +\infty.$$

Of course this fact gives extra information if the sequence u_ϵ has a weak limit $u_0 \ne 0$.

A challenging problem on Trudinger-Moser inequalities was whether the extremal functions exist or not. It has remained open for quite some years. In 1986, Carleson and Chang [21] proved that the supremum is attained in (1) if $\Omega = B_1$ is the unit disk in $\mathbb{R}^2$, by using a symmetrization argument. Successively Struwe [78] showed that the supremum is attained in (1) if Ω is close to a ball in the sense of measure. Flucher [44] extended this result to arbitrary bounded domains in $\mathbb{R}^2$ using conformal deformations.

In 2004 Adimurthi and Druet [3] highlighted that the result of Lions [57] suggested

$$C_\alpha(\Omega) := \sup_{u \in H_0^1(\Omega), \|\nabla u\|_2 \le 1} \int_\Omega e^{4\pi u^2 (1 + \alpha \|u\|_2^2)} dx < +\infty \tag{2}$$

for some $\alpha > 0$. In fact, they proved that if Ω is a smooth bounded domain in $\mathbb{R}^2$, denoted by $\lambda_1(\Omega)$ the first eigenvalue of the Laplacian with respect to the Dirichlet boundary condition, it results that for any $0 \le \alpha < \lambda_1(\Omega)$, $C_\alpha(\Omega) < +\infty$. On the contrary, for any $\lambda \ge \lambda_1(\Omega)$, $C_\alpha(\Omega) = +\infty$. These results in [3] add new information when the weak limit of u_ϵ is 0. A stronger version of (2) was obtained by Tintarev [84] and existence of extremal functions was proved by Yang [88]. We point out that the existence of extremal functions for these improved inequalities is an extremely delicate problem. For example, the supremum in (2) is attained for small values of α, but it is not attained when α is close to $\lambda_1(\Omega)$, see [64]. In general, even small perturbations of the exponential term in (1) may affect the attainability of the supremum, as it is pointed out by [76] and clarified by [61] and [46]. We refer to [83] for recent developments on the subject.

As regards Trudinger-Moser inequality (1) on unbounded domains, we mention for the subcritical growth $\alpha < 4\pi$ the results by Cao [20] and Adachi, Tanaka [1] (see also [38]). In the critical case $\alpha = 4\pi$, the inequality with the sharp exponent (proved by Moser) fails on the whole plane. Li and Ruf [77] and Ruf [55] proved that there exists a constant $d > 0$ such that, for any Ω unbounded domain in $\mathbb{R}^2$

$$\sup_{u\in H_0^1(\Omega),\|u\|\le 1} \int_\Omega \left(e^{4\pi u^2} - 1\right)dx \le d, \tag{3}$$

where $\|u\| = \left(\int_\Omega (|\nabla u|^2 + u^2)dx\right)^{1/2}$ denotes the usual Sobolev norm on $H_0^1(\Omega)$.

The supremum is attained if $\Omega = \mathbb{R}^2$ (see [47, 55]) and the inequality is sharp: for any growth $e^{\alpha u^2}$ with $\alpha > 4\pi$, the corresponding supremum equals $+\infty$. We mention [55] for a general dimension and [45] for a more recent improvement of the inequality on the whole plane $\mathbb{R}^2$. Furthermore an equivariant version of (3) has been obtained in [22] and sharp weighted Trudinger-Moser type inequalities were proved in [18, 41]. For unbounded domains with $\lambda_1(\Omega) > 0$, bounds on the full Sobolev norm are unnecessary, as one can prove that

$$\sup_{u\in\mathcal{H}} \int_\Omega (e^{4\pi u^2} - 1)\, dx < \infty,$$

where $\mathcal{H}$ is defined as in Theorem 1, see [63] and [9].

Even if in this survey we focus on inequalities for $H_0^1(\Omega)$, it is important to mention the work of Adams [2], who proved similar inequalities for domains $\Omega \subseteq \mathbb{R}^n$ and the Sobolev spaces $W^{k,p}(\Omega)$ whenever $kp = n$. We also refer to [65] for Trudinger-Moser-Adams type inequalities on fractional order Sobolev spaces. Concerning the study of extremal functions for this more general family of inequalities, we refer to [37, 58, 62] and references therein.

Trudinger-Moser type inequalities have also been widely studied for Sobolev spaces on Riemannian surfaces and manifolds. Indeed, as we discuss in the next section, they have important applications to prescribed curvature problems in conformal geometry. A general formulation of Trudinger-Moser-Adams inequalities for compact Riemannian manifolds was given by Fontana [43] (see also [52, 53] for related results). Recently subcritical and critical Trudinger-Moser inequalities are proved for complete noncompact Riemannian manifolds in [54].

The singular Trudinger-Moser inequality and its extremal functions were considered in [4, 26, 34, 48, 85, 89] and generalizations to the whole space in [56, 88].

The aim of the present survey is two-fold. On the one hand we discuss some connections between different notions of Trudinger-Moser inequality in 2D, announcing some forthcoming results in [15]. In particular, in Sect. 2, starting from geometrical and physical applications, we introduce an inequality due to Onofri [70] (see Eq. (8)) for Sobolev spaces on the unit sphere and we discuss its connection with similar inequalities for domains in $\mathbb{R}^2$.

On the other hand we present some recent results and new insights into nonlocal interaction energy functionals in two dimension, arising from mathematical models for chemotaxis [40], in the statistical mechanics of selfgravitating clouds [82, 87] and in the description of vortices in turbulent Euler flows [17]. In particular, in Sect. 3, we state some Trudinger-Moser type inequalities with logarithmic kernel in [32] and announce some forthcoming developments in [19, 33].

2 Onofri-Type Inequalities and Geometric Connections

The Trudinger-Moser inequality has applications to several mathematical fields ranging from mathematical physics to differential geometry and calculus of variations. Among the reasons that led Moser to improve Trudinger's result, it is important to mention Nirenberg's problem on $\mathbb{S}^2$, which consists in describing the set of smooth functions that can be obtained as Gaussian curvature of a Riemannian metric conformally equivalent to the standard spherical metric.

Let Σ be a smooth surface, we recall that two Riemannian metrics g_0 and g on Σ are said to be pointwise conformally equivalent if there exists $u \in C^\infty(\Sigma)$ such that $g = e^u g_0$ on Σ. In this setting, the following identity holds:

$$\Delta_{g_0} u + 2K_g e^u = 2K_{g_0},$$

where Δ_{g_0} denotes the Laplace-Beltrami operator on Σ, and K_g and K_{g_0} are the Gaussian curvatures of Σ corresponding respectively to g and g_0. When $\Sigma = \mathbb{S}^2$ and g_0 is the standard spherical metric, then a function $K \in C^\infty(\mathbb{S}^2)$ is the Gaussian curvature of a metric which is pointwise conformally equivalent to g_0 if and only if there exists $u \in C^\infty(\mathbb{S}^2)$ satisfying

$$\Delta_{g_0} u + 2Ke^u = 2 \quad \text{on} \quad (\mathbb{S}^2, g_0). \tag{4}$$

It is possible to give a variational formulation of (4) by considering the functional

$$\Phi(u) = \frac{1}{2} \int_{\mathbb{S}^2} |\nabla_{g_0} u|_{g_0}^2 \, dv_{g_0} + 2 \int_{\mathbb{S}^2} u \, dv_{g_0} - 8\pi \ln \left(\frac{1}{4\pi} \int_{\mathbb{S}^2} K e^u \, dv_{g_0} \right),$$

which is well defined in

$$\mathcal{A} := \left\{ u \in H^1(\mathbb{S}^2, g_0) | \int_{\mathbb{S}^2} K e^u \, dv_{g_0} > 0 \right\},$$

provided $\max_{\mathbb{S}^2} K > 0$. A function $u \in H^1(\mathbb{S}^2, g_0)$ is a weak solution to (4) if and only if it is a critical point of Φ and

$$\int_{\mathbb{S}^2} K e^u dv_{g_0} = 4\pi.$$

As an analog of Theorem 1 for the Sobolev space $H^1(\mathbb{S}^2, g_0)$, Moser proves that there exists a constant $S > 1$ such that for any $u \in H^1(\mathbb{S}^2, g_0)$ with $\|\nabla_{g_0} u\|_{L^2(\mathbb{S}^2, g_0)} \leq 1$:

$$\frac{1}{4\pi} \int_{\mathbb{S}^2} e^{4\pi(u-\overline{u})^2} dv_{g_0} \leq S, \quad \text{where } \overline{u} := \frac{1}{4\pi} \int_{\mathbb{S}^2} u \, dv_{g_0}. \tag{5}$$

As in the Euclidean inequality (1), the exponent 4π in (5) is sharp. As a straightforward consequence of (5) and Young's inequality, one can prove the weaker inequality:

$$\ln \left(\frac{1}{4\pi} \int_{\mathbb{S}^2} e^u \, dv_{g_0} \right) \leq \frac{1}{16\pi} \int_{\mathbb{S}^2} |\nabla_{g_0} u|^2_{g_0} \, dv_{g_0} + \frac{1}{4\pi} \int_{\mathbb{S}^2} u \, dv_{g_0} + \ln S, \tag{6}$$

for any $u \in H^1(\mathbb{S}^2, g_0)$, where S is the constant appearing in (5), and the coefficient $\frac{1}{16\pi}$ is sharp. In particular, (6) shows that Φ is bounded from below in $\mathcal{A}$. While this is not sufficient to apply standard minimization techniques (actually, it can be shown that minimum points for Φ in $\mathcal{A}$ do not exist unless K is constant), the existence of critical points for Φ can be obtained via improved versions of (6) under suitable assumptions on K. For example, in [69] Moser proves that if $u \in H^1(\mathbb{S}^2, g_0)$ is even, then (6) can be replaced by the stronger inequality:

$$\ln \left(\frac{1}{4\pi} \int_{\mathbb{S}^2} e^u \, dv_{g_0} \right) \leq \frac{1}{32\pi} \int_{\mathbb{S}^2} |\nabla_{g_0} u|^2_{g_0} \, dv_{g_0} + \frac{1}{4\pi} \int_{\mathbb{S}^2} u \, dv_{g_0} + C, \tag{7}$$

for some $C \in \mathbb{R}$. In particular, if $K \in C^\infty(\mathbb{S}^2)$ is even, the restriction of Φ to the space of even functions in $H^1(\mathbb{S}^2, g_0)$ with zero average is coercive, and one can find a critical point of Φ by means of Palais symmetric criticality principle. This leads to one of the first general existence results for the Nirenberg problem: if $K \in C^\infty(\mathbb{S}^2)$ is an even function and $\max_{\mathbb{S}^2} K > 0$, then there exists a solution of (4). Over the years, many improved versions of (6) have been discovered. For example, we mention that (6) can be improved for functions with vanishing barycenter (see [6, 60]) or for functions such that the mass of the measure $e^u dv_{g_0}$ is spread in at least two disjoint open subsets of $\mathbb{S}^2$ (see [27, 59]). We refer to [12, 23, 24, 50, 79] and references therein for further results concerning Nirenberg's problem on $\mathbb{S}^2$.

Another interesting aspect of (6) is that while the coefficient $\frac{1}{16\pi}$ is sharp, the constant $\ln S$ is not. Indeed, Onofri in [70] proved that

$$\inf_{H^1(\mathbb{S}^2, g_0)} \Phi = 0,$$

and he classified all minimum points of Φ. Namely, we have

$$\ln\left(\frac{1}{4\pi}\int_{\mathbb{S}^2} e^u \, dv_{g_0}\right) \le \frac{1}{16\pi}\int_{\mathbb{S}^2} |\nabla_{g_0} u|^2_{g_0} \, dv_{g_0} + \frac{1}{4\pi}\int_{\mathbb{S}^2} u \, dv_{g_0}, \tag{8}$$

for any $u \in H^1(\mathbb{S}^2, g_0)$, and equality holds if and only if $u = \ln|J_\psi| + C$, where $\psi : \mathbb{S}^2 \to \mathbb{S}^2$ is a conformal diffeomorphism of $(\mathbb{S}^2, g_0)$, J_ψ denotes the Jacobian of ψ, and $C \in \mathbb{R}$. In [39], Eq. (8) is obtained as limit of Sobolev's inequalities for $W^{1,p}(\mathbb{R}^2)$ as $p \nearrow 2$. Onofri's inequality (8) plays a role in spectral analysis for the Laplace-Beltrami operator. Indeed, thanks to Polyakov's formula (see [71, 72, 74, 75]), we have

$$\frac{\det(-\Delta_g)}{\det(-\Delta_{g_0})} = e^{-\frac{1}{24}\Phi(u)},$$

for any metric $g = e^u g_0$ with $u \in C^\infty(\mathbb{S}^2)$. In particular, Onofri's inequality (8) implies that, in the conformal class of g_0, the determinant of the Laplace-Beltrami operator is maximized by g_0.

An analog of (8) for the unit disk in $\mathbb{R}^2$ can be deduced from the work of Carleson and Chang [21]. As a crucial step in their proof of the existence of extremal functions for (1), they prove an inequality which is equivalent to:

$$\ln\left(\frac{1}{\pi}\int_{B_1} e^u \, dx\right) < \frac{1}{16\pi}\int_{B_1} |\nabla u|^2 dx + 1, \tag{9}$$

for any $u \in H_0^1(B_1)$. Again, this inequality is sharp in the sense that

$$\inf_{u \in H_0^1(B_1)} \left(\frac{1}{16\pi}\int_{B_1} |\nabla u|^2 dx - \ln\left(\frac{1}{\pi}\int_{B_1} e^u \, dx\right)\right) = -1,$$

and

$$\inf_{u\in H_0^1(B_1)}\left(\alpha\int_{B_1}|\nabla u|^2dx-\ln\left(\frac{1}{\pi}\int_{B_1}e^u\,dx\right)\right)=-\infty,$$

for any $\alpha<\frac{1}{16\pi}$. More generally, it was proved in [17] (see also [25]) that if $\Omega\subseteq\mathbb{R}^2$ is a bounded domain with smooth boundary, then for any $u\in H_0^1(\Omega)$

$$\ln\left(\frac{1}{|\Omega|}\int_\Omega e^u\,dx\right)\le\frac{1}{16\pi}\int_\Omega|\nabla u|^2dx+1+4\pi\sup_\Omega\gamma(x)+\ln\frac{\pi}{|\Omega|},\tag{10}$$

where γ is Robin's function of Ω, which is defined for $x\in\Omega$ as $\gamma(x)=H_x(x)$, where H_x is the unique solution of

$$\begin{cases}\Delta H_x=0 & \text{in } \Omega,\\ H_x(y)=\frac{1}{2\pi}\ln|x-y| & \text{on } \partial\Omega.\end{cases}$$

Differently from (9), (10) is not always sharp. Indeed, denoting

$$I(\Omega):=\inf_{u\in H_0^1(\Omega)}\left(\frac{1}{16\pi}\int_\Omega|\nabla u|^2dx-\ln\left(\frac{1}{|\Omega|}\int_\Omega e^u\,dx\right)\right),$$

there are domains Ω (see [16, Sect. 7]) for which

$$I(\Omega)>-1-4\pi\sup_\Omega\gamma(x)-\ln\frac{\pi}{|\Omega|}.\tag{11}$$

The domains for which (10) is sharp are known in the literature as domains of first kind, while the ones for which (11) holds are known as domains of second kind. Inequality (10) plays a crucial role in the study of the mean field equation

$$-\Delta u=\frac{\rho e^u}{\int_\Omega e^u\,dx}\quad\text{in } \Omega,\tag{12}$$

which appears in the statistical mechanics description of vortex formations in 2D-models for turbulent flows in fluid dynamics (see e.g. [17, 49]). We refer to [7, 8, 28, 29, 59] for existence results for (12) and related problems.

In [68] Moser explicitly states that it seems impossible to deduce (1) from (5) or (5) from (1). The proofs of the two inequalities are somehow independent, although they are based on similar arguments relying on symmetrization techniques. It turns out that a more explicit connection exists between (8) and (9). Indeed, one can show that (8) and (9) are equivalent by exploiting the conformal equivalence between $\mathbb{R}^2$ and $\mathbb{S}^2$ minus one point. In the remaining part of this section, we sketch a simple argument to deduce (9) from (8), which is inspired by [48], where the authors consider generalized versions of (9) and (8) involving singular weights. In a forthcoming paper [15], we will give a complete proof of the equivalence between (8) and (9) and investigate extensions to dimension $n \geq 3$ (see [5, 36, 51]).

In the following, we assume that (8) holds in $H^1(\mathbb{S}^2, g_0)$. We take a function $u \in H_0^1(B_1)$, and we prove that (9) holds. We denote by $N = (0, 0, 1)$ the north pole of $\mathbb{S}^2$ and we let $\pi : \mathbb{S}^2 \setminus \{N\} \to \mathbb{R}^2$ be the standard stereographic projection. We shall consider the function

$$v_0(x) = 2 \ln \left(\frac{2}{1 + |x|^2} \right).$$

We stress that v_0 satisfies

$$-\Delta v_0 = 2e^{v_0} \quad \text{on } \mathbb{R}^2. \tag{13}$$

Moreover, for any $f : \mathbb{S}^2 \to \mathbb{R}$, $f \in L^1(\mathbb{S}^2, g_0)$, we have that

$$\int_{\mathbb{S}^2} f \, dv_{g_0} = \int_{\mathbb{R}^2} (f \circ \pi^{-1}) \, e^{v_0} \, dx.$$

For any $r > 0$, we define $u_r : \mathbb{R}^2 \to \mathbb{R}$ as

$$u_r(x) = \begin{cases} u\left(\dfrac{x}{r}\right) - v_0(x) & \text{if } |x| \leq r, \\ -2 \ln \left(\frac{2}{1+r^2}\right) & \text{if } |x| > r, \end{cases}$$

and $\widehat{u}_r : \mathbb{S}^2 \to \mathbb{R}$ as

$$\widehat{u}_r(x) = \begin{cases} u_r \circ \pi(x) & \text{if } x \neq N \\ -2 \ln \left(\frac{2}{1+r^2}\right) & \text{if } x = N. \end{cases}$$

Observe that $\widehat{u}_r$ is constant in a neighborhood of N and $\widehat{u}_r \in W^{1,2}(\mathbb{S}^2, g_0)$. Then, we can apply (8) to get:

$$\ln \left(\frac{1}{4\pi} \int_{\mathbb{S}^2} e^{\widehat{u}_r} \, dv_{g_0} \right) \leq \frac{1}{16\pi} \int_{\mathbb{S}^2} |\nabla_{g_0} \widehat{u}_r|_{g_0}^2 \, dv_{g_0} + \frac{1}{4\pi} \int_{\mathbb{S}^2} \widehat{u}_r \, dv_{g_0}.$$

A direct computation shows that

$$\frac{1}{4\pi} \int_{\mathbb{S}^2} e^{\widehat{u}_r} \, dv_{g_0} = \frac{r^2}{4\pi} \int_{B_1} e^u \, dx + \frac{1 + r^2}{4}.$$

Similarly, we have

$$\frac{1}{4\pi}\int_{\mathbb{S}^2} \widehat{u}_r \, dv_{g_0} = \frac{1}{4\pi}\int_{B_r} u_r e^{v_0}\, dx + 2 - \ln 4 + o_r(1), \tag{14}$$

and

$$\frac{1}{16\pi}\int_{B_r} |\nabla v_0|^2 \, dx = 2\ln r - 1 + o_r(1), \tag{15}$$

where $o_r(1) \to 0$ as $r \to +\infty$. Using (13)–(15), we find

$$\begin{aligned}
&\frac{1}{16\pi}\int_{\mathbb{S}^2} |\nabla_{g_0}\widehat{u}_r|_{g_0}^2 \, dv_{g_0} \\
&= \frac{1}{16\pi}\int_{B_r} |\nabla u_r|^2 \, dx - \frac{1}{4\pi}\int_{B_r} u_r e^{2v_0} dx + 2\ln r - 1 + o_r(1) \\
&= \frac{1}{16\pi}\int_{B_1} |\nabla u|^2 \, dx - \frac{1}{4\pi}\int_{\mathbb{S}^2} \widehat{u}_r \, dv_{g_0} + \ln\left(\frac{r^2}{4}\right) + 1 + o_r(1).
\end{aligned}$$

Summing up, we have shown that

$$\ln\left(\frac{1}{\pi}\int_{B_1} e^u \, dx + \frac{1+r^2}{r^2}\right) \le \frac{1}{16\pi}\int_{B_1} |\nabla u|^2 \, dx + 1 + o_r(1).$$

Passing to limit as $r \to +\infty$, we conclude

$$\ln\left(\frac{1}{\pi}\int_{B_1} e^u \, dx\right) < \ln\left(\frac{1}{\pi}\int_{B_1} e^u \, dx + 1\right) \le \frac{1}{16\pi}\int_{B_1} |\nabla u|^2 \, dx + 1.$$

Hence u satisfies (9).

3 Trudinger-Moser Type Inequalities with Logarithmic Kernels

In this section we consider some nonlocal interaction functionals, which naturally arise in 2D mathematical models for chemotaxis and in other physical contexts, as statistical dynamics, vortex theory.

These two-dimensional problems have remained for a long time a quite open field of study, due to the difficulties to handle changing sign kernels, not bounded from above and below. In literature the first results are just numerical. Recently variational results have been employed to treat integro-differential equations, involving logarithmic convolution potentials, where classical PDE theory fails because of the presence of the non-locality (see [10, 13, 14, 30, 31, 42, 66, 67, 80, 81]).

In particular we consider the following class of free energy functionals

$$u \mapsto \int_{\Omega}\int_{\Omega} \ln \frac{1}{|x-y|} F(u(x))F(u(y))\,dxdy,$$

where $\Omega = B_1$ is the unit ball of $\mathbb{R}^2$ and $F : \mathbb{R} \to \mathbb{R}$ grows exponentially.

In [32] we faced the following challenging questions:

- *Moser-Trudinger type inequalities in presence of a logarithmic convolution potential;*
- *characterization of the critical nonlinear growth rates for these inequalities;*
- *associated Euler-Lagrange equation;*
- *qualitative properties of the maximizers;*
- *symmetry and uniqueness results.*

Here we give some outlines of the results and the proofs in [32]. Firstly we consider the problem of maximizing the quantity

$$\Phi(u) := \int_{B_1}\int_{B_1} \ln \frac{1}{|x-y|} F(u(x))F(u(y))\,dxdy \tag{16}$$

among functions in the set

$$\mathcal{H} := \{u \in H_0^1(B_1) \ : \ |\nabla u|_2 \le 1\}.$$

We make the following general assumption on F:

(A_0) $F : \mathbb{R} \to [0, \infty)$ is even, continuous, and strictly increasing on $[0, \infty)$. Moreover, there exist constants $\alpha, c > 0$ with

$$F(t) \le c e^{\alpha t^2} \qquad \text{for } t \in \mathbb{R}. \tag{17}$$

Since the logarithmic kernel changes sign, it is not a priori clear if the double integral has a well-defined value. To clarify this point, we split the kernel $\ln \frac{1}{|\cdot|}$ into its positive and negative part and define the functionals

$$\Phi_+(u) = \int_{B_1}\int_{B_1} \ln^+ \frac{1}{|x-y|} F(u(x))F(u(y))\,dxdy \in [0, +\infty],$$

$$\Phi_-(u) = \int_{B_1}\int_{B_1} \ln^+|x-y|\, F(u(x))F(u(y))\,dxdy \in [0, +\infty],$$

where $u \in H_0^1(B_1)$, and $\ln^+ = \max\{\ln, 0\}$. By the classical Trudinger-Moser inequality,

$$\int_{B_1} e^{\alpha u^2} dx < \infty \qquad \text{for any } u \in H_0^1(B_1) \text{ and any } \alpha > 0.$$

By the growth condition

$$F(t) \le ce^{\alpha t^2} \qquad \forall t \in \mathbb{R},$$

we infer $F(u) \in L^s(\mathbb{R}^2)$ for any $u \in H_0^1(B_1)$, $s \in [1, \infty)$. Since

$$\ln^+ \frac{1}{|\cdot|} \in L^s(\mathbb{R}^2) \qquad \text{for } s \in [1, \infty),$$

choosing $s = 2$, we deduce by Young convolution inequality that $\ln^+ \frac{1}{|\cdot|} * F(u) \in L^\infty(\mathbb{R}^2)$ and thus $\Phi_+(u) < \infty$, for every $u \in H_0^1(B_1)$.

Moreover, we have

$$[\ln^+ |\cdot| * F(u)](x) = \int_{B_1} \ln^+ |x-y| F(u(y))\,dy \le \ln 2 |F(u)|_1 \qquad \forall x \in B_1$$

and therefore $\ln^+ |\cdot| * F(u) \in L^\infty(B_1)$ and thus $\Phi_-(u) < \infty$ for every $u \in H_0^1(B_1)$.

Taking into account that $\ln \dfrac{1}{|\cdot|} = \ln^+ \dfrac{1}{|\cdot|} - \ln^+ |\cdot|$, we have

$$\Phi(u) := \int_{B_1}\int_{B_1} \ln \frac{1}{|x-y|} F(u(x))F(u(y))\,dxdy = \Phi_+(u) - \Phi_-(u) \in]-\infty, \infty[$$

for every $u \in H_0^1(B_1)$.

Now in order to study the maximization problem related to the quantity

$$\Phi(u) := \int_{B_1}\int_{B_1} \ln \frac{1}{|x-y|} F(u(x))F(u(y))\,dxdy$$

within the set

$$\mathcal{H} := \{u \in H_0^1(B_1) \ : \ |\nabla u|_2 \le 1\},$$

we need to distinguish different forms of asymptotic growth of the nonlinearity F. As in Trudinger-Moser inequality, the value 4π will play a key role.

Firstly we notice that if F has a subcritical exponential growth $\alpha < 4\pi$, an immediate proof of the claim

$$m_1(F) := \sup_{u\in\mathcal{H}} \Phi(u) < \infty$$

can be given using the logarithmic Hardy-Littlewood-Sobolev inequality

$$\int_{B_1}\int_{B_1} \ln\frac{1}{|x-y|} f(x)f(y)\,dxdy \le \frac{|f|_1}{2}\Big(|f|_1\big(c_0 + \big|\ln|f|_1\big|\big) + \int_{B_1} f\ln f\,dx\Big),$$

which holds for any a.e. non-negative function $f \in L^1(B_1)$ with $f \ln f \in L^1(B_1)$, see [11, Theorem 2].

Indeed, let $u \in \mathcal{H} \cap L^\infty(B_1)$, and let $v := F(u)$. Then $v \le c_1 e^{\alpha u^2}$, $\alpha < 4\pi, c_1 > 0$ and

$$v\ln v \le c_1 e^{\alpha u^2}\Big(\alpha u^2 + \ln c_1\Big) \le c_2 e^{4\pi u^2}$$

with $c_2 > 0$. By the classical Trudinger-Moser inequality, we have

$$|v|_1 \le c_1\int_{B_1} e^{4\pi u^2}dx \le c_1 c(B_1) =: c_2, \qquad \int_{B_1} v\ln v\,dx \le c_2 c(B_1) =: c_3.$$

Combining with the logarithmic Hardy-Littlewood-Sobolev inequality we infer

$$\begin{aligned}\Phi(u) &= \int_{B_1}\int_{B_1} \ln\frac{1}{|x-y|} v(x)v(y)\,dxdy \\ &\le \frac{|v|_1}{2}\Big(|v|_1\big(c_0 + \big|\ln|v|_1\big|\big) + \int_{B_1} v\ln v\,dx\Big) \le \frac{c_2}{2}\big(c_2\big(c_0 + |\ln c_2|\big) + c_3\big) < \infty.\end{aligned}$$

For general $u \in \mathcal{H}$, the same inequality follows by approximation and $m_1(F) < \infty$. This argument does not apply if F is critical, so it does not allow for sharp critical growth rates.

In [32] we define some critical rates for the Log-TM inequality.

Definition 1 Let $\beta \in \mathbb{R}$, and let $g : \mathbb{R} \to \mathbb{R}$ be an arbitrary function. We say that g has

(i) at most β-critical growth if $|g(s)| \le c\, e^{4\pi s^2}(1 + |s|)^\beta$ for $s \in \mathbb{R}$ with some constant $c > 0$;
(ii) at least β-critical growth if there exist $s_0, c > 0$ with the property that

$$|g(s)| \ge c\, e^{4\pi s^2}|s|^\beta \qquad \text{for } |s| \ge s_0.$$

In [32] we obtained the following result.

Theorem 2 *Suppose that F satisfies (A_0).*

1. *If F has at most β-critical growth for some $\beta \leq -1$, then*

$$m_1(F) := \sup_{u \in \mathcal{H}} \Phi(u) < \infty.$$

2. *If F has at most β-critical growth for some $\beta < -1$, then $m_1(F)$ is attained, and every maximizer for Φ in $\mathcal{H}$ is, up to sign, a radial and radially decreasing function in B_1.*
3. *If F has at least β-critical growth for some $\beta > -1$, then $m_1(F) = \infty$.*

In order to maximize Φ among the functions in $\mathcal{H} := \{u \in H_0^1(B_1) \ : \ |\nabla u|_2 \leq 1\}$, we consider

$$\mathcal{H}^* := \{u^* \ : \ u \in \mathcal{H}\}$$

the corresponding Schwartz symmetrized set. Here u^* is nonnegative, radial and nonincreasing in the radial variable.

By the Pólya-Szegö inequality, for any $u \in H^1(\mathbb{R}^2)$ we have $u^* \in H^1(\mathbb{R}^2)$ and

$$|u^*|_p = |u|_p, \quad |\nabla u^*|_2 \leq |\nabla u|_2, \qquad \forall p \in [2, +\infty[.$$

Consequently, we have $u^* \in \mathcal{H}$ if $u \in \mathcal{H}$, and so

$$\mathcal{H}^* \subset \mathcal{H}.$$

Now by Riesz's rearrangement inequality, for every $u \in \mathcal{H}$

$$\begin{aligned} \Phi_+(u) &= \int_{B_1}\int_{B_1} \ln^+ \frac{1}{|x-y|} F(u(x))F(u(y))\, dxdy \\ &\leq \int_{B_1}\int_{B_1} \ln^+ \frac{1}{|x-y|} [F(u(x))]^*[F(u(y))]^*\, dxdy \\ &= \int_{B_1}\int_{B_1} \ln^+ \frac{1}{|x-y|} [F(u^*(x))][F(u^*(y))]\, dxdy = \Phi_+(u^*). \end{aligned}$$

Therefore it is immediate that

$$m_1(F) := \sup_{\mathcal{H}} \Phi(u) \leq m_1^+(F) := \sup_{\mathcal{H}} \Phi_+ = \sup_{u \in \mathcal{H}^*} \Phi_+(u).$$

In [32] we prove that

$$\sup_{u\in\mathcal{H}^*} \Phi_+(u) < +\infty,$$

and thus the the following proposition holds.

Proposition 1 *Suppose that F satisfies (A_0) and has at most β-critical growth for some $\beta \le -1$. Then we have*

$$m_1(F) := \sup_{u\in\mathcal{H}} \Phi(u) < \infty.$$

We emphasize the optimality of the growth exponent $\beta = -1$ for the finiteness of $m_1(F)$. Indeed we can define the Moser type functions $u_n \in H_0^1(B_1) \cap L^\infty(B_1)$, for any $n \in \mathbb{N}$, $n \ge 2$

$$u_n(x) = \frac{1}{\sqrt{2\pi}} \frac{\ln(1/|x|)}{(\ln n)^{1/2}} \Big(1 - \frac{1}{4\ln n}\Big)^{1/2}, \quad \frac{1}{n} \le |x| \le 1, \tag{18}$$

$$u_n(x) = \frac{1}{\sqrt{2\pi}} (\ln n)^{1/2} \Big(1 - \frac{1}{4\ln n}\Big)^{1/2}, \quad 0 \le |x| \le \frac{1}{n}. \tag{19}$$

A standard computation gives the following blow-up estimates.

Proposition 2 *Suppose that F satisfies (A_0) and has at least β-critical growth for some $\beta > -1$. Then, if u_n defined as in (18) and (19), we have $u_n \in \mathcal{H} \cap L^\infty(B_1)$ and $\Phi(u_n) \to \infty$ as $n \to \infty$.*

Finally, using Concentration and Compactness Principle by Lions [57], we can recognize the existence of extremals for the Log-TM maximization problem when the parameter $\beta < -1$.

Proposition 3 *Suppose that F satisfies (A_0) and has at most β-critical growth with $\beta < -1$. Then the value $m_1(F) < \infty$ is attained by a function $u \in \mathcal{H}^*$.*

Let $(u_n)_n$ be a maximizing sequence in $\mathcal{H}$ for $m_1(F)$. We may assume that $u_n \in \mathcal{H}^*$ for $n \in \mathbb{N}$. Since $\mathcal{H}$ is bounded in $H_0^1(B_1)$, we may also assume that

$$u_n \rightharpoonup u \in H_0^1(B_1) \qquad \text{with } u \in \mathcal{H}^*.$$

Proof For reader's convenience we give a sketch of the proof in [32]. By Lions [57], we have two possibilities. Either

(i) $u = 0$ or
(ii) $u \ne 0$, and $\int_{B_1} e^{(4\pi+t)u_n^2}\,dx$ is bounded for some $t > 0$ and thus

$$e^{4\pi u_n^2} \to e^{4\pi u^2} \quad \text{in } L^1(B_1).$$

Assume first that (i) holds. In this case, we can prove that

$$m_1(F) = \lim_{n\to\infty} \Phi(u_n) = \Phi(0).$$

On the other hand, if F satisfies (A_0), it follows from that

$$\begin{aligned}\Phi(u) &= 2(2\pi)^2 \int_0^1 rF(u(r)) \ln\frac{1}{r} \int_0^r \rho F(u(\rho)) d\rho dr \\ &> 2(2\pi)^2 \int_0^1 rF(u(0)) \ln\frac{1}{r} \int_0^r \rho F(u(0)) d\rho dr = \Phi(0)\end{aligned}$$

for every $u \in H_0^1(B_1)\backslash\{0\}$. Hence

$$m_1(F) \in \big(\Phi(0), \infty\big]$$

and therefore $m_1(F)$ is not attained at $u = 0$. So this case does not occur.

It remains to consider the case where $(u_n)_n$ satisfies (ii).

Set $v_n := F(u_n)$ for $n \in \mathbb{N}$ and $v := F(u)$.

By (ii), $\int_{B_1} e^{(4\pi+t)u_n^2}\, dx$ is bounded for some $t > 0$ and thus v_n is bounded in $L^{s_0}(\mathbb{R}^2)$ with $s_0 = 1 + \frac{t}{4\pi} > 1$. Moreover, since

$$v_n \to v \qquad \text{in } L^1(\mathbb{R}^2),$$

interpolation yields that $v_n \to v$ in $L^s(\mathbb{R}^2)$ for $1 \le s < s_0$. Finally we conclude that

$$m_1(F) = \lim_{n\to\infty} \Phi(u_n) = \Phi(u),$$

so $m_1(F)$ is attained at $u \in \mathcal{H}^*$.

Furthermore it is possible to derive qualitative property of the maximizer.

Lemma 1 *Let F satisfy (A_0), and let $u \in \mathcal{H}$ be a maximizer for $\Phi\big|_{\mathcal{H}}$. Then, up to a change of sign, we have $u \in \mathcal{H}^*$, and u is strictly positive in B_1.*

Moreover we can prove that the extremals are solutions of an Euler-Lagrange equation under some additional regularity properties on the function F.

Theorem 3 *Suppose that $F \in C^1(\mathbb{R})$ satisfies (A_0), and suppose that $f := F'$ satisfies*

$$f(t) \le ce^{\alpha t^2} \qquad \textit{for } t \in \mathbb{R} \textit{ with constants } c, \alpha > 0.$$

Suppose furthermore that $u \in \mathcal{H}$ is a maximizer of $\Phi\big|_{\mathcal{H}}$. Then there exists $\theta > 0$ such that u satisfies the Euler-Lagrange equation in weak sense, i.e.,

$$\int_{B_1} \nabla u \nabla \phi \, dx = \theta \int_{B_1} \ln \frac{1}{|\cdot|} * \left(1_{B_1} F(u)\right) f(u) \phi \, dx$$

for all $\phi \in H_0^1(B_1)$. Moreover, $u \in W^{2,p}(B_1) \cap C^{1,\sigma}(\overline{B_1})$ for all $p < \infty$, $\sigma \in (0, 1)$.

In the recent paper [33], we focus on the case F has at most $\beta = -1$ critical growth and we turn to the question whether the supremum is attained in the case of critical growth. In some sense we tackle the open problem if there is a Log-Counterpart of Carleson-Chang paper.

We recall that a sequence $(u_n) \subset H_0^1(\Omega)$ is a normalized concentrating sequence if

(i) $\|\nabla u_n\|_2 = 1$
(ii) $u_n \rightharpoonup 0 \in H_0^1(\Omega)$
(iii) there exists $x_0 \in \Omega$ s.t. $\forall \rho > 0$, $\int_{\Omega \setminus B_\rho(x_0)} |\nabla u_n|^2 dx \to 0$.

We also remark that $(u_n) \subset H_0^1(B_1)$ is a normalized concentrating sequence, we can suppose that $x_0 = 0$ and by symmetrization $u_n = u_n^*$ is radially symmetric decreasing function.

In [33] we give some answers to the above open problem, studying the behaviour of the nonlocal interaction functional on the concentrating sequences in the spirit of the celebrated paper [35].

Remark 1 In the case of unbounded domains, the Trudinger-Moser type inequalities in [32] are also proved for the entire domain $\mathbb{R}^2$ under an additional assumption

(A_1) $F : \mathbb{R} \to [0, \infty)$ satisfies (A_0), and $F(t) = O(|t|)$ as $t \to 0$.

Remark 2 As written in [32], it is still an open problem to derive Trudinger-Moser inequalities with Log-potentials for different geometry of bounded domains in critical cases. The existence of maximizers may also depend on the domain. It is still open the uniqueness, up to sign and translation, of maximizers of $\Phi|_{\mathcal{H}}$.

Acknowledgements The authors are supported by PRIN PNRR P2022YFAJH "*Linear and Non-linear PDEs: New directions and applications*", and partially supported by INdAM-GNAMPA. The third author is supported by the INdAM-GNAMPA Project "Fenomeni di concentrazione in PDEs non locali", CUP E53C22001930001. The authors thank PNRR MUR project CN00000013 HUB—National Centre for HPC, Big Data and Quantum Computing (CUP H93C22000450007).

References

1. Adachi, S., Tanaka, K.: Trudinger type inequalities in $\mathbb{R}^N$ and their best exponents. Proc. Amer. Math. Soc. **128**(7), 2051–2057 (2000)
2. Adams, D.R.: A sharp inequality of J. Moser for higher order derivatives. Ann. Math. **128**(2), 385–398 (1988)
3. Adimurthi, A., Druet, O.: Blow-up analysis in dimension 2 and a sharp form of Moser-Trudinger inequality. Commun. Partial Differ. Eqn. **29**(1–2), 295–322 (2004)
4. Adimurthi, A., Sandeep, K.: A singular Moser-Trudinger embedding and its applications. NoDEA Nonlinear Differ. Eqn. Appl. **13**(5–6), 585–603 (2007)
5. Agueh, M., Boroushaki, S., Ghoussoub, N.: A dual Moser-Onofri inequality and its extensions to higher dimensional spheres. Ann. Fac. Sci. Toulouse Math. **26**(2), 217–233 (2017)
6. Aubin, T.: Meilleures constantes dans le théorème d'inclusion de Sobolev et un théorème de Fredholm non linéaire pour la transformation conforme de la courbure scalaire. J. Funct. Anal. **32**(2), 148–174 (1979)
7. Bartolucci, D.: Non-degeneracy, mean field equations and the onsager theory of 2D turbulence. Arch. Rational Mech. Anal. **230**, 397–426 (2018)
8. Bartolucci, D., Malchiodi, A.: Mean field equations and domains of first kind. Rev. Mat. Iberoam. **38**(4), 1067–1086 (2022)
9. Battaglia, L., Mancini, G.: Remarks on the Moser-Trudinger inequality. Adv. Nonlinear Anal. **2**(4), 389–425 (2013)
10. Battaglia, L., Van Schaftingen, J.: Groundstates of the Choquard equations with a sign-changing self-interaction potential. Z. Angew. Math. Phys. **69**(3), 1–16 (2018)
11. Beckner, W.: Sharp Sobolev inequalities on the sphere and the Moser-Trudinger inequality. Ann. Math. 138(1), 213–242 (1993)
12. Berger, M.S.: Riemannian structures of prescribed Gaussian curvature for compact 2-manifolds. J. Differ. Geom. **5**, 325–332 (1971)
13. Bonheure, D., Cingolani, S., Secchi, S.: Concentration phenomena for the Schrödinger-Poisson system in $\mathbb{R}^2$. Discrete Contin. Dyn. Syst. Ser. S **14**(5), 1631–1648 (2021)
14. Bonheure, D., Cingolani, S., Van Schaftingen, J.: The logarithmic Choquard equation: sharp asymptotics and nondegeneracy of the groundstate. J. Funct. Anal. **272**, 5255–5281 (2017)
15. Borgia, N., Cingolani, S., Mancini, G.: On the equivalence between an Onofri-type inequality by Del Pino–Dolbeault and the sharp logarithmic Moser–Trudinger inequality. Calc. Var. **64**, 80 (2025). https://doi.org/10.1007/s00526-025-02935-5
16. Caglioti, E., Lions, P.L., Marchioro, C., Pulvirenti, M.: A special class of stationary flows for two-dimensional Euler equations: a statistical mechanics description. Commun. Math. Phys. **143**, 501–525 (1992)
17. Caglioti, E., Lions, P.L., Marchioro, C., Pulvirenti, M.: A special class of stationary flows for two-dimensional Euler equations: a statistical mechanics description. Part II. Commun. Math. Phys. **174**, 229–260 (1995)
18. Calanchi, M., Ruf, B.: On Trudinger–Moser type inequalities with logarithmic weights. J. Differ. Eqn. **258**(6), 1967–1989 (2015)
19. Cannone, A., Cingolani, S.: A sharp threshold for Trudinger–Moser type inequalities with logarithmic kernels in dimension N. In: Proceedings of the Royal Society of Edinburgh: Section A Mathematics (2025). https://doi.org/10.1017/prm.2025.9
20. Cao, D.M.: Nontrivial solution of semilinear elliptic equation with critical exponent in $\mathbb{R}^2$. Commun. Partial Differ. Eqn. **17**, 407–435 (1992)
21. Carleson, L., Chang, A.: On the existence of an extremal function for an inequality of J. Moser. Bull. Sci. Math. **110**, 113–127 (1986)
22. Cassani, D., Sani, F., Tarsi, C.: Equivalent Moser type inequalities in $\mathbb{R}^2$ and the zero mass case. J. Funct. Anal. **267**, 4236–4263 (2014)
23. Chang, S.-Y.A., Yang, P.C.: Prescribing Gaussian curvature on S^2. Acta Mathematica **159**(1), 215–259 (1987)

24. Chang, S.-Y.A., Yang, P.C.: Conformal deformation of metrics on S^2. J. Differ. Geom. **27**(2), 259–296 (1988)
25. Chang, S.-Y.A., Chen, C.-C., Lin, C.-S.: Extremal functions for a mean field equation in two dimension. In: Lectures on Partial Differential Equations in Honor of Louis Nirenberg's 75th Birthday. International Press (2003)
26. Chen, W.: A trudinger inequality on surfaces with conical singularities. Proc. Am. Math. Soc. **108**(3), 821–832 (1990)
27. Chen, W., Li, C.: Prescribing gaussian curvatures on surfaces with conical singularities. J Geom Anal **1**, 359–372 (1991)
28. Chen, C.-C., Lin, C.-S.: Sharp estimates for solutions of multi-bubbles in compact Riemann surfaces. Commun. Pure Appl. Math. **55**(6), 728–771 (2002)
29. Chen, C.-C., Lin, C.-S.: Topological degree for a mean field equation on Riemann surfaces. Commun. Pure Appl. Math. **56**(12), 1667–1727 (2003)
30. Cingolani, S., Jeanjean, L.: Stationary solutions with prescribed L^2-norm for the planar Schrödinger -Poisson system. SIAM J. Math. Anal. **51**, 3533–3568 (2019)
31. Cingolani, S., Weth, T.: On the planar Schrödinger-Poisson system. Ann. Inst. H. Poincaré Anal. Non Linéaire **33**, 169–197 (2016)
32. Cingolani, S., Weth, T.: Trudinger-Moser-type inequality with logarithmic convolution potentials. J. Lond. Math. Soc. **105**(3), 1897–1935 (2022)
33. Cingolani, S., Weth, T., Yu, M.: Extremal functions for the critical Trudinger-Moser inequality with logarithmic kernels. ESAIM: Control Optimisation Calc. Var. **30** (2024). https://doi.org/10.1051/cocv/2024061
34. Csató, G., Roy, P.: Extremal functions for the singular Moser-Trudinger inequality in 2 dimensions. Calc. Var. Partial Differ. Eqn. **54**, 2341–2366 (2015)
35. de Figueiredo, D.G., do O M., Ruf, B.: On an inequality by N. Trudinger and J. Moser and related elliptic equations. Commun. Pure Appl. Math. **55**, 135–152 (2002)
36. del Pino, M., Dolbeault, J.: The Euclidean Onofri inequality in higher dimensions. Int. Math. Res. Not. IMRN **2013**(15), 3600–3611 (2013)
37. DelaTorre, A., Mancini, G.: Improved Adams-type inequalities and their extremals in dimension $2m$. Commun. Contemp. Math. **23**(05), 2050043 (2021)
38. do Ó, J.M.: N-Laplacian equations in with critical growth. Abstr. Appl. Anal. **2**, 301–315 (1997)
39. Dolbeault, J., Esteban, M.J., Jankowiak, G.: The Moser-Trudinger-Onofri Inequality. Chinese Ann. Math. Ser. B **36**(5), 777–802 (2015)
40. Dolbeault, J., Perthame, B.: Optimal critical mass in the two dimensional Keller-Segel model in $\mathbb{R}^2$. C. R. Acad. Sci. Paris, Ser. I **339**, 611–616 (2004)
41. Dong, M., Lam, N., Lu, G.: Sharp weighted Trudinger-Moser and Caffarelli-Kohn-Nirenberg inequalities and their extremal functions. Nonlinear Anal. **173**, 75–98 (2018)
42. Du, M., Weth, T.: Ground states and high energy solutions of the planar Schrödinger-Poisson system. Nonlinearity **30**, 3492–3515 (2017)
43. Fontana, L.: Sharp borderline Sobolev inequalities on compact Riemannian manifolds. Comment. Math. Helv. **68**, 415–454 (1993)
44. Flucher, M.: Extremal functions for the Trudinger-Moser inequality in 2 dimensions. Comm. Math. Helv. **67**, 471–479 (1992)
45. Ibrahim, S., Masmoudi, N., Nakanishi, K.: Trudinger-Moser inequality on the whole plane with the exact growth condition. J. Eur. Math. Soc. **17**, 819–835 (2015)
46. Ibrahim, S., Masmoudi, N., Nakanishi, K., Sani, F.: Sharp threshold nonlinearity for maximizing the Trudinger-Moser inequalities. J. Funct. Anal. **278**(1), 108302, 52 pp (2020)
47. Ishiwata, M.: Existence and nonexistence of maximizers for variational problems associated with Trudinger-Moser type inequalities in $\mathbb{R}^N$. Math. Ann. **351**(4), 781–804 (2011)
48. Iula, S., Mancini, G.: Extremal functions for singular Moser-Trudinger embeddings. Nonlinear Anal. **156**, 215–248 (2017)
49. Joyce, G., Montgomery, D.: Negative temperature states for the two-dimensional guiding-cente plasma. J. Plasma Phys. **10**, 107–121 (1973)

50. Kazdan, J.L., Warner, F.W.: Curvature Functions for Compact 2-Manifolds. Annals of Mathematics, Second Series, vol. 99, issue 1, pp. 14–47 (1974)
51. Lam, N., Lu, G.: Weighted Moser-Onofri-Beckner and Logarithmic Sobolev Inequalities. J. Geometrical Analysis, J Geom Anal. **28**, 1687–1715 (2018)
52. Li, Y.X.: Moser-Trudinger inequality on compact Riemannian manifolds of dimension two. J. Partial Differ. Eqn. **14**(2), 163–192 (2001)
53. Li, Y.X.: Extremal functions for the Moser-Trudinger inequalities on compact Riemannian manifolds. Sci. China Ser. A **48**(5), 618–648 (2005)
54. Li, J., Lu, G.: Critical and subcritical Trudinger-Moser inequalities on complete noncompact Riemannian manifolds. Adv. Math. **389**(8), 107915 (2021)
55. Li, Y., Ruf, B.: A sharp Trudinger-Moser type inequality for unbounded domains in $\mathbb{R}^N$. Indiana Univ. Math. J. **57**(1), 451–480 (2008)
56. Li, X., Yang, Y.: Extremal functions for singular Trudinger-Moser inequalities in the entire Euclidean space J. Differ. Eqn. **264**(8), 4901–4943 (2018)
57. Lions, P.L.: The concentration-compactness principle in the calculus of variations. The limit case, Part 1. Riv. Mat. Iberoamericana **1**, 145–201 (1985)
58. Lu, G., Yang, Y.: Adams' inequalities for bi-Laplacian and extremal functions in dimension four. Adv. Math. **220**, 1135–1170 (2009)
59. Malchiodi, A.: Topological methods for an elliptic equation with exponential nonlinearities. Discrete Contin. Dyn. Syst. **21**(1), 277–294 (2008)
60. Malchiodi, A., Ruiz, D.: New improved Moser-Trudinger inequalities and singular Liouville equations on compact surfaces. Geom. Funct. Anal. **21**, 1196–1217 (2011)
61. Mancini, G., Martinazzi, L.: The Moser-Trudinger inequality and its extremals on a disk via energy estimates. Calc. Var. **56**(4), Paper No. 94, 26 pp (2017)
62. Mancini, G., Martinazzi, L.: Extremals for fractional Moser-Trudinger inequalities in dimension 1 via harmonic extensions and commutator estimates. Adv. Nonlinear Stud. **20**(3), 599–632 (2020)
63. Mancini, G., Sandeep, K.: Moser-Trudinger inequality on conformal discs. Commun. Contemp. Math. **12**(6), 1055–1068 (2010)
64. Mancini, G., Thizy, P.-D.: Non-existence of extremals for the Adimurthi-Druet inequality. J. Diff. Eqn. **266**(2–3), 1051–1072 (2019)
65. Martinazzi, L.: Fractional Adams-Moser-Trudinger type inequalities. Nonlinear Anal. **127**, 263–278 (2015)
66. Masaki, S.: Local existence and WKB approximation of solutions to Schrödinger-Poisson system in the two-dimensional whole space. Commun. Partial Differ. Eqn. **35**(12), 2253–2278 (2010)
67. Masaki, S.: Energy solution to a Schrödinger-Poisson system in the two-dimensional whole space. SIAM J. Math. Anal. **43**(6), 2719–2731 (2011)
68. Moser, J.: A sharp form of an inequality by N. Trudinger. Indiana Univ. Math. J. **20**(11), 1077–1092 (1971)
69. Moser, J.: On a nonlinear problem in differential geometry. In: Dynamical Systems, pp. 273–280. Academic Press (1973)
70. Onofri, E.: On the positivity of the effective action in a theory of random surfaces. Comm. Math. Phys. **86**(3), 321–326 (1982)
71. Osgood, B., Phillips, R., Sarnak, P.: Extremals of determinants of Laplacians. J. Funct. Anal. **80**(1), 148–211 (1988)
72. Osgood, B., Phillips, R., Sarnak, P.: Compact isospectral sets of surfaces. J. Funct. Anal. **80**(1), 212–234 (1988)
73. Pohozaev, S.I.: On the eigenfunctions of the equation $\Delta u + \lambda f(u) = 0$. Dokl. Akad. Nauk SSSR **165**, 36–39 (1965) (in Russian)
74. Polyakov, A.M.: Quantum geometry of bosonic strings. Phys. Lett. B **103**(3), 207–210 (1981)
75. Polyakov, A.M.: Quantum geometry of fermionic strings. Phys. Lett. B **103**(3), 211–213 (1981)
76. Pruss, A.R.: Nonexistence of maxima for perturbations of some inequalities with critical growth. Can. Math. Bull. **39**, 227–237 (1996)

77. Ruf, B.: A sharp Trudinger-Moser type inequality for unbounded domains in $\mathbb{R}^2$. J. Funct. Anal. **219**(2), 340–367 (2005)
78. Struwe, M.: Critical points of embeddings of $H_0^{1,n}$ into Orlicz spaces. Ann. Inst. H. Poincaré Anal. Non Lineaire **5**(5), 425–464 (1988)
79. Struwe, M.: A flow approach to Nirenberg's problem. Duke Math. J. **128**(1), 19–64 (2005)
80. Stubbe, J.: Bound states of two-dimensional Schrödinger-Newton equations (2008). ArXiv arXiv:0807.4059v1
81. Stubbe, J., Vuffray, M.: Bound states of the Schrödinger-Newton model in low dimensions. Nonlinear Anal. **73**(10), 3171–3178 (2010)
82. Suzuki, T.: Free energy and self-interacting particles. In: Progress in Nonlinear Differential Equations and their Applications, vol. 62. Birkhäuser, Boston (2005)
83. Thizy, P.D.: When does a perturbed Moser-Trudinger inequality admit an extremal? Anal. PDE. **13**(5), 1371–1415 (2020)
84. Tintarev, C.: Trudinger-Moser inequality with remainder terms. J. Funct. Anal. **266**(1), 55–66 (2014)
85. Troyanov, M.: Prescribing curvature on compact surfaces with conical singularities. Trans. Amer. Math. Soc. **324**(2), 793–821 (1991)
86. Trudinger, N.S.: On imbeddings into Orlicz spaces and some applications. J. Math. Mech. **17**(5), 473–483 (1967)
87. Wolansky, G.: On steady distributions of self-attracting clusters under friction and fluctuations. Arch. Rational Mech. Anal. **119**(4), 355–391 (1992)
88. Yang, Y.: Extremal functions for Trudinger-Moser inequalities of Adimurthi-Druet type in dimension two. J. Differ. Eqn. **258**(9), 3161–3193 (2015)
89. Yang, Y., Zhu, X.: Blow-up analysis concerning singular Trudinger-Moser inequalities in dimension two. J. Funct. Anal. **272**(8), 3347–3374 (2017)
90. Yudovich, V.I.: Some estimates connected with integral operators and with solutions of elliptic equations. Dokl. Akad. Nauk SSSR **138**(4), 805–808 (1961). (in Russian)

Singularity: A Seventh Memo?

Matteo Gallone and Sandra Lucente

Abstract In this paper we explore the relationships between Calvino's memos and Mathematics. In the first part, we discuss how Lightness, Quickness, Exactitude, Visibility, Multiplicity are present in the mathematical language, reasoning and in the work of the mathematician. In addition, we follow a similar path for the topics of Calvino's lecture of which we only know the title or some notes. In the final part, we explain why 'Singularity' could be chosen as topic for a possible Calvino's seventh lecture.

Keywords Italo Calvino · Mathematical language · Six memos

1 Introduction

In 1985, Italo Calvino was to have held a series of lectures at Harvard entitled *Six Memos for the Next Millennium*, but a fatal stroke left only five transparent folders in a hard folder on his desk. Each folder contained the text of a lecture with the following titles: *Lightness*, *Quickness*, *Exactitude*, *Visibility*, and *Multiplicity*. In the Introduction of the Italian version [4], we find that from Calvino's point of view the memos he discusses are "certain values, qualities, or peculiarities of Literature that are very close to my heart." But are these values only literary? What would be the qualities of science to be developed in this era? In particular, can we seek specific characteristics of those who research in Mathematics nowadays?

In the Preface that the author's wife writes to accompany the posthumously published Italian text titled *American Lessons*, Esther Judith Singer Calvino tells us about the period of preparation for the lectures: "They became an obsession, and one

M. Gallone (✉)
Scuola Internazionale Superiore di Studi Avanzati (SISSA), Trieste, Italy
e-mail: matteo.gallone@sissa.it; mgallone@sissa.it

S. Lucente
Dipartimento Interateneo di Fisica Università degli Studi di Bari Aldo Moro, Bari, Italy
e-mail: sandra.lucente@uniba.it

B. Cassano et al. (eds.), *Singularities, Asymptotics, and Limiting Models*, Springer INdAM Series 64, https://doi.org/10.1007/978-981-96-3584-9_10

day he announced to me that he had ideas and material for eight lectures. I know the title for what might have been an eighth lecture: *Sul cominciare e sul finire* (On the beginning and the ending [of novels])." Notes are found for this lecture. It is also known that the sixth lecture would have been titled *Consistency*. However, there is no reference to the seventh lecture.

As part of the *Puglia Summer Trimester 2023*, the authors of this article held a lecture on the "American Lessons" and, recognizing in the sub-title of the trimester "Singularities, Asymptotics, and Limiting Models" three mathematical memos, they sought others in Calvino's text.

This article is divided into three parts. Sections 2–6 explore the relationships between Mathematics and the five memos chosen by Calvino: Lightness, Quickness, Exactitude, Visibility, Multiplicity. Sections 7–8 try to follow a similar path for the lessons of which we only know the titles or some notes. In Sect. 9, we allow ourselves to play with the absence of the seventh lesson by choosing for it one of the memos from the conference: Singularity.

This attempt to apply to science the values of the American Lessons is not a new idea. The most successful mirror between the methods of Literature proposed by Calvino and that of Mathematics is the book by Lolli [13] where a different vision of our Sects. 2–6 can be found. Regarding Sect. 9, i.e., the creation of a seventh scientific lesson in Calvino's style, in 2010 Nicola Cabibbo held a lecture at the Accademia dei Lincei titled "Science as the search for clarity" and indeed he imagines a seventh Calvino-style lesson on "Clarity" (see [3]). In this article, therefore, we do not want to perform any mathematical exegesis of the American Lessons, we only take up Calvino's invitation to be protagonist readers of the text with our characteristic of being mathematicians. This invitation in "If on a winter's night a traveler" [6] becomes imperative, the reader becomes Reader, here the mathematical reader becomes Author of a paper. Perhaps we will remain on the surface of Mathematics and Literature, but as Palomar says

> Only after knowing the surface of things can one venture to seek what is beneath. But the surface of things is inexhaustible. [7]

2 The Lightness of Mathematical Models

Calvino chooses *Lightness* as the first of the "Memos" and we do the same in this first paragraph, showing that is central to Mathematics. This value is an essential characteristic that leads to the birth of all mathematical entities. In other words, we would like to convince the reader that the objects with which Mathematics works are *light* in Calvino's sense. In the cases where Mathematics is the language of an applied science, such as Physics, there is an unavoidable lightening during the process of mathematization of reality through the construction of a model. As an example, let us look at the problem of the motion of celestial bodies, ingeniously solved by Newton with the law of universal gravitation. Today we know that the simplest way to describe the motion of the planets in the Solar System is in a reference frame where the Sun is fixed and the planets orbit around it. But let us pretend we don't know Newton's answer and put ourselves in the shoes of a supposed astronomer from the past aware of the fact that the Earth is a planet. If we want to describe the motion of the Earth around the Sun, we would probably have to start with the description of the Earth itself. We would then realize that the Earth is extremely difficult to describe: there are oceans and continents; on the continents, there are flat areas, mountains, rivers, and lakes. The flat areas can then be green, arid, or icy, and so on. In the attempt to construct the most accurate model of Earth, we would end up getting lost in the wonderful details of the planet we live in, losing sight of the goal we had set at the beginning of our investigation. In a somewhat colloquial jargon, we might comment by saying that this model, with all this wealth of details, is *heavy* to handle. This is where the first of the quotations from Calvino's *Memos* comes into play:

> My working method has more often than not involved the subtraction of weight. I have tried to remove weight, sometimes from people, sometimes from heavenly bodies, sometimes from cities; above all I have tried to remove weight from the structure of stories and from language. [5]

The path suggested by reading Calvino is to remove weight, that is, to deprive the model of superfluous details. But what are the superfluous details? The question is legitimate and it is impossible to give a general answer. To give an idea of what guidelines to follow, Calvino quotes of the myth of Perseus and Medusa.

Briefly, Medusa is a Gorgon, that is, a woman with a terrifying appearance with snakes instead of hair who can petrify instantly anyone who crosses her gaze. The myth tells that a young hero named Perseus had to bring as a bridal gift the head of Medusa. To kill the Gorgon, he uses a very ingenious strategy and, instead of confronting her openly, prefers to use a highly polished shield to observe the reflected image of the Gorgon. After beheading Medusa, Perseus collects her head and carries it in a bag with him, making it a very powerful weapon to be used only "with enemies who deserve to become the statue of themselves."[1] It is interesting to note that Calvino represents in the fight between Perseus and Medusa a clash between the *lightness* (Perseus arrives flying on Pegasus, a winged horse) and the *heavyness* (Medusa petrifies with her gaze). The clash is won by the Lightness embodied by Perseus, who carries with him a sign of heaviness, represented by the decapitated head of the Gorgon which gives him the ability to petrify, that is, to make heavy. To quote Calvino:

> Perseus succeeds in mastering that horrendous face by keeping it hidden, just as in the first place he vanquished it by viewing it in a mirror. Perseus's strength always lies in a refusal to look directly, but not in a refusal of the reality in which he is fated to live; he carries the reality with him and accepts it as his particular burden. [5]

This description fits perfectly when looking at the work of lightening a model. In fact, every detail that is decided to be neglected to lighten the model, could actually

---

[1] Here, we prefer to translate the Italian version. In the English version it reads: "against those who deserve the punishment of being turned into statues".

be an essential one. Out of metaphor, what is said in this passage is that only by carrying with us in reasoning all the details that we have neglected we can perform the operation of lightening in a correct way. For this reason, Calvino quotes Ovid to emphasize with how much care these superfluous details should be treated:

> So that the rough sand should not harm the snake-haired head he makes the ground soft with a bed of leaves, and on top of that he strews little branches of plants born under water, and on this he places Medusa's head, face down. [5]

Let's return to the operation of simplification in the case of analyzing the Earth's motion around the Sun. First, we eliminate all elevations, all seas, and all land masses, turning our planet into a sphere. Then, we reduce the size of this sphere until it becomes a point. Reducing the Earth to a dimensionless mathematical object, such as a point, is sufficient to fully understand its orbit around the Sun. Surprisingly, this simplification, at first glance, seems to be a "loss of information" because of the many neglected details. In fact, not only can the Earth be idealized as a point, but every planet in the Solar System can be idealized as a point. Therefore, in this context, in the simplification where it seemed we had lost information, we actually gain it. We gain that the laws we write for the motion of the Earth are the same that we would write for the motion of any other planet.

A similar discussion can be had for many objects in Mathematics. Abstract geometric entities capture the essence of a vast and diverse set of objects and situations from our daily lives. Circles, triangles, lines, as well as functions and more abstract objects like Banach spaces or differential manifolds arise from the "simplification" of reality in the sense of Calvino. After the simplification process, each such object becomes more versatile and general than the original ones. The relationships established between them form an intricate web of Lemmas, Propositions, and Theorems that enter into Calvino's "thoughtful" Lightness:

> There is such a thing as a lightness of thoughtfulness, just as we all know that there is a lightness of frivolity. In fact, thoughtful lightness can make frivolity seem dull and heavy. [5]

3 Geodetic and Zigzagging Quickness

It is surprising to think that Calvino chooses *Quickness* as a value for Literature of the new millennium, and even more surprising is the idea of rapid Mathematics. What is too rapid today is the time dedicated to reading texts and even scientific articles. In the last ten years, 2.5 million scientific articles have been produced; to read them at a rate of one line per minute without interruptions would take the same amount of time that *Homo Sapiens* has lived on this planet. We do not read; we scroll. And here Calvino helps us understand that sometimes this is a value.

> "Discoursing," or "discourse," for Galileo means reasoning, and very often deductive reasoning. "Discoursing is like coursing": this statement could be Galileo's declaration of faith—style as acmethod of thought and as literary taste. For him, good thinking means Quickness, agility increasoning, economy in argument, but also the use of imaginative examples. [5]

For Calvino, Galileo is the greatest writer of the Italian Literature of every century. He wrote this clearly in the *Corriere della Sera* on December 24th, 1967, sparking not a few controversies (see [10]). The greatest writer of the Italian Literature prefers to write his treatises as dialogues, discourses. And to discourse is to run. Galileo is also the man who introduces the experimental method and the scientist of *objects* like the telescope or the spheres he is said to have dropped from the Tower of Pisa. This lesson from Calvino is full of sought objects (Charlemagne's ring), proposed objects (the horse in a novella by Boccaccio), prepared objects (the drawing of Chuang-Tzu) and even non-existent objects (the book by Borges). The object aids the mental speed that Calvino assures it is a value. Even in this lesson, he emphasizes that

> Each value or virtue I chose as the subject for my lectures does not exclude its opposite. Implicit in my tribute to lightness was my respect for weight, and so this apologia for quickness does not presume to deny the pleasures of lingering. Literature has worked out various techniques for slowing down the course of time. I have already mentioned repetition, and now I will say a word about digression. [5]

In this phrase appears the crucial element for speed: time. Confusing speed and velocity, it appears obvious that time is crucial being velocity a function of time, derivative of space with respect to time. But speed and velocity are not the same thing in this lesson. The desire is for the speed of mental calculation, not the haste of the result. Even though the objects move quickly in the mathematician's mind, they remain static when the mathematician is not looking at them. Time applied to mathematical objects generates variations in the continuous as well as successions in the discrete. Calvino uses the words "continuity" but also "discontinuity" of time in this lesson. He talks about the succession of events like the rhymes of a poem. Let's revisit the objects of the Calvinian lesson in this temporal and mathematical key.

The ring sought by Charlemagne is magical, in the legend it connects all the characters searching for it. The mathematical object most similar to this ring is a vertex of a graph. Calvino clearly writes that the scheme of a story is a graph, well knowing the combinatorial topics that his friends from Oulipo are passionate about.

> The events, however long they last, become puncti-form, connected by rectilinear segments, in a zigzag pattern that suggests incessant motion. [5]

This zigzag alludes to a random walk that requires a pace similar to that of the horse in Boccaccio's novella, from which Madonna Oretta wants to dismount because there is speed without rhythm.

> In an age when other fantastically speedy, widespread media are triumphing, and running the risk of flattening all communication onto a single, homogeneous surface, the function of literature is communication between things that are different simply because they are different, not blunting but even sharpening the differences between them, following the true bent of written language. [5]

Science too has always had this function: to highlight differences. For example, the notion of curvature can distinguish among different geometric worlds. It is precisely in this lesson that Calvino talks about geodesics, about shorter paths, about straight lines in a labyrinth but also about routes that connect distant points in space and time.

He must have been influenced by the ideas of the theory of relativity, a noun that Calvino precisely uses in this lesson.

Borges pretends that his book has already been written, constructs imaginary libraries that open to infinity. Here is the true speed, the swift thought with which we imagine infinity compared to the effort of counting.

On the other hand, the design of a crab made in 10 years with perfection in a single gesture can correspond in Mathematics to the distance between a question about infinity and the corresponding elegant answer. For example, Galileo lets his characters Salvati and Simplicio discuss the correspondence between whole numbers and square numbers. In a single gesture, two hundred years later, Cantor responds to the paradox with a paradise, that of transfinite numbers.

The squares and roots are in Salvati's phrase "just as the products are called squares, the factors, that is, those that are multiplied, are called sides or roots; the other [numbers], which do not arise from numbers multiplied in themselves, are not squares otherwise. Hence, if I say, all numbers, including squares and non-squares, are more than just squares alone, I say something true: isn't it?" [9]. The square and the root are taken up in Calvino's description of Borges' writing "With Borges, a literature raised to the square and at the same time a literature as the square root extraction of itself is born." [5]

Ideas quickly draw a gigantic graph between disciplines. The random walk in knowledge seems confused, but it is rapid, like this paragraph, like the third American lecture.

4 Euclidean Exactitude

In common language, precision and Mathematics are closely intertwined, often exemplified by the expression: "mathematical" as a synonym for "exact". In the Treccani's Italian dictionary, the definition of "mathematical" as an adjective is as follows: "having the character of objective, clear, and consequent precision typical of mathematical calculation" [8]. This definition implicitly identifies Mathematics with "mathematical calculation".

Anyone who studies Mathematics even briefly experiences a division between Mathematics and calculation. For instance, in Italy during the 1990s, a popular booklet of a hundred pages by Giuliano Spirito titled "Matematica senza numeri" ("Mathematics without numbers") [20] was widely read, even by high school students who struggled to relate the core subjects of the book (sets, logic, relations) with the concept of Mathematics learned in primary school. The takeaway message was that Mathematics encompassed more than just calculations, numbers, and polygons. Those who study Mathematics at a university level cannot simply equate the discipline with the science of calculations.

In reading Italo Calvino's third lecture titled "Exactitude", we find not only this dichotomy but also a phrase that could serve as an alternative definition of Mathematics:

> My search for exactitude was branching out in two directions: on the one side, the reduction of secondary events to abstract patterns according to which one can carry out operations and demonstrate theorems; and on the other, the effort made by words to present the tangible aspect of things as precisely as possible.

The point is that if Mathematics is defined as the language of abstract schemes with which operations can be performed and theorems proved, then it is necessary to start from the definition of the entities themselves: its language must define every object of the discipline without ambiguity.

This is not a common feature in the language we usually use to communicate, and indeed, nuances are a powerful tool that poets use in their compositions. This type of ambiguity is the first enemy that a mathematician, in the work, must defeat so that a mathematical entity is the same for all who inhabit a certain abstract universe. For example, in Euclidean geometry, for everyone, a square is a polygon with four equal sides and four equal angles. The omission of any detail would lead to the construction of another figure: omitting only that the polygon has four equal angles leads to the construction of a rhombus, omitting only that the lengths of the sides are the same leads to the construction of a rectangle, and so on. Every word is essential to define the square exactly. At that point, we will need an example of a square, which we will find in an image, use in a building, or in a constructed object. The Lightness of the abstract entity, combined with the Exactitude of its definition, avoids emptiness. There is no opposition between Lightness and Exactitude.

We can conclude that in 1963, Italo Calvino turned to the use of science in the novel, for escaping a fear he reveals in this chapter of the Lectures:

> It sometimes seems to me that a pestilence has struck the human race in its most distinctive faculty—that is, the use of words. It is a plague afflicting language, revealing itself as a loss of cognition and immediacy, an automatism that tends to level out all expression into the most generic, anonymous, and abstract formulas, to dilute meanings, to blunt the edge of expressiveness, extinguishing the spark that shoots out from the collision of words and new circumstances. [5]

Beyond abstract schemes, Calvino talks about theorem proofs. First, a theorem must be stated. The statement of a theorem requires the same precision used in definitions. But a theorem has something more than a definition. A theorem is a statement in which, given some premises called hypotheses, some consequences called theses are asserted.

Consider Proposition 32 from Book I of Euclid's Elements: in his axiomatic system, the sum of the internal angles of a triangle equals a straight angle. A theorem is accompanied by its proof, which necessarily leads from the hypotheses to the theses. The difference between a philosophical discussion and the proof of a theorem is that in the theorem proof, every single step must be justified precisely. The order of the steps is crucial. In Fig. 1 (taken from [15]), there is a directed graph with vertices representing all the propositions of Book I of the Elements up to the 32nd (numbered with P), the common notions (CN), and the five postulates (Pos). The edges of the graph indicate which result is necessary to deduce a certain proposition. We understand that the Fifth Postulate is necessary from Proposition 29 onwards,

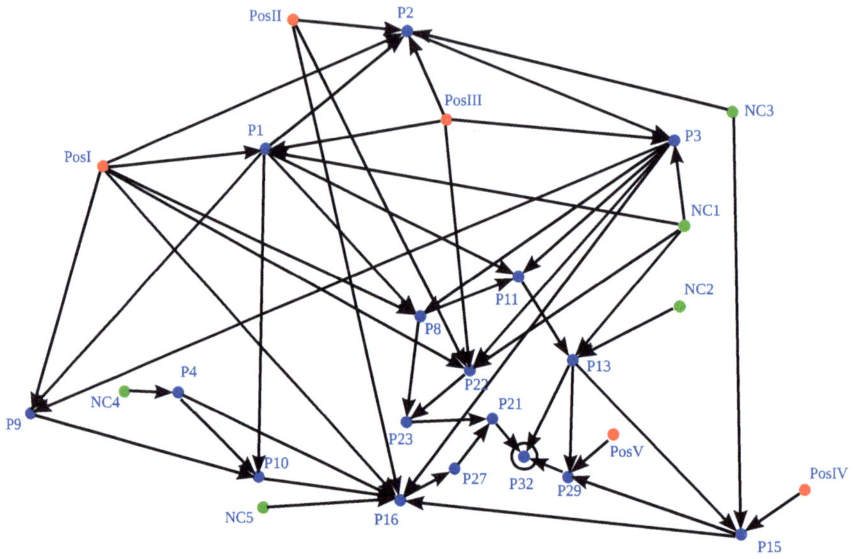

Fig. 1 All Propositions (P), Common Notions (CN) and Postulates (Pos) used in Euclid's elements to prove Proposition 32. An arrow going from a point A to a point B indicates that A is used to prove B

particularly to prove the above-mentioned Proposition 32. Moreover, the theorem on the sum of the internal angles of a triangle could not have been stated for example before Proposition 13, which is used in its proof. The deduction of a result from the previous ones is therefore similar to the process Calvino describes in this lecture on writing: the detail of the detail.

There are, however, several possible proofs for the same theorem: some more elegant and concise, others more cumbersome and convoluted.

Let's go back to to the non-contradictory relationship between Exactitude and Lightness. For Calvino, Exactitude is figuratively represented by the feather which in Egyptian tradition serves as a counterweight for the souls of the deceased reaching the afterlife. It's interesting the analogy of this point with the proof of a theorem which, in everyone's dreams, should be as beautiful and simple as a feather.

There's an interesting story about this. According to Paul Erdös, the most prolific mathematician of all time, God wrote all the most elegant mathematical proofs in a book, to which he referred as "the Book". To convey how important this pursuit of perfection, exactitude, lightness, and elegance was to him, let's quote what he said at a conference in 1985: "you may not believe in God, but you must believe in the Book". An attempt at "the Book" has been made by Aigner and Ziegler (see [1]), but every mathematician must choose the proofs they would put in the book and ask themselves if their proof has the elegance to aspire some pages.

5 Quantum Visibility

We are inhabitants of the third millennium, we regard the American Lectures also with the curiosity to understand if the values Calvino chose have really been preserved. Coming to *Visibility*, we can only assert, meanwhile we receive an Instagram notification, that visibility is the heart of our time, but here Calvino does not talk about shots, about selecting among them, he talks about how the images of the mind become visible.

> I have in mind some possible pedagogy of the imagination that would accustom us to control our own inner vision without suffocating it or letting it fall, on the other hand, into confused, ephemeral daydreams, but would enable the images to crystallize into a well-defined, memorable, and self-sufficient form, the *icastic* form. [5]

Icastic means with effectiveness, with immediacy, with concreteness. An icastic description is not only to remember, it becomes the concept itself. Crystallizing an idea is a process that modern science had somehow rejected, but rereading Calvino's sentence we realize that every part of twentieth-century Physics has proposed itself through the crystallization of images. This is not a paradox, there are many crystals and put together they enhance their reverberation.

Icastic images are Newton's apple, Einstein's twins, Schrödinger's cat, Hilbert's beer mugs. These images were born with the theory, they are not just a didactic expedient for its explanation. Calvino wonders how images are born. We can ask ourselves how an entire theory is born. Instead of writing a treatise on neuroscience, to explain this *memo*, he talks about fantasy:

- Fantasy is a place where it rains inside [5]
- From where do the images rain into fantasy? [5]
- Fantasy is a kind of electronic machine [5]

These three sentences serve to argue on whether imagination is or is not a tool of knowledge. In this work, we lean towards yes and make the example of Quantum Mechanics. This theory is never mentioned in the lessons, but Darconza and Bischi in [2] highlight the sentence found in the lesson on *Multiplicity* "Even before science had officially recognized the principle that observation somehow intervenes to modify the observed phenomenon", Gadda knew that "to know is to insert something into the real: it is, therefore, to deform". In Sect. 9 we will return to Gadda's influence on Calvino, here we are interested in highlighting that Calvino hints at the *Heisenberg Uncertainty Principle*:

$$\Delta x \Delta p \geq \frac{\hbar}{2} \tag{1}$$

This principle has generated many images later used in science fiction stories: microscopic tunnels, collapsing particles, entanglements to be tattooed, up to teleportation. In these narratives, the conjugate variables are increasingly different: not only position-velocity, but for example rigor and clarity in popularization of science (see [11]) have correlated measures.

The uncertainty principle has entered the popular culture of this millennium precisely thanks to its Visibility despite initially appearing fanciful. Let's return to Calvino's lesson: from where have quantum ideas rained down? The very word "quantum" applied by Einstein to light is icastic. If there is a rain of images, then there must be, like for water, a cycle of ideas: they are the ideas of the scientists who preceded that image. Einstein's quantum would not have been there without Planck and Maxwell. Speaking of Maxwell, the formalization of the concept of field is also icastic. We don't need to explain the similarity between vectors and grass, it's immediate.

The icastic image of Schrödinger's cat to describe quantum entanglement is a mental experiment, so if fantasy rains inside literature to generate stories, it also rains in science to generate questions before explanations. For example, the cat led us to reflect on the macroscopic–microscopic transition. Schrödinger's cat is not a metaphor, but an analogy, and if Calvino in this lesson talks precisely about "fields of analogies", Lolli [14] reminds us of the Langlands program to first connect Galois Groups, Algebraic Number Theory and then Harmonic Analysis up to Logic and Quantum Physics. Here again, the key concept is that of field, but we are interested in the attitude of those who are still today looking for a sort of unified theory even in Mathematics. The scientist is not so much the magician who activates the rain, but the scholar who is able to inhabit places where it rains and collect the rain: he inhabits laboratories without a roof of imposed choices, he opens windows on doubts and begins to observe the effect of the wind on the drops, he turns umbrellas upside down to collect the ideas of his predecessors. In a non-metaphorical sense, the imaginative scientist is attentive.

If for Calvino the *imagination is the place of potential and hypothetical*, for a theoretical physicist hypotheses invent potentially real places. The effort in building a quantum computer today is not only linked to speed, efficiency, new applications, but because it must be a new electronic machine, overturning what Calvino said, there must be a kind of fantasy that actuates it.

6 Algebraic Multiplicity and More

Multiplicity, in Mathematics, is an omnipresent concept, and is the second question one tries to answer when faced with a problem. The first question is the problem of existence which at first glance seems to be a naive question but is often anything but trivial. For example, an expert of differential equations can not avoid to ask whether the equation he uses to model reality has a solution or not. Indeed, a negative answer to this question implies that the model is completely useless from the point of view of modelization. But even a simple second-degree equation in the field of reals may or may not have solutions. But, assuming it does, how many are there? Practicing with decompositions and Delta, every high school student distinguishes algebraic equations with one or more solutions. Many also know different types of multiplicity (algebraic, geometric). The fundamental theorem of algebra due to Gauss in the complex field gives the number of these solutions.

For models that translate into differential equations, the same questions of existence and multiplicity depend on the meaning given to the word "solution". Therefore, the answer to the multiplicity of solutions depends on the spaces in which solutions are sought. To illustrate how these problems are of interest, it is useful to remember that one of the Millennium Prize Problems concerns the existence of solutions for equations describing fluid dynamics: the Navier-Stokes equations. In order to stay closer to the Calvinian text, we prefer to return to the example of algebraic equations starting from the following quotation.

> To show that in our own times literature is attempting to realize this ancient desire to represent the multiplicity of relationships, both in effect and in potentiality. [5]

Calvino's lesson on *Multiplicity*, therefore, focuses on *relationships*; we want to identify the world of relationships between various parts of Mathematics that determine the way in which mathematical knowledge progresses. In fact, Calvinian memos have a non-empty intersection: in the paragraph of *Exactitude* Calvino reports the following quotation from "The Man Without Qualities" by Musil.

> He mentions the fact that mathematical problems do not admit of a general solution, but that particular solutions, taken all together, can lead to a general solution. [5]

In Mathematics, there are very general results that are true milestones used daily to discover new things. In the romantic image of the mathematician (of the scientist in general), results come from brilliant and instantaneous intuitions; in the reconstruction that can be made of the history of Mathematics, this is not how discoveries are made. A result dealing with general problems is typically produced after years and years of work carried out independently by many people who, by tackling specific problems with very different techniques, go on to build bricks of knowledge. With many bricks available, it is possible to understand the possible relationships between these elements, that is, to arrange all the bricks coherently. Thus, the formulation of a general result is born.

An example of this process is given by the history of solutions to third-degree equations. We are between the 15th and 16th centuries, and at that time, it was usual for mathematicians to compete in duels by solving mathematical problems. The consequences of a defeat could be dramatic, such as losing a job or going bankrupt; the consequences of a victory could be fame and glory until the next duel. It was therefore particularly important to find methods to solve a large number of problems in a general sense and to play this method time after time without revealing it. Initially, only third-degree equations of a very specific form were known to be solvable, and the first more general result is due to Scipione del Ferro, who managed to find a solution formula for third-degree equations missing the quadratic term. Later, another mathematician, Niccolò Fontana (nicknamed "Tartaglia" for his difficulty in speaking), discovered a way to solve all third-degree equations missing the linear term. It was then Gerolamo Cardano who discovered how to find the general solution of third-degree equations, showing that a cubic equation can always be reduced to an equation missing the quadratic term and thus, quoting Musil, he succeeded by combining individual solutions in a way to obtain the general solution.

In modern Mathematics, the challenge is made by many more competitors on different themes. The lesson on *Multiplicity* begins with a long quotation from Gadda and with the challenge posed by the encyclopedic novel. In the mathematical field, the encyclopedia of twentieth-century results lives in huge databases (ArXiv, ISI, Scopus, MathSciNet, Zentralblatt). The dissemination of results is itself a challenge. Carlo Emilio Gadda's quotation is used by Calvino to talk about a "system of systems" in literature. The mathematical counterpart is the intertwining of various disciplinary interests and the keywords of different scientific articles. According to Calvino, a second aspect of Multiplicity is related to what, according to Calvino, unites writers like Musil and Gadda, namely:

> the inability to conclude. [5]

Imagine a situation where a group of mathematicians works for a certain amount of time on a problem until they find new, perhaps even non-trivial aspects. At that moment, the discussion begins on whether it is appropriate to spend part of the time to properly write down the result obtained and publish it or whether to focus all efforts on finding a more general result. On one hand, the advantage of having only sufficiently general results would allow a streamlining of the mathematical literature. On the other hand, the publication of minor results allows for easier understanding by the public and also a greater dissemination of what Musil calls individual solutions.

This discussion hides, not too subtly, the fact that the definition of the conclusion of a scientific work is not unanimously shared. Learning to conclude a creative work is a complicated act so much so that "Starting and finishing" was among Calvino's memos (see Sect. 8).

Is it possible to say that a result is concluded? Probably this a Multiplicity of scientific relationships in which the newer one does not prevail over the older one but, using a Newtonian image, allows us to see further by "standing on the shoulders of giants".

7 The Missing Lesson: Consistency

No one can say what meaning Calvino would have given to the lesson on *Consistency*; conjectures are made in [19]. In [13] there is the autograph index with the word "consistency" written by Calvino with pencil compared to the previous five memos written with ink. Besides the syntactic aspect (Calvino would almost certainly have gone over the word with ink after writing the lecture), there is a semantic difference between consistency and the other memos: this strongest demand for a novel or a theorem had in the early twentieth century suffered a double attack. Well known to everyone is the literary Decadentism that has its scientific counterpart in the studies of Kurt Gödel.

Consistency is a key word in logic: the property of a system of axioms for which it is not possible to deduce a contradiction from it.

Would a consistent literature be valuable in the sense that it gives us unambiguous outcomes of truth or falsehood? Mathematical consistency, on the other hand, refers to axioms and therefore, anyone wanting to answer such a question should consider the problem of "axiomatic" literature. Tracing back to literary, the axiomatic method would have been daring much more than doing combinatorial literature that Calvino had learned in Oulipo. In combinatorial literature, all the ingredients are already there and need to be mixed. In the axiomatic context, instead, one cannot predict a theorem just by looking at the axioms. *Consistency* is linked not only to the idea of truth but also to that of *proof*. Mathematicians use the terms *consistency* and *logical coherence* interchangeably. But in literary analyses, coherence is more related to the *connection of the parts* of a whole, while consistency is related to the idea of *solidity*. A mathematical proof is always solid and relative to what has been learned up to that moment, therefore what is consistent is always coherent. (see again Fig. 1 as example.)

To reinforce the hypothesis that Calvino would have consulted the logical-mathematical results, we recall the quote from Hofstadter present in the lecture on Visibility. The book *Gödel, Escher, Bach* by Douglas R. Hofstadter [12] had been published in Italy only the year before the writing of the Lectures, but it had struck Calvino, who was investigating imaginative processes. From this quote, we deduce that Calvino was familiar with the incompleteness theorems.

The word consistent appears almost as a constraint in Gödel's First Incompleteness Theorem. Let's rewrite it in a simplified manner:

> Every consistent formalization of Mathematics that is sufficiently powerful to be able to axiomatize the elementary theory of natural numbers is capable of constructing a syntactically correct proposition that cannot be proved or refuted within the same system.

Incompleteness, therefore, is the possibility that paradoxes arise. Let us recall that Calvino cites undecidability in the lesson on Visibility. The paradoxical is of great interest to the literate. The Gödel Theorem itself is based on the arithmetic transposition of the paradox "this statement is not true".

In the Preface to the Garzanti Italian edition of the American Lectures, Esther Calvino informs us that in the lecture on consistency the author would have consulted Herman Melville, in particular the work *Bartleby the Scrivener: A Story of Wall Street* [17]. The character of this story is itself paradoxical: he refuses to perform tasks using the phrase "I would prefer not to". Denial is certainly not a value for the third millennium.

Mathematics overcame the impasse generated by Gödel's Theorem similarly to how literature has overcome negatives for not stop working (which happens in *Bartleby*). The second incompleteness theorem, also by Gödel, asserts that "*Considering a mathematical theory sufficiently expressive to contain arithmetic; if it is consistent, then it is not possible to prove the consistency of the theory within the theory.*"

This too would have been a value for Calvino: distinguishing literature from the argument of literature. Perhaps this is why the novel is today extremely abundant as are the mathematical results: in the majority of cases the novel doesn't talk about itself, mathematicians don't do metatheory. But there is also an incompleteness of the narrated compared to reality. Let's reread the incipit of Melville's story in which the first-person narrator warns the reader: "it doesn't exist material—I am convinced—to compose a complete and satisfactory biography of this man." We, mathematicians, do not preface every work with the theorems of incompleteness, but if we have to bet on why consistency is a value for us, it is precisely by virtue of Gödel's theorems: the mathematical creativity is contained from the broad horizon of metatheory and the distant glimmer of paradoxes. What's beyond this mathematical and literary horizon? Calvino writes in the lesson on *Exactitude*:

> But maybe this lack of substance is not to be found in images or in language alone, but in the world itself. [5]

8 The Incomplete Appendix: On the Beginning and the Ending

The lesson "*On the Beginning and the Ending*" which appears in the appendix of [4] was probably meant to be an introduction to the six Harvard lectures. If these represent not only the values of literature but those of all knowledge, then these notes read as a response to the following questions: Why does one decide to write a certain story? When is a physical theory developed? How is the idea of a theorem generated? Calvino speaks of the world as a "sum of information" and language as something to be extracted from the languages developed by various disciplines. But immediately after, he writes that the beginnings lead us into a world different from the one that generated the idea. The mathematical parallel in this case could be the different geometries. For example, any version of the fifth axiom posed at the beginning of Euclid's masterpiece leads us into the new worlds of elliptic or hyperbolic geometry. Gauss's Theorema Egregium marks the beginning of differential geometry. The problem of the bridges of Königsberg initiates the topological universes. The ancient gasket of Apollonius continues in fractal geometry. These are examples of preambles in Mathematics. Obviously, they are in relation, one with the others: differential geometry includes Euclidean and non-Euclidean; without the concept of topological dimension, there would be no description of fractal dimensions. As Calvino intimates with his literary examples, the preamble is a connection in the continuous connective tissue of human events. Calvino explains the choice of a single story (for us mathematicians, a problem to be addressed) with the words "from the universal to the particular", a story is "isolated" from others to be told. In the same way, in a scientific theory, an experiment or a problem becomes an object of great interest. Having spoken of topology, we cannot but give the example of the Poincaré Conjecture, which has led to great developments in the discipline of which Poincaré himself is considered the founder. Certainly, it is easier to discuss topology starting from Euler's result on graphs than from the concept of dimension in Poincaré. Here, not only those who study, but also those who disseminate Mathematics face the dilemma that Calvino expresses well in this appendix/introduction:

> The moment of choice: we are given the possibility to say everything, in all possible ways; and we must come to say one thing, in a particular way. [5]

As in [16], we draw a parallel between this quote and the axiom of choice stated in 1904 by Zermelo: "Given a non-empty family of non-empty sets, there exists a function that assigns an element from each set of the family to it". We are in set theory, we have an infinity of sets, and we want to call a representative from each set, a kind of symbolic reunion, is it possible? Do the other axioms of set theory already assure us of this? The history of this axiom, its acceptance, its consequences, and equivalences would make a beautiful story, but having to "come to say one thing, in a particular way" we limit ourselves to the concept of function. This has in its definition the notion of uniqueness (of the chosen representative) which in Calvino's manuscript appears as a challenge for each of us: "in the multiplicity of possible experiences [...] the uniqueness of the days we have to live" [4]. We might add the uniqueness of the theorems we have to prove and the stories we have to tell. The axiom does not provide an algorithm on how to choose; it is the creative zone of the mathematician and the writer. Calvino includes oblivion in the set of things to narrate (he writes precisely so, *the set*), for mathematicians, however, the void is not a value of functions: one starts from the void to build for example, the natural numbers, but does not find the void at the conclusion. This asymmetry is also present in the literature. In this lesson, Calvino asks: "Can we make symmetrical considerations for the ending as we have done for the beginning?" It seems so because if the beginning is connected to human events, the ending (even of this incomplete lesson) is "to continue telling", i.e., the story returns to human events. However, everything is transformed, how to recognize the action of the story on these events?

Returning to Mathematics, let's see the beginning and conclusion of an IFS fractal, i.e., obtained by iteration. If Calvino remembers that "Dante ends the three parts of his poem with the word stars", we can consider precisely in the arrangement of the stars a geometry of voids. What is the initial rule? What is the first verse of this astronomical poem? If Calvino hints at "an indeterminate ending", we can take that square (defined with Exactitude in Sect. 4) and start dividing it into nine equal squares, coloring the central square. Repeating this operation in the remaining eight squares, we get 9 colored squares (1 large, 8 small) and 64 uncolored small squares. We can start over, and the "indeterminate ending" after infinite steps will be Sierpinski's carpet square. It is one of the first fractals to appear on the mathematical scene after having made its presence known in finite steps in the floors of churches and castles. If Calvino writes of "an ending that dissolves the illusion", we could look at the Peano or Hilbert curve that seems like a thread but fills the square, therefore, it is two-dimensional, in contrast to the previous carpet that seems a rigid object of the plane but is actually one-dimensional. Finally, Calvino talks about an ending that "questions the entire narration," and we can not avoid to think of the sensation one feels after the beautiful lessons on inductive topological dimension in which the topological dimension is defined in such a way that the previous topological dimension is assigned to the border of an object of the next topological dimension; the reassuring sensation contrasts with the appearance of the Mandelbrot set in the plane, which has a border of dimension 2. Clearly, the notion of dimension has changed, on the other hand, we have also written "finally" in the previous paragraph, we thus have what Calvino calls the "cosmic ending" or how every novel, just like every mathematical theory, is a universe in which the beginning are the axioms and the conclusion is just a passage to a new universe.

9 Singularity: A Seventh Lesson?

A lesson on *Singularity* would seem to be in apparent opposition to the lesson on *Multiplicity*, somewhat like the relationship between *Lightness* and *Exactitude*; but this does not surprise us: as we have already seen, Calvino writes "every value I choose does not intend to exclude the opposite value." [5] From a literary perspective, this opposition is clearly described in [18] also in relation to Gadda's writings: hence, to write, to begin a novel moves from the Multiplicity of stories that could be told to the single story that will become the subject. This choice will inevitably leave out an infinity of possibilities. But in hindsight, after the story is ready for the reader (or the Reader), the act of writing is a break from Multiplicity in favor of a Singularity, represented by the novel itself.

> How is it possible to isolate a singular story if it implies other stories that cross and "condition" it, and these others still, extending to the entire universe? And if the universe cannot be contained in a story, how can one detach from this impossible story other stories that make complete sense? [5]

This way of seeing is not too far from how Singularity is discussed in the sciences. In both Mathematics and the natural sciences, at the point of Singularity, descriptive ambiguities are created, and thus, a large number of different coherent solutions can be proposed. Typically, most of them are not used in the surrounding context.

In Mathematics, the word singularity indicates a point in a certain space where a certain mathematical object is no longer well-defined as in all other points, making it special and interesting. In the more general scientific language, the word singularity indicates a typically extreme situation in which a fairly general theory stops working. Even in Mathematics and the natural sciences, Singularity and Multiplicity—now understood in a different sense from Calvino's—are two concepts inextricably linked.

A typical example, which nowadays is perhaps the most well-known, is the theory of black holes. A black hole is an object where all the mass is concentrated at a point and has such a strong gravity that, beyond a limit called the "event horizon," not even light can escape. The equations of General Relativity that describe this phenomenon, the Einstein equations, predict a mathematical singularity at the point where all the mass is concentrated, and to date, there is no theory on how to describe what happens in such confined spaces and under such strong gravitational fields. This situation triggers an infinite multiplicity of possible descriptions among which, to date, none is experimentally verifiable.

There are also *singular phenomena* closer to our daily lives that are just less famous but no less interesting than black holes. An example is the phase transitions of substances and the existence of so-called critical points, where the usual phases of matter lose their meaning. At these special points, under very precise conditions of pressure, volume, and temperature, wonderful properties occur. In short, near a critical point, there are quantities that exhibit universal behavior, meaning they do not depend on all the details of the interaction between atoms and molecules, but only a tiny part of this information is retained, and therefore, very different substances exhibit the exact same behavior. Conceptually, here it is quite clear how the problem of singularity emerges: when dealing with the critical point, it must happen that the equations describing different substances become identical, and therefore, while it is possible to move towards the critical point, it is not possible to uniquely go back from the critical point. This is, in simple terms, one of the most beautiful ideas in the Theoretical Physics of the 20th-century: the *renormalization group*.

An example of universal behavior is seen considering the singular limit of different equations such as the zero-viscosity limit of the hydrodynamics equations, the zero-range limit for the Schrödinger equation with potential, or the limits leading to the justifications of effective equations for the description of many-body systems.

Today, the study of these singular models is central both in Physics and Mathematics and represents not so much a value to carry into the future, but a starting point for the Mathematics of tomorrow.

Acknowledgements This work for S.L. is part of the PNRR MUR project PE0000023-NQSTI NQSTI Spoke 9. M.G. was supported by the ERC StG project MaMBoQ, no. 80290 and the GNFM (National Group for Mathematical Physics). Both authors thank the INdAM for funding the outreach activity that led to this publication.

References

1. Aiger, M., Ziegler, G.: Proofs from the Book. Springer (2009)
2. Bischi, G.I., Darconza, G.: Calvino e la limpidezza della complessità. Tra Palomar e Parisi, Aras Edizioni (2023)
3. Cabibbo, N.: Tra fossili e particelle, la grande sfida della luce e dell'ombra. Tutto Scienze e Tecnologie. La Stampa (2010)
4. Calvino, I.: Lezioni Americane, Sei proposte per il prossimo millennio. Garzanti 1988. Mondadori (1993)
5. Calvino, I.: Six Memos for the Next Millennium. Harvard University Press (1988)
6. Calvino, I.: Se una notte d'inverno un viaggiatore. Einaudi 1979. Mondadori (1994)
7. Calvino, I.: Palomar. Einaudi 1983. Mondadori (1994)
8. Dizionario Treccani. https://www.treccani.it/vocabolario/matematico/. Accessed 20 Feb. 2024
9. Galilei, G.: Discorsi e dimostrazioni matematiche intorno a due nuove scienze. Boringhieri (1958)

10. Greco, P.: Calvino, dalla Terra alla Luna. Il Bo Live, 21 January 2019
11. Greco, P.: Quale comunicazione della scienza per i paesi emergenti. J. Sci. Comm. **4**(3), 1–6 (2005)
12. Hofstadter D.R.: Gödel, Escher, Bach: un'eterna Ghirlanda Brillante. Adelphi (1984)
13. Lolli, G.: Discorso sulla matematica. Una rilettura delle Lezioni americane di Italo Calvino, Bollati Boringhieri (2011)
14. Lolli, G.: Molteplicità potenziale e creatività al tempo del computer: un matematico del,: legge Calvino. Californian Ital. Stud. **12**, 2023 (2000)
15. Lucente, S.: Elementi dei grafi e grafi negli Elementi. Paginaria Edizioni (2020)
16. Lucente, S.: Il momento della scelta. Prisma n **55**, 40–41 (2023)
17. Melville, H.: Bartleby, the Scrivener: A Story of Wall-Street. Melville House Publishing (2004)
18. Niccolai, S.: Il "Sistema del mondo". Calvino e l'eredità di Gadda. Italianistica: Rivista di letteratura Italiana, vol. 34, issue 3, pp. 29–43
19. Piacentini, A.: Consistenza, l'inesplorata sesta lezione di Calvino. Edizioni Progetto Cultura (2004)
20. Spirito, G.: Matematica senza numeri. Newton Compton Editori (1995)

Resolvent Limits of Exterior Boundary Value Problems and Singular Perturbation of Laplace Operator in 3D

Vladimir Georgiev and Mario Rastrelli

Abstract In this work we study the Helmholtz equation with boundary condition in the exterior of a ball of radius ε in dimension 3, in the radial case. The boundary condition is the Robin condition with a power type weight. Taking the limit $\varepsilon \to 0$, we show that the solution of the problem doesn't feel the initial hole, unless if the power of the weight is 1. In this case a singular perturbation appears and the resolvent behaves exactly like the resolvent of Δ_α, the Laplace operator with point-like interaction.

Keywords Singular perturbation of Laplace operator · Resolvent formula · Helmholtz equation

2020 Mathematics Subject Classification 46E35 · 47A60 · 81Q15 · 35Q41

1 Introduction

The Helmholtz equation and the exterior boundary value problem of the Laplacian are classical, but also studied recently [6, 16, 17]. In this work we want to use these tools to give a new way to understand the operator Δ_α and its resolvent. We are talking about the singular perturbed Laplacian, also called Laplacian with point like interaction, the non trivial self-adjoint extension of $\Delta|_{C_0^\infty(\mathbb{R}^3\setminus\{0\})}$.

V. Georgiev · M. Rastrelli (✉)
Department of Mathematics, University of Pisa, Pisa, Italy
e-mail: rastrelli.mario@outlook.com

V. Georgiev
e-mail: georgiev@dm.unipi.it

V. Georgiev
Faculty of Science and Engineering, Waseda University, Okubo, Shinjuku-ku, Tokyo, Japan

Institute of Mathematics and Informatic, Bulgarian Academy of Sciences, Sofia, Bulgaria

M. Rastrelli
Departmentof Applied Physics, Waseda University, Tokyo 169-8555, Japan

B. Cassano et al. (eds.), *Singularities, Asymptotics, and Limiting Models*, Springer INdAM Series 64, https://doi.org/10.1007/978-981-96-3584-9_11

The literature about Δ_α is wide: we have the first characterizations in [4, 5]; Albeverio et al. [3], Scarlatti and Teta [21], we can find explicit formulas for the linear propagators $e^{t\Delta_\alpha}$ and $e^{it\Delta_\alpha}$; Georgiev et al. [13] and Michelangeli et al. [18] we have more recent results about the fractional domains. The research continues in the direction of the perturbed Schrödinger equation involving Δ_α, with papers regarding dispersive $L^{p'} - L^p$ estimates and Strichartz estimates [8–11, 15], well-posedness [7] and stability of standing waves [1, 2, 12]. Also perturbed Hatree equation is studied in [14].

We give the basic definitions of Δ_α like the domain

$$\mathscr{D}(-\Delta_\alpha) = \left\{ \phi \in L^2(\mathbb{R}^3); \phi = g_\lambda + \frac{g_\lambda(0)}{\frac{\sqrt{\lambda}}{4\pi} + \alpha} \mathbb{G}_\lambda, g_\lambda \in H^2(\mathbb{R}^3) \right\},$$

with $\mathbb{G}_\lambda(x) = \frac{e^{-\sqrt{\lambda}|x|}}{4\pi|x|}$ and its action

$$(\lambda - \Delta_\alpha)\phi = (\lambda - \Delta) g_\lambda.$$

The resolvent is given by Krein formula

$$(\lambda - \Delta_\alpha)^{-1} f = (\lambda - \Delta)^{-1} f + \frac{1}{\frac{\sqrt{\lambda}}{4\pi} + \alpha} \langle \mathbb{G}_\lambda, f \rangle \mathbb{G}_\lambda,$$

that holds for every $\lambda > 0$ if $\alpha \geq 0$ and for every $\lambda \in \mathbb{R}^+ \backslash \{(4\pi\alpha)^2\}$ if $\alpha < 0$.

In [4, Paragraph I.1.3] the operator $-\Delta_\alpha$ is defined also as the limit for $\varepsilon \to 0$ in the resolvent sense of a sequence of perturbed Shrödinger operators $-\Delta + V_\varepsilon$, where V_ε are potential shrinking at zero.

Recently, this approach has been used very often like in [19], where the fractional powers of $-\Delta_\alpha$ where constructed by the limits $(-\Delta)^{\frac{s}{2}} + V_\varepsilon$ as $\varepsilon \to 0$ of fractional Laplace operator with a shrinking potential, or in the context of physics, like [?]. In [?], the approach consists on the Gamma convergence of $H_0 + V_\varepsilon$, where $V_\varepsilon \in C^1$ and is negative, and H_0 is the Hamiltonian of a system describing the interaction of three particles. The Gamma convergence implies the strong resolvent convergence [?]. In [?], the Schrödinger equation with Hatree non linearity is studied. The equation considered is perturbed with a potential V_ε that has a variable magnitude of the scaling σ, when it shrinks around the origin. In Theorem 3.1 they prove that $-\Delta + V_\varepsilon$ tends to $-\Delta$ in the norm resolvent sense, unless if $\sigma = 2$ and V_ε satisfy some resonant condition. In this case we have again $-\Delta + V_\varepsilon \to -\Delta_\alpha$ for a certain α.

Our approach is instead different. We do not have an explicit potential, but a particular Robin condition as a boundary condition for an exterior boundary value problem. In fact. in the next sections we will prove that the resolvent of $-\Delta_\alpha$ can be obtained as a limit of solutions of Helmholtz equation in the exterior of balls, when the hole shrinks.

2 Main Result

We consider the following Helmholtz equation in the exterior of ball of radius $\varepsilon > 0$ in $\mathbb{R}^3$

$$\begin{aligned}(\lambda - \Delta)u(x) &= f(x), \quad |x| > \varepsilon \\ \partial_r(r^k u(x)) &= 4\pi\alpha r^k u(x), \quad r = |x| = \varepsilon.\end{aligned} \tag{1}$$

Here and below $k \geq 0, \alpha \in \mathbb{R}$, and $\lambda \in \mathbb{C}\backslash(-\infty, 0]$. We shall assume $f \in L^2_{rad}(|x| > \varepsilon)$ and we shall look for $u \in H^2_{rad}(|x| > \varepsilon)$ satisfying the boundary condition

$$ku(\varepsilon) + \varepsilon u'(\varepsilon) = \varepsilon 4\pi\alpha u(\varepsilon). \tag{2}$$

It is clear that $u(r)$ satisfies

$$(\lambda - \partial_r^2 - \frac{2}{r}\partial_r)u(r) = f(r), \ r > \varepsilon,$$

where

$$-\partial_r^2 - \frac{2}{r}\partial_r$$

with boundary condition (2) is a self adjoint operator in $L^2_{rad}(r > \varepsilon)$.[1] Indeed, we can rewrite the boundary condition in the form

$$u'(\varepsilon) = \left(4\pi\alpha - \frac{k}{\varepsilon}\right)u(\varepsilon)$$

and see that

$$\langle \Delta u, v\rangle_{L^2(|x|>\varepsilon)} = \langle u, \Delta v\rangle_{L^2(|x|>\varepsilon)}$$

provided $u \in H^2_{rad}(|x| > \varepsilon)$ and (2) holds. Hence the Laplace operator with boundary condition (2) is a symmetric operator. It is easy to see that it is a closed operator. We denote this operator by $\mathscr{L}_{\alpha,\varepsilon,k}$ and it is easy to see that the kernel of $\mathscr{L}_{\alpha,\varepsilon,k} \pm i$ is trivial so this is a self-adjoint operator. The domain of $\mathscr{L}_{\alpha,\varepsilon,k}$ is

$$D_{rad}(\mathscr{L}_{\alpha,\varepsilon,k}) = \{u \in H^2_{rad}(|x| > \varepsilon); \text{(2) holds}\}.$$

[1] Here $L^2_{rad}(r > \varepsilon)$ is equipped with norm

$$\|f(r)\|^2_{L^2_{rad}} = 4\pi \int_\varepsilon^\infty |f(r)|^2 r^2 dr.$$

We immediately note that if $\alpha < 0$, for every $\varepsilon > 0$ the operator $\mathscr{L}_{\alpha,\varepsilon,1}$ has an eigenfunction $\psi_\alpha(r) = \frac{e^{4\pi\alpha r}}{4\pi r}$ with eigenvalue

$$E_\alpha = (4\pi\alpha)^2. \tag{3}$$

Theorem 2.1 *The resolvent*

$$(\lambda + \mathscr{L}_{\alpha,\varepsilon,k})^{-1}$$

has strong L^2 limit

$$\lim_{\varepsilon \searrow 0}(\lambda + \mathscr{L}_{\alpha,\varepsilon,k})^{-1} f = (\lambda + \mathscr{L}_{\alpha,k})^{-1} f, \quad f \in L^2_{rad}(|x| > 0) \tag{4}$$

for every $\lambda \in \mathbb{C}\backslash(-\infty, 0]$, if $k \neq 1$ or $\alpha > 0$, and for every $\lambda \in \mathbb{C}\backslash\big((-\infty, 0] \cup \{E_\alpha\}\big)$, if $k = 1$ and $\alpha < 0$, with E_α defined above in (3).

Remark 2.1 We have a natural inclusion

$$i_\varepsilon : L^2_{rad}(\mathbb{R}^3) \hookrightarrow L^2_{rad}(|x| > \varepsilon)$$

so the relation (4) means that

$$\lim_{\varepsilon \searrow 0}(\lambda + \mathscr{L}_{\alpha,\varepsilon,k})^{-1} i_\varepsilon(f) = (\lambda + \mathscr{L}_{\alpha,k})^{-1} f, \quad f \in L^2_{rad}(\mathbb{R}^3)$$

Theorem 2.2 *The strong L^2 resolvent limit has the following properties:*

(a) The resolvent of the operator $\mathscr{L}_{\alpha,k}$ has a pole if and only if $k = 1$.
(b) If $k = 1$, then

$$(\lambda + \mathscr{L}_{\alpha,1})^{-1} f = (\lambda - \Delta)^{-1} f + \frac{1}{\frac{\sqrt{\lambda}}{4\pi} + \alpha} \mathbb{G}_\lambda \langle \mathbb{G}_\lambda, f \rangle_{L^2(\mathbb{R}^3)}, \tag{5}$$

where $\mathbb{G}_\lambda(r) = \frac{e^{-\sqrt{\lambda} r}}{4\pi r}$;
c) If $k \neq 1$, then

$$(\lambda + \mathscr{L}_{\alpha,k})^{-1} f = (\lambda - \Delta)^{-1} f. \tag{6}$$

3 Proof of Theorem 2.1 and Theorem 2.2

First we discuss the pointwise limit of resolvent $(\lambda + \mathscr{L}_{\alpha,\varepsilon,k})^{-1} f$ by construction the Green function for the corresponding Sturm problem on half line.

The substitution

$$u(r) = \frac{\mathbb{U}(r)}{r},\ f(r) = \frac{\mathbb{F}(r)}{r}$$

transforms the boundary—value problem (1) into

$$\begin{aligned} &\lambda\mathbb{U}(r) - \mathbb{U}''(r) = \mathbb{F}(r), \quad r > \varepsilon \\ &(k-1)\mathbb{U}(\varepsilon) + \varepsilon\mathbb{U}'(\varepsilon) = \varepsilon 4\pi\alpha\mathbb{U}(\varepsilon) \ \ r = |x| = \varepsilon. \end{aligned} \tag{7}$$

Remark 3.1 If $k = 1$, then the boundary condition becomes

$$\mathbb{U}'(\varepsilon) = 4\pi\alpha\mathbb{U}(\varepsilon)$$

that is the condition (when $\varepsilon \searrow 0$) in (1.1.12) in [4].

Lemma 3.1 *The solution* $\mathbb{U}(r)$ *has the representation*

$$\begin{aligned} \mathbb{U}(r) = \frac{1}{2\sqrt{\lambda}}\Bigg(&\int_{\varepsilon}^{r} e^{-\sqrt{\lambda}r}\big(e^{\sqrt{\lambda}s} - e^{-\sqrt{\lambda}s}\big)\mathbb{F}(s)ds \\ &+ \int_{r}^{\infty} e^{-\sqrt{\lambda}s}\big(e^{\sqrt{\lambda}r} - e^{-\sqrt{\lambda}r}\big)\mathbb{F}(s)ds\Bigg) \\ &+ \frac{1}{\sqrt{\lambda} + 4\pi\alpha - \frac{k-1}{\varepsilon}} e^{-\sqrt{\lambda}r}\int_{\varepsilon}^{\infty} e^{-\sqrt{\lambda}s}\mathbb{F}(s)ds. \end{aligned}$$

Proof The solution can be represented via the Green function

$$G(r,s) = \begin{cases} C\mathbb{U}_1(r)\mathbb{U}_2(s) \text{ if } s \geq r, \\ C\mathbb{U}_2(r)\mathbb{U}_1(s) \text{ if } s < r. \end{cases}$$

Here and below $\mathbb{U}_j$, $j = 1, 2$ are two linearly independent solutions of the homogeneous equation

$$\lambda\mathbb{U}(r) - \mathbb{U}''(r) = 0,$$

such that $\mathbb{U}_1(r)$ satisfies the boundary condition

$$(k-1)\mathbb{U}_1(\varepsilon) + \varepsilon\mathbb{U}_1'(\varepsilon) = \varepsilon 4\pi\alpha\mathbb{U}_1(\varepsilon)$$

at $r = \varepsilon$, while $\mathbb{U}_2$ satisfies the boundary condition

$$\lim_{r \to \infty} \mathbb{U}_2(r) = 0$$

at $r = \infty$. These solution can be obtained as linear combinations of solutions

$$\mathbb{U}_\pm(r) = e^{\pm\sqrt{\lambda}r}$$

Obviously, we have

$$\mathbb{U}_2 = \mathbb{U}_-(r) = e^{-\sqrt{\lambda}r}. \tag{8}$$

Then we look for $\mathbb{U}_1$ in the form

$$\mathbb{U}_1(r) = C_+\mathbb{U}_+ + C_-\mathbb{U}_-$$

and using the boundary condition at $r = \varepsilon$ we find

$$\begin{aligned}(k-1)\left(C_+\mathbb{U}_+(\varepsilon) + C_-\mathbb{U}_-(\varepsilon)\right) + \varepsilon\left(C_+\sqrt{\lambda}\mathbb{U}_+(\varepsilon) - \sqrt{\lambda}C_-\mathbb{U}_-(\varepsilon)\right)\\ = \varepsilon 4\pi\alpha\left(C_+\mathbb{U}_+(\varepsilon) + C_-\mathbb{U}_-(\varepsilon)\right),\end{aligned}$$

so

$$\left(\varepsilon\sqrt{\lambda} + (k-1) - \varepsilon 4\pi\alpha\right)C_+\mathbb{U}_+(\varepsilon) = \left(\varepsilon\sqrt{\lambda} - (k-1) + \varepsilon 4\pi\alpha\right)C_-\mathbb{U}_-(\varepsilon).$$

Hence we can choose

$$\begin{aligned}C_+ &= \left(\sqrt{\lambda} + 4\pi\alpha - \frac{k-1}{\varepsilon}\right)\mathbb{U}_-(\varepsilon),\\ C_- &= \left(\sqrt{\lambda} - 4\pi\alpha + \frac{k-1}{\varepsilon}\right)\mathbb{U}_+(\varepsilon),\end{aligned}$$

and we conclude that

$$\begin{aligned}\mathbb{U}_1(r) &= \left(\sqrt{\lambda} + 4\pi\alpha - \frac{k-1}{\varepsilon}\right)\mathbb{U}_-(\varepsilon)\mathbb{U}_+(r)\\ &\quad + \left(\sqrt{\lambda} - 4\pi\alpha + \frac{k-1}{\varepsilon}\right)\mathbb{U}_+(\varepsilon)\mathbb{U}_-(r)\\ &= \left(\sqrt{\lambda} + 4\pi\alpha - \frac{k-1}{\varepsilon}\right)e^{\sqrt{\lambda}(r-\varepsilon)} + \left(\sqrt{\lambda} - 4\pi\alpha + \frac{k-1}{\varepsilon}\right)e^{-\sqrt{\lambda}(r-\varepsilon)}.\end{aligned}$$

Note that

$$\begin{aligned}
\mathbb{U}_1(\varepsilon) &= 2\sqrt{\lambda}, \\
\mathbb{U}_1'(\varepsilon) &= \sqrt{\lambda}\left(\sqrt{\lambda} + 4\pi\alpha - \frac{k-1}{\varepsilon}\right) - \sqrt{\lambda}\left(\sqrt{\lambda} - 4\pi\alpha + \frac{k-1}{\varepsilon}\right) \\
&= 2\sqrt{\lambda}\left(4\pi\alpha - \frac{k-1}{\varepsilon}\right).
\end{aligned} \tag{9}$$

Then we have

$$\mathbb{U}(r) = C\int_{\varepsilon}^{r} \mathbb{U}_2(r)\mathbb{U}_1(s)\mathbb{F}(s)ds + C\int_{r}^{\infty} \mathbb{U}_1(r)\mathbb{U}_2(s)\mathbb{F}(s)ds.$$

The constant C can be found by using the equation for $\mathbb{U}_j$. Indeed, we have

$$\mathbb{U}'(r) = C\int_{\varepsilon}^{r} \mathbb{U}_2'(r)\mathbb{U}_1(s)\mathbb{F}(s)ds + C\int_{r}^{\infty} \mathbb{U}_1'(r)\mathbb{U}_2(s)\mathbb{F}(s)ds$$

and

$$\begin{aligned}
\mathbb{U}''(r) = C\int_{\varepsilon}^{r} \mathbb{U}_2''(r)\mathbb{U}_1(s)\mathbb{F}(s)ds + C\int_{r}^{\infty} \mathbb{U}_1''(r)\mathbb{U}_2(s)\mathbb{F}(s)ds \\
+ C\left(\mathbb{U}_2'(r)\mathbb{U}_1(r) - \mathbb{U}_1'(r)\mathbb{U}_2(r)\right)\mathbb{F}(r),
\end{aligned}$$

so, taking

$$C = -\frac{1}{W(\mathbb{U}_1, \mathbb{U}_2)},$$

where $W = W(\mathbb{U}_1, \mathbb{U}_2)$ is the Wronskian

$$W = W(\mathbb{U}_1, \mathbb{U}_2) = \mathbb{U}_2'(r)\mathbb{U}_1(r) - \mathbb{U}_1'(r)\mathbb{U}_2(r),$$

we find the Eq. (7) is satisfied.

From (8) and (9), we can compute the Wronskian.

$$\begin{aligned}
W &= \mathbb{U}_2'(\varepsilon)\mathbb{U}_1(\varepsilon) - \mathbb{U}_1'(\varepsilon)\mathbb{U}_2(\varepsilon) \\
&= -\sqrt{\lambda}e^{-\sqrt{\lambda}\varepsilon}2\sqrt{\lambda} - 2\sqrt{\lambda}\left(4\pi\alpha - \frac{k-1}{\varepsilon}\right)e^{-\sqrt{\lambda}\varepsilon} \\
&= -2\sqrt{\lambda}\left(\sqrt{\lambda} + 4\pi\alpha - \frac{k-1}{\varepsilon}\right)e^{-\sqrt{\lambda}\varepsilon}.
\end{aligned}$$

In this way we arrive at the following representation of the solution for (7)

$$\mathbb{U}(r) = C\int_{\varepsilon}^{r} \mathbb{U}_2(r)\mathbb{U}_1(s)\mathbb{F}(s)ds + C\int_{r}^{\infty} \mathbb{U}_1(r)\mathbb{U}_2(s)\mathbb{F}(s)ds,$$

$$C = \frac{e^{\sqrt{\lambda}\varepsilon}}{2\sqrt{\lambda}\left(\sqrt{\lambda} + 4\pi\alpha - \frac{k-1}{\varepsilon}\right)}.$$

Now we switch to $\mathbb{U}_{\pm}$ in the representation of $\mathbb{U}$

$$\begin{aligned}
\mathbb{U}(r) &= C\int_{\varepsilon}^{r} \mathbb{U}_-(r)\left(C_+\mathbb{U}_+(s) + C_-\mathbb{U}_-(s)\right)\mathbb{F}(s)ds \\
&\quad + C\int_{r}^{\infty} \left(C_+\mathbb{U}_+(r) + C_-\mathbb{U}_-(r)\right)\mathbb{U}_-(s)\mathbb{F}(s)ds \\
&= CC_+\left(\int_{\varepsilon}^{r} \mathbb{U}_-(r)\mathbb{U}_+(s)\mathbb{F}(s)ds + \int_{r}^{\infty} \mathbb{U}_+(r)\mathbb{U}_-(s)\mathbb{F}(s)ds\right) \\
&\quad + CC_-\mathbb{U}_-(r)\int_{\varepsilon}^{\infty} \mathbb{U}_-(s)\mathbb{F}(s)ds,
\end{aligned} \tag{10}$$

where

$$CC_+ = \frac{\left(\sqrt{\lambda} + 4\pi\alpha - \frac{k-1}{\varepsilon}\right)}{2\sqrt{\lambda}\left(\sqrt{\lambda} + 4\pi\alpha - \frac{k-1}{\varepsilon}\right)} = \frac{1}{2\sqrt{\lambda}}$$

$$CC_- = \frac{\left(\sqrt{\lambda} - 4\pi\alpha + \frac{k-1}{\varepsilon}\right)}{2\sqrt{\lambda}\left(\sqrt{\lambda} + 4\pi\alpha - \frac{k-1}{\varepsilon}\right)} = -\frac{1}{2\sqrt{\lambda}} + \frac{1}{\sqrt{\lambda} + 4\pi\alpha - \frac{k-1}{\varepsilon}}$$

and this concludes the proof of the Lemma, remembering that $\mathbb{U}_{\pm}(r) = e^{\pm\sqrt{\lambda}r}$. □

The above Lemma enables us to find that the solution for (1) is

$$u_\varepsilon(r) = \frac{1}{2\sqrt{\lambda}r}\left(\int_\varepsilon^r e^{-\sqrt{\lambda}r}\left(e^{\sqrt{\lambda}s} - e^{-\sqrt{\lambda}s}\right)sf(s)ds\right.$$
$$\left. + \int_r^\infty e^{-\sqrt{\lambda}s}\left(e^{\sqrt{\lambda}r} - e^{-\sqrt{\lambda}r}\right)sf(s)ds\right)$$
$$+ \frac{1}{\sqrt{\lambda} + 4\pi\alpha - \frac{k-1}{\varepsilon}}\frac{e^{-\sqrt{\lambda}r}}{r}\int_\varepsilon^\infty e^{-\sqrt{\lambda}s}sf(s)ds.$$

Assuming $f \in L^2((0,\infty), sds)$ and f has a compact support it is not difficult to obtain the pointwise bound

$$|u_\varepsilon(r)| \le \frac{C}{|x|} \tag{11}$$

where $C = C_{\lambda,k} > 0$ is independent of $\varepsilon > 0$.

In order to take the pointwise limit of $(\lambda + \mathscr{L}_{\alpha,\varepsilon,k})^{-1}f(r) = u_\varepsilon(r)$ it is sufficient to compute:

$$\lim_{\varepsilon \searrow 0} CC_- = \begin{cases} -\dfrac{1}{2\sqrt{\lambda}} & \text{if } k \neq 1, \\ -\dfrac{1}{2\sqrt{\lambda}} + \dfrac{1}{\sqrt{\lambda} + 4\pi\alpha} & \text{if } k = 1. \end{cases}$$

In particular we have proved the following Lemma.

Lemma 3.2 *We have the following poinwise limits*

(a) If $k = 1$, then

$$(\lambda + \mathscr{L}_{\alpha,1})^{-1}f = (\lambda - \Delta)^{-1}f + \frac{1}{\frac{\sqrt{\lambda}}{4\pi} + \alpha}\mathbb{G}_\lambda\langle \mathbb{G}_\lambda, f\rangle_{L^2(\mathbb{R}^3)},$$

where $\mathbb{G}_\lambda(r) = \frac{e^{-\sqrt{\lambda}r}}{4\pi r}$;

(b) If $k \neq 1$, then

$$(\lambda + \mathscr{L}_{\alpha,k})^{-1}f = (\lambda - \Delta)^{-1}f.$$

We only need to show that

$$\frac{1}{2\sqrt{\lambda}r}\left(\int_0^r e^{-\sqrt{\lambda}r}\left(e^{\sqrt{\lambda}s} - e^{-\sqrt{\lambda}s}\right)sf(s)ds + \int_r^\infty e^{-\sqrt{\lambda}s}\left(e^{\sqrt{\lambda}r} - e^{-\sqrt{\lambda}r}\right)sf(s)ds\right) \tag{12}$$

is equal to $(\lambda - \Delta)^{-1} f$ in L^2_{rad}. This is possible thanks to the Newton's shell theorem:

$$\int_{\mathbb{S}^2} f(|x+\rho\omega|)d\omega = \frac{2\pi}{\rho|x|} \int_{\left||x|-\rho\right|}^{|x|+\rho} f(s)sds,$$

where $\mathbb{S}^2$ is the unit sphere in $\mathbb{R}^3$. We have that

$$\begin{aligned}
(\lambda - \Delta)^{-1} f(x) &= \int_{\mathbb{R}^3} \frac{e^{-\sqrt{\lambda}|y|}}{4\pi|y|} f(|x-y|)dy \\
&= \int_0^\infty e^{-\sqrt{\lambda}\rho} \rho \frac{1}{4\pi} \int_{\mathbb{S}^2} f(|x+\rho\omega|)d\omega d\rho \\
&= \frac{1}{2|x|} \int_0^\infty e^{-\sqrt{\lambda}\rho} \int_{\left||x|-\rho\right|}^{|x|+\rho} sf(s)dsd\rho.
\end{aligned}$$

We note that also $u = (\lambda - \Delta)^{-1} f$ is radial and the domain of integration is symmetric with respect to $s = \rho$, so we can swap the integrals, obtaining

$$\begin{aligned}
(\lambda - \Delta)^{-1} f(r) &= \frac{1}{2r} \int_0^\infty sf(s) \int_{|r-s|}^{r+s} e^{-\sqrt{\lambda}\rho} d\rho ds \\
&= \frac{1}{2\sqrt{\lambda}r} \int_0^\infty sf(s)(e^{-\sqrt{\lambda}|r-s|} - e^{-\sqrt{\lambda}(r+s)})ds.
\end{aligned} \tag{13}$$

And it is clear that (13) is equal to (12).

We have proved the limits (5) and (6) pointwise.Our next step is to deduce the limit in L^2. We can also rewrite (10) as

$$\begin{aligned}
u_\varepsilon(x) = &\frac{1}{2\sqrt{\lambda}|x|} \int_{\mathbb{R}^3 \setminus B_\varepsilon} \frac{e^{-\sqrt{\lambda}|y|}}{4\pi|y|} f(|x-y|)dy \\
&+ \frac{1}{\sqrt{\lambda} + 4\pi\alpha - \frac{k-1}{\varepsilon}} \frac{e^{-\sqrt{\lambda}|x|}}{|x|} \int_{\mathbb{R}^3 \setminus B_\varepsilon} \frac{e^{-\sqrt{\lambda}|y|}}{4\pi|y|} f(|y|)dy
\end{aligned}$$

and the limit as

$$u(x) = \begin{cases} \dfrac{1}{2\sqrt{\lambda}|x|} \displaystyle\int_{\mathbb{R}^3} \dfrac{e^{-\sqrt{\lambda}|y|}}{4\pi|y|} f(|x-y|)dy & \text{if } k \neq 1, \\ \dfrac{1}{2\sqrt{\lambda}|x|} \displaystyle\int_{\mathbb{R}^3} \dfrac{e^{-\sqrt{\lambda}|y|}}{4\pi|y|} f(|x-y|)dy & \\ + \dfrac{1}{\sqrt{\lambda}+4\pi\alpha} \dfrac{e^{-\sqrt{\lambda}|x|}}{|x|} \displaystyle\int_{\mathbb{R}^3} \dfrac{e^{-\sqrt{\lambda}|y|}}{4\pi|y|} f(|y|)dy & \text{if } k = 1, \end{cases}$$

where B_ε is the ball of radius ε centered at the origin. Thanks to the regularity of ∂B_ε, we can extend u_ε at zero inside B_ε. In order to apply the dominate convergence theorem in $L^2(|x| \geq \varepsilon_0)$ we can use the pointwise estimate (11) and use the pointwise convergence.

To be more precise, we have the following estimate

$$|u_\varepsilon(x) - u(x)| \leq C \frac{\mathbb{1}_{|x|<1}}{|x|} \|\mathbb{G}_\lambda\|_{L^2} \|f\|_{L^2} + \int_{\mathbb{R}^3} \frac{e^{-\sqrt{\lambda}|y|}}{4\pi|y|} \big|f(|x-y|)\big| dy$$
$$+ C \frac{e^{-\sqrt{\lambda}|x|}}{|x|} \|\mathbb{G}_\lambda\|_{L^2} \|f\|_{L^2},$$

this implies, thanks to the properties $f, \mathbb{G}_\lambda \in L^2$ and the dominate convergence theorem that

$$\lim_{\varepsilon\to 0} \|u_\varepsilon - u\|_{L^2(|x|>\varepsilon_0)} = 0$$

for any $\varepsilon_0 > 0$. Combining the estimate (11) and the above pointwise estimate of $|u_\varepsilon - u|$ we see that $u \in L^2_{rad}(\mathbb{R}^3)$.

Remark 3.2 We can discuss the possible extensions of functions defined in the exterior of a ball of radius ε. The zero extension gives one possibility. We can consider the interior of the ball as a "black box" as it is often assumed in scattering and inverse scattering problems. In fact for the case of obstacle moving in time a natural space in $L^2(\Omega(t))$, where $\Omega(t)$ is the exterior of an obstacle $K(t)$ that is a compact set with sufficiently smooth boundary. We have a natural embedding $i_t : L^2(\mathbb{R}^3) \to L^2(\Omega(t))$. The existence of wave operators requires to define limits of type

$$\lim_{t\to\infty} \|u(t) - i_t U_0(t) f\|_{L^2(\Omega(t))}$$

Here $u(t) \in L^2(\Omega(t))$ for any $t \in [0, T]$ and such that for any $g \in L^2(\mathbb{R}^3)$ we have

$$\int_{\Omega(t)} u(t, x) g(x) dx \in C([0, T]).$$

We refer to [20] and the references therein for the scattering for moving obstacles. In our case, we have a family u_ε of functions in $L^2_{rad}(|x| > \varepsilon)$ so from the pointwise convergence of $u_\varepsilon(|x|)$ for any fixed $|x|$ we can define a function $u(|x|)$ and then to prove

$$\lim_{\varepsilon \to 0} \|u_\varepsilon - u\|_{L^2(|x|>\varepsilon)} = 0 \tag{14}$$

that implies $u \in L^2_{rad}(\mathbb{R}^3)$. Indeed, using the zero extension $\widetilde{u_\varepsilon}(x)$ of u_ε we take $0 < \varepsilon < \varepsilon_0$ and from (14) we see that $\widetilde{u_\varepsilon}(x)$ is a Cauchy sequence in $L^2(|x| > \varepsilon_0)$ and moreover $f \in L^2(|x| > \varepsilon_0)$.

Acknowledgements Vladimir Georgiev was partially supported by Gruppo Nazionale per l'Analisi Matematica, by the project PRIN 2020XB3EFL with the Italian Ministry of Universities and Research, by Institute of Mathematics and Informatics, Bulgarian Academy of Sciences, by Top Global University Project, Waseda University, the Project PRA 2022 85 of University of Pisa and the MIUR Excellence Department Project awarded to the Department of Mathematics, University of Pisa, CUP I57G22000700001. Mario Rastrelli was partially supported by Gruppo Nazionale per l'Analisi Matematica.

References

1. Adami, R., Boni, F., Carlone, R., Tentarelli, L.: Existence, structure, and robustness of ground states of a NLSE in 3D with a point defect. J. Math. Phys. **63**(7), 071501 (2022)
2. Adami, R., Boni, F., Carlone, R., Tentarelli, L.: Ground states for the planar NLSE with a point defect as minimizers of the constrained energy. Calc. Var. Partial Differ. Eqn. **61**(5), Paper No. 195, 32 (2022)
3. Albeverio, S., Brzeźniak, Z., Dpolhkabrowski, L.: Fundamental solution of the heat and Schrödinger equations with point interaction. J. Funct. Anal. **130**(1), 220–254 (1995)
4. Albeverio, S., Gesztesy, F., Höegh-Krohn, R., Holden, H.: Solvable Models in Quantum Mechanics, 2nd edn. AMS Chelsea Publishing, Providence, RI (2005). With an appendix by Pavel Exner
5. Berezin, F.A., Faddeev, L.D.: Remark on the Schrödinger equation with singular potential. Dokl. Akad. Nauk SSSR **137**, 1011–1014 (1961)
6. Burq, N., Gérard, P., Tzvetkov, N.: On nonlinear schrödinger equations in exterior domains. Annales de l'Institut Henri Poincaré C, Analyse non linéaire **21**(3), 295–318 (2004)
7. Cacciapuoti, C., Finco, D., Noja, D.: Well posedness of the nonlinear schrödinger equation with isolated singularities. J. Differ. Eqn. **305**, 288–318 (2021)
8. Cornean, H.D., Michelangeli, A., Yajima, K.: Two-dimensional Schrödinger operators with point interactions: threshold expansions, zero modes and L^p-boundedness of wave operators. Rev. Math. Phys. **31**(4), 1950012, 32 (2019)
9. Cornean, H.D., Michelangeli, A., Yajima, K.: Erratum: two-dimensional Schrödinger operators with point interactions: threshold expansions, zero modes and L^p-boundedness of wave operators. Rev. Math. Phys. **32**(4), 2092001, 5 (2020)
10. D'Ancona, P., Pierfelice, V., Teta, A.: Dispersive estimate for the Schrödinger equation with point interactions. Math. Methods Appl. Sci. **29**(3), 309–323 (2006)
11. Dell'Antonio, G., Michelangeli, A., Scandone, R., Yajima, K.: L^p-boundedness of wave operators for the three-dimensional multi-centre point interaction. Ann. Henri Poincaré **19**(1), 283–322 (2018)

12. Fukaya, N., Georgiev, V., Ikeda, M.: On stability and instability of standing waves for 2d-nonlinear schrödinger equations with point interaction. J. Differ. Eqn. **321**, 258–295 (2022)
13. Georgiev, V., Michelangeli, A., Scandone, R.: On fractional powers of singular perturbations of the Laplacian. J. Funct. Anal. **275**(6), 1551–1602 (2018)
14. Georgiev, V., Michelangeli, A., Scandone, R.: Standing waves and global well-posedness for the 2d Hartree equation with a point interaction (2022). arXiv:2204.05053
15. Iandoli, F., Scandone, R.: Dispersive estimates for Schrödinger operators with point interactions in $\mathbb{R}^3$. In: Advances in Quantum Mechanics, Volume 18 of Springer INdAM Series, pp. 187–199. Springer, Cham (2017)
16. Melrose, R.B., Sjöstrand, J.: Singularities of boundary value problems. I. Comm. Pure Appl. Math. **31**(5), 593–617 (1978)
17. Melrose, R.B., Sjöstrand, J.: Singularities of boundary value problems. II. Commun. Pure Appl. Math. **35**(2), 129–168 (1982)
18. Michelangeli, A., Olgiati, A., Scandone, R.: Singular Hartree equation in fractional perturbed Sobolev spaces. J. Nonlinear Math. Phys. **25**(4), 558–588 (2018)
19. Michelangeli, A., Scandone, R.: Point-like perturbed fractional Laplacians through shrinking potentials of finite range. Complex Anal. Oper. Theory **13**(8), 3717–3752 (2019)
20. Petkov, M., Georgiev, V.: RAGE theorem for power bounded operators and decay of local energy for moving obstacles. Ann. Inst. H. Poincaré Phys. Théor. **51**(2), 155–185 (1989)
21. Scarlatti, S., Teta, A.: Derivation of the time-dependent propagator for the three-dimensional Schrodinger equation with one point interaction. J. Phys. A: Math. Gen. **23**(19), L1033 (1990)

The Search for NLS Ground States on a Hybrid Domain: Motivations, Methods, and Results

Riccardo Adami, Filippo Boni, Raffaele Carlone, and Lorenzo Tentarelli

Abstract We discuss the problem of establishing the existence of the Ground States for the subcritical focusing Nonlinear Schrödinger energy on a domain made of a line and a plane intersecting at a point. The problem is physically motivated by the experimental realization of hybrid traps for Bose-Einstein Condensates, that are able to concentrate the system on structures close to the domain we consider. In fact, such a domain approximates the trap as the temperature approaches the absolute zero. The spirit of the paper is mainly pedagogical, so we focus on the formulation of the problem and on the explanation of the result, giving references for the technical points and for the proofs.

Keywords Hybrids · Standing waves · Nonlinear schrödinger · Ground states · Delta interaction

AMS Subject Classification 35R02 · 81Q35 · 35Q55 · 35Q40 · 35B07 · 35B09 · 35R99

R. Adami (✉) · L. Tentarelli
Politecnico di Torino, Dipartimento di Scienze Matematiche "G.L. Lagrange", Torino, Italy
e-mail: riccardo.adami@polito.it

L. Tentarelli
e-mail: lorenzo.tentarelli@polito.it

F. Boni
Scuola Superiore Meridionale, Napoli, Italy
e-mail: f.boni@ssmeridionale.it

R. Carlone
Universitá degli Studi di Napoli Federico II, Dipartimento di Matematica e Applicazioni "Renato Caccioppoli", Napoli, Italy
e-mail: raffaele.carlone@unina.it

B. Cassano et al. (eds.), *Singularities, Asymptotics, and Limiting Models*, Springer INdAM Series 64, https://doi.org/10.1007/978-981-96-3584-9_12

1 Introduction

Modeling Bose-Einstein Condensation has been a major challenge in the Mathematical Physics of the last decades. The phenomenon consists in a phase transition undergone by systems made of a large number, i.e. around 10^4, of identical bosons at low temperature, namely 10^{-7} K and below. In the new phase all bosons of the system share the same quantum state called Ground State of the condensate, and such state occupies all the available space. Notice that the fact that every boson possesses a quantum state is not a natural property: indeed, even assuming that the whole system lies in a quantum state, its parts may not. In the mathematical description, this means that a wave function of the whole system may not be the product of single-particle wave functions. This is what happens in the presence of entanglement among the particles. Therefore, in the transition to a Bose-Eintein condensate every particle acquires an individual quantum state, i.e. bosons in a Bose-Einstein condensate lying in its Ground State are not entangled to one another.

Such a new physical phase was first foreseen one hundred years ago [13, 20] and experimentally realized seventy years later [8, 18], when the techinques of laser and evaporative cooling made possible to reach the necessary temperature. Since then, the mathematical investigation on condensates has become topical and has involved several approaches: operator theory, nonlinear evolution equation, calculus of variation, analysis on Fock spaces and the formalism of second quantization. An important part of such a research consisted in singling out the conditions under which the so-called Gross-Pitaevskii regime [12, 29, 35] holds. In such a regime it is possible to approximate the fundamental linear N-body dynamics of the bosons with an effective, nonlinear one-body dynamics: if, on the one hand, the problem becomes more difficult for the presence of the nonlinearity, on the other hand there is a great simplification in passing from an N-body to a one-body problem, and the simplification greatly overcomes the additional difficulty.

Let us be more specific. First, a quantum mechanical description of the N-boson system is made through the N-boson wave function $\Psi_{N,t}(x_1, \ldots x_N)$, that represents the quantum state of the system at time t. By x_j we denote the three-dimensional coordinate of the jth boson. The function $\Psi_{N,t}$ is symmetric under permutation of its variables, as it describes the state of N identical bosons. The evolution of the system is described by the N-body Schrödinger equation

$$i\partial_t \Psi_{N,t} = -\sum_{j=1}^{N} \Delta_j \Psi_{N,t} + \sum_{1 \leqslant i < j \leqslant N} w(x_i - x_j)\Psi_{N,t} + \sum_{j=1}^{N} V(x_j)\Psi_{N,t}, \quad (1)$$

where w represents the two-body interaction between the bosons and V is the potential of an external field whose role is typically to confine the system.

The already mentioned Gross-Pitaevskii regime is the physical setting in which the dynamics of a Bose-Einstein condensate is effectively described by the one-body nonlinear Gross-Pitaevskii equation

$$i\partial_t\psi_t = -\Delta\psi_t + V(x)\psi_t + 8\pi\alpha|\psi_t|^2\psi_t, \tag{2}$$

where α is the scattering length of the two-body interaction w between the particles in the condensate. It turns then out that the nonlinearity is the mark of the interaction between the bosons that are the elementary constituents of the condensate. Of course, the problem of deducing (2) from (1) is of paramount importance in understanding the underlying physics. It is nowadays well understood that the transition to the Gross-Pitaevskii regime happens if the interacting potential scales with respect to the number N of bosons as

$$w(x) \to N^2 w(Nx) \tag{3}$$

and if the initial data fulfil at least approximately a bosonic Stosszahlansatz, namely $\Psi_{N,0} = \psi_0^{\otimes N}$. The scaling law (3) means that the Gross-Pitaevskii regime is a good approximation of the fundamental dynamics (1) if the interacting potential is strong, due to the factor N^2, and concentrated, as expressed by the argument Nx that shrinks the range of the potential to a scale N^{-1}. The Stosszahlansatz hypothesis states in turn that the correlations between the particles are small, which is more likely the case if the system is dilute. Notice that at the level of the dynamics the process that makes the system uncorrelated is not described, as one has to impose the Stosszahlansatz on the initial data. Under such hypothesis, one has that in the Gross-Pitaevskii regime every particle possesses an individual state, so particles are not entangled with one another, and that at every time t such a state is represented by a complex-valued function ψ_t to be interpreted according the usual Born's rule: the modulus square of ψ_t gives the probability density of finding one particle of the condensate in a specified spatial region. Such interpretation imposes ψ_t to be normalized at 1, but in fact it is often preferable to normalize it to the number N of particles in the condensate, so that

$$\int_{\mathbb{R}^3} |\psi_t|^2\, dx = N,$$

thus $|\psi_t|^2$ is better understood as a particle density rather than as a probability density.

Besides their relevance in the physical understanding of the phenomenon, the rigorous proofs of Bose-Einstein condensation in the static framework [32] and of the stability of the condensates under time evolution [4, 21, 22, 31] greatly enriched the mathematical techniques employed in the derivation of nonlinear one-body equations from linear N-body quantum dynamics of interacting particles. In particular, the scaling (3) is highly singular, so a main breakthrough was accomplished as it was understood how to rigorously deal with it.

Later, a major challenge was to get a good estimate of the error made in approximating the fundamental linear N-body description with the effective nonlinear one-body evolution equation [9, 11, 30, 34, 38].

In the following we focus on the existence and the shape of the Ground State for a condensate in the Gross-Pitaevskii regime under the action of a trap whose shape can be modeled as the union of a plane and a line orthogonal to it. Since the Bose-Einstein condensation occurs at low energy, it does not involve neither the creation nor the annihilation of particles, therefore the natural notion of Ground State is that of the minimizer of the energy functional E_{GP} whose value is conserved by the flow of the Gross-Pitaevskii equation, under the constraint that the number of particles, i.e. the L^2-norm, is fixed. In the following the L^2-norm will be called mass and will be denoted by μ. We will refer to the prescription that it has to be a fixed quantity as to the mass constraint.

We then get to the problem of finding the minimizers of the Gross-Pitaevskii energy

$$E_{GP}(\psi) = \frac{1}{2}\|\nabla\psi\|^2_{L^2(\mathbb{R}^3)} + 2\pi\alpha\|\psi\|^4_{L^4(\mathbb{R}^3)} + \frac{1}{2}\int_{\mathbb{R}^3} V(x)|\psi(x)|^2\,dx \tag{4}$$

with the constraint

$$\int_{\mathbb{R}^3} |\psi(x)|^2\,dx = \mu,$$

provided they exist.

Usually the potential V is referred to as the trap. Of course, the actual shape of a condensate depends on its particular state, thus it has become customary to think of the spatial distribution of the Ground State as the shape of the trap.

Early condensates [8, 18] were subject to an external harmonic and isotropic potential, so that their shape was approximately spherical, but the subsequent impressive technological advances in magnetic and optical confinement made it possible to build up condensates with various shapes: disc-shaped, cigar-shaped, branched, and others.

We assume that the trapping potential V is so effective that the Ground State of the condensate can be approximated as supported in a region $\Omega \subset \mathbb{R}^3$, so that

$$E(\psi) = \frac{1}{2}\|\nabla\psi\|^2_{L^2(\Omega)} + 2\pi\alpha\|\psi\|^4_{L^4(\Omega)}.$$

Clearly, the Ground State ψ_t evolves in time just by periodically changing its phase, namely

$$\psi_t(x) = e^{i\omega t}\psi_0(x),$$

so that its spatial profile and therefore the associated density of particles remain unaltered. It is a stationary state.

1.1 Hybrid Traps

As already mentioned, we are interested in describing an experimental arrangement in which the trap has the shape of a plane with a line attached to it, as in Fig. 1. This structure results from gluing together two components of different dimensionality and will be referred to as the hybrid structure $\mathcal{H}$. Traps of this kind have been realized by employing a magnetic trap together with an optical trap [33]. Because of the use of two different physical mechanisms, in the physical literature such traps are called hybrid, while for us the same word refers to the resulting multi-dimensional geometry of the system. Serendipitously, it happens that hybrid traps give rise to hybrid geometry, although the two terms were introduced independently with different meaning.

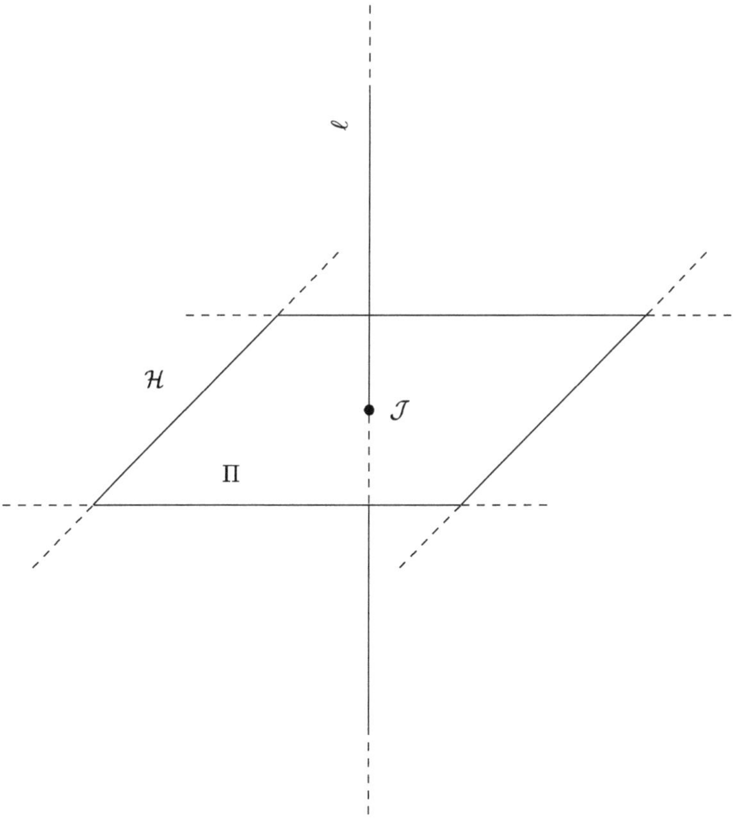

Fig. 1 The hybrid $\mathcal{H}$. The origin of the coordinates on both the line ℓ and the plane Π is set at the junction $\mathcal{J}$

Let us describe more closely the effect of a hybrid trap. The confining potential V decomposes into a term V_M, describing the magnetic trap, and a term V_O, related to the optical trap, so that

$$V(\mathbf{x}) = V_M(\mathbf{x}) + V_O(\mathbf{x}), \qquad \mathbf{x} = (x, y, z) \in \mathbb{R}^3,$$

where, according to the description given in [33], the effective shape of the two potentials can be modeled as

$$V_M(\mathbf{x}) = 2\nu B|z|, \qquad V_O = \frac{1}{2} m\omega_O^2(x^2 + y^2).$$

Here ν is the magnetic moment of the particles in the condensate and m their mass, B is the magnetic field generated by the trap, and ω_O is the frequency of the optical trap. The net effect of the double trap is that the resulting spatial distribution of the condensate is

$$n(\mathbf{x}) \;=\; n_0\, e^{-\frac{2\nu B}{kT}|z|} e^{-\frac{m\omega_O^2}{2kT}(x^2+y^2)}, \tag{5}$$

where k is the Boltzmann constant and T is the temperature of the system. Notice that the first factor concentrates the condensate on the plane $z = 0$, while the second factor on the z axis.

One can then distinguish three regions in the hybrid, according to the size of the density:

(1) the origin, where both exponentials in (5) equal one, so the density is of order 1;
(2) the (x, y)-plane, where the density n is dumped by the optical trap, namely by the second exponential factor in (5), and the z-axis, where the first exponential factor, representing the magnetic trap, attenuates the density of particles;
(3) the rest of the space, where the density is shrunk by both traps.

Moreover, notice that as the temperature goes to the absolute zero, the density concentrates at the origin, with an exponential tail on $\mathcal{H}$, and a double exponentially tiny rest outside $\mathcal{H}$. Our aim is to describe such profile at least qualitatively.

Hybrids were first introduced in [23], then further investigated in [24] and later in [15, 16]. In such works the physical context ranges from classical electromagnetism to quantum mechanics, and the dynamics is always linear. The hybrid structure considered there is made of a plane with a halfline attached at its origin. On the same hybrid manifold we studied the problem of the Ground State of condensates in [3]. The problem we investigate here is a simple variant of that, so we refer to that work for the technical aspects of the proofs. Here we aim at discussing the meaning and the formulation of the problem. Therefore, in Sect. 2 we construct the energy functional on $\mathcal{H}$ and compare it to the classical Nonlinear Schrödinger energies. In Sect. 3 we give the Gagliardo-Nirenberg estimates, that are crucial for the existence of a Ground State, and present the main result. Technically it deals with the lack of compactness of minimizing sequences, but it can be told quite simply, by saying that a Ground State exists if the hybrid is energetically more convenient than the line. In other words, the hybrid must compete, at an energetical level, with the line and *not* with the plane. Finally, we give sufficient conditions to make the hybrid energetically convenient, namely, conditions on the parameters of the system in order to get this convenient.

One can sketch such conditions as follows. If one aims at constructing an effective trap for a Bose-Einstein condensate with a shape close to $\mathcal{H}$, namely a trap able to capture the system in a geometry of that kind, then the junction must unfold an attractive effect on the condensate. Now, since the most general junction can produce a contact interaction on the line, another contact interaction on the plane, and a coupling between the two, one has to consider the possible attractive feature of the three of them. More precisely, to trap a condensate on a region modeled by $\mathcal{H}$, one must ensure that:

(1) the junction between the line and the plane deploys an attractive action on the part of the condensate located on the line, towards the origin of the line; indeed, this makes the system "line + junction" energetically convenient with respect to the line alone, and so the hybrid prevails on the line;
(2) if the previous condition is not fulfilled, then one has to make the plane more convenient than the line, so the hybrid will be more convenient too;
(3) if both the previous conditions are not fulfilled, then one has to provide that the coupling between the plane and the line is strong: indeed its effect always proves energetically convenient and, above a certain threshold, it makes the hybrid more convenient than the line.

These prescriptions are stated more precisely in Corollary 3.7.

On the other hand, if none among the conditions (1)–(3) is satisfied, then it may happen that the condensate tends to escape through the line far from the junction. According to our analysis, this runaway phenomenon cannot take place on the plane.

We stress here that in our model we are not able to treat the fourth power nonlinearity for the component of the energy on the plane, for technical reasons that will be clarified in the next section. We can of course approach such power arbitrarily, but this does not guarantee that our result can be transposed by continuity to the physical case with the power four.

Propositions 3.2 and 3.5 are proved in some detail since their direct proof is not in [3].

2 The Problem

We assume that the spatial domain in which the condensate is confined by the trap is represented by the hybrid domain $\mathcal{H}$ and aim at investigating whether, given the number of particles N that compose the condensate, there is a Ground State. Let us translate such problem in the terms of the Calculus of Variations.

2.1 Heuristic Construction of the Energy Functional

First, we write the hybrid domain $\mathcal{H}$ as

$$\mathcal{H} = \ell \cup \Pi,$$

where ℓ is the z-axis and Π is the (x, y)-plane. Second, we introduce a mathematical object that plays the role of a wave function defined on $\mathcal{H}$. It is natural to assume that the state of the condensate is represented by a two-component function $U = (u, v)$, where $u : \ell \to \mathbb{C}$ is the portion of the wave function supported on the line, while $v : \Pi \to \mathbb{C}$ is the part supported on the plane. In order to establish to which spaces the functions u and v belong, we have to introduce an energy functional E that adapts the energy (4) to the structure of $\mathcal{H}$. It is natural to assume that such energy functional is composed by three terms:

$$E(U) = E_\alpha(u, \ell) + E_\sigma(v, \Pi) + E_{\mathcal{J}}(U),$$

where

- $E_\alpha(\cdot, \ell)$ is the energy of the portion of U on ℓ. It comprises the kinetic energy of the one-dimensional wave packet u, the self-consistent nonlinear term, and the contact interaction between the line and the junction. Then, we write it as

$$E_\alpha(u, \ell) = \frac{1}{2}\|u'\|^2_{L^2(\mathbb{R})} - \frac{1}{p}\|u\|_p^p + \frac{\alpha}{2}|u(0)|^2, \qquad 2 < p < 6,\ \alpha \in \mathbb{R}. \qquad (6)$$

  Notice that the contact interaction corresponds here to the action of a Dirac's delta potential.
- $E_\sigma(v, \Pi)$ is the energy of the portion of U on Π. It includes a nonlinear term and a contact interaction at the junction. Now, as widely known [7, 36], a pointwise contact interaction in dimension two cannot be described by a distributional delta potential as in dimension one, i.e. it cannot be written as $|v(0)|^2$, since in two dimensions a generic function in H^1 has no pointwise evaluation. More precisely, contrarily to the one-dimensional case, in two dimensions such a pointwise perturbation is not small with respect to the kinetic term in the sense of the quadratic forms, therefore the definition of the point interaction cannot be achieved in the same way. It is nowadays well understood that the construction of such kind of interaction can be equivalently realized by following two different paths: a renormalization procedure [10] or the use of the theory of self-adjoint extensions of hermitian operators [37]. Even though the two approaches end up with the same result, they are complementary to each other, as the renormalization procedure highlights the interpretation of contact interactions as limits of spatially extended potentials, while the self-adjoint extension theory provides a simple algorithm that directly gives all possible point interactions [7].

Eventually the introduction of a contact interaction modifies the energy domain by introducing a singularity at the site of the interaction. More specifically such a domain results in

$$\mathcal{D}_2 = \left\{ u = \phi + q\frac{K_0}{2\pi},\ \phi \in H^1(\mathbb{R}^2),\ q \in \mathbb{C} \right\}, \tag{7}$$

where K_0 is a function of the Bessel family, more precisely the MacDonald function of order zero whose asymptotic behaviour at the origin is $K_0(x) \sim -\log|x|$ and that decays exponentially fast for large $|x|$ [28]. Notice that the presence of a point interaction in dimension two results in the formation of a logarithmic singularity. The complex factor q is usually called the charge and represents the size of the singularity of the function v. Notice that this prescription concentrates a significant portion of the state around the origin unless $q = 0$.
The energy functional $E_\sigma(\cdot, \mathbb{R}^2)$ acts as follows:

$$\begin{aligned} E_\sigma(v, \Pi) &= \frac{1}{2}\|\phi\|^2_{H^1(\mathbb{R}^2)} - \frac{1}{2}\|v\|^2_{L^2(\mathbb{R}^2)} \\ &\quad + \frac{\sigma}{2}|q|^2 - \frac{1}{r}\|v\|^r_{L^r(\mathbb{R}^2)}, \quad 2 < r < 4,\ \sigma \in \mathbb{R}. \end{aligned} \tag{8}$$

We stress that here the coupling constant σ does not coincide with its analogous ρ employed in [3] as well as with the parameter σ previously used in [1]. Indeed, one has

$$\sigma = \rho + \frac{\gamma - \log 2}{2\pi},$$

where γ is the Euler-Mascheroni constant.

- $E_{\mathcal{J}}(U)$ is the energy of the interaction between ℓ and Π that takes place at the junction. It describes exclusively the exchange of energy between the line and the plane. In other words, it is the only term which, under a dynamical point of view, realizes the transmission of a signal from ℓ to Π and vice versa. For its precise shape we choose the simplest possible option, namely a quadratic coupling:

$$E_{\mathcal{J}}(U) = -\beta \mathrm{Re}(\overline{q}u(0)), \qquad \beta \geqslant 0. \tag{9}$$

Summarizing, we constructed the energy functional

$$\begin{aligned} E(U) &= E_\alpha(u, \ell) + E_\sigma(v, \Pi) + E_{\mathcal{J}}(U) \\ &= \frac{1}{2}\|u'\|^2_{L^2(\mathbb{R})} + \frac{\alpha}{2}|u(0)|^2 - \frac{1}{p}\|u\|^p_{L^p(\mathbb{R})} + \frac{1}{2}\|\phi\|^2_{H^1(\mathbb{R}^2)} - \frac{1}{2}\|v\|^2_{L^2(\mathbb{R}^2)} \\ &\quad + \sigma\frac{|q|^2}{2} - \frac{1}{r}\|v\|^r_{L^r(\mathbb{R}^2)} - \beta\mathrm{Re}(\overline{q}u(0)), \\ &\quad \text{with } 2 < p < 6,\ 2 < r < 4,\ \alpha, \sigma \in \mathbb{R}, \beta \geqslant 0, \end{aligned} \tag{10}$$

defined on the energy domain

$$\mathcal{D} := \left\{ U = (u, v) \text{ s.t. } u \in H^1(\mathbb{R}),\ v = \phi + q\frac{K_0}{2\pi},\ \phi \in H^1(\mathbb{R}^2),\ q \in \mathbb{C} \right\}.$$

2.2 Some Comments on the Energy Functional E

Let us explain the meaning of the choice of the parameters in the construction of the energy functional (10).

Concerning the term $E_\alpha(u, \ell)$ notice that in (6) we consider an arbitrary nonlinearity power p and limit our analysis to the focusing case, namely the case in which the nonlinearity has a negative sign and thus, dynamically, an attractive effect. The upper bound $p < 6$ means that we restrict to nonlinearities that can be controlled by the kinetic term, as described by the one-dimensional Gagliardo-Nirenberg estimate

$$\|u\|^p_{L^p(\mathbb{R})} \leqslant C\|u'\|^{\frac{p}{2}-1}_{L^2(\mathbb{R})}\|u\|^{\frac{p}{2}+1}_{L^2(\mathbb{R})}, \qquad u \in H^1(\mathbb{R}), \tag{11}$$

showing that, while for $p < 6$ the nonlinearity grows slower than the kinetic term, for $p = 6$ it may grow at the same rate of the kinetic term, making the energetic balance more delicate. Eventually, it is well-known [17] that if $\alpha = 0$ then the choice $p = 6$ results in a phenomenology that radically differs from that of the case $p < 6$, both in the statics and in the dynamics: if $p = 6$ then existence of Ground States depends on the chosen value of μ, and in the related evolution problem blow-up solutions appear whose existence cannot be extended to arbitrarily large time. Both features are not present if $p < 6$, for which a Ground State for $E_0(u, \ell)$ is present at every value of the mass and all solutions to the dynamical problem in the energy space are defined globally in time.

For its peculiar character, the power nonlinearity with $p = 6$ is called critical, and here we treat subcritical nonlinearities only.

Let us point out that for $\alpha = 0$ the resulting energy functional $E_0(\cdot, \ell)$ is the standard NLS energy functional with focusing subcritical power nonlinearity. It is a long standing result [17, 40] that for every value of the mass there is a family of ground states, called solitons, that are obtained by translating and multiplying by a phase the unique positive, even Ground State

$$\varphi_\mu(x) = \mu^{\frac{2}{6-p}}\varphi(\mu^{\frac{p-2}{6-p}}x), \tag{12}$$

where

$$\varphi(x) = C_p \operatorname{sech}^{\frac{2}{p-2}}(c_p x),$$

with C_p and c_p constants depending on the power p only.

Furthermore, the energy of the solitons scales as

$$\mathcal{E}_{0,\ell}(\mu) = E_0(\varphi_\mu, \ell) = -\theta_p \mu^{\frac{p+2}{6-p}},$$

where θ_p is a positive constant that depends on p only, and we introduced the symbol

$$\mathcal{E}_{\alpha,\ell}(\mu) = \inf_{u \in H^1(\mathbb{R}),\, \int_\ell |u|^2 = \mu} E_\alpha(u, \ell).$$

The problem of the Ground States for $E_\alpha(\cdot, \ell)$ was studied in [26, 27]. For an exhaustive review see [39].

Concerning the term $E_\sigma(v, \Pi)$ in (10), we introduced the strength of the contact interaction σ with either sign, restricted to the focusing nonlinearity by choosing a negative sign for the last term, and moved from the physical fourth power to a generic power r. In fact, in this paper we do not allow the values $r \geqslant 4$ because $r = 4$ is the critical power in dimension two, as described by the two-dimensional Gagliardo-Nirenberg inequality:

$$\|v\|_{L^r(\mathbb{R}^2)}^r \leqslant C \|\nabla v\|_{L^2(\mathbb{R}^2)}^{r-2} \|v\|_{L^2(\mathbb{R}^2)}^2,$$

that states that from $r = 4$ the kinetic term in $E_\sigma(\cdot, \Pi)$ ceases to control the nonlinear term. From the construction of the two-dimensional point interaction follows that in order to recover the standard focusing two-dimensional NLS energy, one has to set $\sigma = \infty$ instead of the more intuitive $\sigma = 0$. This delicate point can be heuristically explained by considering that for $\sigma = \infty$ in order to get a finite energy it must be $q = 0$, that reduces the domain $\mathcal{D}_2$ to $H^1(\mathbb{R}^2)$ through $v = \phi$ and makes the energy $E_\sigma(\cdot, \Pi)$ equal to

$$E_\infty(v, \Pi) = \frac{1}{2} \|\nabla v\|_{L^2(\mathbb{R}^2)}^2 - \frac{1}{r} \|v\|_{L^r(\mathbb{R}^2)}^r,$$

that is the standard NLS energy.

We will use the symbol

$$\mathcal{E}_{\sigma,\Pi}(\mu) = \inf_{v \in \mathcal{D}_2,\, \int_\Pi |v|^2 = \mu} E(v, \ell).$$

The problem of the Ground States for $E_\sigma(\cdot, \Pi)$ was analyzed in [1, 25] for the action functional of the same model.

Concerning the term $E_{\mathcal{J}}(U)$ in (10) we point out that considering a non-negative strength β is not restrictive, since for every complex β it is possible to reduce to (9) by suitably modulating the relative phase between q and $u(0)$. The negative sign corresponds to the choice for which having the same constant phase on ℓ and Π is energetically convenient.

2.3 Formulation of the Problem

As anticipated, the problem we investigate is the existence of Ground States. Among the several notions of Ground States present in the literature and consistently with the physical features of the Ground States in a condensate, we choose the one which generalizes the ordinary quantum mechanical notion of Ground State to a nonlinear setting.

Definition 2.1 Given $\mu > 0$, a Ground State at mass μ for the energy functional E defined by (10) on the domain $\mathcal{D} \subset L^2(\mathcal{H})$, is an element of $\mathcal{D}$ that satisfies the mass constraint

$$\int_{\mathcal{H}} |U|^2 = \int_{\ell} |u|^2 + \int_{\Pi} |v|^2 = \mu \tag{13}$$

and minimizes E among all functions in $\mathcal{D}$ satisfying the same constraint.

It is convenient to introduce the notation

$$\mathcal{D}^\mu := \{U \in \mathcal{D} \,\text{s.t.}\, U \text{ fulfils the constraint (13)}\}$$

and to denote

$$\mathcal{E}(\mu) = \inf_{U \in D^\mu} E(U).$$

We are then led to the following formulation of the problem of the Ground States for the Nonlinear Schrödinger Equation on the hybrid $\mathcal{H}$.

Let $\alpha, \sigma \in \mathbb{R}, \beta \geqslant 0, 2 < p < 6, 2 < r < 4,$ *and* $\mu > 0$. *Establish whether there exist Ground States at mass* μ *of the energy* (10).

Notice that without imposing the mass constraint there would be no Ground State, since $E(\lambda U) \to -\infty$ as $\lambda \to \infty$ for every non-zero $U \in \mathcal{D}$. However the presence of the constraint alone does not ensure the existence for the following reasons:

- first, the energy, although constrained to a fixed mass, may not be lower bounded. This happens for instance in the problem on the line ℓ in the supercritical case $p > 6$ for every value of the mass. To realize it, consider a function u of a real variable and the family of functions $u_\lambda(x) = \sqrt{\lambda} u(\lambda x)$. Then,

$$E_\alpha(u_\lambda, \ell) = \frac{\lambda^2}{2} \|u'\|^2_{L^2(\mathbb{R})} - \frac{\lambda^{\frac{p}{2}-1}}{p} \|u\|^p_{L^p(\mathbb{R})} + \frac{\lambda^2 \alpha}{2} |u(0)|^2 \to -\infty, \quad \lambda \to +\infty.$$

  Therefore, since $\|u_\lambda\|_{L^2(\mathbb{R})} = \|u\|_{L^2(\mathbb{R})}$ we have that the energy $E_\alpha(\cdot, \ell)$ constrained to the arbitrary mass $\|u\|^2_{L^2(\mathbb{R})} > 0$ is not lower bounded. Notice that the assumption $p > 6$ is crucial to this conclusion;
- on the other hand, consider the subcritical case $p < 6$ on the line; then the Gagliardo-Nirenberg inequality (11) applied to $E_\alpha(\cdot, \ell)$ with $\alpha > 0$ gives the uniform lower bound

$$E_\alpha(u,\ell) \geqslant \frac{1}{2}\|u'\|^2_{L^2(\mathbb{R})} - C\|u'\|^{\frac{p}{2}-1}_{L^2(\mathbb{R})}\mu^{\frac{p}{4}+\frac{1}{2}} \geqslant M > -\infty,$$

where we crucially used $p < 6$ and denoted $\mu = \|u\|^2_{L^2(\mathbb{R})}$. Therefore, owing to the constraint the energy functional $E_\alpha(\cdot,\ell)$ is lower bounded for every mass, so that

$$\inf_{u\in H^1(\mathbb{R}),\int_\ell |u|^2=\mu} E_\alpha(u,\ell) > -\infty.$$

On the other hand, for any $u \in H^1(\mathbb{R})$ such that $\int_{\mathbb{R}} |u|^2 dx = \mu$, due to the choice $\alpha > 0$ it holds

$$E_\alpha(u,\ell) \geqslant E_0(u,\ell) \geqslant E_0(\varphi_\mu,\ell), \tag{14}$$

where φ_μ is the soliton introduced in (12). One then gets

$$\mathcal{E}_{\alpha,\ell}(\mu) \geqslant \mathcal{E}_{0,\ell}(\mu). \tag{15}$$

On the other hand, consider the family of translations $\varphi_{\mu,\lambda}(x) = \varphi_\mu(x-\lambda)$ of the Ground State φ_μ at mass μ for $E_0(\cdot,\mathbb{R})$, so

$$E_\alpha(\varphi_{\mu,\lambda},\ell) = E_0(\varphi_\mu,\ell) + \frac{\alpha}{2}|\varphi_\mu(\lambda)|^2 \to E_0(\varphi_\mu,\ell), \quad \lambda\to\infty,$$

$$\mathcal{E}_{\alpha,\ell}(\mu) \leqslant \mathcal{E}_{0,\ell}(\mu). \tag{16}$$

Then, from (15) and (16) one concludes

$$\mathcal{E}_{\alpha,\ell}(\mu) = \mathcal{E}_{0,\ell}(\mu).$$

Now, the first inequality in (14) is an equality if and only if $u(0) = 0$, while the second is an equality if and only if $u = \varphi_\mu$ up to translations and multiplications by a phase. But as φ_μ is strictly positive, the two inequalities cannot be both equalities, so that

$$E_\alpha(u,\ell) > E_0(\varphi_\mu,\ell), \quad \forall u \in H^1(\mathbb{R}) \text{ s.t. } \int_{\mathbb{R}} |u|^2 = \mu,$$

from which it follows that there is no function u that attains the infimum $\mathcal{E}_{\alpha,\ell}(\mu)$, so that there is no ground state at mass μ.

Summing up, the existence of a Ground State at mass μ is equivalent to two conditions: first, the infimum of the constrained energy must be finite. Second, it must be attained by some function. We stress that, contrarily to what happens in standard Quantum Mechanics, when a nonlinearity is involved these two conditions are not guaranteed.

3 The Main Result

Here we give a condition equivalent to the existence of Ground States at mass μ. For the proof we refer to the proof of the analogous result given in [3] for the hybrid plane $\mathcal{I}$.

3.1 *Lower Boundedness of the Constrained Energy*

First, like in the case of classical domains, the lower boundedness of the constrained energy is guaranteed by subcriticality through the Gagliardo-Nirenberg estimates, which establish that every subcritical nonlinearity can be controlled by the kinetic term of the energy. Now, such estimates are widely known for functions in Sobolev spaces, but here one needs them to hold in the energy domain $\mathcal{D}$, that strictly contains the space $H^1(\mathbb{R}) \oplus H^1(\mathbb{R}^2)$. In [3] it was shown how to extend Gagliardo-Nirenberg estimates to the hybrid plane $\mathcal{I}$, and the very same method can be applied for the hybrid manifold $\mathcal{H}$.

First of all notice that, in the nonlinear term

$$-\frac{1}{p}\|u\|^p_{L^p(\mathbb{R})} - \frac{1}{r}\|v\|^2_{L^r(\mathbb{R}^2)},$$

the first term can be treated by the classical Gagliardo-Nirenberg estimate (11). The same is not true for the second term, which lies outside $H^1(\mathbb{R}^2)$ unless $q = 0$. Following [1, 3], for the case $q \neq 0$ we rewrite $v \in \mathcal{D}_2$, $v = \phi + q\frac{K_0}{2\pi}$, as

$$v(x) = \phi_q(x) + q\frac{K_0(|q|x)}{2\pi}, \qquad \phi_q \in H^1(\mathbb{R}^2), \tag{17}$$

which is possible since $K_0(|q|x) \sim K_0(x) \sim -\log|x|$ as $x \to 0$. Then

$$\phi_q(x) = \phi(x) + \frac{q}{2\pi}(K_0(x) - K_0(|q|x)). \tag{18}$$

One has the following proposition:

Proposition 3.1 (Gagliardo-Nirenberg estimate) Let $U = (u, v) \in \mathcal{D}^\mu$. Write

$$v(x) = \begin{cases} \phi_q(x) + q\frac{K_0(|q|x)}{2\pi}, & q \neq 0 \\ \phi_0(x), & q = 0 \end{cases}$$

Then,

$$\begin{aligned} \|u\|^p_{L^p(\mathbb{R})} &\leqslant C\|u'\|^{\frac{p}{2}-1}_{L^2(\mathbb{R})} \\ \|v\|^r_{L^r(\mathbb{R}^2)} &\leqslant C(\|\nabla\phi_q\|^{r-2}_{L^2(\mathbb{R}^2)} + |q|^{r-2}), \end{aligned} \tag{19}$$

where the mass μ was absorbed in the constants C.

As in the standard case, Gagliardo-Nirenberg estimates are useful in the proof of the lower boundedness of the constrained energy. To show this, we first rewrite the energy of the planar component v in terms of the decomposition (17). We get the following result.

Proposition 3.2 *Let $v \in \mathcal{D}_2$ with $q \neq 0$ and consider the decomposition (17). Then*

$$\begin{aligned} E_\sigma(v,\Pi) = {} & \frac{1}{2}\|\nabla\phi_q\|^2_{L^2(\mathbb{R}^2)} + \frac{|q|^2}{2}\|\phi_q\|^2_{L^2(\mathbb{R}^2)} - \frac{|q|^2}{2}\|v\|^2_{L^2(\mathbb{R}^2)} \\ & + \left(\frac{\sigma}{2} + \frac{\log|q|}{4\pi}\right)|q|^2 - \frac{1}{r}\|v\|^r_{L^r(\mathbb{R}^2)}. \end{aligned} \tag{20}$$

Proof We define the Fourier transform as

$$\widehat{f}(k) = \frac{1}{2\pi}\int_{\mathbb{R}^2} f(x)e^{-ik\cdot x}\,dx$$

for $f \in L^1(\mathbb{R}^2)$, then extend it to $L^2(\mathbb{R}^2)$ by continuity. With such a definition the Fourier transform is a unitary operator in all Sobolev spaces $H^s(\mathbb{R}^2)$, $s \in \mathbb{R}$. Now, by properties of the Bessel functions it is known that, in the space of distributions,

$$(-\Delta + 1)K_0 = 2\pi\delta$$

and, since in Fourier space the gradient is mapped into the multiplication by the vector ik, one has

$$\widehat{K_0}(k) = \frac{1}{k^2+1} \qquad k^2 := |k|^2,$$

then by (18) one gets

$$\widehat{\phi}(k) = \widehat{\phi}_q(k) + \frac{q}{2\pi}\frac{1-|q|^2}{(k^2+|q|^2)(k^2+1)}.$$

The two first terms of (8) can be rewritten as

$$
\begin{aligned}
&\frac{1}{2}\|\phi\|^2_{H^1(\mathbb{R}^2)} - \frac{1}{2}\|v\|^2_{L^2(\mathbb{R}^2)} \\
&= \frac{1}{2}\int_{\mathbb{R}^2} (k^2+1)\left|\widehat{\phi}_q(k) + \frac{q}{2\pi}\frac{1-|q|^2}{(k^2+|q|^2)(k^2+1)}\right|^2 dk \\
&\quad - \frac{1}{2}\int_{\mathbb{R}^2} \left|\widehat{\phi}_q(k) + \frac{q}{2\pi(k^2+|q|^2)}\right|^2 dk \qquad (21)\\
&= \frac{1}{2}\int_{\mathbb{R}^2} k^2|\widehat{\phi}_q(k)|^2 dk - \mathrm{Re}\int_{\mathbb{R}^2} \frac{|q|^2 q\,\overline{\widehat{\phi}_q(k)}}{2\pi(k^2+|q|^2)} dk \\
&\quad + \frac{|q|^2}{8\pi^2}\int_{\mathbb{R}^2} \frac{dk}{k^2+|q|^2}\left(\frac{(1-|q|^2)^2}{k^2+1} - 1\right)
\end{aligned}
$$

On the other hand, the three first terms in (20) can be rewritten as

$$
\begin{aligned}
&\frac{1}{2}\|\nabla\phi_q\|^2_{L^2(\mathbb{R}^2)} + \frac{|q|^2}{2}\|\phi_q\|^2_{L^2(\mathbb{R}^2)} - \frac{|q|^2}{2}\|v\|^2_{L^2(\mathbb{R}^2)} \\
&= \frac{1}{2}\int_{\mathbb{R}^2} (k^2+|q|^2)|\phi_q(k)|^2 dk - \frac{|q|^2}{2}\int_{\mathbb{R}^2} \left|\widehat{\phi}_q(k) + \frac{q}{2\pi(k^2+|q|^2)}\right|^2 dk \qquad (22)\\
&= \frac{1}{2}\int_{\mathbb{R}^2} k^2|\phi_q(k)|^2 dk - |q|^2\mathrm{Re}\, q\int_{\mathbb{R}^2} \frac{\overline{\widehat{\phi}_q(k)}\,dk}{2\pi(k^2+|q|^2)} - \frac{|q|^4}{8\pi^2}\int_{\mathbb{R}^2} \frac{dk}{(k^2+|q|^2)^2}.
\end{aligned}
$$

Then, taking the difference between the quantities computed in (21) and (22) we conclude

$$
\begin{aligned}
&\frac{1}{2}\|\phi\|^2_{H^1(\mathbb{R}^2)} - \frac{1}{2}\|v\|^2_{L^2(\mathbb{R}^2)} - \left(\frac{1}{2}\|\nabla\phi_q\|^2_{L^2(\mathbb{R}^2)} + \frac{|q|^2}{2}\|\phi_q\|^2_{L^2(\mathbb{R}^2)} - \frac{|q|^2}{2}\|v\|^2_{L^2(\mathbb{R}^2)}\right) \\
&= \frac{|q|^2}{8\pi^2}\int_{\mathbb{R}^2} \frac{dk}{(k^2+|q|^2)^2}\left(\frac{(1-|q|^2)^2}{k^2+1} - 1 + |q|^2\right) \qquad (23)\\
&= \frac{|q|^2}{8\pi^2}\int_{\mathbb{R}^2} \frac{|q|^2-1}{(k^2+|q|^2)(k^2+1)} dk = \frac{|q|^2}{4\pi}\log|q|.
\end{aligned}
$$

Therefore, from (8) and (23)

$$E_\sigma(v,\Pi) = \frac{1}{2}\|\nabla\phi_q\|^2_{L^2(\mathbb{R}^2)} + \frac{|q|^2}{2}\|\phi_q\|^2_{L^2(\mathbb{R}^2)} - \frac{|q|^2}{2}\|v\|^2_{L^2(\mathbb{R}^2)}$$
$$+ \frac{|q|^2}{4\pi}\log|q| + \frac{\sigma}{2}|q|^2 - \frac{1}{r}\|v\|^2_{L^2(\mathbb{R}^2)},$$

and the proof is complete. □

Plugging estimate (19) in the expression of the energy (10) one has the following

Proposition 3.3 *Let* $2 < p < 6$ *and* $2 < r < 4$*. Then the functional (10) with the constraint (13) is lower bounded.*

The proof given in [3] for the hybrid plane $\mathcal{I}$ can be closely repeated for $\mathcal{H}$. Here we recall the main passage. After estimating the coupling term by

$$-\beta\mathrm{Re}\,(\overline{q}u(0)) \geqslant -C(|q|^2 + \sqrt{\mu}\|u'\|_{L^2(\mathbb{R})}),$$

through (19) we have

$$E(U) \geqslant \frac{1}{2}\|u'\|^2_{L^2(\mathbb{R})} - C\left(\|u'\|^{\frac{p}{2}-1}_{L^2(\mathbb{R})} + \|u'\|_{L^2(\mathbb{R})}\right) + \frac{1}{2}\|\nabla\phi_q\|^2_{L^2(\mathbb{R}^2)}$$
$$- C\|\nabla\phi_q\|^{r-2}_{L^2(\mathbb{R}^2)} + \frac{|q|^2}{2}(\log|q| - C) - C|q|^{r-2},$$

where the mass was absorbed in the constants and the contributions of $\|u'\|_{L^2(\mathbb{R})}$, $\|\nabla\phi_q\|_{L^2(\mathbb{R}^2)}$, and q are separated from one another, so that one can introduce three functions f, g, h that are lower bounded due to the subcritical character of the nonlinearities. Then one gets

$$E(U) \geqslant f(\|u'\|_{L^2(\mathbb{R})}) + g(\|\nabla\phi_q\|_{L^2(\mathbb{R}^2)}) + h(|q|), \tag{24}$$

and it follows that the constrained energy is lower bounded too.

3.2 *Minimizing Sequences*

Let $\{U_n\} \subset D^\mu$ be a minimizing sequence for the energy E at mass μ. This means that every element of the sequence has mass μ and that

$$\lim E(U_n) = \mathcal{E}(\mu).$$

We aim at singling out the Ground States as limits of the minimizing sequences.

Writing $U_n = (u_n, v_n)$ and considering the decomposition

$$v_n(x) = \phi_{n,q_n}(x) + q_n \frac{K_0(|q_n|x)}{2\pi}$$

by (24) one has

$$E(U_n) \geqslant f(\|u_n'\|_{L^2(\mathbb{R})}) + g(\|\nabla\phi_{n,q_n}\|) + h(|q_n|).$$

Now, since $E(U_n)$ is a convergent real sequence, it has to be bounded, so none of the three terms can diverge, which implies that the three quantities $\|u_n'\|_{L^2(\mathbb{R})}$, $\|\nabla\phi_{n,q_n}\|$, $|q_n|$ are bounded too. As a consequence, u_n is a bounded sequence in $H^1(\mathbb{R})$, ϕ_{n,q_n} is a bounded sequence in $H^1(\mathbb{R}^2)$, q_n is a numerical bounded sequence.

Remark 3.4 Therefore, up to subsequences,

- By Banach-Alaoglu theorem u_n converges to some u weakly in $H^1(\mathbb{R})$.
- q_n converges to some q in $\mathbb{C}$.
- $q_n \frac{K_0(|q_n|x)}{2\pi}$ converges strongly to $q\frac{K_0(|q|x)}{2\pi}$ in $L^p(\mathbb{R}^2)$ for $2 \leqslant p < \infty$ if $q \neq 0$. It converges to zero weakly in $L^2(\mathbb{R}^2)$ if $q = 0$.
- By Banach-Alaoglu theorem ϕ_{n,q_n} converges to some ϕ_q weakly in $H^1(\mathbb{R}^2)$. From this one can easily see that ϕ_n, that appears in the decompositon (7) of v_n, converges weakly in $H^1(\mathbb{R}^2)$ to the function ϕ that decomposes v according to (7), namely

$$v = \phi + q\frac{K_0}{2\pi}.$$

- v_n converges weakly to some v in $L^p(\mathbb{R}^2)$, $2 \leqslant p < \infty$.

As a consequence, U_n converges weakly to U in $L^2(\mathcal{H})$. Now, U may not belong to the minimizing domain $\mathcal{D}^\mu$ since, as a weak limit, its mass could be strictly smaller than μ. However, it turns out that, if the mass is conserved in the weak limit so that $U \in \mathcal{D}^\mu$, then U is a Ground State.

Proposition 3.5 *Let us consider a minimizing sequence $\{U_n\} \subset \mathcal{D}^\mu$, and denote by U a weak limit of it in the sense specified by Remark 3.4.*

If $\|U\|^2_{L^2(\mathcal{H})} = \mu$, then U is a Ground State at mass μ for the energy functional E.

Proof By weak convergence,

$$\|U_n - U\|^2_{L^2(\mathcal{H})} = 2\mu - 2\mathrm{Re}(U_n, U)_{L^2(\mathcal{H})} \to 0,$$

so conservation of mass in the weak limit ensures strong convergence. This immediately entails that u_n, v_n and ϕ_n converge strongly to u, v and ϕ in the respective L^2 spaces, while, as already seen, q_n converges to q. Now, since Banach-Steinhaus' Theorem guarantees boundedness in L^2 for the weakly convergent sequences $\{u_n'\}$, $\{\nabla\phi_n\}$, then by Gagliardo-Nirenberg estimates one gets

$$\|u_n - u\|^p_{L^p(\mathbb{R})} \leqslant C\|u_n' - u'\|^{\frac{p}{2}-1}_{L^2(\mathbb{R})}\|u_n - u\|^{\frac{p}{2}+1}_{L^2(\mathbb{R})}$$

thus u_n converges strongly to u in $L^p(\mathbb{R})$. On the other hand

$$\begin{aligned}\|v_n - v\|^r_{L^r(\mathbb{R}^2)} &\leqslant C\|\phi_n - \phi\|^r_{L^r(\mathbb{R}^2)} + C|q_n - q|^r \\ &\leqslant C\|\phi_n - \phi\|^2_{L^2(\mathbb{R}^2)}\|\nabla\phi_n - \nabla\phi\|^{r-2}_{L^2(\mathbb{R}^2)} + C|q_n - q|^r\end{aligned}$$

thus v_n converges strongly to v in $L^r(\mathbb{R}^2)$. Therefore, the nonlinear terms in $E(U_n)$ converge to the corresponding terms in $E(U)$. All other terms converge except the two kinetic terms $\frac{1}{2}\|u_n'\|^2_{L^2(\mathbb{R})}$ and $\frac{1}{2}\|\nabla\phi_n\|^2_{L^2(\mathbb{R}^2)}$. Then, since $U \in D^\mu$, one has the following double inequality

$$\begin{aligned}0 &\geqslant \mathcal{E}(\mu) - E(U) = \lim E(U_n) - E(U) \\ &= \frac{1}{2}(\lim(\|u_n'\|^2_{L^2(\mathbb{R})} + \|\nabla\phi_n\|^2_{L^2(\mathbb{R}^2)}) - \|u\|^2_{L^2(\mathbb{R})} - \|\nabla\phi\|^2_{L^2(\mathbb{R}^2)}) \\ &\geqslant 0\end{aligned}$$

by weak convergence. Then $\mathcal{E}(\mu) = E(U)$ so U is a Ground State at mass μ and the proof is complete. □

Now it is necessary to give conditions that guarantee that the weak limit of a minimizing sequence preserves the mass, so to apply Proposition 3.5. It turns out that there is a simple necessary and sufficient condition.

Theorem 3.6 *A Ground State at mass μ for the energy functional $E(U)$ defined in (10) exists if and only if there is a function $U \in \mathcal{D}^\mu$ such that*

$$E(U) \leqslant \mathcal{E}_{0,\ell}(\mu). \tag{25}$$

Such a result is relevant for two reasons: first, it provides a practical criterion for proving the existence of a Ground State by exhibiting a function U that fulfils (25); second, it states that the infimum of the constrained energy can be reached in two ways only: either by constructing a Ground State or by escaping at infinity through the line. This alternative greatly simplifies the options foreseen by the concentration-compactness theory. Let us further examine this point.

According to the concentration-compactness theory (see e.g. [17]), a weakly convergent sequence can preserve its mass or lose it. Furthermore, the loss can be partial or complete. For the present model, we proved in Proposition 3.5 that if for a given minimizing sequence there is no loss of mass, then the sequence converges to a Ground State. One can show that the partial loss of mass is never energetically convenient due to the subadditivity of the energy, so that it never applies to minimizing sequences (see the proof of Theorem 1 in [3]). The complete loss of mass can happen through three mechanisms: dispersion of the sequence, with vanishing $\|u_n\|_{L^p(\mathbb{R})}$ and $\|v_n\|_{L^r(\mathbb{R}^2)}$, escape at infinity through the plane or escape at infinity through the line. It turns out that [3]:

(1) if among the three mechanisms involved by the complete loss of mass the most convenient is the dispersion, then

$$E(U_n) \to E_{\text{lin}}(\Psi_\mu) = E(\Psi_\mu) + \frac{1}{p}\|\psi_\mu\|^p_{L^p(\mathbb{R})} + \frac{1}{r}\|\varsigma_\mu\|^r_{L^r(\mathbb{R}^2)},$$

with $\Psi_\mu = (\psi_\mu, \varsigma_\mu)$ the Ground State at mass μ of the quadratic part E_{lin} of E. Therefore, $\mathcal{E}(\mu) = E_{\text{lin}}(\Psi_\mu)$ and

$$\mathcal{E}(\mu) \leqslant E(\Psi_\mu) = E_{\text{lin}}(\Psi_\mu) - \frac{1}{p}\|\psi_\mu\|^p_{L^p(\mathbb{R})} - \frac{1}{r}\|\varsigma_\mu\|^r_{L^r(\mathbb{R}^2)} < \mathcal{E}(\mu),$$

which is absurd. So there cannot be complete loss of mass by dispersion;

(2) if the most convenient mechanism is the escape at infinity through the plane, then $E(U_n) \to E_\infty(\xi_\mu, \Pi)$, where ξ_μ is the soliton of the standard nonlinear Schrödinger equation in dimension two. But in this case one has $\Xi_\mu = (0, \xi_\mu) \in \mathcal{D}^\mu$ and so Ξ_μ is a Ground State and one proves existence.

(3) If, finally, the most convenient mechanism is the escape at infinity through the line, then $E(U_n) \to E_0(\varphi_\mu, \ell)$. Of course the state $\Phi_\mu = (\varphi_\mu, 0)$ belongs to $\mathcal{D}^\mu$, but

$$E(\Phi_\mu) = E_0(\varphi_\mu, \ell) + \frac{\alpha}{2}|\varphi_\mu(0)|^2,$$

thus if $\alpha > 0$, then the escape at infinity is a convenient choice for the soliton.

The previous discussion is aimed at clarifying why the only hindrance to the existence of a Ground State is the possibility, for a minimizing sequence, to run to infinity through the line. In other words, the plane is not dangerous since it can always host the two-dimensional soliton centered at the origin, equalling the optimal energy $E_\infty(\xi_\mu, \Pi)$ reached when escaping through the plane. In fact, it has been proved in [1] that on the plane it is possible to make better: for every $\sigma \in \mathbb{R}$ there exist a Ground State at mass μ whose energy is strictly less than $E_\infty(\xi_\mu, \Pi)$. Then one can say that the plane is strictly attractive, regardless of the sign of σ. An analogous result for the analogous three-dimensional problem was proved in [2].

Gathering together the previous comments, one ends up with the following

Corollary 3.7 *Inequality (25) and therefore the existence of a Ground State at mass μ for the energy functional (10) is guaranteed by one of the following conditions:*

(1) *The contact interaction between the plane and the line is non-repulsive on the line, i.e. $\alpha \leqslant 0$. This entails that a soliton at the infinity on the line is not less energetic than the same soliton centred at the origin, and therefore the escape at infinity along the line is not strictly convenient.*

(2) *Given the mass μ, the nonlinearity powers p and r are such that the plane is energetically more convenient than the line, in the sense that*

$$E_\sigma(\varsigma_\mu, \Pi) \leqslant E_\alpha(\varphi_\mu, \ell).$$

More precisely, this is accomplished by the following condition: there exists a threshold value $\mu^\star$ for the mass, that depends on p and r, such that

$$\text{if } \mu > \mu^\star \text{ and } \frac{2}{4-r} > \frac{p+2}{6-p}$$

$$\textit{or}$$

$$\text{if } \mu < \mu^\star \text{ and } \frac{2}{4-r} < \frac{p+2}{6-p},$$

then there is a Ground State at mass μ for (10).

(3) *The contact interaction between the plane and the line is attractive enough on the plane, which means that σ lies below a threshold $\sigma^\star$. Indeed, since $\mathcal{E}_{\sigma,\Pi}$ is a monotonically increasing function of σ and*

$$\mathcal{E}_{\sigma,\Pi}(\mu) \to -\infty, \qquad \sigma \to -\infty,$$

then either

$$\mathcal{E}_{+\infty,\Pi}(\mu) < \mathcal{E}_{0,\ell}(\mu)$$

or there exists $\sigma^\star \in \mathbb{R}$ such that

$$\mathcal{E}_{\sigma^\star,\Pi} = \mathcal{E}_{0,\ell}(\mu),$$

Then, for $\sigma \leqslant \sigma^\star$ possibly infinite, there exists a Ground State.

(4) *The junction energy $E_{\mathcal{J}}$ is strong enough, namely $\beta \geqslant \beta^\star$ for a certain threshold $\beta^\star$. In this case, $E_{\mathcal{H}}$ is a monotonically decreasing function of β, so one has*

$$\mathcal{E}(\mu) \to -\infty, \qquad \beta \to +\infty$$

and then the result immediately follows.

Notice that the previous Corollary makes more precise the prescriptions given in Sect. 1.1 in order to construct a trapping hybrid plane.

We point out that the criterion given at (1) is simpler than the analogous condition given in [1] for the hybrid $\mathcal{I}$. This reflects the fact that the problem of the existence of

Ground States on a halfline with a delta interaction at the origin is more complicated than the same problem on the line, as detailed in [14].

3.3 Further Results

We list here for the sake of completeness some results that immediately translate from [3] to this context.

(1) A Ground State always exists for small mass. Indeed, the linear ground state Ψ_μ is energetically more convenient than the one-dimensional soliton, as it scales linearly with the mass while $\mathcal{E}_{0,\ell}$ scales superlinearly.
(2) Fixed $\mu > 0$, if α and σ are large enough, while β is small enough, then there is no Ground State. Here we are in fact denying the hypotheses of existence theorem. Notice that the threshold depends on μ.
(3) If $\beta = 0$ and there exists a Ground State, then it is supported either on ℓ or on Π and it coincides with a Ground State of the problem on ℓ or on Π.
(4) If $\beta \neq 0$, then every Ground State $U = (u, v)$ has non-trivial components u and v. Moreover, $u \in H^2(\mathbb{R}\backslash\{0\}) \cap H^1(\mathbb{R})$, while $v = \phi + q\frac{K_0}{2\pi}$ with $\phi \in H^2(\mathbb{R}^2)$. More precisely, u is an even function made by glueing together two chunks of a soliton, while v is a radially symmetric function whose shape is that of a Ground State for the Nonlinear Schrödinger Equation on the plane with a point interaction at the origin. They are connected by the matching conditions

$$\begin{cases} u'(0+) - u'(0-) = \alpha u(0) - \beta q \\ \phi(0) = -\beta u(0) + \sigma q \end{cases}$$

Remark 3.8 In our search for Ground States we excluded the cases $\alpha = \infty$ and $\sigma = \infty$. This was done because such cases lead to modifications of the functional E and of the energy domain $\mathcal{D}$ and including them would make the formulation of the problem too cumbersome. However, since they are not completely trivial, we quickly summarize the results.

The condition $\alpha = \infty$ amounts to imposing Dirichlet boundary conditions at the origin of the line. As a consequence, both the contact interaction on the line and the coupling term vanish. Therefore one has

$$E(U) = E_\infty(u, \ell) + E_\sigma(v, \Pi) = E_0(u, \ell) + E_\sigma(v, \Pi)$$

on the domain

$$\mathcal{D}_{\text{dir}} = \{U = (u, v),\ u \in H^1(\mathbb{R}) \text{ s.t. } u(0) = 0,\ v \in \mathcal{D}_2\}.$$

The problem reduces then to a competition between ℓ and Π. If p, r, σ and μ are such that the line turns out to be more convenient, then minimizing sequences leave the plane and seek the lowest energy on the line, which is given by the energy of the soliton that cannot be attained because of the Dirichlet boundary condition at the origin. Then there is no Ground State. If vice versa the chosen parameters make the plane energetically convenient with respect to the line, then minimizing sequences concentrate on the plane and converge to the two-dimensional Ground State according to the results of [1].

In the case with $\sigma = \infty$ the line and the plane are decoupled too since, as discussed in Sect. 2.2, such conditions entail $q = 0$. Therefore minimizing sequences concentrate on the most convenient component among ℓ and Π. In the first case, the issue of the existence of a Ground State reduces to that of the existence for $E_\alpha(u, \ell)$. In the second case, a Ground State exists.

If $\alpha = \sigma = \infty$, then the existence of Ground States is determined by the competition between the free line and the free plane.

Lastly, we stress that the energy functional (10) does not exhaust all possible choices for the contact interactions. In fact, if one starts from the linear case and seeks for all possible interactions taking place at the junction, then one is led to consider a nine-parameter family of possible energy functionals. This family includes several rich dynamics as the one generated by the so-called delta-prime interaction, that on ℓ would give rise to singular phenomena like symmetry breaking bifurcation on the Ground State [5, 6]. We plan to investigate such dynamics in a forthcoming study. A further point that we left untouched concerns the uniqueness of Ground States. This is in general a delicate point that requires ad hoc techniques. For instance, for the analysis on uniqueness on metric graphs instead of hybrids see [19].

Acknowledgements F.B. and L.T. have been partially supported by the INdAM GNAMPA project 2023 "Modelli nonlineari in presenza di interazioni puntuali" (CUP E53C22001930001).
R.A., R.C. and L.T. acknowledge that this study was carried out within the project E53D23005450006 "Nonlinear dispersive equations in presence of singularities"—funded by European Union —Next Generation EU within the PRIN 2022 program (D.D. 104-02/02/2022 Ministero dell'Universitá e della Ricerca).

References

1. Adami, R., Boni, F., Carlone, R., Tentarelli, L.: Ground states for the planar NLSE with a point defect as minimizers of the constrained energy. Calc. Var. Partial Differ. Eqn. **61**(5), Paper No. 195, 32 pp (2022)
2. Adami, R., Boni, F., Carlone, R., Tentarelli, L.: Existence, structure, and robustness of ground states of a NLSE in 3D with a point defect. J. Math. Phys. **63**(7), Paper No. 071501, 16 pp (2022)
3. Adami, R., Boni, F., Carlone, R., Tentarelli, L.: NLS Ground states on a hybrid plane (2024). Preprint arXiv:2401.09888

4. Adami, R., Golse, F., Teta, A.: Rigorous derivation of the cubic NLS in dimension one. J. Stat. Phys. **127**, 1193–1220 (2007)
5. Adami, R., Noja, D.: Stability and symmetry-breaking bifurcation for the ground states of a NLS with a δ' interaction. Comm. Math. Phys. **318**(1), 247–289 (2013)
6. Adami, R., Noja, D., Visciglia, N.: Constrained energy minimization and ground states for NLS with point defects. Discrete Contin. Dyn. Syst. Ser. B **18**(5), 1155–1188 (2013)
7. Albeverio, S., Gesztesy, F., Høegh-Krohn, R., Holden, H.: Solvable Models in Quantum Mechanics. Texts and Monographs in Physics. Springer, New York (1988)
8. Anderson, M.H., Ensher, J.R., Matthews, M.R., Wieman, C.E., Cornell, E.A.: Observation of Bose-Einstein condensation in a dilute atomic vapor. Science **269**(5221), 198–201 (1995)
9. Benedikter, N., de Oliveira, G., Schlein, B.: Quantitative derivation of the Gross-Pitaevskii equation. Commun. Pure Appl. Math. **68**(8), 1399–1482 (2015)
10. Berezin, F.A., Faddeev, L.D.: A remark on Schrödinger's equation with a singular potential. Dokl. Akad. Nauk SSSR **137**(5), 1011–1014 (1961)
11. Boccato, C., Brennecke, C., Cenatiempo, S., Schlein, B.: Optimal rate for Bose-Einstein condensation in the Gross-Pitaevskii regime. Commun. Math. Phys. **376**(3), 1311–1395 (2020)
12. Bogoliubov, N.N.: On the theory of superfluidity. Izv. Akad. Nauk. USSR **11** (1947). Engl. Transl. J. Phys. USSR **11**, 23 (1947)
13. Bose, S.N.: Plancks Gesetz und Lichtquantenhypothese. Zeitschrift für Physik **26**(1), 178–181 (1924)
14. Boni, F., Carlone, R.: NLS ground states on the half-line with point interactions. NoDEA Nonlinear Differ. Eqn. Appl. **30**(4), Paper No. 51, 23 pp. (2023)
15. Carlone, R., Exner, P.: Dynamics of an electron confined to a "hybrid plane" and interacting with a magnetic field. Rep. Math. Phys. **67**(2), 211–227 (2011)
16. Carlone, R., Posilicano, A.: A quantum hybrid with a thin antenna at the vertex of a wedge. Phys. Lett. A **381**(12), 1076–1080 (2017)
17. Cazenave, T.: Semilinear Schrödinger Equations, Courant Lecture Notes 10. American Mathematical Society, Providence, RI (2003)
18. Davis, K.B., Mewes, M.-O., Andrews, M.R., van Druten, N.J., Durfee, D.S., Kurn, D.M., Ketterle, W.: Bose-Einstein condensation in a gas of sodium atoms. Phys. Rev. Lett. **75**(22), 3969–3973 (1995)
19. Dovetta, S., Serra, E., Tilli, P.: Uniqueness and non-uniqueness of prescribed mass NLS ground states on metric graphs. Adv. Math. **374**, Paper No. 107352, 41 pp. (2020)
20. Einstein, A.: Quantentheorie des einatomigen idealen Gases, Königliche Preussische Akademie der Wissenschaften. Sitzungsberichte, pp. 261–267 (1924)
21. Erdős, L., Schlein, B., Yau, H.-T.: Derivation of the Gross-Pitaevskii equation for the dynamics of the Bose-Einstein condensation (2010)
22. Erdős, L., Schlein, B., Yau, H.-T.: Derivation of the cubic nonlinear Schrödinger equation from quantum dynamics of many-body systems. Inv. math. **167**(3), 515–614 (2007)
23. Exner, P., Šeba, P.: Quantum motion on a half-line connected to a plane. J. Math. Phys. **28**(2), 386–391 (1987). Erratum: Quantum motion on a half-line connected to a plane [J. Math. Phys. **28**, 386 (1987)], J. Math. Phys. **28**(9), 2254 (1987)
24. Exner, P., Šeba, P.: Mathematical models for quantum point-contact spectroscopy. Czechoslovak J. Phys. B **38**(1), 1–11 (1988)
25. Fukaya, N., Georgiev, V., Ikeda, M.: On stability and instability of standing waves for 2d-nonlinear Schrödinger equations with point interaction. J. Differ. Eqn. **321**, 258–295 (2022)
26. Fukuizumi, R., Jeanjean, L.: Stability of standing waves for a nonlinear Schrödinger equation with a repulsive Dirac delta potential. Discrete Contin. Dyn. Syst. **21**(1), 121–136 (2008)
27. Fukuizumi, R., Ohta, M., Ozawa T., Nonlinear Schrödinger equation with a point defect. Ann. Inst. H. Poincaré C Anal. Non Linéaire **25**(5), 837–845 (2008)
28. Gradshteyn, I.S., Ryzhik, I.M., Geronimus, Y.V., Tseytlin, M.Y.: Table of Integrals, Series, and Products (A. Jeffrey, Ed.), Translated by Scripta Technica, Inc., 3rd edn. (1966)
29. Gross, E.P.: Structure of a quantized vortex in boson systems. Il Nuovo Cimento **20**(3), 454–457 (1961)

30. Knowles, A., Pickl, P.: Mean-field dynamics singular potential and rate of convergence. Commun. Math. Phys. **298**, 101–138 (2010)
31. Kirkpatrick, K., Schlein, B., Staffilani, G.: Derivation of the two-dimensional nonlinear Schrödinger equation from many-body quantum dynamics. Amer. J. Math. **133**(1), 91–130 (2011)
32. Lieb, E.H., Seiringer, R., Solovey, J.-P., Yngvason, J.: The Mathematics of the Bose Gas and its Condensation, Oberwolfach Seminar Series, vol. 34. Birkhäuser (2005)
33. Mishra, H.P., Flores, A.S., Vassen, W., Knoop, S.: Efficient production of an 87Rb, $F = 2$, $m_F = 2$ Bose-Einstein condensate in a hybrid trap. Eur. Phys. Jour. D, **69**(52), 8 pp. (2015)
34. Nam, P.T., Napiórowski, M.: Bogoliubov correction to the mean-field dynamics of interacting bosons. Adv. Theor. Math. Phys. **21**(3), 683–738 (2017)
35. Pitaevskii, L.: Vortex lines in an imperfect Bose gas. Sov. Phys. JETP. **13**(2), 451–454 (1961)
36. Reed, M., Simon, B.: Methods of Modern Mathematical Physics. I. Functional Analysis, 2nd edn. Academic Press Inc., New York (1980)
37. Reed, M., Simon, B.: Methods of Modern Mathematical Physics. II. Fourier Analysis and Self-adjointness, 2nd edn. Academic Press Inc., New York
38. Rodninanski, I., Schlein, B.: Quantum fluctuations and rate of convergence towards mean field dynamics. Commun. Math. Phys. **291**(1), 31–61 (2009)
39. Tentarelli, L.: A general review on the NLS equation with point-concentrated nonlinearity. Commun. Appl. Ind. Math. **14**(1), 62–84 (2023)
40. Zakharov, V.E., Shabat, A.B.: Exact theory of two-dimensional self-focusing and one-dimensional self-modulation of waves in nonlinear media. Zh. Exp. Teor. Fiz. **61**, 118–134 (1971)

From Microscopic to Macroscopic: The Large Number Dynamics of Agents and Cells, Possibly Interacting with a Chemical Background

Thierry Paul

Abstract In this note we review some recent results dealing with the transition between microscopic and macroscopic scales in different fields, including kinetic theory, cells movement in biology, chemotaxis, flocking phenomena and agent systems. The methodology of the mean-field approach of this study uses the concept of marginals instead of the empirical measure paradigm. Numerical computations showing some theoretically unexpected features are presented at the end of the paper.

Keywords Mean field · Chemiotaxis · Agent systems · Biomathematics · Mathematical physics · PDEs

1 Introduction

The passage from micro to macro represents nowadays an important feature when considering large scale situations built up out of small scales ones. Initiated with a great success in the kinetic theory of gases and thermodynamics in physics, as macroscopic descriptions of large number microscopic particle, it has gained lately an increasing importance in biology for the study of large number of moving cells. If in these two situations just mentioned the "entry" of the passage consists in microscopic entities (particle, cells) and the "exit" macroscopic media, macroscopic objects like birds or agent have been considered as entering the passage, leading as output consisting in large scale population movements and social behaviors.

It is quite astonishing that the same mathematical modeling methods can be used to handle situations as diverse as kinetic theory, biology including chemotaxis and, say, social or "animal" dynamics.

One plausible reason of this (sort of) universality might reside in the fact that already in each of theses thematically different fields (physics, biology, sciences of collective behavior) the passage connects very different paradigms: large systems of ODEs for the entrance, non-linear PDEs on phase or configuration spaces for the exit.

T. Paul (✉)
Laboratoire Ypatia des Sciences Mathématiques (LYSM), CNRS, Rome, Italy
e-mail: thierry.paul@cnrs.fr

B. Cassano et al. (eds.), *Singularities, Asymptotics, and Limiting Models*, Springer INdAM Series 64, https://doi.org/10.1007/978-981-96-3584-9_13

These tunnels are therefore more than just innocent conduits, they contains obviously subtle algorithms able to transform paradigms involving trajectories to paradigms endowed with non-linear analysis situations. An these complex structures associated, independently, to each field of interest seem to be rich enough to be exported quasi verbatim from one field to another.

In this article we propose a short review of recent results in the micro-macro (say) transition around several emblematic examples: the standard (Hamiltonian or not) two body interaction, the so-called Cucker-Smale model of flacking and its generalizations, the topological interaction system involving more than two-body interactions, the biological chemotaxis and multi-agents (not indistinguishable) systems.

In order to avoid technicalities and heavy notations so that the format of this review can be kept short, the main results are presented sometimes in a kind of rough way. But they all refer to precises theorems in the quoted references.

Part 1. Overview (chi va piano va sano)

2 Micro Versus Macro (Lagrange Versus Euler)

In this paper I address the following question: how to recover the macroscopic scale, OUR scale, from a microscopic one, the PARTICLES or CELLS or AGENT's one.

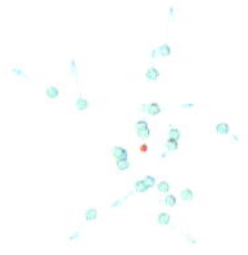

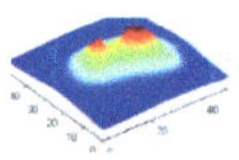

PARTICLES, CELLS, AGENTS US

At the microscopic scale, the dynamic will be given by a type of Newton law, possibly coupled with a chemical scale force deriving from a potential involving a chemotaxis interaction with the continuous chemical media. The model will be therefore a hybrid model ODE/PDE.

The total panorama can be dynamically described as follows.

The first scale is the cells (particles) one:

$$N\ cells$$

The second scale is the mean-field (self consistent) limit on phase space:

$$N\ cells \xrightarrow[\to\infty]{N} \begin{array}{c}(Vlasov)\\ equation\\ phase\ space\end{array}$$

The third scale is the hydrodynamical e with densities and velocity field:

$$N\ cells \underset{\to\infty}{\overset{N}{\longrightarrow}} \begin{matrix} (Vlasov) \\ equation \\ phase\, space \end{matrix} \underset{hier.}{\overset{closing}{\longrightarrow}} \begin{matrix} (Euler) \\ position \\ velocity \\ physical\, space \end{matrix}$$

The fourth one the description with densities only:

$$N\ cells \underset{\to\infty}{\overset{N}{\longrightarrow}} \begin{matrix} (Vlasov) \\ equation \\ phase\, space \end{matrix} \underset{hier.}{\overset{closing}{\longrightarrow}} \begin{matrix} (Euler) \\ position \\ velocity \\ physical\, space \end{matrix} \longrightarrow \begin{matrix} (biological) \\ position \\ physical\, space \\ Keller-Segel \end{matrix}$$

Finally a pre-chemical scale completes the panorama:

$$\boxed{\begin{matrix} chemical \\ scale \end{matrix} \longrightarrow \begin{matrix} N\ cells \\ coupled \\ to \\ chemestry \end{matrix} \underset{\to\infty}{\overset{N}{\longrightarrow}} \begin{matrix} (Vlasov) \\ equation \\ phase\, space \end{matrix} \underset{hier.}{\overset{closing}{\longrightarrow}} \begin{matrix} (Euler) \\ position \\ velocity \\ physical\, space \end{matrix} \longrightarrow \begin{matrix} (biological) \\ position \\ physical\, space \\ Keller-Segel \end{matrix}}$$

But the dichotomy micro/macro can also refer simply to a difference of scale between (macroscopic) animals, birds or agents, and crowd formed by a large number of them. What we called micro becomes now macro seen form far.

3 Newton Law

It is quite astonishing to notice that all the (generalized as just mentioned) microscopic situations considered in the previous section can fall under the scope of Newton type equations.

We will therefore consider a continuous family of mapping in a configuration space:

$$x_i(t) \in \mathbb{R}^d,\ i = 1, \dots, N,\ d = 1, 2, 3, t \in \mathbb{R},$$

subject to an equation of the form:

$$m_i \frac{d^2 x_i}{dt^2}(t) = F_i\Bigg(t, x_1(t), \ldots, x_N(t), \frac{dx_1}{dt}(t), \ldots, \frac{dx_N}{dt}(t),$$
$$\{x_j(s), j = 1, \ldots, N, 0 \le s < t\}\Bigg)$$

giving rise to a trajectory $(t \to x_1(t), \ldots, x_N(t))$ in $\mathbb{R}^{dN}$, possibly NOT given by a flow because of the dependence of F in the full history $\{x_j(s), j = 1, \ldots, N, 0 \le s < t\}$.

Let us recall the (usual) trick consisting in defining $v_i := m_i \dot{x}_i \ \Rightarrow$, the second order Newton ODE becomes a system of two first order ODE:

$$\begin{cases} \dot{x}_i(t) = \frac{v_i(t)}{m_i} \\ \dot{v}_i(t) = F_i\left(t, x_1(t), \ldots, x_N(t), v_1(t), \ldots, v_N(t), \{x_j(s), j = 1, \ldots, N, 0 \le s < t\}\right). \end{cases} \quad (1)$$

4 Examples

Let us start by giving several examples we will be concerned with through this articles.

4.1 *Examples I: Two Body Interaction* ($m_i = 1$)

In this case, the dynamics is given by a two-particle function G of the positions only, possibly the gradient of a potential, i.e. the situation given by (1) with $F_i(\ldots) = \frac{1}{N}\sum_{j=1}^N F(x_i(t), x_(t))$ for a certain function F:

$$\dot{v}_i(t) = \frac{1}{N}\sum_{j=1}^N F(x_i(t), x_j(t)) \qquad i = 1, \ldots, N$$
$$(\text{e.g.}) = \frac{1}{N}\sum_{j=1}^N \nabla V(x_i(t) - x_j(t))$$

It is well know that, even in this simple case, the dynamics can be very complicated, exhibiting chaotic behavior.

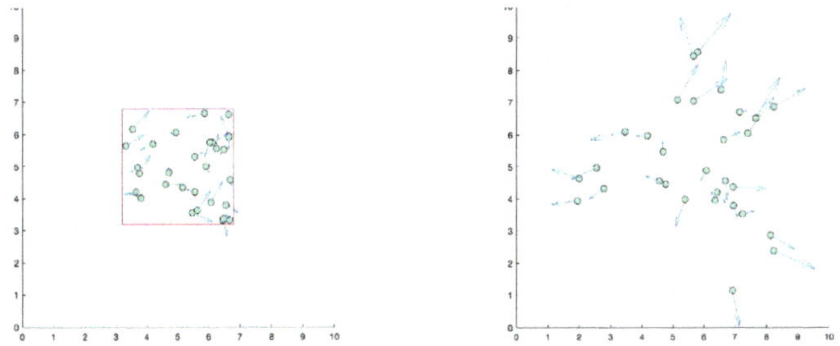

$t = 0$ $\qquad$ $t \gg 0$

4.2 Examples II: Cucker-Smale and All that

In this case, the dynamics of the velocities involves also a dependence in themselves of the following form

$$\dot{v}_i(t) = \frac{\lambda}{N} \sum_{j=1}^{N} \frac{1}{\left(1 + \frac{\|x_i(t) - x_j(t)\|^2}{R^2}\right)^{\beta}} \big(v_i(t) - v_j(t)\big)$$

leading to flocking phenomena

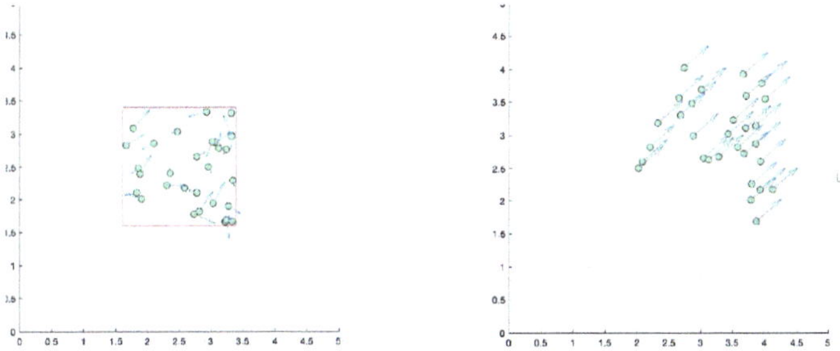

$t = 0$ $\qquad$ $t \gg 0$

as seen by looking at birds flying in the sky of Rome in the fall.

4.3 Examples III: Topological Interaction

Another example of dynamics leading to flocking, with this time a topological (not metric) interaction:

$$\dot{v}_i(t) = \frac{1}{N}\sum_{j=1}^{N} K(M(x_i, |x_i - x_j|)(v_i(t) - v_j(t))$$

with K smooth (Lipschitz) decreasing function and $M(x, r) = \frac{1}{N}\#\{x_k,\ |x_k - x| \leq r\}$.

4.4 Examples IV: Chemotaxis

This model consists in coupling particles with interactions with chemistry

$$\dot{v}_i(t) = \frac{1}{N}\sum_{j=1}^{N} F(x_i(t) - x_j(t), v_i(t) - v_j(t)) + \eta\nabla_x\varphi^t(x_i(t))$$

where φ satisfies the parabolic equation

$$\partial_t\varphi^t(x) = \Delta\varphi^t(x) - \kappa\varphi^t + \frac{1}{N}\sum_{j=1}^{N}\chi(x - x_i(t)),$$

Here η, κ are positive constants and χ is the (Lipschitz mollified) characteristic function of a ball.

4.5 Examples V: Multi-agents

All the four examples above share the property of preserving the indistinguishability of the involved particles or cells. A multi-agents system doesn't share this property anymore. It takes the form (see [13, 14])

$$\dot{\xi}_i(t) = \frac{1}{N}\sum_{j=1}^{N} G_{i.j}(\xi_i(t), \xi_j(t)), \quad \xi_i(t) \in \mathbb{R}^d,\ i = 1, \ldots, N \tag{2}$$

namely the one of a two-body index-dependent interaction.

An example of multi-agents system is the opinion systems (Krause):

$$\dot{\xi}_i(t) = \frac{1}{N}\sum_{j=1}^{N} \sigma_{i.j}(xi_i(t) - xi_j(t)), \quad \xi_i(t) \in \mathbb{R}^d,\ i = 1, \ldots, N.$$

which leads to consensus à la Cucker-Smale.

4.6 Examples V Bis: the Inverse Way

Is there always a microscopic system associated, through the large number of agents limit to any macroscopic system?

A quasilinear PDE

$$\partial_t u^t(x) = \sum_{j=0}^{p} a_j(u^t(x), x) D_x^l u^t(x),$$

can bee seen as the hydrodynamics mean-field limit of a (singular) agent system (see [15]).

Part 2. Methodology (empirical-free)

5 Microscopic Probability (Liouville)

The approach of the mean-field limit presented in this paper relies fully on the study of the propagation of probability measures on phase-space by the 'flow" generated by (1).

Indeed, instead of considering all the trajectories solutions of (1), we consider the mapping on phase-space $\Phi_N^t : \mathbb{R}^{2dN} \mapsto \mathbb{R}^{2dN}$ which sends an initial condition

$$(x_1(0), \ldots, x_N(0), v_1(0), \ldots, v_N(0))$$

to

$$\Phi_N^t(x_1(0), \ldots, x_N(0), v_1(0), \ldots, v_N(0)) := (x_1(t), \ldots, x_N(t), v_1(t), \ldots, v_N(t))$$

solution of (1).

We consider moreover a probabilistic distribution $\rho_N^0(x_1, \ldots, x_N; v_1, \ldots, v_N)$ on $\mathbb{R}^{2dN}$ whose interpretation is the following:

> $\rho_N^0(x_1, \ldots, x_N; v_1, \ldots, v_N)$ is the
> probability that each particle i is originally at position x_i with velocity v_i.

We can now propagate ρ_N^0 by "pushing it forward"[1] to

$$\rho_N^t = (\Phi_N^t)_* \rho_N^0$$

Obviously

$\rho_N^t(x_1, \ldots, x_N; v_1, \ldots, v_N)$ is the probability that each particle i is at time t at position x_i with velocity v_i.

That way we do not consider the trajectories one-by-one (Lagrangian view) but handle all of them in a row(Euler view).

Of course, ρ_N^t becomes intractable when N becomes large, but part of it remains at the limit $N \to \infty$, after loosing part of its information. This is what we will see i the next section.

6 None But One: Marginals (Vlasov)

The key idea in order to control the large N behavior of microscopic N-particle probabilities consist in averaging over all particle but one. That is to integrate the density over all variable but one: $\int \ldots \int dx_2 \ldots dx_N dv_2 \ldots dv_N$. Note that in case of densities invariant by permutations of the particles, that is invariant by any change $(x_i, v_i) \leftrightarrow (x_j, v_j)$, the choice of the selected remaining variable, the "one", is obviously inoperative. This is indeed the case in physical situations where the particles/cells are indistinguishable.

We define the (first) marginal of ρ_N^t by

$$\rho_{N;1}^t(x, v) := \int \dot{\int} \rho_N^t(x, x_2, \ldots, x_N; v, v_2, \ldots, v_N) dx_2 \ldots dx_N dv_2 \ldots dv_N$$

There are three important, and not obvious, facts concerning this definition:

(1) as $N \to \infty$, $\rho_{N;1}^t(x, v)$ has a limit $\rho^t(x, v)$
(2) $\rho^t(x, v)$ is a probability measure
(3) when the initial condition ρ_N^0 has the tensorial power factorized form $\rho_N^0 = (\rho^0)^{\otimes N}$, with ρ^0 probability on $\mathbb{R}^{2d}$, ρ^t is a solution to a *mean-field type (Vlasov)* **nonlinear** equation of the following general form in the case of Sect. 4.4

[1] We recall that the pushforward of a measure μ by a measurable function Φ is $\Phi_*\mu$ defined by $\int \varphi d(\Phi_*\mu) := \int (\varphi \circ \Phi) d\mu$ for every test function φ.

$$\partial_t \rho^t(x,v) + v\partial_x \rho^t(x,v) =$$

$$\partial_v \left(\left(\int G(x,x';v,v')\rho(x',v')dx'dv' + \eta\nabla_x\varphi^t(x) \right) \rho(x,v) \right) \quad (3)$$

$$\partial_s \psi^s(x) = \Delta\psi^s(x) - \kappa\psi^s + \chi * \rho^s(x) \quad 0 \leq s \leq t,$$

with initial condition ρ^0.

The meaning of (4) is easy to catch when no chemical interaction is present, i.e. when $\eta = 0$. In this case the equation reduces to the equation satisfied by the push forward of the initial condition ρ^0 by the one particle dynamical system obtained by replacing the two body interaction G by the average on the second of the two bodies by the solution itself (self-consistent field). It therefore reflect the fact that only one of the particle is kept moving, all the other being replaced by an average pondered by the density itself.

7 Propagation of Chaos (Recovering Independence)

The tensorial power property of the initial condition $\rho_N^0 = (\rho^0)^{\otimes N}$ is of fundamental importance for the derivation of the mean-field equation. Its meaning is that one starts with N independent particles.

Of course this property immediately breaks down after the dynamics starts, i.e. $\rho_N^t \neq (\rho^t)^{\otimes N}$ for $t > 0$ (except for trivial examples of dynamics): the dynamics couples particles and kills the independence property.

Nevertheless some trace of it remains valid when considering higher order marginals, namely j particles probabilities of the form

$$\begin{aligned}&\rho_{N;j}^t(x_1,\ldots,x_j,v_1,\ldots,v_j)\\&:= \int\int \rho_N^t(x_1,\ldots,x_j,x_{j+1},\ldots,x_N;v_1,\ldots,v_j,v_{j+1},\ldots,v_N)dx_2\ldots dx_N dv_2\ldots dv_N\end{aligned}$$

for $j \leq N$ kept fixed.

Propagation of chaos states that, at any time $t > 0$ and for any j,

$$\lim_{N\to\infty} \rho_{N;j}^t = (\rho^t)^{\otimes j}.$$

8 Velocity and Density On (Our) Physical Space (Euler)

Marginals again: in order to pass from a (single) one-particle function on phase-space, such as the solution ρ^t of the Vlasov equation, to two functions on physical space, we define

$$\underbrace{\mu^t(x) = \int \rho^t(x, v)dv}_{\text{density}} \qquad \underbrace{t(x) = \int v\rho^t(x, v)dv}_{\text{velocity field}}$$

Associated to these two quantities arises naturally by induction from hydrodynamics the following Euler system of partial differential equations on physical space:

$$\begin{cases} \partial_t \mu^t + \nabla(u^t \mu^t) = 0 \\ \partial_t(\mu^t u^t) + \nabla(\mu^t (u^t)^{\otimes 2}) = \int f(\cdot - x'; u^t(\cdot) - u^t(x')\mu^t(x')dx' + \eta \nabla \psi^t \\ \partial_s \psi^s = \Delta \psi^s - \kappa \psi^s + \chi * \rho^s \end{cases} \tag{4}$$

A natural question then pops out of (4); wondering in which situations the density and velocity fields of a solution for the Vlasov equation (3) are solution for the Euler system (4).

And indeed this is true:

- when $\rho^t(x, v) = \mu^t(x)\delta(v - u^t(x))$ monokinetic ansatz
- for the case of the Krause model for agents systems (see below)
- sometimes numerically, with both Cucker-Smale and chemotaxis (see Sect. 15).

9 Density Only On (Our) Physical Space (Keller-Segel)

The phenomenological Keller-Segel describes the collective motion of cells by using the density only, with no use of the velocity field [8, 16, 17]. It reads

$$\begin{cases} \partial_t \mu^t - \Delta \mu^t - \nabla(\mu^t \varphi^t) = 0 \\ \alpha \partial_s \varphi^s = \Delta \varphi^s - \kappa \varphi^s + \mu^s, \ s \in [0, t]. \end{cases} \tag{5}$$

In [17, Sect. 5.2] was considered the case $\alpha = 0$ and in [17, Sect. 5.5.1] the case where the Laplacian in (5) is replaced, e.g., by $\nabla(\mu^t \nabla \cdot)$. With these changes, the system (5) becomes

$$\begin{cases} \partial_t \mu^t = \nabla(\mu^t \nabla \mu^t) + \nabla(\mu^t \varphi^t) \\ \Delta \varphi^t = \kappa \varphi^t - \mu^t. \end{cases} \tag{6}$$

This system can be derived from a two-body chemotaxis system with friction of the form

$$\begin{cases} \dot{x}i(t) = & \frac{v_i(t)}{m} \\ \dot{v}_i(t) = & \frac{1}{N}\sum_{j=1}^{N} \nabla V^\varepsilon(x_i(t)-x_j(t)) + \eta\nabla_x\varphi^t(x_i(t)) - v_i \\ \partial_t\varphi^t(x) = & \Delta\varphi^t(x) - \kappa\varphi^t + \frac{1}{N}\sum_{j=1}^{N} \chi^\varepsilon(x-x_i(t)), \end{cases} \tag{7}$$

Having in mind a large mass limit, we set $m = \varepsilon^{-1}$.

A simple computation shows that, after a rescaling of time by ε, the particle system (7) leads to the following Euler system

$$\begin{cases} \partial_t\mu^t + \nabla(u^t\mu^t) = 0 \\ \varepsilon\partial_t(\mu^t u^t) + \nabla(\mu^t(u^t)^{\otimes 2}) = \mu^t \int \nabla V^\varepsilon(\cdot - y)\mu^t(y)dy + \eta\mu^t\nabla\psi^t - u^t\mu^t \\ \varepsilon\partial_s\psi^s = \Delta\psi^s - \kappa\psi^s + \chi^\varepsilon * \rho^s. \end{cases} \tag{8}$$

Taking V^ε and χ^ε as approximations of the Dirac mass and letting $\varepsilon \to 0$ in (8) gives

$$u^t\mu^t = \eta\mu^t\nabla\psi^t - +\mu^t\nabla\mu^t \tag{9}$$

and

$$\Delta\psi^t = \kappa\psi^t - \rho^t \tag{10}$$

so that we get (6) after u^t has been eliminated from the first equation in (8) thanks to (9), and having replaced the third equation in (8) by (10).

A rigorous proof of this derivation, obtained by linking the limit $\varepsilon \to 0$ to $N \to \infty$ in (7), together with other Keller-Segel type non-local equations corresponding to other asymptotic behaviors of the different parameters entering in (7), can be found in [12].

Part 3: Theorems (fini de rire ! (J. Lacan))

We will now present rigorous results concerning the five examples already described.

As we saw in Sect. 4, all these examples are particular cases of the system (1):

$$\begin{cases} \dot{x}_i(t) = & \frac{v_i(t)}{m_i},\ m_i \in (0,+\infty],\ i = 1,\dots,N \\ \dot{v}_i(t) = & F_i\left(t, x_1(t),\dots,x_N(t), v_1(t),\dots,v_N(t), \{x_j(s), j=1,\dots,N, 0 \le s < t\}\right). \end{cases}$$

In the sequel we will denote by $\rho_{N;s}$, $s = 0,\dots,N$ the s-marginal of a probability ρ on $\mathbb{R}^{2dN}$ defined by

$$\begin{aligned} &\rho_{N,s}(x_1,\dots,x_s,v_1,\dots,v_s) \\ &:= \int_{\mathbb{R}^{2d(N-s}} \rho(x_1,\dots,x_N,v_1,\dots,v_N)dx_{s+1}dv_{s+1}\dots dx_N dv_N) \end{aligned} \tag{11}$$

with the convention $\rho_{N;0} = \rho$ and $\rho_{N,N} = 1$.

Let us recall that we denote in Sect. 5 the mapping Φ_N^t which associates to any initial condition the corresponding solution of (1).

In the sequel we will use the two Wasserstein distances defined as follow ([18, 19]).

Definition 9.1 (*Wasserstein distances*) For all probability measures μ_1, μ_2 on $\mathbb{R}^m$

$$W_1(\mu_1,\mu_2) = \sup \left\{ \int_{\mathbb{R}^m} f\, d(\mu_1-\mu_2) \ \mid\ f \in \mathrm{Lip}(\mathbb{R}^m),\ \mathrm{Lip}(f) \le 1 \right\}.$$

For all probability measures μ, ν on $\mathbb{R}^m$ with finite second moments

$$W_2(\mu,\nu)^2 = \inf_{\gamma\in\Gamma(\mu,\nu)} \int_{\mathbb{R}^m\times\mathbb{R}^m} |x-y|^2\gamma(dx,dy)$$

where $\Gamma(\mu,\nu)$ is the set of probability measures on $\mathbb{R}^m \times \mathbb{R}^m$ whose marginals on the two factors are μ and ν.

In the sequel

- Φ_N^t will be the mapping defined through (1):

$$\Phi_N^t(x_1(0),\ldots,x_N(0)t; v_1(0),\ldots,v_N(0))$$
$$:= (x_1(t),\ldots,x_N(t); v_1(t),\ldots,v_N(t))\text{solution to 1}$$

- ρ^t will be solution of the Vlasov equation (3) with initial condition ρ^{in}
- ψ^t ill be the potential part of the solution to the Vlasov equation (3) with initial condition $\psi^{in} = 0$.

10 Bounded Two-Body (Hamiltonian) Interactions

In this case, all $m_i = 1$ and $F_i(\ldots) = -\frac{1}{N}\sum_{j=1}^N \nabla V(x_i - x_j)$, ∇V Lipschiz.

The Vlasov equation in this case is the "standard" one and reads

$$\partial_t\rho(x,v) = -v\cdot\nabla_x\rho(x,v) + \int \nabla V(x-y)\rho(y,w)dydw\cdot\nabla_v\rho(x,v). \tag{12}$$

The result is the following.

Theorem 10.1 ([6]) For each j,

$$W_2\left((\Phi^t_N\#(\rho^{in})^{\otimes N})_{N;j}, (\rho^t)^{\otimes j}\right) \le e^{\Lambda t} N^{-\frac{1}{2}}$$

for some $\alpha, \Lambda > 0$.

11 Cucker-Smale Type Systems

In this case all the m_i are equal to 1 and

$$F_i(t, X, V) = \frac{1}{N}\sum_{j=1}^{N} \gamma(v_i - v_j, x_i - x_j) \tag{13}$$

γ is the collective interaction function supposed to be Lipschitz continuous.

The Vlasov equation, first stated in [7] reads

$$\partial_t \rho(x, v) = -v \cdot \nabla_x \rho(x, v) + \nabla_v \cdot \left(\left(\int \gamma(x - y, v - w)\rho(y, w) dy dw \rho(x, v) \right) \right). \tag{14}$$

Note that when $\gamma(x, v) = \nabla V(x)$ (14) reduces to (12).

The result holds true.

Theorem 11.1 ([10]) For each j,

$$W_2\left((\Phi^t_N\#(\rho^{in})^{\otimes N})_{N;j}, (\rho^t)^{\otimes j}\right) \le C_t N^{-\alpha}$$

for some $\alpha > 0$.

12 Topological Models

In this case one takes $m_i = 1, i = 1, \ldots, N$ and

$$F_i(\ldots) = \tfrac{1}{N}\sum_{j=1}^{N} K\big(M(x_i, |x_i - x_j|)\big)(v_i - v_j), \tag{15}$$

where $K\colon [0, 1] \to \mathbb{R}^+$ is a positive Lipschitz continuous non-increasing function, and, for $r > 0$, the rank function

$$M(x_i, r) = \frac{1}{N}\sum_{k=1}^{N} \mathcal{X}\{|x_k - x_i| \leq r\} \tag{16}$$

counts the number of agents at a distance less than or equal to r from x_i, normalized with N. Note that in this case p_{ij} is a stepwise function of the positions of all the agents and F_i is not of type "two body" anymore, it depends on all agents.

The Vlasov equation was introduced in [1] and reads

$$\partial_t \rho(x, v) = -v \cdot \nabla_x \rho(x, v) + \tag{17}$$

Note that it is not quadratic in ρ anymore.

The result is the following (see also [3–5]).

Theorem 12.1 ([2]) For each j,

$$W_1\left((\Phi_N^t \#(\rho^{in})^{\otimes N})_{N;j}, (\rho^t)^{\otimes j}\right) \leq C_t N^{-\alpha}$$

for some $\alpha > 0$.

13 Chemotaxis

In this case all the m_i are equal to 1 and

$$F_i(t, X, V) = \frac{1}{N}\sum_{j=1}^{N} \gamma(v_i - v_j, x_i - x_j) + \eta \nabla_x \varphi^t(x_i) + F_{ext}(x_i), \tag{18}$$

γ is the collective interaction function supposed to be Lipschitz continuous, F_{ext} is an external force and φ satisfies the equation

$$\partial_s \varphi^s(x) = D\Delta_x \varphi - \kappa\varphi + f(x, X(s)),\ s \in [0, t],\ \varphi^{s=0} = \varphi^{in} \tag{19}$$

for some $\kappa, D, \eta \geq 0$ and function f of the form

$$f(x, X) = \frac{1}{N}\sum_{j=1}^{N} \chi(x - x_i),\ \ \chi \in C_c^1. \tag{20}$$

The Vlasov equation, in fact a system when considering the potential φ introduced in [11] reads

$$\begin{cases} \partial_t \rho^t(x,v) + v\partial_x \rho^t(x,v) \\ = \partial_v \left(\left(\int F(x-x'; v-v')\rho(x',v')dx'dv' + \eta \nabla_x \varphi^t(x) \right) \rho(x,v) \right) \\ \rho^{t=0} := \rho^{in} \\ \partial_s \psi^t(x) = \Delta\psi(x) + \chi * \rho^s, \quad 0 \le s \le t, \quad \psi^{s=0} = 0 \end{cases} \tag{21}$$

In order to state the result in this case, we need to fix the notation.

With the notation $Z^{in} : (X_N^{in}, V_N^{in}) \to Z^t = (X^t, V^t) = \Phi_N^t(Z^{in})$, $X^t = (x_1(t), \cdot, x_N(t))$ and the same for V and Z,

The *non-Markovian* particle system reads then

$$\begin{aligned} \dot{x}_i = v_i, \ \dot{v}_i(t) = & \quad \frac{1}{N}\sum_{j=1}^N F(x_i(t)-x_j(t), v_i(t)-v_j(t)) + \eta \nabla_x \varphi^t(x_i(t)) \\ Z^{t=0} & := Z^{in} \\ \partial_t \varphi^t_{Z^{on}}(x) = & \quad \Delta \varphi^t_{Z^{on}}(x) + \frac{1}{N}\sum_{j=1}^N \chi(x - x_i(t)), \end{aligned} \tag{22}$$

since the solution of (22) at time t depends on the whole trajectory $\{X^s, \ 0 \le s \le t\}$ that is on Z^{in}. By the same argument, the last equation of the Vlasov system (21) should be stated this way:

$$\partial_s \psi^t_{\rho^{in}}(x) = \Delta \psi^t_{\rho^{in}}(x) + \chi * \rho^s, \quad 0 \le s \le t, \quad \psi^{s=0} = 0$$

The question which naturally arises is then: how to compare $\varphi^t_{Z^{in}}$ (particles) which depends on Z^{in} to $\psi^t_{\rho^{in}}$ (Vlasov) which depends on ρ^{in}?

The answer is given in our next result.

Theorem 13.1 ([11])

Let $\varphi^t_{Z^{in}}$ the chemical potential of the particle system starting at X^{in}, and let $\psi^t_{\rho^{in}}$ the chemical potential solution to the Vlasov system (21).

Then

$$W_2\left((\Phi_N^t \# (\rho^{in})^{\otimes N})_{N;1}, \rho^t \right)^2 \le \tau(t) N^{-\frac{1}{d}} \quad (d > 2)$$

and

$$\int_{\mathbb{R}^{2dN}} \|\nabla \varphi^t_{Z^{in}} - \nabla \psi^t_{\rho^{in}}\|^2_\infty (\rho^{in})^{\otimes N}(dZ^{in}) \le \tau_c(t) N^{-\frac{1}{d}} \quad (d > 2)$$

14 Hydrodynamics Limit

In the case of Vlasov evolution with a monokinteic initial condition $\mu^{in}(x)\delta(v - u^{in}(x))$, the following result holds true in the chemotaxis case.

Theorem 14.1 ([11]) Let μ^t, u^t solves the Euler system:

$$\begin{cases} \partial_t \mu^t + \nabla(u^t \mu^t) = 0 \\ \partial_t(\mu^t u^t) + \nabla(\mu^t (u^t)^{\otimes 2}) = \int f(\cdot - x'; u^t(\cdot) - u^t(x')\mu^t(x')dx' + \eta \nabla \psi^t \\ \partial_s \psi^s = \Delta \psi^s - \kappa \psi^s + \psi^s * \rho^s \end{cases}$$

Then $\mu^t(x)\delta(v - u^t(x))$ solves the Vlasov system (21) with initial condition $\mu^{in}(x)\delta(v - u^{in}(x))$.

A similar result is valid in the topological interactions type situation, see [2].

15 Numerical Intermezzo [9]

In this review we have considered three different ways of handling large number of particle systems: the ODE original (and true) vision (1), its mean-field limit Vlasov equation (3) and its hydrodynmaics picture Euler equation (4).

Let us briefly show in this section some numerical results (therefore done for a finite number N of particles) showing first that the kinetic (Vlasov) description is quite faithful to the original particle one, and second how the hydrodynamical picture (Euler) can be compared to the kinetic one.

15.1 Particles $\longrightarrow$ Vlasov

Here we compare the particle evolution with the corresponding Vlasov one for three different times. The two first simulations correspond to a flocking situation (with zero momentum of the center of gravity), and the third one a non-flocking one.

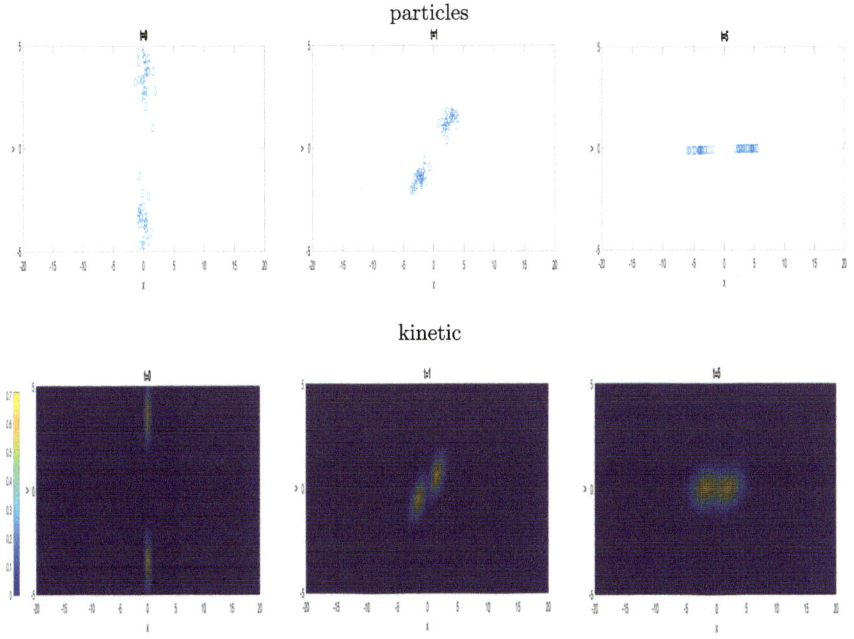
particles
kinetic

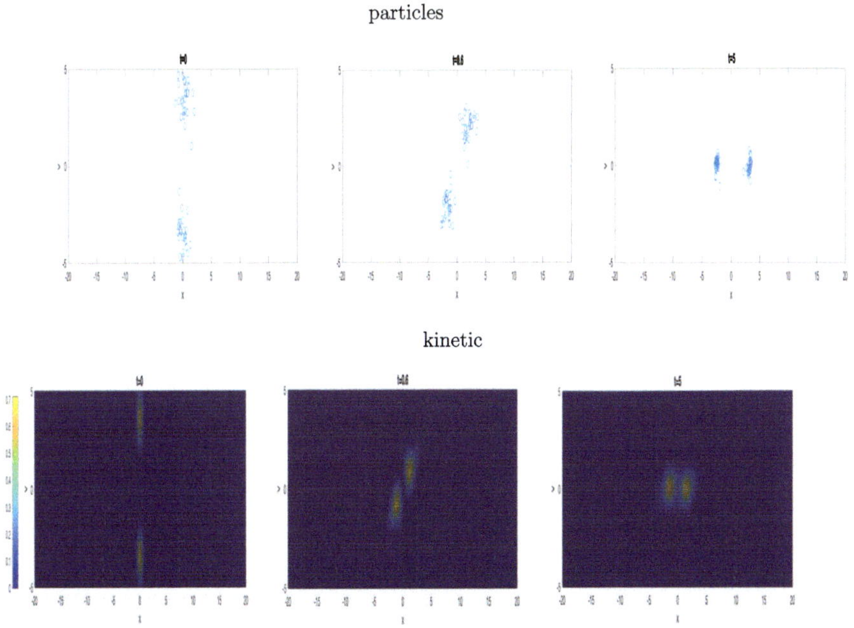
particles
kinetic

particles

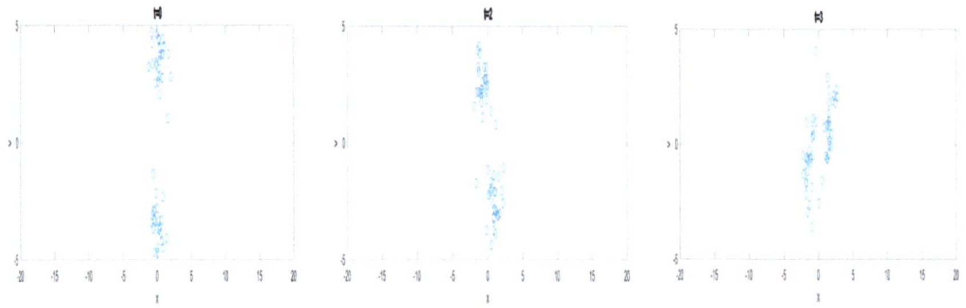

kinetic

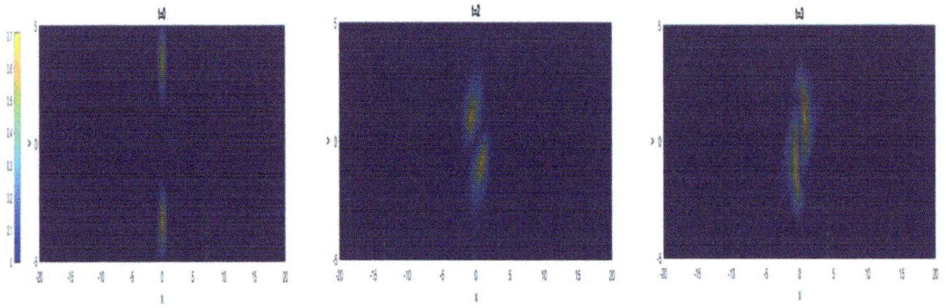

15.2 Vlasov ⟶ Euler

Let us compare now the evolution through Vlasov to the one of the density and velocity fields through Euler. More precisely, we compare the density and the velocity field of the solution to the Vlasov equation (blue curve) with the evolution through Euler equation of the density and velocity field of the same initial condition to Vlasov (dashed curves).

The first line shows the initial condition, the second and third ones the evolution (at time $t = 2$) without chemotaxis and with chemotaxis respectively.

15.3 Moral

Instead of commenting each figure, let us exhibit the main features of these numerical computations (see [9] for (much) more details):

- in many situations the most striking features of the finite particle level are preserved by passing to the one of Vlasov

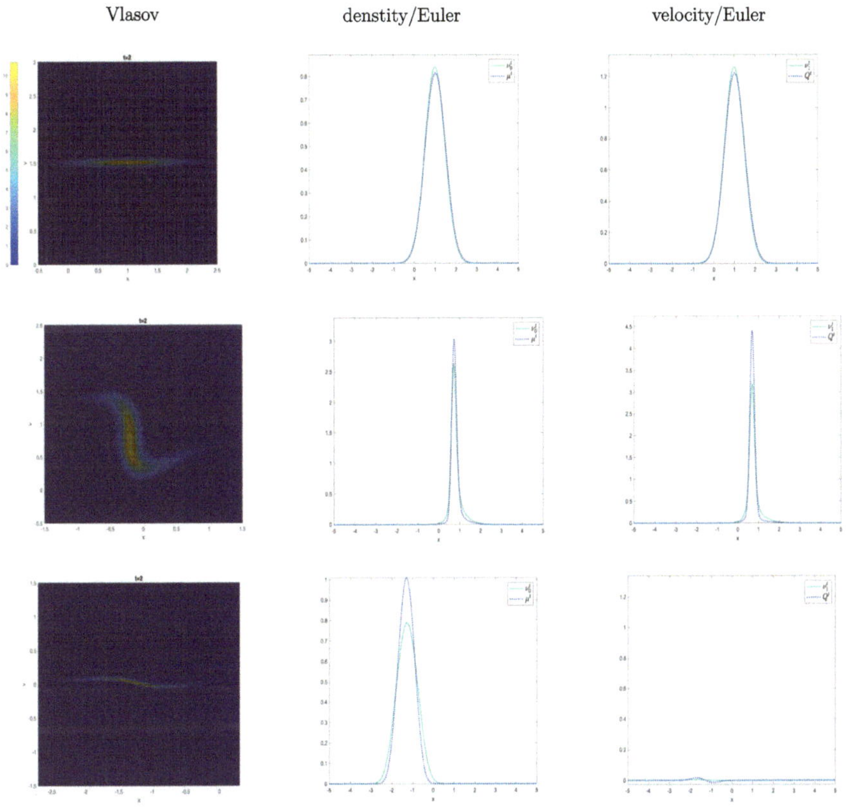

- the fidelity of Euler versus Vlasov is increased by the presence of the chemical interaction
- and this even without monokineticity hypothesis (therefore outside of the paradigm of strict equivalence Vlasov $\sim$ Euler
- the nonlocal integral Euler system keeps memory of the interactions at the microscopic level
- adding an additional pressure term of size ϵ, still keeping the nonlocal integral term but with no equivalent term in Vlasov, helps
- there exists of an optimal value ϵ realizing an improved correspondence between Vlasov moments and Euler solutions.

References

1. Benedetto, D., Caglioti, E., Rossi, S.: Mean-field limit for particle systems with topological interactions. Math. Mech. Complex Syst. **9**, 423–440 (2021)
2. Benedetto, D., Paul, T., Rossi, S.: Propagation of chaos and hydrodynamic description for topological models, preprint hal-04102006v1

3. Blanchet, A., Degond, P.: Kinetic models for topological nearest-neighbor interactions. J. Stat. Phys. **169**, 929–950 (2017)
4. Degond, P., Pulvirenti, M.: Propagation of chaos for topological interactions. Ann. Appl. Prob. **29**, 2594–2612 (2019)
5. Degond, P., Pulvirenti, M., Rossi, S.: Propagation of chaos for topological interactions by a coupling technique (2022). arXiv:2212.08884
6. Golse, F., Mouhot, C., Paul, T.: On the mean field and classical limits of quantum mechanics. Commun. Math. Phys. **343**, 165–205 (2016)
7. Ha, S.-Y., Tadmor, E.: From particle to kinetic and hydrodynamic description of flocking. Kinet. Relat. Models **1**, 415–435 (2008)
8. Keller, E.F., Segel, L.A.: Model for chemotaxis. J. Theor. Biol. **30**, 225–234 (1971)
9. Menci, M., Natalini, R., Paul, T.: Microscopic, kinetic and hydrodynamic hybrid models of collective motions with chemotaxis: a numerical study, to appear in MeMoCs
10. Natalini, R., Paul, T.: On the mean field limit for Cucker-Smale models. Discrete Continuous Dyn. Syst. B **27**(5), 2873–2889 (2022)
11. Natalini, R., Paul, T.: The mean-field limit for hybrid models of collective motions with chemotaxis. SIAM J. Math. Anal. **55**(2), 900–928 (2023)
12. Paul, T.: From Liouville to Keller-Segel (In preparation)
13. Paul, T., Trélat, E.: From microscopic to macroscopic scale equations: mean field, hydrodynamic and graph limits, hal-03779694
14. Paul, T., Trélat, E.: From microscopic to macroscopic dynamics: mean field, hydrodynamic and graph limits, hal-03779694
15. Paul, T., Trélat, E.: Universal approximation of PDEs by finite distinguishable particle systems, hal-04895207
16. Patlak, C.S.: Random walk with persistence and external bias. Bull. Math. Biol. Biophys. **15**, 311–338 (1953)
17. Perthane, B.: Transport Equations in Biology. Birhaüser (2000)
18. Villani, C.: Topics in Optimal Transportation. Graduate Studies in Mathematics, vol. 58. American Mathematical Society, Providence, RI (2003)
19. Villani, C.: Optimal Transport. Old and New. Grundlehren der mathematischen Wissenschaften [Fundamental Principles of Mathematical Sciences], vol. 338. Springer, Berlin (2009)

Open Problems and Perspectives on Solving Friedrichs Systems by Krylov Approximation

Noè Angelo Caruso and Alessandro Michelangeli

Abstract We set up, at the abstract Hilbert space setting, the general question on when an inverse linear problem induced by an operator of Friedrichs type admits solutions belonging to (the closure of) the Krylov subspace associated to such operator. Such Krylov solvability of abstract Friedrichs systems allows to predict when, for concrete differential inverse problems, truncation algorithms can or cannot reproduce the exact solutions in terms of approximants from the Krylov subspace.

Keywords Inverse linear problems · Krylov subspace methods · Krylov solvability · Friedrichs systems of differential equations · Abstract Friedrichs systems

2010 Mathematics Subject Classification 35R30 · 47N20 · 47N40 · 47A58

1 Introduction

At the intersection of the two general fields of numerical solutions to partial differential equations, and of iterative approximations of inverse linear problems, an interesting class of questions has been recently emerging, with particularly rich devel-

---

N. A. Caruso (✉)
Dipartimento di Scienze Umane, Università degli Studi "Link Campus University", Rome, Italy
e-mail: n.caruso@unilink.it

A. Michelangeli
Mathematics and Science Department, American University Bulgaria AUBG, Blagoevgrad, Bulgaria
e-mail: amichelangeli@aubg.edu

Hausdorff Center for Mathematics, University of Bonn, Bonn, Germany

TQT Trieste Institute for Theoretical Quantum Technologies, Trieste, Italy

B. Cassano et al. (eds.), *Singularities, Asymptotics, and Limiting Models*, Springer INdAM Series 64, https://doi.org/10.1007/978-981-96-3584-9_14

opments and impact both in an abstract setting and in concrete applications. In this note we would like to briefly review this subject and, in a sense, to advocate for it.

In abstract terms, an *inverse linear problem* in some linear space $\mathcal{H}$ consists of the following: an element $g \in \mathcal{H}$ (the 'datum') and a linear map A on $\mathcal{H}$ are given, and one is to find, if any, those element(s) $f \in \mathcal{H}$ such that $Af = g$.

This clearly encompasses a huge variety of concrete classes of inverse problems, including those where A is a differential operator. On the one hand, it is customary to study each such class developing ad hoc methods and tools. This falls within the realm of numerical analysis, and includes the issue of reproducing the 'exact' solution(s) f to $Af = g$ as a limit of approximants $f_n \to f$ obtained by solving (easier) truncated problems $A_n f_n = g_n$, typically in an iterative way.

Another perspective is to disregard the specific features and setting of inverse linear problems and to investigate them at a sufficiently high level of abstract generality, namely to study the solvability of $Af = g$ and the approximability $f_n \to f$ depending on: the finite or infinite dimensionality of the vector space $\mathcal{H}$, the additional analytic or topological structure of $\mathcal{H}$ (e.g., a Banach space, a Hilbert space), the boundedness or unboundedness of the operator A on $\mathcal{H}$ and further features of A such as its compactness or normality or self-adjointness and the like, and the class of data g with respect to the given operator A.

Obviously, there is a natural interplay between the two perspectives above. For instance, the actual convergence $f_n \to f$ from solutions to suitably truncated inverse problems obtained in a concrete setting, as well as the convergence rate and the quantitative control (say, the norm) of the displacement $f - f_n$, are often established by means of abstract functional-analytic and operator theoretic methods. We mention, in this respect, excellent classical references such as [8, 24, 28, 33, 40, 41, 45, 46, 50–52, 54, 56], and also, to quickly grasp the flavour of the abstract point of view, [17, Sects. 1.1 and 1.2, and Appendix A].

Our interest here concerns the abstract analysis of a class of approximation schemes for inverse linear problems known as *Krylov subspace methods*. Their efficiency and popularity places them among the 'Top 10 Algorithms' of the 20th century [25, 27].

In a nutshell (Sect. 2 provides a concise and somewhat more detailed 'crash course' on the subject), one assumes that $\mathcal{H}$ is a Hilbert space (clearly, the richest scenario of problems and results occurs when $\dim \mathcal{H} = \infty$), and given the operator A acting on $\mathcal{H}$ and the vector $g \in \operatorname{ran} A$ one suitably truncates the inverse linear problem $Af = g$ to the finite-dimensional subspaces $\operatorname{span}\{g, Ag, A^2g, \dots, A^{n-1}g\}$ for larger and larger integers n. If f_n is a solution of the n-truncated problem, then under favourable circumstances a solution f to $Af = g$ is approximated in norm by the f_n's, i.e., $\|f - f_n\|_{\mathcal{H}} \to 0$ as $n \to \infty$, which means, in particular, that f belongs to the closure of the subspace generated by $g, Ag, A^2g, \dots$

Thus, Krylov subspace methods are (typically iterative) algorithms where a solution f to an inverse linear problem $Af = g$ is obtained by closer and closer approximants that are expressed by convenient finite linear combinations of the vectors $g, Ag, A^2g, \dots$. The subspace

$$\mathcal{K}(A, g) := \operatorname{span}\{g, Ag, A^2g, \dots\} \subset \mathcal{H} \tag{1.1}$$

is called in this context the *Krylov subspace* associated with A and g (in other contexts it is also called the *cyclic subspace* for A and g). When the inverse problem $Af = g$ admits a solution $f \in \overline{\mathcal{K}(A, g)}$, the problem is said to be *Krylov-solvable*.

In a sense, for inverse problems that are Krylov-solvable, the advantage is that the trial space for the approximants is spanned by vectors (the $A^n g$'s) that are explicit and typically easy to calculate in the applications. This makes Krylov subspace methods particularly versatile.

The study of the Krylov solvability of inverse linear problems at an abstract Hilbert space setting, and of the asymptotic convergence of the Krylov approximants, is a fascinating subject that only very recently has been set up in a thorough and systematic way. We refer on this to the monograph [17] and to the 'crash course' of Sect. 2.

As a matter of fact, the theoretical apparatus developed so far in the subject of Krylov solvability is sufficiently articulated to make it natural to specialise it for distinguished classes of inverse problems. The one we consider here is the type of inverse problems customarily referred to as *Friedrichs systems.*

Friedrichs systems are a wide class of differential linear inverse problems essentially characterised by the fact that the sum of the differential operator A and its formal adjoint is bounded and coercive. They have recently attained particular relevance in numerical analysis [33, 34, 42, 43] as a convenient unified framework for numerical solutions to partial differential equations of different types, including elliptic, parabolic, and hyperbolic.

Starting from concrete, differential formulations of Friedrichs-type inverse problems, it is possible to re-formulate them in an abstract Hilbert space setting, so as to exploit general operator-theoretic methods and then apply the results to concrete versions [1, 3, 7, 29, 30, 33, 37]. Section 3 revisits the basic aspects of this construction.

This naturally leads to the notion of a *joint pair of abstract Friedrichs operators* in a Hilbert space $\mathcal{H}$, namely a pair $(A_0, \widetilde{A}_0)$ of two densely defined and closed linear operators $A_0 : \mathcal{W}_0 \to \mathcal{H}$ and $\widetilde{A}_0 : \mathcal{W}_0 \to \mathcal{H}$ such that

$$\begin{aligned} & A_0 \subset (\widetilde{A}_0)^* =: A_1, \qquad \widetilde{A}_0 \subset (A_0)^* =: \widetilde{A}_1, \\ & A_1 + \widetilde{A}_1 \ \text{ is a bounded self-adjoint operator in } \mathcal{H} \\ & \text{with strictly positive bottom.} \end{aligned} \tag{1.2}$$

At this abstract level, A_0 and A_1 model what in a typical concrete example of inverse problem of Friedrichs type are, respectively, the minimal and the maximal realisation of a differential operator on a bounded domain $\Omega \subset \mathbb{R}^n$. Any closed operator A with $A_0 \subset A \subset A_1$ models instead a specific realisation of the differential operator, identified by specific boundary conditions at the boundary $\partial\Omega$.

At the overlap of the two seemingly distant subjects of abstract Krylov solvability and abstract Friedrichs systems we finally make the following observation: *investigating in the abstract Hilbert space setting the Krylov solvability of an inverse* *problem $Af = g$ governed by an operator A of Friedrichs type would yield extremely valuable a priori information on the possibility of approximating numerically the solution to a concrete differential inverse problem within the vast Friedrichs class.*

This note is conceived to serve both as a self-contained introduction to the set of problems we outlined above, including the current state of the art and the recent achievements, and as a survey of relevant questions on top of the research agenda in this subject.

As mentioned already, Sects. 2 and 3 provide background materials. In Sect. 4 we outline a few of the most relevant and natural questions that are worth addressing.

2 A Very Short 'Crash Course' on Krylov Solvability in Abstract Hilbert Space

For the concise review materials of this section we refer to the corpus of recent works [16–22].

2.1 *Krylov Approximation Methods as Truncation Algorithms*

We begin by presenting Krylov subspace methods as a family of approximation methods based on suitable compressions, or 'truncations', of the operator A and the vector g of the inverse linear problem. Certainly one may think of other ways to represent various Krylov based methods, such as the more well-known iterative or variational formulations, yet these viewpoints have the drawback that they somewhat mask the relevance of the Krylov subspace and the concept of 'Krylov solvability'.

Let us recall from the Introduction that the inverse linear problem, in the complex Hilbert space $\mathcal{H}$,

$$Af = g \tag{2.1}$$

in the unknown $f \in \mathcal{H}$ and for given densely defined (and possibly closed) linear operator A, and $g \in \operatorname{ran} A \subset \mathcal{H}$—which ensures the existence of solution(s)—is said to be *Krylov-solvable* if there is a solution f belonging to the closure in $\mathcal{H}$ of the Krylov subspace

$$\mathcal{K}(A, g) := \operatorname{span}\{g, Ag, A^2 g, \dots\} \subset \mathcal{H} \tag{2.2}$$

associated to A and g. This expresses the notion of *Krylov solvability* of (2.1) and f in this case is referred to as a *Krylov solution*.

Krylov methods work by constructing a sequence of approximants $(f_n)_{n\in\mathbb{N}}$ of f that are solutions to a *truncated* problem obtained by 'projecting' the original (2.1) onto an n-dimensional subspace built with a finite amount of vectors of the type $A^k g$.

In practice one considers the subspace

$$\mathcal{K}_n(A, g) := \operatorname{span}\{g, Ag, A^2 g, \ldots, A^{n-1} g\}, \qquad \dim \mathcal{K}_n(A, g) \leqslant n, \tag{2.3}$$

as well as another, possibly distinct, n-dimensional subspace $\mathcal{J}_n(A, g)$ consisting of finite linear combinations of an amount of the $A^k g$'s. Correspondingly, consider the orthogonal projections $P_n : \mathcal{H} \to \mathcal{H}$ and $Q_n : \mathcal{H} \to \mathcal{H}$, respectively, onto $\mathcal{K}_n(A, g)$ and $\mathcal{J}_n(A, g)$. It is customary to refer to $\mathcal{K}_n(A, g) = \operatorname{ran} P_n$ and $\mathcal{J}_n(A, g) = \operatorname{ran} Q_n$ as, respectively, the *solution subspace* and the *trial subspace*.

The n-truncated inverse linear problem is then formulated as

$$\begin{aligned} A_n f_n &= g_n, \\ \text{where } A_n := Q_n A P_n, &\quad g_n := Q_n g, \end{aligned} \tag{2.4}$$

and the $n \times n$ matrix A_n is the *compression* of A to the considered subspaces. It need *not* happen that $f_n = P_n f$, where f is a solution to the 'full' problem (2.1). The finite-dimensional inverse linear problems (2.4) form the basis of many different approximation schemes, in particular the famous Petrov-Galerkin methods.

As we increase the dimension of the subspaces, i.e., as n becomes large, ideally an approximation scheme produces solutions to the truncated problem $(f_n)_{n \in \mathbb{N}}$ that tend to an exact solution of the original inverse linear problem (2.1). The convergence $f_n \to f$ in the norm of $\mathcal{H}$ is tantamount of the Krylov solvability of (2.1) defined above.

We remark that the choice of Q_n (and hence of $\mathcal{J}_n(A, g)$) may vary depending on the precise method used. For example, the celebrated GMRES method uses the choice of Q_n as the orthogonal projection operator onto the subspace $A\mathcal{K}_n(A, g)$. On the other hand, for the class of problems for which A is positive self-adjoint, the famous conjugate gradient (CG) method uses the choice $Q_n = P_n$.

Beside the above-mentioned recent discussion [21], an excellent overview and general results associated with the formulation and solution to (2.4) may be found in several excellent articles and monographs such as [24, 45, 50, 55–57].

2.2 A Glance at Krylov Solvability for Bounded Inverse Problems

Interesting structural properties emerge in the discussion of Krylov solvability of inverse problems (2.1) with $A \in \mathcal{B}(\mathcal{H})$ (for a more systematic discussion, see [20]).

Let us start with the notion of *Krylov reducibility*. The following inclusions are easily seen:

$$\begin{aligned} A\,\overline{\mathcal{K}(A, g)} &\subset \overline{\mathcal{K}(A, g)}, \\ A^* \mathcal{K}(A, g)^{\perp} &\subset \mathcal{K}(A, g)^{\perp}, \end{aligned} \tag{2.5}$$

which express the invariance of $\overline{\mathcal{K}(A,g)}$, respectively, of $\mathcal{K}(A,g)^{\perp}$, under the action of A, respectively, of A^*. One then says that A is $\mathcal{K}(A,g)$*-reduced* if in addition $\mathcal{K}(A,g)^{\perp}$ is invariant also under the action of A, i.e.,

$$A\,\mathcal{K}(A,g)^{\perp} \subset \mathcal{K}(A,g)^{\perp}. \tag{2.6}$$

Here *Krylov reducibility* indicates that A is reduced (in the standard operator sense, e.g., [53, Sect. 1.4]) into two blocks with respect to the Hilbert space direct sum decomposition

$$\mathcal{H} = \overline{\mathcal{K}(A,g)} \oplus \mathcal{K}(A,g)^{\perp}. \tag{2.7}$$

An evident sufficient condition for Krylov reducibility is self-adjointness: a bounded self-adjoint operator A is *always* $\mathcal{K}(A,g)$-reduced for any $g \in \mathcal{H}$. More generally, bounded normal operators A are $\mathcal{K}(A,g)$-reduced if and only if $A^*g \in \overline{\mathcal{K}(A,g)}$ [20, Proposition 2.1]. Thus, for unitary operators Krylov reducibility is equivalent to Krylov solvability [20, Remark 3.6].

It turns out that the Krylov reducibility of A is intimately connected to the Krylov solvability of the associated inverse linear problem (2.1).

Proposition 1 ([20, Proposition 3.9]) Let $A \in \mathcal{B}(\mathcal{H})$ and let $g \in \operatorname{ran}A$. If A is $\mathcal{K}(A,g)$-reduced, then there exists a Krylov solution to the problem $Af = g$. For example, if $f \in \mathcal{H}$ satisfies $Af = g$ and $P_{\mathcal{K}}$ is the orthogonal projection onto $\overline{\mathcal{K}(A,g)}$, then $f^{\circ} := P_{\mathcal{K}} f$ is a Krylov solution

The immediate corollary of the above proposition is:

Corollary 1 *Let $A \in \mathcal{B}(\mathcal{H})$ be self-adjoint and let $g \in \operatorname{ran}A$. Then there exists a solution to the inverse linear problem $Af = g$ in the space $\overline{\mathcal{K}(A,g)}$.*

Therefore, the bounded-self adjoint inverse problems are *always* Krylov-solvable.

As discovered in [20], the following subspace is inherently linked to the Krylov solvability of an inverse linear problem.

Definition 1 The *Krylov intersection* associated to given $A \in \mathcal{B}(\mathcal{H})$ and $g \in \mathcal{H}$ is the subspace

$$\mathcal{I}(A,g) := \overline{\mathcal{K}(A,g)} \cap A(\mathcal{K}(A,g)^{\perp}). \tag{2.8}$$

The possible triviality of the Krylov intersection is another structural property relevant for the Krylov solvability.

Proposition 2 ([20, Proposition 3.4], [16, Proposition 2.4]) Let $A \in \mathcal{B}(\mathcal{H})$ and let $g \in \operatorname{ran}A$. If $\mathcal{I}(A,g) = \{0\}$, then there exists a Krylov solution $f \in \overline{\mathcal{K}(A,g)}$ to the inverse linear problem $Af = g$. If, additionally, $A^{-1} \in \mathcal{B}(\mathcal{H})$, then $\mathcal{I}(A,g) = \{0\}$ if and only if the solution f to $Af = g$ is in $\overline{\mathcal{K}(A,g)}$.

Observe that, obviously, the $\mathcal{K}(A,g)$-reducibility of an operator A is just a special case of the triviality of the Krylov intersection $\mathcal{I}(A,g)$, which indicates that the latter subspace is more informative concerning the possibility that an inverse problem be Krylov-solvable.

We refer to [20] for an expanded discussion of classes of inverse linear problems that are Krylov-reduced, or have trivial Krylov intersection, and in general are Krylov-solvable, including when the Krylov solution is unique. To that, we add the following interesting recent result:

Proposition 3 ([22, Propositions 3.4 and 3.9]) Let $A \in \mathcal{B}(\mathcal{H})$ be a compact normal operator and let $g \in \mathcal{H}$. Then A is $\mathcal{K}(A, g)$-reduced. If, additionally, $g \in \operatorname{ran} A$, then $Af = g$ has a unique Krylov solution that is also a minimal norm solution to the inverse linear problem.

2.3 *Krylov Solvability and Unbounded Self-Adjoint Operators*

As a matter of fact, the notion of Krylov reducibility and Krylov intersection, in connection to the Krylov solvability of an inverse linear problem, have a counterpart in the case of *unbounded* A, with analogous results to those outlined above: this has been discussed in [16, 23].

The unavoidable domain issues that arise in the unbounded case require suitable adaptations. For one thing, the very notion (2.2) of the Krylov subspace $\mathcal{K}(A, g)$ only makes sense if g belongs to the domain of *every* power of A, an inescapable minimal condition when A is unbounded. When this happens, the generalised notion of Krylov intersection reads

$$\mathcal{I}(A, g) := \overline{\mathcal{K}(A, g)} \cap A(\mathcal{K}(A, g)^{\perp} \cap \operatorname{dom} A), \tag{2.9}$$

and a discussion of its properties and relevance for the Krylov solvability of $Af = g$ may be found in [16, Sects. 5 and 6].

In fact, distinguished classes of data g turn out to play a role in the analysis of the Krylov solvability of $Af = g$, and the following classical definition collects some of them.

Definition 2 Let A be a closed operator acting in a complex Hilbert space $\mathcal{H}$ and let $g \in \mathcal{H}$.

(i) g belongs to the class $C^{\infty}(A)$ of *smooth vectors* for A if

$$g \in \operatorname{dom} A^n \qquad \forall n \in \mathbb{N}.$$

(ii) g belongs to the class $\mathcal{D}^b(A)$ of *bounded vectors* for A if g is smooth for A and there is a constant $B_g > 0$ such that

$$\|A^n g\|_{\mathcal{H}} \leqslant (B_g)^n \qquad \forall n \in \mathbb{N}.$$

(iii) g belongs to the class $\mathcal{D}^a(A)$ of *analytic vectors* for A if g is smooth for A and there is a constant $C_g > 0$ such that

$$\|A^n g\|_{\mathcal{H}} \leqslant (C_g)^n \, n! \qquad \forall n \in \mathbb{N}.$$

(iv) g belongs to the class $\mathcal{D}^{qa}(A)$ of *quasi-analytic vectors* for A if g is smooth for A and

$$\sum_{n=0}^{\infty} \|A^n g\|_{\mathcal{H}}^{-\frac{1}{n}} = +\infty.$$

Clearly,

$$\mathcal{D}^b(A) \subset \mathcal{D}^a(A) \subset \mathcal{D}^{qa}(A) \subset C^\infty(A). \tag{2.10}$$

The self-adjoint case exhibits in this respect distinguished properties. For one thing, as well known (see, e.g., [53, Lemma 7.13]), if A is self-adjoint, then $\mathcal{D}^b(A)$ is dense in $\mathcal{H}$, and so too are, consequently, all the other classes above. Moreover, a noticeable conclusion can be drawn concerning the Krylov solvability of possibly *unbounded* self-adjoint inverse problems.

Theorem 1 ([16, Theorems 4.1 and 4.2]) Let A be a self-adjoint or skew-adjoint operator on a Hilbert space $\mathcal{H}$, and let $g \in \operatorname{ran} A \cap \mathcal{D}^{qa}(A)$. Then there exists a unique Krylov solution to the inverse linear problem $Af = g$.

It is worth commenting that Theorem 1 relies on an ingenious and somewhat laborious analysis carried out in the framework of the convergence of the conjugate gradient algorithm on an infinite-dimensional Hilbert space. This was initially developed in the *bounded* case [48, 49] and recently re-designed in the *unbounded* setting [18, Theorem 2.4 and Corollary 2.5]. The full proof obtained in [18] in fact reveals that *the sequence of approximants from the conjugate-gradients algorithm converges to the minimal norm solution of the inverse problem* (2.1) when A is positive self-adjoint and $g \in \mathcal{D}^{qa}(A) \cap \operatorname{ran} A$. In other words: the truncation scheme and truncated inverse problems (2.4) defined by setting the trial space $Q_n = P_n$ for P_n the orthogonal projection onto $\mathcal{K}_n(A, g)$, produce a well-defined sequence of approximants $(f_n)_{n\in\mathbb{N}}$ and $f_n \xrightarrow{n\to\infty} f^\circ$ in $\mathcal{H}$, where f° is the minimal norm solution to (2.1).

2.4 Krylov Subspaces and Solvability Under Perturbations

A further approach to the question of determining the Krylov solvability (or lack-of) of an inverse linear problem $Af = g$ is to analyse possibly 'easier' *perturbed* problems $\widetilde{A}\widetilde{f} = \widetilde{g}$, where in some suitable sense $\widetilde{A} \sim A$ and $\widetilde{g} \sim g$.

Even elementary toy problems show that the phenomenology may be pretty diverse.

Example 1 ([19, Example 3.3]) Let $R : \ell^2(\mathbb{Z}) \to \ell^2(\mathbb{Z})$ be the right shift operator $e_n \mapsto e_{n+1}$ (where e_n is the nth canonical vector of $\ell^2(\mathbb{Z})$). R is unitary with $R^* = R^{-1} = L$, the left-shift operator $e_n \mapsto e_{n-1}$. R admits a dense set $\mathcal{C} \subset \ell^2(\mathbb{Z})$ of cyclic vectors (i.e., vectors $g \in \mathcal{H}$ such that $\mathcal{K}(R, g)$ is dense in $\ell^2(\mathbb{Z})$) and a dense subset $\mathcal{N} \subset \ell^2(\mathbb{Z})$ of all finite linear combinations of the canonical vectors. Clearly all vectors in $\mathcal{N}$ are non-cyclic for the operator R, and moreover the solution f to $Rf = g$ for any $g \in \mathcal{N}$ does not belong to $\overline{\mathcal{K}(R, g)}$.

(i) (Loss of Krylov solvability). For a datum $g \in \mathcal{N}$, the inverse problem $Rf = g$ admits a unique solution f that is not in $\overline{\mathcal{K}(R, g)}$. Yet, by the density of $\mathcal{C}$ in $\ell^2(\mathbb{Z})$, there exists a sequence $(g_m)_{m\in\mathbb{N}} \subset \mathcal{C}$ such that $\|g - g_m\|_{\ell^2} \xrightarrow{m\to\infty} 0$ and moreover for each $m \in \mathbb{N}$ the inverse linear problem $Rf_m = g_m$ is Krylov-solvable with the unique solution $f_m = R^{-1}g_m$. Here we see that Krylov solvability is lost in the limit $m \to \infty$, although $\|f_m - f\|_{\ell^2} \xrightarrow{m\to\infty} 0$.

(ii) (Gain of Krylov solvability). For a datum $g \in \mathcal{C}$ the inverse linear problem $Rf = g$ is Krylov-solvable as $\overline{\mathcal{K}(R, g)} = \ell^2(\mathbb{Z})$. Yet, by density there exists a sequence $(g_m)_{m\in\mathbb{N}} \subset \mathcal{N}$ such that $\|g_m - g\|_{\ell^2} \xrightarrow{m\to\infty} 0$, and each perturbed problem $Rf_m = g_m$ is not Krylov-solvable. Therefore, Krylov solvability is absent at the perturbed level but appears only in the limit, still with the approximation $\|f_m - f\|_{\ell^2} \xrightarrow{m\to\infty} 0$.

In a perturbative perspective, it is relevant to compare the possibly "close" Krylov subspaces $\mathcal{K}(A, g)$ and $\mathcal{K}(\widetilde{A}, \widetilde{g})$. A promising tool is the *weak gap metric* between subspaces of *separable* Hilbert space, inspired to the standard *norm gap distance* (namely, the Hausdorff distance) between subspaces (see, e.g., [44, Chap. 4, Sect. 2]). Owing to the separability, and also reflexivity, of $\mathcal{H}$, the Hilbert space weak topology restricted to the (norm-closed) unit ball $B_{\mathcal{H}}$ of $\mathcal{H}$ makes the ball a compact and complete metric space $(B_{\mathcal{H}}, \rho_w)$, and the metric ρ_w is induced by a norm, namely $\rho_w(x, y) = \|x - y\|_w$ for some explicitly constructed norm in $B_{\mathcal{H}}$. This allows to define, for any pair of norm-closed linear subspaces U and V of $\mathcal{H}$,

$$\begin{cases} d_w(B_U, B_V) := \sup_{u\in B_U} \inf_{v\in B_V} \rho_w(u, v), \\ \widehat{d}_w(U, V) := \max\{d_w(B_U, B_V), d_w(B_V, B_U)\}, \end{cases} \tag{2.11}$$

where $B_U = B_{\mathcal{H}} \cap U$ and $B_V = B_{\mathcal{H}} \cap V$. As a matter of fact [19, Lemmas 6.1 and 6.2], $\widehat{d}_w$ is indeed a metric for the norm-closed subspaces of $\mathcal{H}$, albeit in general it is *not* complete.

In the context of Krylov subspaces, the distance $\widehat{d}_w$ is effective in a multiplicity of instances, for example in monitoring the convergence

$$\mathcal{K}_N(A, g) \xrightarrow[N\to\infty]{\widehat{d}_w} \overline{\mathcal{K}(A, g)} \tag{2.12}$$

of the N-dimensional truncated Krylov subspace $\mathcal{K}_N(A, g)$ to the (closure of the) full Krylov subspace $\mathcal{K}(A, g)$. Such quite natural *inner Krylov approximability* is actually false when the standard Hausdorff distance is used for an infinite-dimensional $\mathcal{H}$ [19, Lemma 7.1]. Another natural aspect of the inner Krylov approximability that manifests in the $\widehat{d}_w$-sense is the following [19, Lemma 7.6]:

$$\left.\begin{array}{l}(g_n)_n \subset \overline{\mathcal{K}(A,g)} \\ \|g_n - g\|_{\mathcal{H}} \xrightarrow{n\to\infty} 0\end{array}\right\} \quad \Rightarrow \quad \overline{\mathcal{K}(A,g_n)} \xrightarrow[n\to\infty]{\widehat{d}_w} \overline{\mathcal{K}(A,g)}. \tag{2.13}$$

Other natural approximations, instead, do not hold in general in the $\widehat{d}_w$-sense, which suggests that further conditions need be added to the perturbation picture. For example [19, Lemma 7.3, Examples 7.4 and 7.5],

$$\|g_n - g\|_{\mathcal{H}} \xrightarrow{n\to\infty} 0 \quad \Rightarrow \quad \left\{\begin{array}{c} d_w\big(\overline{\mathcal{K}(A,g)}, \overline{\mathcal{K}(A,g_n)}\big) \xrightarrow{n\to\infty} 0 \\ \text{but in general} \\ \overline{\mathcal{K}(A,g_n)} \ \stackrel{\widehat{d}_w}{\not\longrightarrow}\ \overline{\mathcal{K}(A,g)}. \end{array}\right. \tag{2.14}$$

The study of the behaviour of the Krylov solvability along $\widehat{d}_w$-limits is at an early stage. Let us conclude this concise overview with this type of results.

Proposition 4 ([19, Proposition 7.7]) Let $\mathcal{H}$ be a separable Hilbert space, and let the following be given:

(i) an operator $A \in \mathcal{B}(\mathcal{H})$ with inverse $A^{-1} \in \mathcal{B}(\mathcal{H})$,
(ii) a sequence $(g_m)_{m\in\mathbb{N}} \subset \mathcal{H}$ such that for each $m \in \mathbb{N}$ the unique solution $f_m := A^{-1}g_m$ to the inverse linear problem $Af_m = g_m$ is a Krylov solution,
(iii) a vector $g \in \mathcal{H}$ such that

$$\overline{\mathcal{K}(A,g_m)} \xrightarrow{\widehat{d}_w} \overline{\mathcal{K}(A,g)} \quad \text{as } n \to \infty.$$

Then the unique solution $f := A^{-1}g$ to the inverse linear problem $Af = g$ is a Krylov solution. If additionally $g_m \to g$ (resp. $g_m \rightharpoonup g$), then $f_m \to f$ (resp. $f_m \rightharpoonup f$).

3 Overview on Abstract Friedrichs Systems

Let us condense in this section a short overview on Friedrichs systems from their concrete formulation as differential problems to their lifting to abstract inverse linear problems on Hilbert space.

Friedrichs systems were initially introduced by Friedrichs [38] in his research on symmetric positive systems, and became relevant in numerical analysis [33, 34, 42, 43] as a unified and versatile scheme for solving partial differential equations of different types, including elliptic, parabolic, and hyperbolic.

They are currently being investigated in a variety of directions, including well-posedness [1, 3, 7, 37], homogenisation [11, 12], representations of boundary conditions [1], connection with the classical theory [2–4, 6], applications to various initial or boundary value problems of elliptic, hyperbolic, and parabolic type [5, 13, 26, 34, 47], development of different numerical schemes [9, 10, 14, 34–36], classification of abstract realisations of Friedrichs systems [7, 29–32], the Friedrichs system structure of contact interactions [29], and many others.

The classical Friedrichs operator on a bounded domain $\Omega \subset \mathbb{R}^d$ with Lipschitz boundary $\partial\Omega$ is the first-order differential operator A acting as the $L^2(\Omega, \mathbb{C}^r) \to \mathcal{D}'(\Omega)^r$ map

$$Af = \sum_{k=1}^{d} \partial_k(\mathbf{B}_k f) + \mathbf{C}f \tag{3.1}$$

for given matrix functions $\mathbf{B}_k \in W^{1,\infty}(\Omega, M_r)$, $k \in \{1, \dots, d\}$, and $\mathbf{C} \in L^\infty(\Omega, M_r)$ satisfying, for some $\mu > 0$,

$$\mathbf{B}_k = \mathbf{B}_k^* \quad \text{and} \quad \mathbf{C} + \mathbf{C}^* + \sum_{k=1}^{d} \partial_k \mathbf{B}_k \geqslant \mu \mathbb{1} \quad \text{a.e. on } \Omega. \tag{3.2}$$

The inequality in (3.2) is in the sense of expectations on $L^2(\Omega, \mathbb{C}^r)$.

The solvability or also the well-posedness of the inverse problem $Af = g$ induced by the operator (3.1)–(3.2) and by a given $g \in L^2(\Omega, \mathbb{C}^r)$ depend on the kind of realisations of A that are considered in terms of boundary conditions imposed at $\partial\Omega$.

Many other concrete differential problems may be cast in the form $Af = g$ with A of the type (3.1)–(3.2) [1, 2, 5, 6, 13, 34, 35]. In addition, it is possible to reformulate all such concrete problems at an abstract Hilbert space setting: this includes the classification of the distinct operator realisations of A (namely, the counterpart of boundary conditions at $\partial\Omega$), as well as the identification of certain intrinsic abstract conditions ensuring the bijectivity of A and hence the solvability of $Af = g$ [1, 3, 7, 29–32, 37].

This way one naturally comes to the definition of a *joint pair of abstract Friedrichs operators* on a complex Hilbert space $\mathcal{H}$ as a pair $(A_0, \widetilde{A}_0)$ of two closed and densely defined operators acting in $\mathcal{H}$ and satisfying the conditions

$$\begin{aligned} & A_0 \subset (\widetilde{A}_0)^* =: A_1, \qquad \widetilde{A}_0 \subset (A_0)^* =: \widetilde{A}_1, \\ & A_1 + \widetilde{A}_1 \text{ is a bounded self-adjoint operator in } \mathcal{H} \\ & \text{with strictly positive bottom.} \end{aligned} \tag{3.3}$$

(The first condition in (3.3) is in the sense of operator inclusion.)

In the concrete case (3.1)–(3.2) A_0 may be taken as a minimally defined version of the operator A therein, and $\widetilde{A}_0$ is its formal adjoint.

In the abstract setting (3.3) it can be shown [37, Sect. 2.1], [3, Theorem 7] that the graph norms associated with A_0 and $\widetilde{A}_0$ are equivalent, and as a consequence A_0 and $\widetilde{A}_0$ have a common (dense) domain

$$\mathrm{dom}\, A_0 \;=\; \mathrm{dom}\, \widetilde{A}_0 \;=:\; \mathcal{W}_0 \tag{3.4}$$

and so do A_1 and $\widetilde{A}_1$, that is,

$$\mathrm{dom}\, A_1 \;=\; \mathrm{dom}\, \widetilde{A}_1 \;=:\; \mathcal{W}_1 \qquad (\text{with } \mathcal{W}_0 \subset \mathcal{W}_1), \tag{3.5}$$

and moreover,

$$A_1 + \widetilde{A}_1 \;\subset\; \overline{A_0 + \widetilde{A}_0}\,. \tag{3.6}$$

Furthermore, between the minimal and the maximal operators defined above one is led to consider intermediate pairs $(A, \widetilde{A})$ with

$$A_0 \subset A \subset A_1\,, \qquad \widetilde{A}_0 \subset \widetilde{A} \subset \widetilde{A}_1\,, \tag{3.7}$$

equivalently, pairs $(V, \widetilde{V})$ of linear subspaces included between $\mathcal{W}_0$ and $\mathcal{W}_1$, and the restrictions $A := A_1|_V$, $\widetilde{A} := \widetilde{A}_1|_{\widetilde{V}}$. Each choice of V and $\widetilde{V}$ corresponds, in this abstract setting, to a selection of certain boundary conditions for a concrete differential operator such as (3.1)–(3.2).

In particular, a special status is attributed to those V's that are closed in $(\mathcal{W}_1, \|\cdot\|_{A_1})$ and correspondingly those realisations $A_1|_V$'s that are bijective as $(V, \|\cdot\|_{A_1}) \to \mathcal{H}$ maps. Indeed, such operators that are closed and densely defined in $\mathcal{H}$, are also $(V, \|\cdot\|_{A_1}) \to \mathcal{H}$ bounded, and their inverse is $\mathcal{H} \to (V, \|\cdot\|_{A_1})$ bounded, implying that the inverse linear problem induced by each such $A_1|_V$ is well-posed. Incidentally, one can see that if $A := A_1|_V$ is a *closed bijective realisation* of A_0, then its adjoint $A^* = (A_1|_V)^* \equiv \widetilde{A}_1|_{\widetilde{V}}$, with $\widetilde{V} := \mathrm{dom}\,(A_1|_V)^*$, is a closed bijective realisation of $\widetilde{A}_0$.

The investigation and classification of mutually adjoint pairs (A, A^*) of closed bijective realisations of a Friedrichs operator A_0 (and of its partner $\widetilde{A}_0$) has been the object of multiple recent studies [7, 29–32]. Quite noticeably [7, Theorem 13]:

(i) A pair of mutually adjoint closed bijective realisations of $(A_0, \widetilde{A}_0)$ always exists.
(ii) If both $\ker A_1$ and $\ker \widetilde{A}_1$ are non-trivial, then there exist uncountably many such pairs. If instead at least one among A_1 and $\widetilde{A}_1$ is injective, then the only mutually adjoint pair of closed bijective realisations is $(A_1, \widetilde{A}_0)$ when $\ker A_1 = \{0\}$, or $(A_0, \widetilde{A}_1)$ when $\ker \widetilde{A}_1 = \{0\}$.

Part (i) above expresses the fact that any differential operator of Friedrichs type admits at least one set of boundary conditions for which the corresponding inverse problem is well-posed. In abstract terms it provides the existence of a 'reference' pair $(A_{\mathrm{r}}, A_{\mathrm{r}}^*)$ with $A_0 \subset A_{\mathrm{r}} \subset A_1$ where A_{r} is closed and bijective.

4 Perspective on Krylov Solvability of Friedrichs-Type Inverse Problems

As stated already, the study of the Krylov solvability of abstract inverse problems $Af = g$ in a Hilbert space, governed by operators A of Friedrichs type, provides valuable a priori information on the possibility of approximating numerically the solution to a concrete differential inverse problem within the vast class of Friedrichs systems.

At this level of generality the main question is therefore.

Question 1 Under what conditions on the (closed) operator A and the vector $g \in \operatorname{ran} A$ does the inverse problem $Af = g$ admit solution(s) $f \in \overline{\mathcal{K}(A, g)}$?

Obviously, Question 1 has a variety of variants, depending on what features of A are considered, such as its general type (normal, self-adjoint, skew-adjoint, or in some special class), or a priori estimates satisfied by A (including boundedness, unboundedness, semi-boundedness, coercivity, etc.).

It is also relevant to investigate in Question 1 the issue of the uniqueness of the Krylov solution, let alone the issue of convergence rate of the Krylov approximants $f_n \to f$ in one of the abstract algorithms considered.

Work in progress is showing an increasing understanding of the general Question 1 in concrete toy problems as well as special settings that retain a significant amount of instructiveness.

In this spirit, consider for a moment a generic Hilbert space $\mathcal{H}$ and an operator A acting in $\mathcal{H}$ and $g \in \mathcal{H}$ such that

$$\begin{aligned} &A \text{ is densely defined and closed,} \\ &g \in \operatorname{ran} A \cap \operatorname{dom} A^*, \\ &A^* g \in \mathcal{D}^{qa}(A^* A) \end{aligned} \tag{4.1}$$

(the latter condition meaning that $A^* g$ is a quasi-analytic vector for $A^* A$, in the sense of Definition 2). The assumption that $g \in \operatorname{ran} A \cap \operatorname{dom} A^*$ ensures that $g = Au$ for some $u \in \operatorname{dom} A^* A$, and $A^* g = A^* A u \in \operatorname{ran} A^* A$. This leads to consider, instead of the inverse problem $Af = g$, the auxiliary inverse problem

$$A^* A f = A^* g. \tag{4.2}$$

Here $A^* A$ is self-adjoint (see, e.g., [39, Theorem 12.11]), and $A^* g \in \operatorname{ran} A^* A \cap \mathcal{D}^{qa}(A^* A)$. These are precisely the conditions that allow to deduce, via Theorem 1, that there exists a unique solution

$$f \in \overline{\mathcal{K}(A^* A, A^* g)}$$

to the inverse problem (4.2).

On the other hand, one observes that the above solution f also satisfies $Af = g$. Indeed, from $A^*Af = A^*g$ one sees that $Af - g \in \ker A^*$, whereas obviously $Af - g \in \operatorname{ran} A \subset (\ker A^*)^\perp$ because Af and g belongs to ran A. From the triviality of $\ker A^* \cap (\ker A^*)^\perp$ one thus deduces that $Af = g$.

It then remains established that under the present conditions (4.1) there is a unique solution f to $Af = g$ belonging to the (closure of the) Krylov subspace $\overline{\mathcal{K}(A^*A, A^*g)}$.

This is *not* the actual Krylov subspace one would like to have in the ultimate conclusion of Krylov solvability, namely $\overline{\mathcal{K}(A, g)}$, and the very condition of quasi-analiticity of A^*g with respect to A^*A differs from the standard quasi-analiticity property $g \in \mathcal{D}^{qa}(A)$. The following meaningful question then arises naturally.

Question 2 Assume that $(A_0, \widetilde{A}_0)$ is a joint pair of abstract Friedrichs operators on $\mathcal{H}$ satisfying (4.1) and hence such that there is a unique solution f to $A_0 f = g$ belonging to $\overline{\mathcal{K}(A_0^*A_0, A_0^*g)}$. What assumptions need be added in order to obtain the standard Krylov solvability of $Af = g$, where (A, A^*) is one of the closed realisations of $(A_0, \widetilde{A}_0)$?

In particular, it is of interest to consider Question 2 for the *maximal* realisation $(A_1, \widetilde{A}_1)$ of $(A_0, \widetilde{A}_0)$, in the sense of Sect. 3, or also for one of the mutually adjoint pairs of closed *bijective* realisations of $(A_0, \widetilde{A}_0)$.

Current work in progress in collaboration with Erceg (Zagreb) [15] has allowed us to produce instructive partial answers to Question 2. For instance, quite noticeably, if $A_1 + \widetilde{A}_1$ is a multiple of the identity, and if the smooth vectors of A are also smooth vectors of A^* (or vice versa), then indeed

$$\overline{\mathcal{K}(A^*A, A^*g)} \subset \overline{\mathcal{K}(A, g)},$$

thereby closing in the affirmative the issue of the Krylov solvability for this class of Friedrichs systems.

This has naturally brought attention to those joint pairs $(A_0, \widetilde{A}_0)$ of abstract Friedrichs operators such that

$$A_0 + \widetilde{A}_0 = \alpha \mathbb{1} \qquad \alpha \in \mathbb{R}. \tag{4.3}$$

Consider for instance this very prototypical Friedrichs operator: the complex Hilbert space $\mathcal{H} = L^2(\mathbb{R})$, and

$$\operatorname{dom} A := H^1(\mathbb{R}), \qquad Af := -f' + f. \tag{4.4}$$

A is actually an unbounded Friedrichs operator, as it is closed and its adjoint acts as $\frac{\mathrm{d}}{\mathrm{d}x} + \mathbb{1}$ on $H^1(\mathbb{R})$, thereby implying that $A + A^* = 2\mathbb{1}$ is obviously bounded, self-adjoint, and coercive (see Eq. 3.3).

More generally, one may consider variants where A acts as

$$\begin{aligned} A &= -\frac{\mathrm{d}}{\mathrm{d}x} + c(x) \\ &\text{for } c : \mathbb{R} \to \mathbb{C} \text{ such that } c(x) + \overline{c(x)} \text{ is a} \\ &\text{positive bounded function separated from zero} \end{aligned} \tag{4.5}$$

(The second term in $-\frac{\mathrm{d}}{\mathrm{d}x} + c(x)$ indicates the multiplication by the function c.)

Question 3 To investigate the Krylov-solvability of the inverse problem $Af = g$ induced by (4.4) or (4.5) (or variants) under suitable conditions on g and c.

It is not difficult to argue that the general arguments preceding Question 2, as well as the above-mentioned partial answers to Question 2, may be made applicable to Question 3.

For instance, we could show [15] that in the case (4.4) it suffices to assume further that the $\widehat{g} \in C_c^\infty(\mathbb{R})$ (i.e., the Fourier transform of g is smooth and compactly supported) in order to deduce that A^*g is a *bounded* (and hence quasi-analytic) vector for A^*A, in the sense of Definition 2. This ensures that A^*g is in particular quasi-analytic for A^*A and as a consequence of the facts established before Question 2 and the partial results stated thereafter one deduces that the Friedrichs-type inverse problem

$$-f' + f = g$$

admits a unique Krylov solution. Analogous partial result are currently being produced for the variant (4.5).

In fact, the latter line of reasoning mirrors at the level of concrete differential operators what Questions 1 and 2 deal with at an abstract setting, and could be therefore continued by inquiring the following.

Question 4 To investigate the Krylov-solvability of the inverse problem $Af = g$ induced by the classical Friedrichs operator (3.1) under suitable conditions on the matrix functions matrix functions $\mathbf{B}_k \in W^{1,\infty}(\Omega, M_r)$, $k \in \{1, \ldots, d\}$, and $\mathbf{C} \in L^\infty(\Omega, M_r)$.

Back to the abstract Hilbert space setting, a variety of additional meaningful problems arise naturally. Among them, we find the following two particularly intriguing and promising.

Question 5 To characterise the Krylov intersection $\mathcal{I}(A, g)$ relative to an abstract Friedrichs operator A acting in a Hilbert space $\mathcal{H}$ and a vector $g \in \mathcal{H}$ (see Eqs. 2.8 and 2.9), as well as the Krylov reducibility of A (as in the spirit of Proposition 1). Or, at the concrete level, to characterise the Krylov intersection and the Krylov reducibility of the classical Friedrichs operator (3.1).

Question 5 is a 'structural' question, where the general goal is to investigate when the Krylov intersection is trivial, which under suitable conditions is a sufficient criterion for Krylov solvability, in the spirit of Propositions 2. It is reasonable to expect that the special constraints (3.3) in the abstract Hilbert space setting (or (3.1)–(3.2) in a concrete differential setting) on an operator of Friedrichs type result in a special behaviour concerning its Krylov reducibility or the triviality of its Krylov intersection, which in turn determine the Krylov solvability of the associated inverse problem.

Question 6 Consider a perturbation $\widetilde{A}\widetilde{f} = \widetilde{g}$ of the inverse linear problem $Af = g$ induced by a Friedrichs operator A, where in some suitable sense $\widetilde{A} \sim A$ and $\widetilde{g} \sim g$. When, in the $\widehat{d}_w$-sense (Sect. 2.4), are the Krylov subspaces $\mathcal{K}(A, g)$ and $\mathcal{K}(\widetilde{A}, \widetilde{g})$ close? When does the perturbation preserve or alter the Krylov solvability of the problem? What is the answer to the above questions when, in particular, the perturbation preserves the Friedrichs character of the linear operator?

Concerning Question 6, it was recalled in Sect. 2.4 that the sole vicinity $\widetilde{A} \sim A$ or $\widetilde{g} \sim g$, even in some strong sense, does not guarantee the smallness of

$$\widehat{d}_w\Big(\overline{\mathcal{K}(A, g)}, \overline{\mathcal{K}(\widetilde{A}, \widetilde{g})}\Big),$$

and, typically, it is the term $d_w\big(\overline{\mathcal{K}(\widetilde{A}, \widetilde{g})}, \overline{\mathcal{K}(A, g)}\big)$ in (2.11) to be problematic. It was also recalled (Proposition 4) that if the perturbation leaves the Krylov subspaces $\overline{\mathcal{K}(A, g)}$ and $\overline{\mathcal{K}(\widetilde{A}, \widetilde{g})}$ $\widehat{d}_w$-close, then under favourable conditions the Krylov solvability of the corresponding inverse problems is preserved, whereas in general it may be gained or lost in a perturbation (Example 1). It is plausible to expect, once again, that the constraints for A to be an operator of Friedrichs type select distinguished results in this general scenario. It should be even more so for those perturbations that actually *preserve* the status of Friedrichs system of the inverse problem.

The above selection of questions, be they formulated at an abstract or concrete level, should convey the message, as we strongly believe, that this intriguing subject is ready for very promising developments. For one more time we should like to stress that general criteria of Krylov solvability in the context of inverse problems of Friedrichs type are most valuable, in view of the wide popularity of Krylov approximation methods in numerical analysis, as well as of the treatment of partial differential equations cast in the form of Friedrichs systems.

Acknowledgements For this work the authors gratefully acknowledge the support of the GNFM of the Italian National Instutute for Higher Mathematics INdAM. In addition, A.M. gratefully acknowledges support from the David Flanagan funds at the AUBG, American University Bulgaria, and the Alexander von Humboldt Foundation, Germany.

References

1. Antonić, N., Burazin, K.S.: Intrinsic boundary conditions for Friedrichs systems. Commun. Partial Differ. Eqn. **35**, 1690–1715 (2010)
2. Antonić, N., Burazin, K.S.: Boundary operator from matrix field formulation of boundary conditions for Friedrichs systems. J. Differ. Eqn. **250**, 3630–3651 (2011)
3. Antonić, N., Burazin, K.S., Crnjac, I., Erceg, M.: Complex Friedrichs systems and applications. J. Math. Phys. **58**, 101508, 22 (2017)
4. Antonić, N., Burazin, K.S., Vrdoljak, M.: Connecting classical and abstract theory of Friedrichs systems via trace operator. ISRN Math. Anal. Art. ID 469795, 14 (2011)
5. Antonić, N., Burazin, K.S., Vrdoljak, M.: Heat equation as a Friedrichs system. J. Math. Anal. Appl. **404**, 537–553 (2013)
6. Antonić, N., Burazin, K.S., Vrdoljak, M.: Second-order equations as Friedrichs systems. Nonlinear Anal. Real World Appl. **15**, 290–305 (2014)
7. Antonić, N., Erceg, M., Michelangeli, A.: Friedrichs systems in a Hilbert space framework: solvability and multiplicity. J. Diff. Eqn. **263**, 8264–8294 (2017)
8. Atkinson, K., Han, W.: Theoretical Numerical Analysis: A Functional Analysis Framework, Vol. 39 of Texts in Applied Mathematics, 3rd edn. Springer, Dordrecht (2009)
9. Bui-Thanh, T.: From Godunov to a unified hybridized discontinuous Galerkin framework for partial differential equations. J. Comput. Phys. **295**, 114–146 (2015)
10. Bui-Thanh, T., Demkowicz, L., Ghattas, O.: A unified discontinuous Petrov-Galerkin method and its analysis for Friedrichs' systems. SIAM J. Numer. Anal. **51**, 1933–1958 (2013)
11. Burazin, K., Erceg, M., Waurick, M.: G-convergence of Friedrichs systems revisited. Math. Method. Appl. Sci. **48**, 6081–6091 (2025)
12. Burazin, K., Vrdoljak, M.: Homogenisation theory for Friedrichs systems. Commun. Pure. Appl. Anal. **13**, 1017–1044 (2014)
13. Burazin, K.S., Erceg, M.: Non-stationary abstract Friedrichs systems. Mediterr. J. Math. **13**, 3777–3796 (2016)
14. Burman, E., Ern, A., Fernández, M.A.: Explicit Runge-Kutta schemes and finite elements with symmetric stabilization for first-order linear PDE systems. SIAM J. Numer. Anal. **48**, 2019–2042 (2010)
15. Caruso, N.A., Erceg, M., Michelangeli, A.: Work in progress (2024)
16. Caruso, N.A., Michelangeli, A.: Krylov solvability of unbounded inverse linear problems. Integr. Eqn. Oper. Theory **93**, Paper No. 1 (2021)
17. Caruso, N.A., Michelangeli, A.: Inverse Linear Problems on Hilbert Space and Their Krylov Solvability. Springer Monographs in Mathematics. Springer, Cham (2021)
18. Caruso, N.A., Michelangeli, A.: Convergence of the conjugate gradient method with unbounded operators. Oper. Matrices **16**, 35–68 (2022)
19. Caruso, N.A., Michelangeli, A.: Krylov solvability under perturbations of abstract inverse linear problems. J. Appl. Anal. **29**, 3–29 (2023)
20. Caruso, N.A., Michelangeli, A., Novati, P.: On Krylov solutions to infinite-dimensional inverse linear problems. Calcolo **56**, 32 (2019)
21. Caruso, N.A., Michelangeli, A., Novati, P.: On general projection methods and convergence behaviours for abstract linear inverse problems. Asymptotic Anal. **127**, 167–189 (2022)
22. Caruso, N.A.: A note on the Krylov solvability of compact normal operators on Hilbert space. Complex Anal. Oper. Theory **17** (2023)
23. Caruso, N.A.: Structural properties of Krylov subspaces, Krylov solvability, and applications to unbounded self-adjoint operators. Banach J. Funct. Anal. **19**, Paper No. 7 (2025)
24. Chen, M., Chen, Z., Chen, G.: Approximate Solutions of Operator Equations. Series in Approximations and Decompositions, vol. 9. World Scientific Publishing Co., Inc., River Edge, NJ (1997)
25. Cipra, B.A.: The best of the 20th century: editors name top 10 algorithms. SIAM News, p. 33 (2005)

26. Després, B., Lagoutière, F., Seguin, N.: Weak solutions to Friedrichs systems with convex constraints. Nonlinearity **24**, 3055–3081 (2011)
27. Dongarra, J., Sullivan, F.: The top 10 algorithms (Guest editors' intruduction). Comput. Sci. Eng. **2**, 22–23 (2000)
28. Engl, H.W., Hanke, M., Neubauer, A.: Regularization of Inverse Problems. Mathematics and Its Applications, vol. 375. Kluwer Academic Publishers Group, Dordrecht (1996)
29. Erceg, M., Michelangeli, A.: On contact interactions realised as Friedrichs systems. Complex Anal. Oper. Theory **13**, 703–736 (2019)
30. Erceg, M., Soni, S.K.: Classification of classical Friedrichs differential operators: one-dimensional scalar case. Commun. Pure. Appl. Anal. **21**, 3499–3527 (2022)
31. Erceg, M., Soni, S.K.: The von Neumann extension theory for abstract Friedrichs operators. Z. Anal. Anwend. **44**, 193—218 (2025)
32. Erceg, M., Soni, S.K.: Friedrichs systems on an interval (2024). arXiv:2401.11941
33. Ern, A., Guermond, J.-L.: Theory and Practice of Finite Elements. Applied Mathematical Sciences, vol. 159. Springer, New York (2004)
34. Ern, A., Guermond, J.-L.: Discontinuous Galerkin methods for Friedrichs' systems. I. General theory. SIAM J. Numer. Anal. **44**, 753–778 (2006)
35. Ern, A., Guermond, J.-L.: Discontinuous Galerkin methods for Friedrichs' systems. II. Second-order elliptic PDEs. SIAM J. Numer. Anal. **44**, 2363–2388 (2006)
36. Ern, A., Guermond, J.-L.: Discontinuous Galerkin methods for Friedrichs' systems. III. Multifield theories with partial coercivity. SIAM J. Numer. Anal. **46**, 776–804 (2008)
37. Ern, A., Guermond, J.-L., Caplain, G.: An intrinsic criterion for the bijectivity of Hilbert operators related to Friedrichs' systems. Commun. Partial Differ. Eqn. **32**, 317–341 (2007)
38. Friedrichs, K.O.: Symmetric positive linear differential equations. Commun. Pure Appl. Math. **11**, 333–418 (1958)
39. Grubb, G.: Distributions and Operators. Graduate Texts in Mathematics, vol. 252. Springer, New York (2009)
40. Hanke, M.: Conjugate Gradient Type Methods for Ill-Posed Problems. Pitman Research Notes in Mathematics Series, vol. 327. Longman Scientific & Technical, Harlow (1995)
41. Hansen, P.C.: Rank-Deficient and Discrete Ill-Posed Problems: Numerical Aspects of Linear Inversion. SIAM Monographs on Mathematical Modeling and Computation, Society for Industrial and Applied Mathematics (SIAM), Philadelphia, PA (1998)
42. Houston, P., Mackenzie, J.A., Süli, E., Warnecke, G.: A posteriori error analysis for numerical approximations of Friedrichs systems. Numer. Math. **82**, 433–470 (1999)
43. Jensen, M.: Discontinuous Galerkin methods for Friedrichs systems with irregular solutions. Ph.D. thesis, University of Oxford (2004)
44. Kato, T.: Perturbation Theory for Linear Operators. Classics in Mathematics. Springer, Berlin (1995). Reprint of the 1980 edition
45. Krasnosel′ skiĭ, M.A., Vaĭnikko, G.M., Zabreĭko, P.P., Rutitskii, Y.B., Stetsenko, V.Y.: Approximate Solution of Operator Equations. Wolters-Noordhoff Publishing, Groningen (1972). Translated from the Russian by D. Louvish
46. Liesen, J., Strakoš, Z.E.: Krylov Subspace Methods: Principles and Analysis. Numerical Mathematics and Scientific Computation, Oxford University Press, Oxford (2013)
47. Mifsud, C., Després, B., Seguin, N.: Dissipative formulation of initial boundary value problems for Friedrichs' systems. Commun. Partial Differ. Eqn. **41**, 51–78 (2016)
48. Nemirovskiy, A.S., Polyak, B.T.: Iterative methods for solving linear ill-posed problems under precise information. I. Izv. Akad. Nauk SSSR Tekhn. Kibernet. 13–25, 203 (1984)
49. Nemirovskiy, A.S., Polyak, B.T.: Iterative methods for solving linear ill-posed problems under precise information. II. Eng. Cybern. **22**, 50–57 (1984)
50. Nevanlinna, O.: Convergence of Iterations for Linear Equations. Lectures in Mathematics ETH Zürich, Birkhäuser Verlag, Basel (1993)
51. Quarteroni, A.: Numerical Models for Differential Problems, Vol. 16 of MS&A. Modeling, Simulation and Applications, 3rd edn. Springer, Cham (2017)

52. Saad, Y.: Iterative Methods for Sparse Linear Systems, 2nd edn. Society for Industrial and Applied Mathematics, Philadelphia, PA (2003)
53. Schmüdgen, K.: Unbounded Self-adjoint Operators on Hilbert Space. Graduate Texts in Mathematics, vol. 265. Springer, Dordrecht (2012)
54. Sullivan, T.J.: Introduction to Uncertainty Quantification. Texts in Applied Mathematics, vol. 63. Springer, Cham (2015)
55. Vainikko, G.: Regular convergence of operators and approximate solution of equations. J. Sov. Math. **15**, 675–705 (1981)
56. Vainikko, G.: Multidimensional Weakly Singular Integral Equations. Lecture Notes in Mathematics, vol. 1549. Springer, Berlin (1993)
57. Vainikko, G., Karma, O.: The convergence of approximate methods for solving linear and non-linear operator equations. USSR Comput. Math. Math. Phys. **14**, 9–19 (1974)

MIX
Papier aus verantwortungsvollen Quellen
Paper from responsible sources
FSC® C105338

If you have any concerns about our products,
you can contact us on
ProductSafety@springernature.com

In case Publisher is established outside the EU,
the EU authorized representative is:
Springer Nature Customer Service Center GmbH
Europaplatz 3, 69115 Heidelberg, Germany

Printed by Libri Plureos GmbH
in Hamburg, Germany